Natural Product-Based Synthetic Drug Molecules in Alzheimer's Disease

Abha Sharma • Gyan Prakash Modi
Editors

Natural Product-Based Synthetic Drug Molecules in Alzheimer's Disease

Therapeutic & Theranostic Agents

Springer

Editors
Abha Sharma
Department of Medicinal Chemistry
National Institute of Pharmaceutical Education and Research
Raebareli, Uttar Pradesh, India

Gyan Prakash Modi
Department of Pharmaceutical Engineering and Technology
Indian Institute of Technology BHU
Varanasi, Uttar Pradesh, India

ISBN 978-981-99-6037-8 ISBN 978-981-99-6038-5 (eBook)
https://doi.org/10.1007/978-981-99-6038-5

This Springer imprint is published by the registered company Springer Nature Singapore Pte Ltd.
The registered company address is: 152 Beach Road, #21-01/04 Gateway East, Singapore 189721, Singapore

Paper in this product is recyclable.

Preface

In this book, we present reviews of the most recent findings and information on several critical aspects of Alzheimer's disease (AD), providing the readers with a broad picture of the underlying pathophysiology, biomarkers, potential targets, and their modulators. The book will serve as a resource for various researchers and teachers to understand the current state of the disease and to identify the key research areas that need to be addressed and explored for AD drug development. This book is primarily organized into six parts, each of which is further subdivided into several chapters. The book parts all have their own distinct identities that help readers to deeply understand the various facets of Alzheimer's disease.

The first part covers the fundamentals of AD with an emphasis on the disease's pathophysiology, possible therapeutic options, and how oxidative stress and metal ions play a significant role in its development. Additionally, this part fully explains the significance of neuroinflammation in AD. The second part provides a thorough overview of the numerous diagnostics and illness-related biomarkers for AD. This part primarily focuses on the several fluorescence probes used in the detection and management of AD. Furthermore, the focus of this part is on nanotechnology and the application of nanomedicine to the diagnosis and treatment of AD. The third part contains a variety of therapeutic compounds that primarily target acetylcholine receptors as well as certain memantine-based medication derivatives that have undergone biological review and synthesis. The fourth part contains a variety of treatments that are naturally available for AD. These cutting-edge medications are mostly based on natural sources including huperzine, flavonoids, polyphenols, and vitamins. The fifth part offers details on the metal hypothesis and metal chelators, which may be used as an innovative medicinal agent to treat AD. Apart from that, the focus of this part is on ferroptosis modulators and how they can function as an improved therapeutic agent for the treatment of AD. The topic of several epigenetic modulators in AD is covered in the final part, which also contains inhibitors of histone deacetylase and sirtuin modulators. This part is primarily concerned with the design, production, and biological assessment of diverse pharmacological compounds that operate as epigenetic modulators.

We express our gratitude to all the authors for their meticulous work in contributing to this book. We consider that this book will inspire readers to inspect deeper into this field and expedite the research in this direction for searching new drugs/probes for AD treatment/diagnosis. Further, this book will serve as an excellent teaching tool for undergraduate and postgraduate students.

Raebareli, Uttar Pradesh, India Abha Sharma
Varanasi, Uttar Pradesh, India Gyan Prakash Modi

Contents

Editors and Contributors

About the Editors

Abha Sharma is currently an Associate Professor in the Department of Medicinal Chemistry at the National Institute of Pharmaceutical Education and Research, Raebareli, India.

Gyan Prakash Modi is currently an Associate Professor in the Department of Pharmaceutical Engineering and Technology, at the Indian Institute of Technology Banaras Hindu University, Varanasi, India.

Contributors

Anurag T. K. Baidya Department of Pharmaceutical Engineering and Technology, Indian Institute of Technology (B.H.U.), Varanasi, Uttar Pradesh, India

Rahul Barot Faculty of Pharmacy, Kalabhavan Campus, The Maharaja Sayajirao University of Baroda, Vadodara, Gujarat, India

Sunita Beni Department of Medicinal Chemistry, National Institute of Pharmaceutical Education and Research, Ahmedabad, Gujarat, India

Swati Biswas Epigenetic Research Laboratory, Department of Pharmacy, Birla Institute of Technology and Science-Pilani, Hyderabad, India

Praveen Bodhe Department of Medicinal Chemistry, National Institute of Pharmaceutical Education and Research, Ahmedabad, Gujarat, India

Vedika G. Dadlani Department of Pharmaceutical Chemistry, Dr. L. H. Hiranandani College of Pharmacy, University of Mumbai, Ulhasnagar, Maharashtra, India

Bhanuranjan Das Department of Pharmaceutical Engineering and Technology, Indian Institute of Technology (B.H.U.), Varanasi, Uttar Pradesh, India

Bhaskar Dewangan Department of Medicinal Chemistry, National Institute of Pharmaceutical Education and Research, Ahmedabad, Gujarat, India

Yashasvi Anant Dhamapurkar Department of Medicinal Chemistry, National Institute of Pharmaceutical Education and Research (NIPER)-Raebareli, Lucknow, Uttar Pradesh, India

Vineet Diwakar Department of Pharmacoinformatics, National Institute of Pharmaceutical Education and Research (NIPER), Mohali, Punjab, India

Shaik Ayesha Fathima Department of Regulatory Toxicology/Pharmacology and Toxicology, National Institute of Pharmaceutical Education and Research (NIPER-R), Lucknow, Uttar Pradesh, India

Mareechika Gaddam Department of Medicinal Chemistry, National Institute of Pharmaceutical Education and Research (NIPER)-Raebareli, Lucknow, Uttar Pradesh, India

Abhiram Ganorkar Epigenetic Research Laboratory, Department of Pharmacy, Birla Institute of Technology and Science-Pilani, Hyderabad, India

Prabha Garg Department of Pharmacoinformatics, National Institute of Pharmaceutical Education and Research (NIPER), Mohali, Punjab, India

Balaram Ghosh Epigenetic Research Laboratory, Department of Pharmacy, Birla Institute of Technology and Science-Pilani, Hyderabad, India

Kondakamarla Imran Department of Medicinal Chemistry, National Institute of Pharmaceutical Education and Research (NIPER)-Raebareli, Lucknow, Uttar Pradesh, India

Arun K. Iyer Use-Inspired Biomaterials and Integrated Nano Delivery (U-BiND) Systems Laboratory, Department of Pharmaceutical Sciences, Wayne State University, Detroit, MI, USA
Molecular Imaging Program, Karmanos Cancer Institute, Detroit, MI, USA

Keerti Jain Department of Pharmaceutics, National Institute of Pharmaceutical Education and Research (NIPER)-Raebareli, Lucknow, Uttar Pradesh, India

Karan Joshi Faculty of Pharmacy, Kalabhavan Campus, The Maharaja Sayajirao University of Baroda, Vadodara, Gujarat, India

Kailash Laboratory of Organic and Medicinal Chemistry, Department of Chemistry, Central University of Punjab, Bathinda, Punjab, India

Santanu Kaity Department of Pharmacology, Delhi Pharmaceutical Sciences & Research University (DPSRU), New Delhi, India

Pradnya Kamble Department of Pharmacoinformatics, National Institute of Pharmaceutical Education and Research (NIPER), Mohali, Punjab, India

Sushil Kumar Kashaw Integrated Drug Discovery Research Laboratory, Department of Pharmaceutical Sciences, Dr. Harisingh Gour University (A Central University), Sagar, Madhya Pradesh, India

Varsha Kashaw Sagar Institute of Pharmaceutical Sciences, Sagar, Madhya Pradesh, India

Shivam K. Kori Integrated Drug Discovery Research Laboratory, Department of Pharmaceutical Sciences, Dr. Harisingh Gour University (A Central University), Sagar, Madhya Pradesh, India

Amit Kumar Department of Pharmaceutics, National Institute of Pharmaceutical Education and Research (NIPER)-Raebareli, Lucknow, Uttar Pradesh, India

Amit Kumar Department of Medicinal Chemistry, National Institute of Pharmaceutical Education and Research, Ahmedabad, Gujarat, India

Anoop Kumar Department of Pharmaceutics, National Institute of Pharmaceutical Education and Research (NIPER), Kolkata, West Bengal, India

Madhavi Kumari Department of Pharmacoinformatics, National Institute of Pharmaceutical Education and Research (NIPER), Mohali, Punjab, India

Kunal Kumar Department of Medicinal Chemistry, National Institute of Pharmaceutical Education and Research, Ahmedabad, Gujarat, India

Naveen Kumar Laboratory of Organic and Medicinal Chemistry, Department of Chemistry, Central University of Punjab, Bathinda, Punjab, India

Rajnish Kumar Department of Pharmaceutical Engineering and Technology, Indian Institute of Technology (B.H.U.), Varanasi, Uttar Pradesh, India

Vijay Kumar Laboratory of Organic and Medicinal Chemistry, Department of Chemistry, Central University of Punjab, Bathinda, Punjab, India

Vinay Kumar Laboratory of Organic and Medicinal Chemistry, Department of Chemistry, Central University of Punjab, Bathinda, Punjab, India

Vinod Kumar Laboratory of Organic and Medicinal Chemistry, Department of Chemistry, Central University of Punjab, Bathinda, Punjab, India

Ranika Maurya Department of Regulatory Toxicology/Pharmacology and Toxicology, National Institute of Pharmaceutical Education and Research (NIPER-R), Lucknow, Uttar Pradesh, India

Indubhusan Mishra Department of Medicine, Heritage Hospital, Varanasi, Uttar Pradesh, India

Gyan Prakash Modi Department of Pharmaceutical Engineering and Technology, Indian Institute of Technology (Banaras Hindu University), Varanasi, Uttar Pradesh, India

Esther Rani Motamarri Department of Medicinal Chemistry, National Institute of Pharmaceutical Education and Research (NIPER)-Raebareli, Lucknow, Uttar Pradesh, India

Prashant R. Murumkar Faculty of Pharmacy, Kalabhavan Campus, The Maharaja Sayajirao University of Baroda, Vadodara, Gujarat, India

Saba Naqvi Department of Regulatory Toxicology/Pharmacology and Toxicology, National Institute of Pharmaceutical Education and Research (NIPER-R), Lucknow, Uttar Pradesh, India

Jyoti Pandey Department of Chemistry, School for Physical and Decision Sciences, Babasaheb Bhimrao Ambedkar University, Lucknow, Uttar Pradesh, India

Sandhya Pathak Department of Chemistry, Dr. Harisingh Gour University (A Central University), Sagar, Madhya Pradesh, India

Harshal Ashok Pawar Department of Pharmacognosy, Dr. L. H. Hiranandani College of Pharmacy, University of Mumbai, Ulhasnagar, Maharashtra, India

Sravani Pulya Epigenetic Research Laboratory, Department of Pharmacy, Birla Institute of Technology and Science-Pilani, Hyderabad, India

Nishant Kumar Rana Department of Pharmaceutical Engineering and Technology, Indian Institute of Technology (Banaras Hindu University), Varanasi, Uttar Pradesh, India

Bichismita Sahu Department of Medicinal Chemistry, National Institute of Pharmaceutical Education and Research, Ahmedabad, Gujarat, India

Abha Sharma Department of Medicinal Chemistry, National Institute of Pharmaceutical Education and Research (NIPER)-Raebareli, Lucknow, Uttar Pradesh, India

Gourav Singh Department of Pharmaceutical Engineering and Technology, Indian Institute of Technology (Banaras Hindu University), Varanasi, Uttar Pradesh, India

Vandana Soni Integrated Drug Discovery Research Laboratory, Department of Pharmaceutical Sciences, Dr. Harisingh Gour University (A Central University), Sagar, Madhya Pradesh, India

Teeja Suthar Department of Pharmaceutics, National Institute of Pharmaceutical Education and Research (NIPER)-Raebareli, Lucknow, Uttar Pradesh, India

Pushpendra K. Tripathi Institute of Pharmaceutical Sciences, University of Lucknow, Lucknow, Uttar Pradesh, India

Tanmaykumar Varma Department of Pharmacoinformatics, National Institute of Pharmaceutical Education and Research (NIPER), Mohali, Punjab, India

Satyamshyam Vishwakarma Integrated Drug Discovery Research Laboratory, Department of Pharmaceutical Sciences, Dr. Harisingh Gour University (A Central University), Sagar, Madhya Pradesh, India

Awesh Kumar Yadav Department of Pharmaceutics, National Institute of Pharmaceutical Education and Research (NIPER)-Raebareli, Lucknow, Uttar Pradesh, India

Mange Ram Yadav Centre of Research for Development, Parul University, Vadodara, Gujarat, India

Rasana Yadav Faculty of Pharmacy, Kalabhavan Campus, The Maharaja Sayajirao University of Baroda, Vadodara, Gujarat, India

Part I

Introduction of Alzheimer's Disease

Alzheimer's Disease and Drug Targets

1

Amit Kumar and Awesh Kumar Yadav

Abstract

Alzheimer's disease (AD) is among the neurodegenerative disorders mainly characterized by loss of memory is also termed as dementia. The primary mechanism is centered on the deposition of extracellular plaques, which are mainly composed of amyloid-β (Aβ). Also, the deposition of neurofibrillary tangles of phosphorylated tau proteins inside the neurons affects the longevity of neurons. The other factors like the formation of reactive oxygen species (ROS) and neuroinflammation have crucial roles in the development of AD. Moreover, the improper enzymatic processing of Aβ is thought to be a responsible factor for the deposition of insoluble Aβ plaques. However, the drugs which prolong the half-life of acetylcholine by inhibiting the activity of the acetylcholine esterase enzyme are administered for the management of AD. Unfortunately, these available drugs provide only symptomatic support, cause severe side effects, and may cause the death of patients. Hence, new agents are required to target the disease at different levels or more than one level for the management of AD. The published pieces of literature show that inhibition of Aβ formation, degradation or disaggregation of Aβ plaques, targeting hyperphosphorylation of tau protein, modulation of ROS production, and mitigation of neuroinflammation could be the best strategies to treat AD. Interestingly, the development of agents that may target BACE, APOE, APP, γ-secretase, and other families of gene(s) could set a milestone for the treatment of AD. In this chapter, we have tried our best to explore the possible targets for the management of AD.

A. Kumar · A. K. Yadav (✉)
Department of Pharmaceutics, National Institute of Pharmaceutical Education and Research, (NIPER) Raebareli (An Institute of National Importance under Dept. of Pharmaceuticals, Ministry of Chemicals and Fertilizers, GOI), Lucknow, Uttar Pradesh, India
e-mail: awesh.yadav@niperraebareli.edu.in

A. Sharma, G. P. Modi (eds.), *Natural Product-Based Synthetic Drug Molecules in Alzheimer's Disease*, https://doi.org/10.1007/978-981-99-6038-5_1

Keywords

Neurodegeneration · Amyloid-β · Hyperphosphorylation · Tau protein · Neuroinflammation · BACE · APOE · APP · γ-Secretase

1.1 Introduction

Alzheimer's disease is growing rapidly worldwide, and this type of disease is closely related to age and has a greater incidence after the age of 60; it is a neurodegenerative disease that is characterized by brain cell destruction and dementia (Tam et al. 2018), in which the patient suffers from memory loss and judgment disorders. AD is considered as the fifth most popular reason for death over the age of 65 (Cummings et al. 2021); amyloid beta and tau protein play an important role in the pathophysiology of Alzheimer's disease. APP and AB disposition can lead to the generation of AD. Tau phosphorylation is also a key factor for Alzheimer's disease progression (Aisen et al. 2017). Refer to Fig. 1.1.

Inflammation of neurons leads to abnormal responses within the brain and spinal cord. This neuroinflammation may be caused by Aβ accumulation and tau protein aggregation, which may also lead to the occurrence of oxidative stress, continuous loss of cholinergic synapse, and impairment of cognitive function (Saha et al. 2022). The genetic description of AD is complex; a majority of occurrences of Alzheimer's disease have been seen in the elderly population, and genetically it may due to the

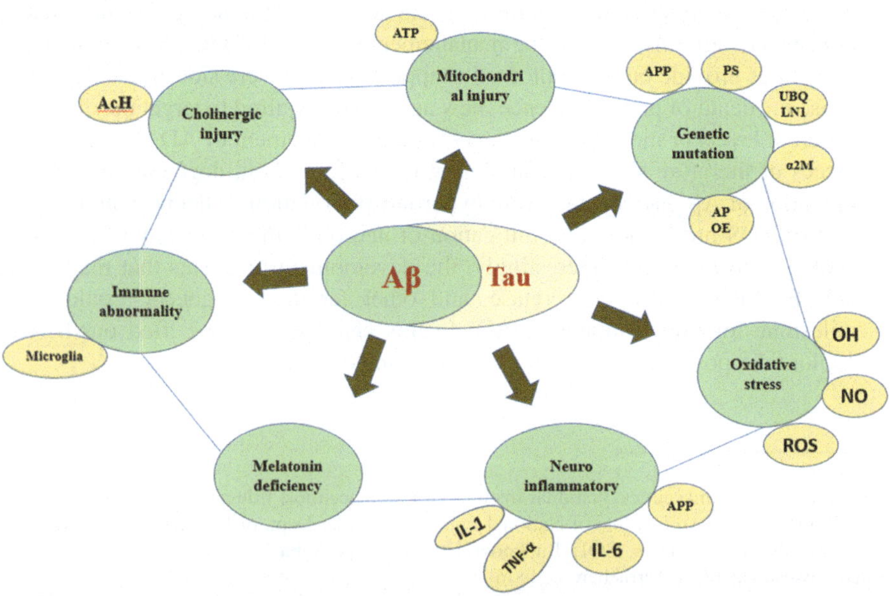

Fig. 1.1 Illustrating several factors which may lead to the generation of Alzheimer's disease

mutation; several mutations have been seen in different genes like amyloid precursor protein (APP), PSEN1 gene, and PSEN2 gene; these three genes are majorly described for the occurrence of this disease (Tanzi and Bertram 2001).

Proteins in the space between nerve cells may be aggregated as misfolded proteins to form amyloid plaque; the main component of amyloid plaque is amyloid-β (Aβ), which is derived from amyloid precursor protein (APP); APP undergoes proteolytic cleavage to form Aβ-amyloid peptide (Koudinov et al. 2002). The care of AD is one of the biggest medical challenges nowadays, and dementia is the main reason for this disease's progression. According to the literatures, in total, 40 million people are suffering from dementia in the whole world population, and by 2050, it will become twice approximately (Yiannopoulou and Papageorgiou 2020). Amyloid precursor protein, which upon cleavage produced Aβ-amyloid peptide, is coded on chromosome 21 by APP gene. This cleavage of APP is done by different cleaving enzymes, generally beta-cleaving enzyme (BACE-1), and ϒ-secretase complex cleaves APP in which presenilin has a catalytic site (Calderon-Garcidueñas and Duyckaerts 2018). The breakdown of amyloid precursor protein may also be due to other enzymatic effects; in these cases, Aβ is not produced.

Patients diagnosed with AD show several symptoms of the disease. Many patients have symptoms of depression and apathy, and these symptoms become more frequent and progressive with age. During the disease progression stage, many patients suffer from neuropsychiatric symptoms, and these neuropsychiatric symptoms progress with disease progression. Higher cognitive impairment has been seen in AD patients as compared to the normal population. Patients with AD symptoms also have physical agitation and uncontrolled movement; with time, disease progression also leads to the development of delusion and hallucinations (Lyketsos et al. 2011).

In Fig. 1.1, various factors have been listed which may contribute to Alzheimer's disease progression. These factors may influence Aβ aggregation and tau protein phosphorylation, which may lead to the generation of Alzheimer's disease.

1.2 Pathological and Therapeutic Target in AD

1.2.1 Amyloid Beta (Aβ)

Amyloid beta plays an important role in the pathogenesis of Alzheimer's disease, and it has also a crucial role in the progression of disease from one cell to another cell because amyloid oligomer has the property to diffuse cellular membrane; thus, AD pathology strongly correlates with the Aβ because of amyloid-β having neural toxicity and folding, and deposition of these peptides and proteins causes this type of neurodegenerative disorder (Breydo et al. 2016). Refer to Fig. 1.2.

Misfolding of several proteins may lead to the occurrence of this type of neurodegenerative disorder; $A\beta_{1-40/42}$ is one of the important identified peptides whose aggregation and misfolding may promote soluble to stable amyloid fiber, and this can be linked to Alzheimer's disease (Protein and Doi 2013). Transport of amyloid

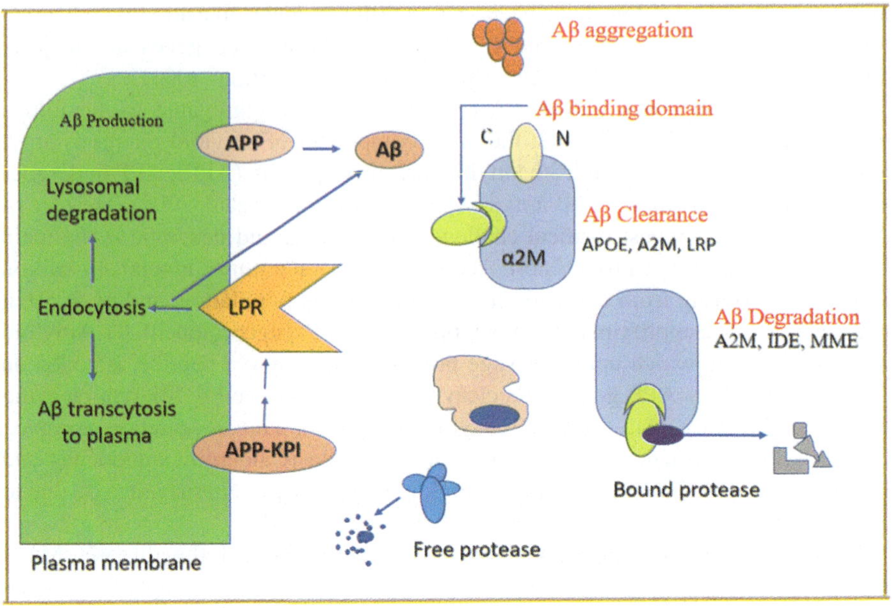

Fig. 1.2 Aggregation of Aβ, clearance, and degradation

Table 1.1 Transport of Aβ

S. No.	Receptor	Role
1	LPR	Aβ transport from brain into the blood
2	P-glycoprotein	Aβ transport from brain into the blood
3	RAGE	Aβ transport from blood into brain
4	gp330/megalin	Aβ transport from blood into brain

beta involves several pathways and many receptors (Table 1.1) like lipoprotein receptor-related protein (LRP), P-glycoprotein, receptor for the advanced glycation end product (RAGE), and gp330/megalin-associated Aβ transport across blood-brain barrier, and degradation of amyloid beta is mediated by a number of enzymes like neprilysin, insulin-degrading enzyme, and endothelin-converting enzymes (Wang et al. 2006).

1.2.2 Tau Phosphorylation

Tau phosphorylation and aggregation are among the major causes of the generation of Alzheimer's disease. Different pathways are there through which tau protein can be phosphorylated; two types of kinases may be involved in phosphorylation: proline-directed kinase and non-proline-directed kinase (Avila 2006). The cytoskeleton consists of microtubules, and these microtubule-associated proteins retain morphology and structure stabilization of neurones (Lee et al. 2005); tau protein is also a microtubule-associated protein whose aggregation and phosphorylation may lead to the generation of AD symptoms.

GSK-3 activity can influence the tau protein; increase in GSK-3 activity can promote phosphorylation of tau protein, and this tau protein phosphorylation can be caused by presenilin-1 (PS-1) mutation (Avila et al. 2012). Presenilin-1 is a multipass transmembrane protein, and the mutation of this protein or deletion of PS-1 leads to an increase in GSK-3 activity which phosphorylates tau protein. It has been studied that PS-1 prevents apoptosis by downregulation of GSK-3 activity; this suppression is via activating PI3K/Akt signaling (Baki et al. 2004). Refer to Fig. 1.3.

Studies reported that cdk-2 and cdk-5 are also involved in the modification of tau into Alzheimer's-like state; cdk-2 is one of the important factors in the disease generation, and cdk-5 facilitates tau into Alzheimer's-like state and is associated with brain microtubules (Baumann et al. 1993); they have reported significant characteristics of tau phosphorylation by using reaction with diagnostic antibodies (Baumann et al. 1993). Many studies have been performed for the measurement of phosphorylated tau proteins in CSF using markers. They developed the first antibody-free mass spectrometric method to measure the level of phosphorylated tau in CSF; in this study, they confirmed that tau phospho-epitopes pT231 and pT217 can be potential markers of Alzheimer's disease in early stage, and in late Alzheimer's disease, pT205 is identified as an important marker (Gobom et al. 2022).

Alzheimer's disease generation may also be due to the brain somatic mutation. The study performed on whole-exome sequencing in 111 postmortem hippocampal formation and AD blood samples matched with normal samples, and they found that mutation rate accumulation in AD sample is significantly higher than in normal sample (Park et al. 2019).

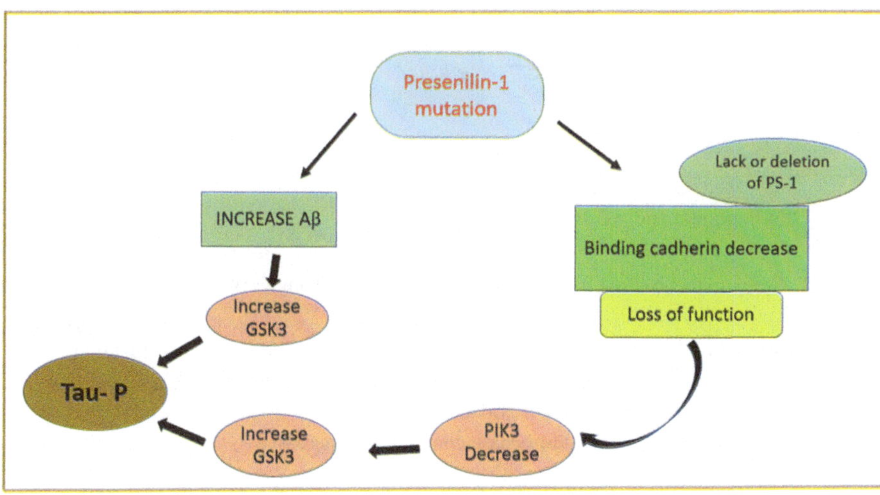

Fig. 1.3 Illustrating tau phosphorylation by mutation in gene

1.3 Mitochondrial Dysfunction and Oxidative Stress in Alzheimer's Disease

Mitochondria is a double-membrane organelle of the body, and it provides energy to the cells. Mitochondria produces almost 93% of all the cellular energy and mitochondrial dysfunction including cellular energy production impairment, oxidative stress, mtDNA damage, and calcium regulation impairment causing symptoms and generation of Alzheimer's disease (Wong et al. 2020). Refer to Fig. 1.4. Also, Refer to Table 1.2.

In mammals, mitochondria can produce reactive oxygen species (ROS) after sequences of biochemical reactions, and this ROS plays an important role in the oxidative damage of mitochondrial protein (Murphy 2009). This can also influence

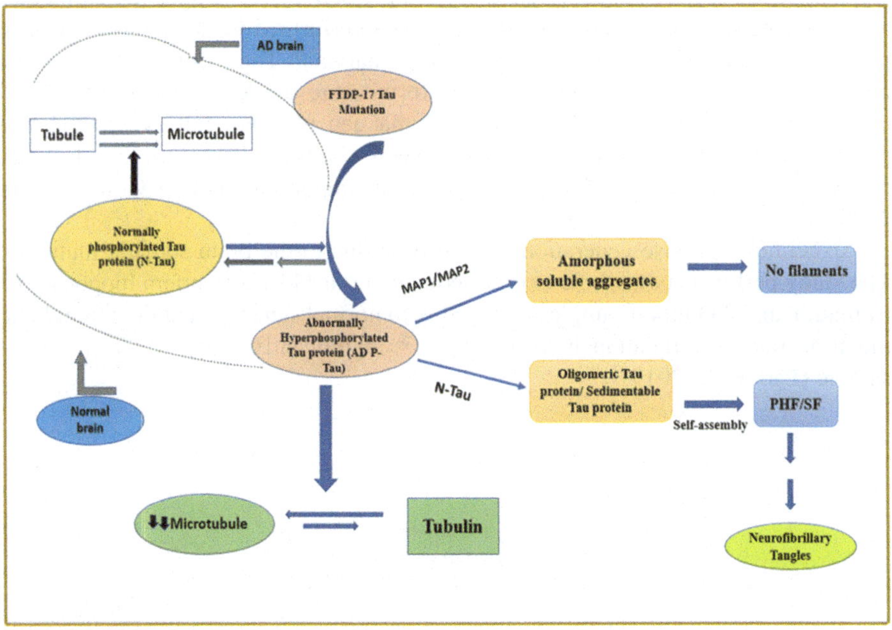

Fig. 1.4 Tau in normal and AD brain

Table 1.2 Some mitochondrial dysfunctions and possible outcomes (Wong et al. 2020)

S. No.	Description	Mitochondrial dysfunction
1	Oxidative scavengers decrease, and elevated level of reactive oxygen species. Increased peroxidation	Oxidative stress
2	Decreased pyruvate dehydrogenase complex and increased succinate dehydrogenase	TCA cycle dysregulation
3	Calcium influx increases from extracellular to cytosol upon excitotoxicity	Calcium dysregulation

the ability of ATP production; mitochondrial oxidative damage can promote cell apoptotic process by secreting cytochrome C to the cytosol by mitochondrial outer membrane permeabilization (Murphy 2009).

1.3.1 Alzheimer's Disease Drug Targets

1.3.1.1 Prevention of Aβ Accumulation and Amyloid Inhibitors

Aβ deposition and its aggregation are among the major causes behind this disease. Its aggregation in the brain leads to the disturbance of function and promotes Alzheimer's disease symptoms. Currently, various approaches are utilized to control the unnecessary accumulation of amyloid beta, and prevention of deposition of the same may become a valuable approach to control disease progression. Currently, many molecules or compounds are being investigated for their amyloid inhibitory action; molecules with amyloid accumulation inhibition properties can be used as a therapeutic approach in the treatment and management of Alzheimer's disease by correcting the misfolding and aggregation of amyloid beta.

Neurotoxicity and inflammation can also be reduced or prevented by the modification and regulation of APP and its expression (Estrada and Soto 2006). Synthesis of molecules for amyloid beta inhibition can be done by molecule library and further assessing them for their activity whether a molecule has Aβ inhibition properties or not.

Fibril formation is required for Aβ-mediated neuronal toxicity. A study has been performed for evaluating the connection of the physical form of Aβ and neurotoxicity by examining amorphous Aβ against fibrillar Aβ (Lorenzo and Yankner 1994), and it has been concluded that fibrillar Aβ is responsible for neurotoxicity. Congo dye has been investigated for its fibrillar inhibiting properties, and they found that Congo dye inhibits fibrillar Aβ complex and prevents Alzheimer's symptoms (Lorenzo and Yankner 1994). Nordihydroguaiaretic acid and rosmarinic acid are the phenolic compounds, and they are reported as Aβ aggregation inhibitors using AD transgenic mice in vitro performance study (Hamaguchi et al. 2009). Tanshinones, which are derived from herbal source, are examined for its amyloid beta aggregation inhibitory action; in vitro studies have been done on their two forms TS1 and TS2, and they found that both compounds have the ability of amyloid fibril inhibition (Wang et al. 2013). The outcome of AFM and ThT suggested that TS1 had better inhibitory action than TS2 (Wang et al. 2013).

Nicotine has also been studied for its anti-amyloid plaque formation activity. Proteolytic cleavage of APP produced an alpha-helical structure, which is monomeric in nature. Nicotine binding with alpha helix leads to inhibition of β-sheet formation. This slow or inhibition of the conversion of alpha helix into β-sheet prevents amyloid aggregation (Salomon et al. 1996).

Suppression of vitamin D receptor (VDR) by effects of Aβ leads to neurodegeneration and cytotoxicity (Dursun et al. 2011). Supplement of vitamin prevents neural toxicity and Aβ-mediated neural damage in the model, and this novel approach may become a potential target to prevent AD.

Resveratrol has properties to interfere with amyloid aggregation. The study performed on transgenic strain expressing $A\beta_{1-42}$ paralysis with amyloid toxicity has shown that resveratrol has properties to decrease lysosomes by activating proteasomal degradation (Regitz et al. 2016).

Mitochondrial function and health have an important contribution to the progression of Alzheimer's disease; neural damage, cytotoxicity, and oxidative stress are closely related to the mitochondrial abnormalities. Level of oxidation stress and dysfunction in ATP production can lead to neurodegeneration disorder. Oxidative stress can be reduced by several mitochondrial targeted antioxidants to control AD symptoms. Antioxidant SS31 has been studied for preventing mitochondrial dysfunction induced by Aβ deposition and aggregation (Calkins et al. 2012). In some kinds of literature, it has been reported that SS31 antioxidants inhibit lipid peroxidase accumulation and scavenge ROS, and mitochondrial swelling is also reduced by SS31 antioxidant (Zhao et al. 2004).

Cholinesterase inhibitors can be used to treat AD. The cholinergic system is mainly involved in learning and memory functions; cholinesterase drug, tacrine, improves cognitive impairment (Lahiri et al. 2000).

Graphene quantum dots have been studied for their ability to treat Alzheimer's disease due to its low cytotoxicity properties as compared to other carbon compounds; their small size favors to cross blood-brain barrier (Liu et al. 2015), and results show that graphene quantum dots possess the ability to inhibit $A\beta_{1-42}$ and prevent amyloid beta-mediated cytotoxicity (Liu et al. 2015). A cholesterol-lowering agent atorvastatin can be used to prevent amyloid beta deposition; it has been reported that Aβ aggregation can be influenced by atorvastatin in a calcium-dependent manner; in the absence of calcium, atorvastatin inhibits Aβ aggregation, but in the presence of calcium, it shows opposite action (Nedaei et al. 2022).

Clusterin is a glycoprotein (Spatharas et al. 2022), heterodimeric in nature, and has the ability to inhibit amyloid aggregation. Clusterin is a chaperone protein having five aggregation-sensitive regions in its alpha chain, and these sites have the ability to prevent β-amyloid fibril formation (Spatharas et al. 2022).

1.3.1.2 Prevention of Tau Protein Phosphorylation

One of the important ways to prevent Alzheimer's disease and its symptoms is by inhibition of tau protein aggregation. Many tau protein inhibitors may be used in the prevention of phosphorylation and pathways of phosphorylation.

Tau protein could be an important potential target in the treatment of Alzheimer's disease. Lithium is investigated for its inhibition activity against the tau protein (Muñoz-Montaño et al. 1997). Abnormal tau phosphorylation or uncontrolled hyperphosphorylation can cause neuron damage. Lithium has been reported as a GSK-3 inhibitor, and GSK-3 has a role in hyperphosphorylation. In vivo study has been performed to evaluate the activity of lithium in rats upon 7 days of intraperitoneal administration, and results show dephosphorylation of tau protein at the Tau-1 and PHF-1 epitope site (Muñoz-Montaño et al. 1997). This indicates prevention of tau protein hyperphosphorylation activity of lithium.

Phenothiazine was reported as an inhibitor of tau protein deposition (Wischik et al. 1996). A study revealed that diaminophenothiazine has selective inhibition activity against tau protein aggregation to prevent neuron damage and symptoms of AD by a proteolytic breakdown of tau aggregates (Wischik et al. 1996).

Beta-carboline compounds were reported as tau aggregation inhibitors; these alkaloids have a high affinity to inhibit kinase-dependent tau aggregation. It has been reported that beta-carboline prevents tau aggregation via inhibition of DYRK1A in cell culture assay (Frost et al. 2011). In that study, harmine and its derivative with some structural modifications were studied for their inhibition activity, and results show that fully aromatic ring-containing compounds have greater inhibition activity against tau protein aggregation (Frost et al. 2011). Exploring a number of possible therapeutic molecules may be beneficial in the management and treatment of Alzheimer's disease. An in vitro research was conducted on protein-capped metal nanoparticles for evaluating therapeutic potency in preventing tau protein deposition, and they found that PC-Fe3O4 and PC-Cds nanoparticles (Sonawane et al. 2019) have the ability to inhibit tau protein. Disaggregation of tau and efficient clearance of tau aggregates by the action of PC-Cd nanoparticles were demonstrated by SDS-PAGE (Sonawane et al. 2019). And the study also revealed that a concentration of PC-Cd nanoparticles below 70 μg/mL for 24 h of duration exhibits no cytotoxicity in neuron cells during MTT assay (Sonawane et al. 2019). Several herbal compounds are explored to yield potential molecules, which can be effective in the treatment of Alzheimer's disease. Cinnamon aqueous extract was tested for evaluating its tau protein aggregation prevention activity, and the study performed in vitro revealed that aqueous *Cinnamomum zeylanicum* extract can break aggregates of tau protein and change the morphology of paired-helical filaments (PHFs) (Peterson et al. 2009).

Several fungal products were investigated for their tau protein aggregation inhibition property; this aggregation cleavage property of the natural products is a potential strategy for the treatment of Alzheimer's disease. A fungal derived compound asperbenzaldehyde was tested in vitro for tau aggregation reversal property; asperbenzaldehyde is an intermediate compound formed in the biosynthesis of azaphilone (Paranjape et al. 2015). They tested 11 azaphilone derivatives for their activity, and they found that the tau aggregation reduces with azaphilone (Paranjape et al. 2015).

Intracellular insoluble aggregates of MT-associated tau are responsible for the generation of AD and neuron cell toxicity; toluidine blue is derived from methylene blue (Dubey et al. 2019). This toluidine belongs to the phenothiazine group, which on irradiation is converted into photoexcited form and acts as tau aggregation inhibitor and also improves memory impairment (Dubey et al. 2019).

1.3.1.3 MicroRNAs in the Management of AD

MicroRNA may have the property to inhibit phosphorylation. The study was conducted to evaluate the effect of miRNAs on hyperphosphorylation, and results revealed that miR-106b overexpression reduced or inhibited tau protein phosphorylation induced by $A\beta_{1-42}$ peptide (Liu et al. 2016).

1.3.1.4 Pyroglutamate Aβ Cascade as a Potential Target

Pyroglutamate peptide production is associated with AD. Figure 1.5 shows the systematic steps of generation of pyroglutamate peptide, and pyroglutamate-modified Abeta peptide is found in AD brain, thus represented as a potential target for the drug in the management of AD (Bayer 2022). Different events of targeting pyroglutamate Aβ are shown in Fig. 1.6. Also, refer to Table 1.3.

The mAb 9D5 reacted with a low-molecular-weight oligomer of AβpE3-X and not with amyloid plaques in Alzheimer's disease brain tissue. 5XFAD mice reduced amyloid plaques via passive immunization and recovered behavior deficits. Some mAbs like pan-AβpE are designed to target plaques. In an acute application design using APPPS1 mice, passive immunization shows therapeutic effects. A DPP4 inhibitor, vildagliptin, effect was investigated in a clinical study on cognitive dysfunction in 60 diabetes elderly patients with MCI diagnosis; 50% of patients were treated with standard medication control, e.g., metformin, and the other 50% with metformin plus vildagliptin. As a result, in the DPP4 inhibitor group, a significant response to stabilizing cognitive scores was observed (Borzì et al. 2019).

1.3.1.5 Immunotherapy in Alzheimer's Disease

Immunotherapy is one of the valuable approaches to provoke an immune response or use of antibodies to control AD pathogenesis and progression. Anti-Aβ approaches by immunotherapy can be beneficial in the management of Alzheimer's symptoms. Production of long-term antibodies from short-term drug intake at limited cost is the main advantage of active immunotherapy (van Dyck 2018). Unlike active vaccination, passive immunization has some advantages of maintaining consistency of

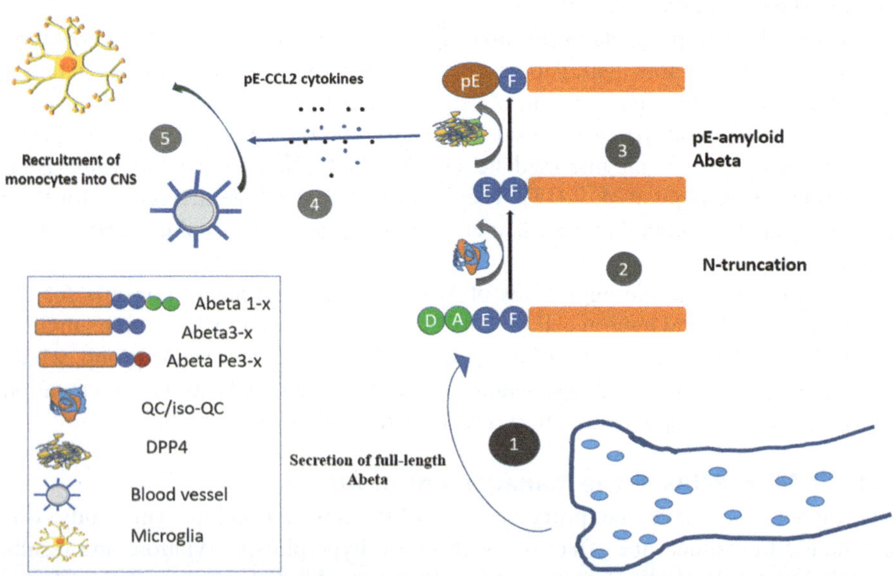

Fig. 1.5 Pyroglutamate peptide in Alzheimer's disease

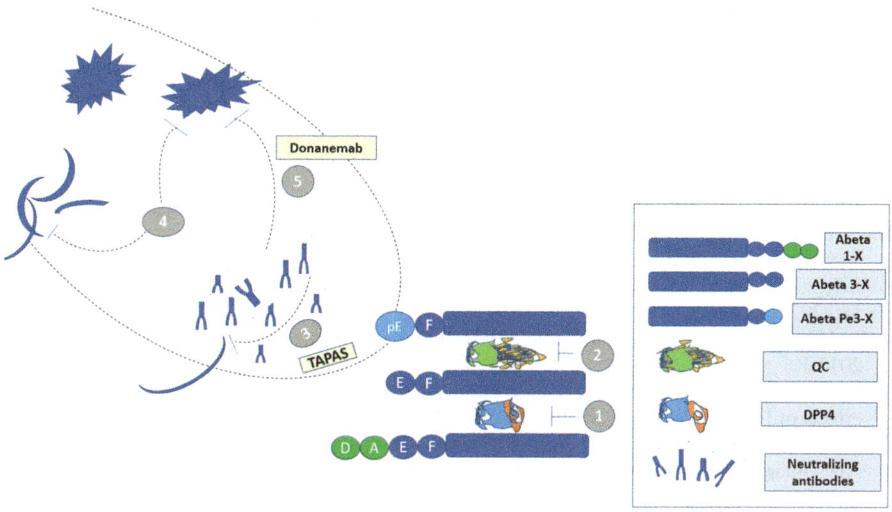

Fig. 1.6 Pyroglutamate Abeta as a potential target against Alzheimer's disease

Table 1.3 Various miRs with their target site in the management of AD

Interaction	Target site	Examples
Direct	Aβ peptide formation	miR-101, miR-106, miR-126
Direct	Tau protein phosphorylation	miR-219, miR-125b, miR-34a, miR-106b, etc.
Direct	α-Secretase and γ-secretase (α-cleavage)	miR-29a, miR-b-1, miR-9, etc.
Direct	β-Secretase-1 activity (β-cleavage)	miR-29a, miR-29b, miR-29c, miR-124, etc.
Indirect	Loss of function in mitochondria	miR-761, miR-30, miR-499, etc.
Indirect	Inflammatory mechanisms (T-cell association immune response)	miR-155, miR-125, miR-146a, miR-7b, etc.
Indirect	Senescence and aging	miR-155, miR-21, miR-34 series, miR-320

antibody titers and permitting management and control of adverse effects by withdrawing treatment.

It has been reported that antibodies are effective in the treatment of AD by targeting amyloid-β when peripherally administered; this is sufficient to reduce amyloid aggregation or burden (Guyen et al. 2000). Passive administration of antibodies was able to cross the central nervous system to prevent plaque and promote clearance of amyloid. Ex vivo assay study shows that anti-amyloid-β antibody stimulates microglial cells to clear plaques through phagocytosis (Guyen et al. 2000). These results indicate the ability of the antibody to cross the blood-brain barrier to act on CNS, and it is considered a potential approach for AD treatment.

A study had been performed to assess the effect of solanezumab antibody binding to soluble amyloid beta peptide, and it was observed that cognitive impairment decreases in the AD population. Solanezumab is an anti-amyloid antibody that has the property to bind with amyloid-β; these findings describe the solanezumab's effect and efficacy in a mild AD population (Siemers et al. 2016).

References

Aisen PS et al (2017) On the path to 2025: understanding the Alzheimer's disease continuum. Alzheimers Res Ther 9:1–10

Avila J (2006) Tau phosphorylation and aggregation in Alzheimer's disease pathology. FEBS Lett 580:2922–2927

Avila J et al (2012) Tau phosphorylation by GSK3 in different conditions. Int J Alzheimers Dis 2012:1

Baki L et al (2004) PS1 activates PI3K thus inhibiting GSK-3 activity and tau overphosphorylation: effects of FAD mutations. EMBO J 23:2586–2596

Baumann K, Mandelkow EM, Biernat J, Piwnica-Worms H, Mandelkow E (1993) Abnormal Alzheimer-like phosphorylation of tau-protein by cyclin-dependent kinases cdk2 and cdk5. FEBS Lett 336:417–424

Bayer TA (2022) Pyroglutamate Aβ cascade as drug target in Alzheimer's disease. Mol Psychiatry 27:1880–1885

Borzì AM et al (2019) Effects of vildagliptin, a DPP-4 inhibitor, in elderly diabetic patients with mild cognitive impairment. Arch Gerontol Geriatr 84:103896

Breydo L et al (2016) Structural differences between amyloid beta oligomers. Biochem Biophys Res Commun 477:700–705

Calderon-Garcidueñas AL, Duyckaerts C (2018) Alzheimer disease. Handb Clin Neurol 145:325–337

Calkins MJ, Manczak M, Hemachandra Reddy P (2012) Mitochondria-targeted antioxidant SS31 prevents amyloid beta-induced mitochondrial abnormalities and synaptic degeneration in Alzheimer's disease. Pharmaceuticals 5:1103–1119

Cummings J, Lee G, Zhong K, Fonseca J, Taghva K (2021) Alzheimer's disease drug development pipeline: 2021. Alzheimer's Dement Transl Res Clin Interv 7:1–24

Dubey T, Gorantla NV, Chandrashekara KT, Chinnathambi S (2019) Photoexcited toluidine blue inhibits tau aggregation in Alzheimer's disease. ACS Omega 4(20):18793–18802. https://doi.org/10.1021/acsomega.9b02792

Dursun E, Gezen-Ak D, Yilmazer S (2011) A novel perspective for Alzheimer's disease: vitamin D receptor suppression by amyloid-β and preventing the amyloid-β induced alterations by vitamin D in cortical neurons. J Alzheimers Dis 23:207–219

Estrada L, Soto C (2006) Disrupting beta-amyloid aggregation for Alzheimer disease treatment. Curr Top Med Chem 7:115–126

Frost D et al (2011) β-Carboline compounds, including Harmine, inhibit DYRK1A and tau phosphorylation at multiple Alzheimer's disease-related sites. PLoS One 6:e19264

Gobom J et al (2022) Antibody-free measurement of cerebrospinal fluid tau phosphorylation across the Alzheimer's disease continuum. Mol Neurodegener 17:1–14

Guyen MINHN et al (2000) F. Bard Nature Medicine. 6

Hamaguchi T, Ono K, Murase A, Yamada M (2009) Phenolic compounds prevent Alzheimer's pathology through different effects on the amyloid-β aggregation pathway. Am J Pathol 175: 2557–2565

Koudinov AR, Berezov TT, Koudinova NV (2002) Cholesterol and Alzheimer's disease: is there a link? Neurology 58:1135

Lahiri DK, Farlow MR, Hintz N, Utsuki T, Greig NH (2000) Cholinesterase inhibitors, β-amyloid precursor protein and amyloid β-peptides in Alzheimer's disease. Acta Neurol Scand Suppl 102: 60–67

Lee HG et al (2005) Tau phosphorylation in Alzheimer's disease: pathogen or protector? Trends Mol Med 11:164–169
Liu Y et al (2015) Graphene quantum dots for the inhibition of β amyloid aggregation. Nanoscale 7: 19060–19065
Liu W, Zhao J, Lu G (2016) miR-106b inhibits tau phosphorylation at Tyr18 by targeting Fyn in a model of Alzheimer's disease. Biochem Biophys Res Commun 478:852–857
Lorenzo A, Yankner BA (1994) Beta-amyloid neurotoxicity requires fibril formation and is inhibited by Congo red. Proc Natl Acad Sci U S A 91:12243–12247
Lyketsos CG et al (2011) Neuropsychiatric symptoms in Alzheimer's disease. Alzheimers Dement 7:532–539
Muñoz-Montaño JR, Moreno FJ, Avila J, Díaz-Nido J (1997) Lithium inhibits Alzheimer's disease-like tau protein phosphorylation in neurons. FEBS Lett 411:183–188
Murphy MP (2009) How mitochondria produce reactive oxygen species. Biochem J 417:1–13
Nedaei H et al (2022) The calcium-free form of atorvastatin inhibits amyloid-β(1–42) aggregation in vitro. J Biol Chem 298:101662
Paranjape SR et al (2015) Azaphilones inhibit tau aggregation and dissolve tau aggregates in vitro. ACS Chem Neurosci. 6(5):751–760. https://doi.org/10.1021/acschemneuro.5b00013
Park JS et al (2019) Brain somatic mutations observed in Alzheimer's disease associated with aging and dysregulation of tau phosphorylation. Nat Commun 10
Peterson DW, George RC, Scaramozzino F, Lapointe NE (2009) Cinnamon extract inhibits tau aggregation associated with Alzheimer's disease in vitro. J Alzheimers Dis 17:585–597
Protein A, Doi S (2013) Article protein science. 2234:1–42
Regitz C, Fitzenberger E, Mahn FL, Dußling LM, Wenzel U (2016) Resveratrol reduces amyloid-beta (Aβ1–42)-induced paralysis through targeting proteostasis in an Alzheimer model of Caenorhabditis elegans. Eur J Nutr 55:741–747
Saha S, Buttari B, Profumo E, Tucci P, Saso L (2022) A perspective on Nrf2 signaling pathway for neuroinflammation: a potential therapeutic target in Alzheimer's and Parkinson's diseases. Front Cell Neurosci 15:1–15
Salomon AR, Marcinowski KJ, Friedland RP, Zagorski MG (1996) Nicotine inhibits amyloid formation by the β-peptide. Biochemistry 35:13568–13578
Siemers ER et al (2016) Phase 3 solanezumab trials: secondary outcomes in mild Alzheimer's disease patients. Alzheimers Dement 12:110–120
Sonawane SK, Ahmad A, Chinnathambi S (2019) Protein-capped metal nanoparticles inhibit tau aggregation in Alzheimer's disease. ACS Omega 4(7):12833–12840. https://doi.org/10.1021/acsomega.9b01411
Spatharas PM et al (2022) Clusterin in Alzheimer's disease: an amyloidogenic inhibitor of amyloid formation? Biochim Biophys Acta Mol Basis Dis 1868:166384
Tam C, Wong JH, Ng TB, Tsui SKW, Zuo T (2018) Drugs for targeted therapies of Alzheimer's disease. Curr Med Chem 26:335–359
Tanzi RE, Bertram L (2001) New frontiers in Alzheimer's disease genetics. Neuron 32:181–184
van Dyck CH (2018) Anti-amyloid-β monoclonal antibodies for Alzheimer's disease: pitfalls and promise. Biol Psychiatry 83:311–319
Wang YJ, Zhou HD, Zhou XF (2006) Clearance of amyloid-beta in Alzheimer's disease: progress, problems and perspectives. Drug Discov Today 11:931–938
Wang Q et al (2013) Tanshinones inhibit amyloid aggregation by amyloid-β peptide, disaggregate amyloid fibrils, and protect cultured cells. ACS Chem Neurosci 4(6):1004–1015
Wischik CM, Edwards PC, Lai RYK, Roth M, Harrington CR (1996) Selective inhibition of Alzheimer disease-like tau aggregation by phenothiazines. Proc Natl Acad Sci U S A 93: 11213–11218
Wong KY et al (2020) Relationships between mitochondrial dysfunction and neurotransmission failure in Alzheimer's disease. Aging Dis 11:1291–1316

Yiannopoulou KG, Papageorgiou SG (2020) Current and future treatments in Alzheimer disease: an update. J Cent Nerv Syst Dis 12:117957352090739

Zhao K et al (2004) Cell-permeable peptide antioxidants targeted to inner mitochondrial membrane inhibit mitochondrial swelling, oxidative cell death, and reperfusion injury. J Biol Chem 279: 34682–34690

2 Oxidative Stress and Metals in Alzheimer's Disease

Shaik Ayesha Fathima, Ranika Maurya, and Saba Naqvi

Abstract

Alzheimer's disease (AD) is one of the progressive neurodegenerative disorders allied with genetic, lifestyle, and environmental factors and mainly affects the elderly population. Key pathological hallmarks of AD are extracellular depositions of beta-amyloid plaques, intracellular accumulation of neurofibrillary tangles (NFTs), and alterations in amyloid precursor protein (APP) gene, and it is clinically characterised by changes in cellular and molecular cascades of synaptic loss, cognitive dysfunctions, and neuronal death. Some evidences support the role of major biomarker candidates to induce AD, which are oxidative stress and metals, as brain is more prone to generate the reactive oxygen species (ROS) by different metal ions. Imbalance between free radicals and antioxidant defence system leads to cell apoptosis and dyshomeostasis of essential endogenous redox active metal ions in brain such as iron (Fe), copper (Cu), chromium (Cr), and cobalt (Co), and redox inactive metals like cadmium (Cd), arsenic (As), and lead (Pb) and inert metal zinc (Zn) show their toxic effects in Alzheimer's disease. To overcome the toxic effects of metal-induced ROS, there must be enhancement of two major antioxidant enzymes, catalase and superoxide dismutase. The chapter emphasises the dual protagonists oxidative stress and metals and their relation in the progression of AD.

Keywords

Oxidative stress · Metal toxicity · Neurodegeneration · Alzheimer's disease

S. A. Fathima · R. Maurya · S. Naqvi (✉)
Department of Pharmacology and Toxicology/Department of Regulatory Toxicology, National Institute of Pharmaceutical Education and Research (NIPER-R), Lucknow, Uttar Pradesh, India
e-mail: saba.naqvi@niperraebareli.edu.in

A. Sharma, G. P. Modi (eds.), *Natural Product-Based Synthetic Drug Molecules in Alzheimer's Disease*, https://doi.org/10.1007/978-981-99-6038-5_2

2.1 Introduction

Alzheimer's disease (AD) is a neurological condition characterised by progressive degeneration of nerve cells as it worsens, causing dementia and other brain abnormalities. It accounts for approximately 60–80% of cases and is mainly observed in the geriatric population as an age-related risk factor (Anonymous 2020). The changes in AD of the brain are structural changes like progressive brain atrophy and damaged neurons of the cerebral cortex and hippocampus regions causing functional changes like difficulty in thinking and concentration, delusions, mental decline, disorientation, forgetfulness, and inability to create new memories.

There are countless propositions to explain AD pathogenesis, involving the amyloid cascade theory, tau hyperphosphorylation theory, cellular excitotoxicity, metal-induced oxidative stress, and apoptosis (Fan et al. 2020). The excessive production of ROS and their further interactions with various metals can cause oxidative stress that alters different biological processes. Oxidative stress and metal-induced pathogenesis of AD are covered in this chapter.

2.2 Oxidative Stress

The difference between the amount of excessive oxidants (such as molecular oxygen and its derivatives, or ROS) produced and the antioxidants released to counteract them is known as oxidative stress. Because of the numerous intracellular metabolisms in the mitochondria, lipoxygenases, peroxisomes, and NADPH oxidase enzyme systems, oxidants are produced endogenously (Mattson 2004). In order to maintain physiological balance, an advanced enzymatic antioxidant system, such as catalase (CAT), superoxide dismutase (SOD), and glutathione peroxidase (GPx), and a non-enzymatic antioxidant system, such as glutathione and vitamins A, C, and E, neutralise and regulate overall ROS levels (Akbar et al. 2016).

It should be emphasised that owing to the high concentration of polyunsaturated fatty acids, excitotoxicity, and a mixed-function oxidase system, brain cells are predominantly vulnerable to oxidative damage (Manisha et al. 2017). As a source of oxidative stress, ROS are largely formed in the nervous system and active in the brain and neurons. As glial cells (post-mitotic cells) are easily vulnerable or prone to free radicals and result in the degeneration of neurons, this stress primarily affects them. Some studies have reported that harmful or toxic effects of ROS on various cells may cause shrinkage of cytoplasm and condensation of the nucleus and nuclear envelope that lead to the formation of phosphatidylserine on the surface region of the cell, i.e. sign of apoptosis and necrosis (Moreira et al. 2009). It provides enough evidence that the aetiology of Alzheimer's disease may be significantly influenced by the oxidative impairment triggered by free radicals (AD).

2.2.1 Oxidative Stress and Neurodegeneration

Neurodegeneration is a process that is characterised by abnormal functioning or complete loss of brain nerve cells commonly known as ataxia or dementia. Some factors like excitotoxicity, abnormal mitochondrial (Mt) functions, and finally programmed cell death are the main pathological changes in ageing and neurodegeneration of Alzheimer's disease (AD). While considering the part oxidative stress plays in Alzheimer's disease, some important points need to be discussed in short:

1. Why nerve cells are especially most sensitive or more vulnerable to oxidative stress?
2. What role do environmental and genetic factors play in the production of free radicals?
3. How do the toxic 'free radicals' cause neuronal death or cell apoptosis?
4. Oxygen is a key element in regulating various metabolic and physiological functions, and in normal physiology, it is exchanged for oxidative phosphorylation in mitochondria. Then how it becomes toxic to neurons?

Reactive oxygen species (ROS) are induced by the formation of free radicals associated with increased pro-oxidants and decreased antioxidants. It consists of nitric oxide (NO), hydrogen peroxide (H_2O_2), hydroxyl radicals (OH), and super-anions, which are highly reactive and damage mitochondrial activated microglial cells, which are present in the brain and turn as a reservoir of reactive oxygen species (ROS). A free radical is an electrically charged atom; one such free radical is superoxide, which is produced by the addition of an electron to an oxygen molecule (Fridovich 1999; Nordberg and Arnér 2001; Leigh 1990). These formed radicals do not get destroyed, so they enter various tissues, cells, cell organelles, organs, and organ systems; alter various cell signalling mechanisms, metabolic dysfunctions, and signal transduction pathways; and generate a phenomenon called oxidative stress. This production of reactive oxygen species (ROS) stimulates the mitogen-activated pathway (MAP), which in turn triggers a cascade of calcium excitotoxicity and apoptosis. Additionally, it was discovered that older rats' front parietal cortex, cortical neurons, and white matter all had much higher levels of RNA oxidation (Liu et al. 2002). Extensive oxidative damage to nucleic acids that results in changes in DNA structure is another indicator of oxidative stress and metals associated with AD (Cioffi et al. 2019).

Humans are continuously exposed to free radicals that are produced by any chemical, physical, or biological pollutants that are generated from the industries and pollutants, which are present in the ecosystem. Some physical sources like ionising radiations, γ-rays, cosmic rays, and X-rays also generate free radical ions in the environment. Living organisms are always attacked by oxygen free radical (OFR), which is present in pollutants with oxygen. Environmental exposures, like air pollution, might increase an organism's ability to create ROS; consequently, air pollution exposure may constitute a risk factor for AD by amplifying oxidative stress mechanisms capable of triggering physiological changes in the central nervous system (Yang and Omaye 2009). Air pollution has been found in clinical tests to impair cognitive performance and cause cerebrovascular damage (Moulton and Yang 2012). ROS-mediated damage seems to be complex in Alzheimer's disease, with interactions between dysfunctional mitochondria, redox transition metals, and other variables.

2.2.2 Relation Between Oxygen and Oxidative Stress

Oxygen, oxidative stress, and hypoxia are some factors that affect the cell's physiology. A constant supply of oxygen is indispensable for nerve survival and physiology. However, the role of oxygen and its associated mechanisms and other cascade processes in the nervous system is complex. As oxygen is essential for cells, whether neuronal or other cells that are involved in tissue formation, excessive oxygen level can be equally hazardous. During oxidative phosphorylation, the breakdown of glucose molecules takes place in the mitochondria (Mt) with the help of oxygen via ATP molecules. Mitochondria itself promotes oxidative stress and increases its intensity. Any genetic alterations like DNA mutation lead to dysfunction in the generation of ATP, leading to neuronal dysfunction and further leading to neuronal death. The primary reason of neuronal cell hypersensitivity to oxidative stress is related to structural and metabolic variables. The brain is predominantly sensitive to oxidative stress (Behl 2005), and glial cells of diverse types are involved in structural support and metabolic requirements. When compared to other endothelial cells in the body, endothelial cells are encircled by glial cells and are impervious to the absorption of different substances and defence cells such as macrophages. Furthermore, the brain requires greater oxygen and glucose intake to make continual ATP pooling in vivo for appropriate brain physiology since it always strives to maintain the other organs functioning, and it must be in normal physiology. Wherever the antioxidant levels are reduced in the aged brain, the consumption of oxygen is highly converted to free radicals and generates reactive oxygen species (ROS).

2.2.3 Mechanism of Generation of Reactive Oxygen Species (ROS)

The creation of free radicals is caused by oxygen molecules, which are hazardous to cells under certain conditions. Cells have regulatory mechanisms for the effective interaction of oxygen and metal ions, which results in reactive oxygen species (ROS) generation and oxidative stress. The metabolic route is useful for interacting with organic molecules in vivo since contact with oxygen is an important necessity owing to reactive oxygen species (ROS) and is actively inappropriate. In any form of reactive oxygen species (ROS) generation, molecular oxygen is needed to be actual and cellular or physiologically functioning, which is evolved in a wide range of metallo-enzymes that facilitate reactive oxygen species (ROS) generation upon the interaction of redox metals with oxygen using catalytic pathways:

$$\mathbf{Fe(III) + O_2^{\cdot -} \quad Fe(II) + O_2} \tag{2.1}$$

$$\mathbf{Fe(II) + H_2O_2 \quad Fe(III) + {}^{\cdot}OH + OH^{-}\ (Fenton\ reaction)} \tag{2.2}$$

The whole reaction of the pooled steps (3) is termed Haber-Weiss reaction:

$$\mathbf{O_2 + H_2O_2 \quad O_2 + {}^{\cdot}OH + OH^{-}} \tag{2.3}$$

In totalling the above reactions, the subsequent reactions may also take place:

$$\mathbf{OH + H_2O_2 \quad H_2O + H^{+} + O_2^{\cdot -}} \tag{2.4}$$

$$\mathbf{OH + Fe(II) \quad Fe(III) + OH^{-}} \tag{2.5}$$

$$\mathbf{LOOH + Fe(II) \quad Fe(III) + LO^{-} + {}^{\cdot}OH} \tag{2.6}$$

Fenton reaction yields hydroxyl radical, which in turn rejoins with various biomolecules, including lipids, proteins, and DNA, and causes their oxidation.

2.2.4 Role of Amyloid Beta (Aβ)-Induced Oxidative Stress in Alzheimer's Disease (AD)

AD is characterised by the existence of disorientated plaques of amyloid proteins, hyperphosphorylation of neurofibrillary tangles of tau proteins, changes in amyloid precursor protein (APP) gene, and clinical symptoms like memory loss, motor deficiency, and cognitive impairment by inducing various mechanistic factors like ROS, neuroinflammation, cerebrovascular damage, Aβ accumulation, and neuronal loss. Stress can also cause vasculitis in the brain, making the brain more susceptible to dementia. Stressful life events have been linked to causing havoc to the brain, which could lead to AD. Numerous studies demonstrate that oxidation processes occur in the AD brain and that Aβ proteins can directly generate free radicals by activating the NADPH oxidase system (Shelat et al. 2008). The condition progresses as the reactive chemicals develop into pro-oxidants, which sustain an ongoing auto-destructive process (Praticò 2005) (Fig. 2.1).

A fragment of the amyloid precursor protein (APP), which is a normal neuron membrane protein synthesised in the brain, aggregates to form extracellular amyloid plaques. Excessive synthesis of Aβ can be neurotoxic and can cause oxidative stress. Direct evidence shows increasing oxidation stress in AD:

1. Generation of free radicals in neurons and ROS formation
2. Decreased neurocytochromes like cytochrome C oxidase in the brain
3. Elevation of oxidation of proteins and DNA in aged brain

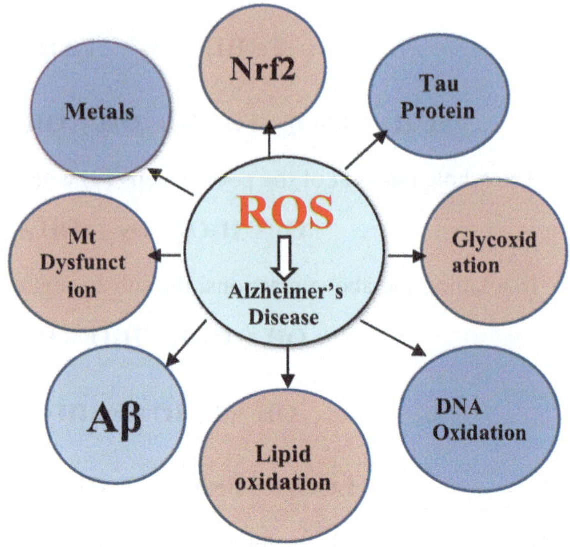

Fig. 2.1 Causative factors of Alzheimer's disease (AD)

4. Increased per oxidation of lipids in AD ventricular fluid due to membrane damage by ROS
5. Amyloid β peptide capable of generating free radicals
6. Decreased energy metabolism due to alterations in metabolic pathways
7. Decreased polyunsaturated fatty acids and increased aldehyde (4-hydroxynonenal) products during the peroxidation of lipids
8. Advanced glycation end products (AGE), malondialdehyde, and SOD-1 neurofibrillary tangles

The enzymes glutamine synthetase and creatine kinase can both be rendered inactive by the direct production of ROS. ROS can control the JNK/stress-activated protein kinase pathways. The instigation of these cascades is related to the hyperphosphorylation of tau proteins and Aβ-induced cell death (Ferrer et al. 2005). Aβ is created when the amyloid precursor protein is successively broken down by the secretases, β-site amyloid precursor protein-cleaving enzyme 1, and γ-secretase protein of the amyloidogenic pathway (APP). BACE 1 initiates the amyloidogenic process by cleaving the APP at two secretase sites. As a result, the plasma membrane secretes APP, which is a long version of APP, as well as CTF-99 or CTF-89. Since Aβ42 is more likely to oligomerise than Aβ40 or Aβ38, which increases its capacity to exert cytotoxic effects, Aβ42 looks to be more neurotoxic than Aβ40 or Aβ38 (Dimitrov et al. 2013), causing oxidative mutilation to synaptic membranes and inducing hyperphosphorylation of tau proteins (Fan et al. 2020). These neurons internalised and degraded the Aβ peptide when exposed to soluble oligomers of it (AβOs), released extracellular vesicles containing the active enzyme catalase (CAT), and selectively secreted interleukins-6 and -10 and vascular endothelial growth factor (VEGF) into the medium. The enzymes glutamine synthetase and creatine kinase can be inactivated directly by ROS, which is produced by the production of Aβ fragment (de Godoy et al. 2018). As a result, the main metabolic variation in AD is the alteration of Aβ into hazardous compounds by ROS, which form senile plaques and promote apoptosis (Ribarič 2018). In fact, a growing body of research indicates that the primary toxic agents in AD rather than amyloid deposits are soluble Aβ oligomers, monomers, and protofibrils (Verma et al. 2015). Fyn is rendered inactive as a result of γ-secretase, with Aβ activating the enzyme (STEP) striatal enriched protein tyrosine phosphatase, which sets off a chain of subsequent events that result in dendritic spine collapse (Mairet-Coello et al. 2013). It is clear that the interaction between Aβ and oxidative stress has a significant role in the pathological changes experienced by AD patients.

2.2.5 Oxidative Damage-Induced Tau Neurofibrillary Tangles

Microtubule-associated protein (MAP), tau protein, is recurrently present in the cytosol as well as in axons of neurons. Overexpression of Tau protein causes the course of stabilisation and destabilisation of microtubules, protein interactions that augment the pathological effects of tau and inhibition of kinase-dependent transport of Golgi complex-derived vesicles into neuritis, transportation defects and APP trafficking into neuron cells causing mutations in genes (apoE4), neurodegenerative histological dysfunctions, increased metal levels, inflammation, and consequent loss of synapses and neurons. Through the triggering of the p38 mitogen-activated protein kinase (MAPK) and c-Jun amino-terminal kinase, oxidative stress enhances the production of β-secretase (Tamagno et al. 2008) and promotes deviant phosphorylated tau through the activation of glycogen synthase kinase 3-(GSK-3-β) (Fang et al. 2000); the inactivation of certain molecules by oxidants may be significant as well. Prolyl isomerase PIN1 was discovered using a proteomic technique to be especially vulnerable to oxidative stress, being significantly downregulated and oxidised in the hippocampus of AD patients (Pastorino et al. 2006). Tau proteins include approximately 30 possible phosphorylation sites because of the abundance of serine and threonine phosphate-accepting residues. One of the many post-translational alterations that tau may experience is phosphorylation by kinases such as JKN, AMPK, and GSK-3. It was discovered that tau-enriched neurofibrillary tangles indicate oxidative damage. Nitric oxide is produced in close vicinity to the tau that creates the neurofibrillary tangles, according to research on the location of the enzyme dimethyl arginase (MacAllister et al. 1996) and regulation of the activity of nitric oxide synthase in hippocampal tissue from AD patients. These conclusions are confirmed by the revelation that an antibody that recognises an HNE-lysine adduct co-localises with endogenously produced paired helical tau filaments from AD brains (Takeda et al. 2000). Furthermore, in AD patients' brain, an aldehyde by-product of lipid peroxidation, i.e. acrolein, was found to co-localise with neurofibrillary tangles. Additionally, the antibody Alz50, which detects a conformational change in tau, corresponds with the antioxidant enzyme heme oxygenase-1 (HO-1), whose levels are noticeably raised in the AD and moderate cognitive impaired (MCI) brain (Carmel et al. 1996; Barone et al. 2012), and perhaps plays a critical role in the harmful development of dementia. Antioxidant therapy, for instance of nitrone and N-acetyl cysteine, inhibited the immune response to tau oligomers, demonstrating a further straight involvement of oxidative tension in tau assembly (Du et al. 2016).

2.3 ROS-Induced Disruption of Calcium Signalling in Alzheimer's Disease

In the AD brain, calcium (Ca^{2+}) dysregulation ensues preceding to the development of Aβ plaques plus neurofibrillary tangles, signifying so as disturbance in cytoplasmic Ca^{2+} may be one of the disease's primary causes. Cellular Ca^{2+} homeostasis is a critical regulator of many aspects of neuronal physiology, including synaptic plasticity, growth and differentiation, action potential properties, knowledge, and memory. The pathogenesis of AD is influenced by plasma membrane Ca^{2+} channels, lysosomal Ca^{2+} signalling, and mitochondrial Ca^{2+} signalling and shows apoptosis, necrosis, degeneration, and poor autophagy, which are also facilitated by aberrant cellular Ca^{2+}. Hydroxyl radicals, superoxide anion, and hydrogen peroxide are recognised ROS that control Ca^{2+} signalling pathways. The ROS generation/Ca^{2+} signalling paradigm was discovered as a consequence of functional impairment of membrane-bound receptors and channels that regulate Ca^{2+} influx or efflux brought on by oxidative stress-induced lipid peroxidation. The primary Ca^{2+} storage organelle in a cell is the ER, which can release Ca^{2+} in response to electrical and chemical cell stimulation (Bootman et al. 2001) through two types of Ca^{2+} release channels, the IP3R and the RyR. Ca^{2+} modulation is indirectly accomplished by causing membrane-damaging oxidative stress through voltage-dependent channels and ionotropic glutamate receptors, Ca^{2+} pumps, and increases Ca^{2+} inflow (Keller et al. 1997; Blanc et al. 1998; Mark et al. 1997). Aβ aggregates can inhibit Ca^{2+} signalling in a number of ways, including by activating the InsP3R and RyR, which release Ca^{2+} from ER storage, and by generating cation permeation pores on the plasma membrane, which enables Ca^{2+} entry. In response to agonists of phospholipase C (PLC)-coupled receptors, the IP3R releases Ca^{2+}. When PLC is activated, PI (4,5) P_2 is cleaved, releasing diacylglycerol and INS (1, 4, 5) P3 (IP3), which binds to IP3R. Calcium-induced calcium release by IP3R is a mechanism that further increases Ca^{2+} release through RyR (CICR). It has previously been demonstrated that IP3-evoked Ca^{2+} transients are amplified by the presence of the AD-linked presenilin 1 (PS1) mutation (Stutzmann et al. 2004). Exaggerated ER Ca^{2+} signals were shown to be caused by RyR activation brought on by Ca^{2+} released by IP3R. Increased intracellular Ca^{2+} overload, specifically its release from the ER, may cause an excessive amount of Ca^{2+} to be taken up by the mitochondria. The closeness of the ER and mitochondria on a physical level may contribute to this impact (Pinton et al. 1998; Csordás et al. 2006).

Increased ROS generation and decreased mitochondrial membrane potential are effects of excessive Ca^{2+} buildup in mitochondria. Cell death results from the large

Ca^{2+} influx into mitochondria that causes the potential of the mitochondrial membrane to collapse (Duchen 2000). The classic store-operated Ca^{2+} entry (SOCE) mechanism is activated by the depletion of ER Ca^{2+}, which then causes the stimulation of plasma membrane Ca^{2+} channels, which eventually stimulate a prolonged extracellular Ca^{2+} influx to the cytosol. By controlling DRD1 that is dopamine receptor D1 to activate CaMKII (which is a Ca^{2+}/calmodulin-dependent protein kinase II) via the non-canonical Gaq-Ca^{2+} signalling pathway, GHSR1a contributes to hippocampus synaptic physiology and memory preservation (Kern et al. 2015; Hsu et al. 2018a; Seminara et al. 2018). In AD, tau is abnormally hyperphosphorylated, which disrupts axonal transport and results in the death of neuronal cells. Given that numerous kinases are triggered through Ca^{2+}, disturbance of Ca^{2+} homeostasis caused via the PS mutation may greatly increase tau phosphorylation and cause neurofibrillary tangle development. Ca^{2+} fluctuations, mitochondrial function, gene expression, and apoptosis are just a few of the cellular activities that store-operated Ca^{2+} entry signalling in AD regulates. Some of the molecular targets include NMDR, which lessens cognitive impairment and behavioural outcomes in patients with moderate Alzheimer's disease by preventing excessive Ca^{2+} influx and sustained glutamate release, which cause excitotoxicity of neurons, rescuing tau hyperphosphorylation and protecting synapse type A. Regulating sodium and calcium permeability, which improves memory and learning, type L VGCC diminishes the synthesis of A-42 and the neurotoxicity caused by A-25-33 in cortical neurons; several of these agents have good blood-brain barrier penetration; T-type voltage-gated calcium channels are controlled in their activation, which improves cognitive performance; RyR stabilises ER calcium release, prevents synaptic loss, and enhances cognitive function, whereas InsP3R guards cells by limiting excessive caspase-3 and calcium activity (Tong et al. 2018). Antioxidants and mechanisms that control Ca^{2+} homeostasis by preventing its release from ER may be effective therapeutic strategies for preventing AD-related neuronal death.

2.4 ROS-Induced Mitochondrial Dysfunction (Mitochondrial Cascade Hypothesis)

Mitochondria control both cellular metabolism and apoptosis. The onset of AD has been related to microglial mitochondrial oxidative stress. AD has been connected, via a number of mechanisms, to elevated reactive oxygen species (ROS) generation and decreased mitochondrial membrane potential. AD interacts with microglial receptors including TREM2, which opens up a cascade of reactions that harm mitochondria and amplify inflammatory and cytotoxic reactions. Microglia's generation of mitochondrial ROS increases as a result of fibrillary Aβ stimulation of NADPH oxidase, which worsens the neurotoxic effects.

The pathophysiology of AD is influenced by the metals, microglia, TREM2, apoptosis, P2X7R, and mitochondrial dysfunction. Damage to the mitochondria impairs their ability to produce energy, causes oxidative stress, and produces mitochondrion-derived damage-associated molecular patterns that harm neurons and promote inflammation. The dynamics of mitochondrial fission and fusion are also out of balance in AD, which results in abnormal mitochondrial dispersion in neurons. Reactive oxygen species can be created when electrons decrease oxygen outside of the electron transport chain (ROS). By coordinating electron transfer with the pumping of protons across the inner mitochondrial membrane, mitochondria make ATP. Membrane potential (MP) and ROS production are related, and a high MP encourages more ROS production. High MP alters the redox potential of ETC carriers and lengthens the half-life of ubisemiquinone, both of which increase the generation of ROS. Additionally, any damage to ETC components could cause decreased intermediates to stall, increasing the chance of an electron sliding and lowering O_2 to produce ROS. To maintain the functioning of neurons, mitochondria produce ATP using the electron transport chain (ETC). Reduced mtDNA copy number from altered mitochondrial dynamics may lead to problems in mitochondrial electron transport function (Readnower et al. 2011). mtDNA oxidation rises in comparison to nuclear DNA, leading to an age-dependent buildup of mtDNA mutations. These mutations would result in a general decrease in the number of copies of the mtDNA, which would lower oxidative phosphorylation. The development of AD is aided by the elevated levels of cyclophilin D and Aβ in synaptic mitochondria. Patients with AD have abnormal mitochondrial dynamics, and it has been found that mitochondrial fission occurs more regularly than fusion in AD. Evidence shows that there are fewer mitochondria in AD, which is consistent with larger mitochondria, which supports this observation (Hirai et al. 2001). Fis1 and other proteins linked to fusion (dynamin-like protein and OPA1) have been found to have higher protein levels in response to APP overexpression through the synthesis of Aβ. Aβ has also been demonstrated to harm Drp1 by oxidative damage, which leads to mitochondrial fission (Cho et al. 2009). There is growing indication from studies that mitochondrial dysfunction is a key influence in the onset and advancement of AD.

2.5 Oxidative Stress and Damage to Biomolecules (Lipids, Proteins, DNA/RNA)

Strong oxidants called reactive oxygen species, also known as peroxynitrite, are produced when nitric oxide and superoxide anion react, and they can harm lipids, proteins, and nucleic acids (DNA and RNA). The pre-existing oxidative stress-induced damage that precedes and may involve and/or contribute to the neurofibrillary deterioration of neurons in the Alzheimer's disease brain may constitute the selective variation of a multitude of intracellular proteins, including key enzymes and structural proteins. Another sign of oxidative stress linked to AD is extensive oxidative injury to nucleic acids that results in modifications to DNA assembly. The nucleotide guanosine is oxidised in DNA/RNA oxidation to create 8-hydroxyguanosine (8-oxoG). High amounts of 8-oxoG were discovered in neurons in the hippocampus, subiculum, entorhinal cortex, and frontal, temporal, and occipital neocortex in apoptotic brain samples from AD patients. Additionally, it was discovered that RNA oxidation was much higher in various locations such as the front parietal cortex, white matter, cortical neurons, and hippocampus of old rats. According to certain reports, oxidative stress causes changes to the proteins tau and Aβ. Tau interacts dynamically with the generated microtubules to aid in microtubule structure, and their organisation's intracellular dynamics were shown to be disturbed in AD patients (Weingarten et al. 1975; Heston and White 1978).

Lipids, which make up the majority of cellular membranes, are essential for preserving the structural integrity of cells. The physical characteristics of cellular membranes are changed by excessive lipid oxidation, which can also lead to the covalent alteration of proteins and nucleic acids (Gaschler and Stockwell 2017). Numerous biological settings result in the assembly of lipid peroxides, which can act as signalling molecules by post-translationally altering proteins through enzymes or non-enzymatic mechanisms to be used in the production of lipid peroxides such as 5-lipoxygenase, 12/15-lipoxygenases, and chemistry of the Fenton-type reactions. The amount of the chromophore produced by the reaction between MDA and thiobarbituric acid can be evaluated by measuring the absorbance. This chromophore is utilised to detect lipid peroxides and the by-products of their breakdown. Similar to this, the degree of protein carbonylation has been determined using the reaction between the aldehyde moiety of 4-HNE and 2,4-dinitrophenylhydrazine. The role of lipid peroxides in numerous diseases and cell death has also been made clear by the fact that they can produce hazardous secondary messengers. In some situations, it has been demonstrated that the lipid breakdown product 4-HNE can cause apoptosis (Dalleau et al. 2013). Lipid peroxides' capacity to produce harmful secondary messengers has also served to emphasise their significance in a number of diseases and the part that cell death plays in the development and control of inflammation (Ackermann et al. 2017). Inhibiting the enzymes responsible for their synthesis, or using peroxidation inhibitors, is a very popular method for averting the progress of lipid peroxides.

DNA and RNA, which are made up of proteins involved in the disease, peaked early and remained increased. According to the study, nucleic acid oxidation is a common occurrence in neurodegeneration. Protein synthesis is slowed down or is abnormal when mRNA, rRNA, and tRNA are oxidised. Oxidative stress messes with the regulatory mechanisms of noncoding RNAs, particularly microRNAs, as well as this translational machinery (miRNAs). Oxidised miRNAs can mistakenly recognise target mRNAs. The vulnerable brain areas of AD contain three downregulated miRNAs that are miR-107, miR-210, and miR-485 as well as seven upregulated miRNAs that are miR-125b, miR-146a, miR-200c, miR-26b, miR-30e, miR-34a, and miR-34c, all of which are associated to oxidative stress (Nunomura and Perry 2020).

Transfer RNA (tRNA), microRNA, and ribosomal RNA (rRNA) susceptible to oxidative damage (miRNA): The stimulation of caspase-3 and subsequent apoptosis are caused, in part, by the preferential attachment of poly(C)-binding protein 1 (PCB1) to oxidised mRNA containing two contiguous 8-oxoGua residues (Ishii et al. 2018). Because it is available in the mitochondrial inter-membrane space, cytochrome c (cytc) can catalyse the oxidation of transfer RNA (tRNA), which primes to the formation of a cross-linking complex between tRNA and cytc and encourages its release from mitochondria, which then causes apoptosis (Tanaka et al. 2012). Even though Bcl-xL and Bcl-w are not their natural targets, oxidised miRNA-184 that contains 8-oxo-guanosine binds with them. The eventual apoptosis is caused by a subsequent decrease in Bcl-xL and Bcl-w (Wang et al. 2015). Better early intervention tactics may result from a deeper understanding of the effects and cellular conduct processes of the oxidatively changed RNAs, which may reveal information about the primary causes of neurodegenerative disorders.

DNA oxidative damage may be a significant factor in ageing and neurological illnesses like Alzheimer's disease (AD). Reactive oxygen species, predominantly hydroxyl radicals, can harm DNA by breaking DNA strands, forming DNA-DNA and DNA-protein cross links, exchanging and translocating sister chromatids, and producing at least 20 oxidised base adducts. Altering DNA bases can result in mutation and altered protein synthesis (Markesbery and Lovell 2006). Mild cognitive weakening, the first form of AD, has elevated levels of 8-OHG, 8-OHA, and 5,6-diamino-5-formamidopyrimidine in nuclear and mitochondrial DNA, indicating that DNA oxidation is an early occurrence in AD. Additionally, the interaction of DNA bases with alpha- and beta-unsaturated aldehydic by-products of lipid peroxidation, such as 4-hydroxynonenal and acrolein, can result in the production of large exocyclic adducts (Lovell and Markesbery 2007). Antioxidants as treatments have engrossed a lot of attentiveness due to the role that biomolecule oxidation and overall oxidative stress play in the progress of Alzheimer's disease.

2.6 Role of Metals in AD

Metals are part of the earth's crust and are found in water, the atmosphere, and many other ecosystems. Copper (Cu), chromium, cobalt, magnesium, iron (Fe), lithium, manganese (Mn), selenium (Se), nickel (Ni), and zinc (Zn) are among the essential metals. Metals play a vital role in our daily lives since they are engaged in so many enzymatic activities and serve crucial roles in maintaining cell structure and controlling processes like neurotransmission, antioxidant response, and gene expression. Both endogenous and exogenous metal exposures are known to cause changes in oligo-element homeostasis and can harm the central nervous system (CNS), cause oxidative stress, interfere with mitochondrial function, and inhibit the activity of various enzymes that play crucial catalytic, structural, and regulatory roles in various proteins, transporters, and receptors. These systems may be circumvented by toxic metal compounds, or they may be trapped inside and endanger the BBB. Evidence suggests that metal-induced neurotoxicity may be brought about via chemically induced blood-brain interface damage (Chen et al. 2014). While roughly they are necessary in tiny amounts, undue concentrations in the humanoid body typically cause neurotoxicity. The most prevalent deficiencies linked to metal-induced toxicity embrace mitochondrial malfunction, oxidative stress, and protein misfolding when metals get hoarded in the brain system (Wright and Baccarelli 2007; Angeli et al. 2014; Zhang et al. 2013; Seo et al. 2013). Numerous antioxidant enzymes, which are crucial for the brain and other organ functions, depend heavily on metal homeostasis. Additionally, it has been proposed that AD is caused by changes in critical metal homeostasis, which result in the interaction of metals with proteins and their consequent activation of aggregate formation (Gunter et al. 2010).

In the biological processes of metalloproteins and in neural processes, metal ions in the AD brains can rise up to three times over those in healthy control brains physiologically (Malecki 2001). A positive feedback loop of greater oxidation and higher ROS production is produced when the cations Zn^{2+} and Cu^{2+} bind to the hydrophilic N-terminal ends of Aβ peptides. Here, they are able to undertake continuous redox reactions that generate significant levels of ROS (Bondy et al. 1998; Strong et al. 1996).

2.7 Zinc

The second most common transition metal after Fe, Zn is a crucial trace element needed by humans and several other living organisms. Over 300 enzymes and metalloproteins use it as a cofactor, which controls gene transcription and antioxidant response. The testes, muscle, liver, and brain contain the majority of the body's Zn. Zn deficiency in humans impacts learning capacity as well as mental and physical development.

Researchers have looked into the unusually high Zn^{2+} content in the brains of AD patients to draw conclusions about the relationship between the pathogenesis of AD and an imbalance in Zn^{2+} homeostasis. Zn inhibits amyloid-induced neurotoxicity by preferentially precipitating aggregation intermediates at low doses (a few micromolars) (Garza-Lombó et al. 2018; Spiers et al. 2022).

Zinc's contribution to the pathophysiology of AD is supported by two major lines of evidence: (1) zinc concentrations in the brain, blood serum, and cerebrospinal fluid are frequently used as AD biomarkers and (2) zinc concentrations in brain, blood serum, and cerebrospinal fluid are densely innervated by zinc-containing axons, whereas those less affected by pathology contain few zinc-containing terminals (Chen et al. 2012). The phenomenon known as 'zinc fooding', in which zinc-containing neurons abruptly release free zinc into the brain's extracellular spaces, is a key aspect of this examination. This phenomenon can appear in response to a number of circumstances, including ischaemia, convulsions, and traumatic brain damage. However, an experimental investigation using a tau mouse model revealed that Zn supplementation exacerbated the behavioural and biochemical deficiencies brought on by tau proteins (Barnham et al. 2003).

The high concentration of Zn^{2+} ions binding to β-amyloid may enhance the formation of fibrillar β-amyloid aggregation in the synaptic cleft, where oxidised Aβ protein can build up and combines with the protein to produce a precipitate of poisonous Aβ peptides that are unable to leave the synapses leading to neurodegeneration (Cuajungco et al. 2000). The easier formation of hazardous oligomers and, ultimately, plaques is made possible by the binding of metal ions to Aβ monomers. Microtubule instability caused by intracellular Zn depletion triggers the release of tau, hyperphosphorylation, and development of neurofibrillary tangles. The release of Zn from metallothionines by intracellular Zn excess, which results from Aβ aggregation and ROS production, may disrupt mitochondrial function and trigger apoptosis.

Metal ions can also interact with tau proteins in a similar way. It has been shown that Zn^{2+} can bind tau and encourage its phosphorylation (Garai et al. 2007). We discovered that tau fibrillisation and formation of paired helical filaments are induced and accelerated by even micromolar concentrations of Zn^{2+} ions. This is accomplished by upregulating kinases that phosphorylate tau, such as GSK-3, and inhibiting kinases that de-phosphorylate tau proteins, such as protein phosphatase 2A.

Zn influences key pro-inflammatory signalling pathways by acting as an anti-inflammatory element. Zn inhibits the dissociation of nuclear factor k B (NF-kB) from the inhibitory protein that it is paired with, preventing NF-kB from translocating into the nucleus and thus suppressing inflammation. Zn also prevents STAT3 activation brought on by IL-6 (Suh et al. 2000). Zn levels that are excessively high induce the inhibition of Cu and Fe 83 absorption, an increase in the production of ROS in the mitochondria, and disruption of the activity of metabolic enzymes, all of which lead to the induction of apoptosis. By activating ERK1/2, which can phosphorylate beclin-1 and thereby facilitate the formation of the beclin-1-PI3K complex during the autophagic process, zinc promotes autophagy. Additionally, zinc can facilitate the degradation of mTOR, a negative regulator of autophagy, which results in cell autophagy.

2.8 Iron Toxicity

All living things require iron (Fe), a redox active metal, for healthy physiological functions. Fe is necessary for cell growth at the cellular level, but too much of it (Fe overload) results in oxidative stress and cell death. Fe homeostasis refers to the process of strictly controlling Fe levels. It is frequently found as heme-containing proteins, as a cofactor in proteins that contain Fe-sulphur clusters, and as proteins that contain Fe ions (Craven et al. 2018). A couple of examples of heme-containing proteins are catalases and peroxidases. Fe is very strongly associated with the pathophysiology of AD.

Although free Fe is more likely to form free radicals and exchange electrons with adjacent molecules than bound Fe, bound Fe is regarded as safe because it can lead to additional Fe release from proteins that contain Fe, such as heme proteins, ferritin, and Fe-sulphur clusters. A lack of iron can also make it easier for the body to absorb divalent elements including lead (Pb), cadmium (Cd), aluminium (Al), and manganese (Mn). Even in the lack of excess Mn in the brain, Fe shortage might promote Mn accumulation there (Atwood et al. 2003). There is an inverse link between Mn and Fe because Fe and Mn, along with other important metals, are regulated inside the basal ganglia by influx into the brain via transferrin and TfRs, as well as by DMT-1. Fe-induced oxidative stress is particularly hazardous (Naqvi et al. 2010).

This is because it can create an intracellular positive feedback loop that worsens the toxic effects of brain Fe overload. Fe is taken up by the BBB in the brain. The capillary endothelial cell TfR absorbs Fe in the form of transferrin-$Fe3^{+}$. Additionally, Fe^{3+} can attach to p-tau monomers to cause aggregation and creation of oligomers. NFTs can undergo continuous Fenton redox reactions as a result of the accumulation of Fe^{3+} ions, which produces large levels of ROS and intensifies intracellular oxidative stress. For iron to be absorbed from food, the DMT-1 is a crucial transporter. Intestinal DMT-1 levels are upregulated by Fe deprivation, and Mn absorption and neurotoxicity are also increased (Cristóvão et al. 2016).

2.9 Copper Toxicity

Copper is thought to be the metal ion that is most redox reactive. Redox regulation guards against various forms of oxidative stress and preserves 'redox equilibrium' by managing the redox state. In actuality, the redox characteristics of Cu ions are what cause the neurotoxic effects, increasing ROS generation in the CNS (Gammoh and Rink 2017). The brain is thought to be particularly vulnerable to the harmful effects of ROS due to its high metabolic rate and relatively low capacity for cellular control when compared to other organs. Cu is a necessary element and a crucial part of many enzymes, such as cytochrome c oxidase, dopamine hydroxylase, monoamine oxidase, and superoxide dismutase (Cu/Zn-SOD or SOD1), which are crucial for oxygen and electron transport, protein modification, and production of neurotransmitters (Mezzaroba et al. 2019). ROS manifestations and neurodegeneration have been linked to necessary metals, such as Cu. Cu in excess is neurotoxic and has been associated with the progression of AD.

Because of its weak albumin binding, it is likely for Cu ions to be liberated from the albumin-bound moiety at the BBB and then transported into the brain. The ATP-dependent transport of Cu across the BBB is carried out by two intracellular proteins called Cu-transporting P-type ATPases (ATP7A and ATP7B), which belong to a subclass of ATPases. However, it is unclear how these proteins are distributed within the BBB. The BBB is impermeable to Cu under typical physiological circumstances. Specific Cu transport systems are needed to move Cu between two fluid compartments across the BBB. The BBB-mediated variations in copper ion (Cu^{2+}) levels have a significant impact on Aβ metabolism. Cu can only passively diffuse into the cerebrospinal fluid under specific pathological circumstances where BBB permeability is reduced.

Three steps can be used to describe how copper ions associate with Aβ. Copper is first bound to endogenous reductants, and then Cu (II) is reduced to copper. Copper's reductive state causes molecular oxygen to also undergo reduction, ensuing in the generation of ROS. The toxicity of amyloid oligomers and plaques is exacerbated as a result of copper's direct interactions with Aβ (Mills et al. 2010).

Copper interacts with amyloid and tau proteins to promote aberrant protein accumulation (Bader et al. 2011). Cu may promote amyloid precursor protein and amyloid peptide self-aggregation, and elevated levels of Cu in cerebral fluid have been observed in some AD patients (Mezzaroba et al. 2019).

Cu appears to accelerate the Aβ cascade by promoting Aβ production and accretion in AD plaques, and its deposition in the CNS. Intracellular tangles first form inside neurons when Fe and Cu bind to hyperphosphorylated tau protein. Free Fe^{2+} or Cu^{1+} species will cause harmful Fenton reactions, including the production of ROS and micro-inflammation (Smith et al. 1997). Cu's potent affinity for Aβ, which promotes its aggregation and heightens oxidative stress via the Fenton reaction, has traditionally been thought to be its neurotoxic mechanism of action. Therefore, it has been proposed that Cu build-up mediates neurotoxicity and that its removal from the brain prevents or reverses the load of Aβ plaque. Recent research reveals that dyshomeostasis of Cu and its valency in the body, rather than acculturation and interaction with Aβ, are the primary determinants of either its neurotoxic or helpful effects as an essential metal. Aβ1–42 that has been stabilised by copper interacts with cell membranes to make them more permeable. Lysosomes, the organelles involved in autophagy, have shown an increase in copper. Elevated copper levels in the redox cycle of metal ions encourage autophagy and apoptosis in glioma cells through the activation of JNK and reactive oxygen species.

Deregulated copper ions may initiate and enhance tau hyperphosphorylation and formation of sheet-rich tau fibrils, which in turn lead to synaptic failure, neuronal death, and cognitive impairment found in AD patients.

The fact that genetic mutations on Cu transporters that cause loss of function cause severe neurological symptoms is another argument in AD pathology. Unique pathways of Cu neurotoxicity were postulated, mediated by non-neuronal cell lineages in the brain, like capillary endothelial cells, contributing to the development of AD neuropathology, along with its changed distribution (Acevedo et al. 2019).

2.10 Aluminium

Al is used in many different things, such as food packaging, preservative cans, cookware, automobiles, and vaccine adjuvants, to name a few. Oxidative stress and mitochondrial dysfunction are two potential reasons of the brain damage. In human AD, aluminium increases plaque and tangle pathology, impairing cholinergic neurotransmission and causing similar neural network destruction (Kawamata and Manfredi 2010). The prolonged retention percentage of aluminium in the brain suggests that it may accumulate to hazardous levels over time. The same neurons that form intracellular neurofibrillary clumps also develop aluminium deposits; however, Al hydroxide impairs long-term memory, increases anxiety, and kills neurons in the spinal cord and motor cortex. Aluminium salt produces localised neurodegenerative effects that resemble AD.

The growth of iron-driven oxidative stress events linked to numerous separate but functionally related gain-of-toxicity and loss-of-function processes is the postulated mechanism of aluminium-induced neurotoxicity. The inositol phosphate system and calcium control may be disrupted by aluminium deposition, which could lead to neurodegeneration. Aluminium promotes harmful redox reactions caused by iron, primarily by simultaneously activating superoxide dismutase and inhibiting catalase (Stutzmann et al. 2004). After chronic aluminium exposure, transgenic mice's hippocampal tissues' RNA expression can be examined to indicate redox stress patterns. Because the integrity of biological membranes is compromised by free radical attacks on proteins and lipids, unchecked redox reactions are devastating. By opening the mitochondrial permeability transition pore (MTP) and encouraging cytochrome c translocation into the cytoplasm, aluminium ions indirectly drive mitochondrial dysfunction. Effects of aluminium on biological systems, such as changes in protein accumulation, gene expression, and membrane disruption, have only been observed in quantity that is significantly higher than that found in people (Wang et al. 2020).

2.11 Manganese

Manganese (Mn) is a universally important trace element needed for healthy cellular homeostasis, growth, and development. The chemical forms of Mn include chelates, salts (sulphate and gluconate), and oxidation states (Mn^{2+}, Mn^{3+}, Mn^{4+}, Mn^{6+}, and Mn^{7+}) (aspartate, fumarate, succinate) (Roos et al. 2006). In astrocytes, the abundant manganoprotein glutamine synthetase (GS) expresses itself mostly and produces glutamine by converting glutamate to glutamine. Low levels of Mn in the brain lower GS activity since it has been proposed that Mn regulates GS activity. Enhanced glutamate transport and glutamatergic oxidative stress, mitochondrial malfunction, dysregulation of autophagy, build-up of intracellular hazardous compounds, and apoptosis are some of the underlying mechanisms. Alzheimer's disease (AD) and other neurological illnesses associated with ageing depend on mitochondria.

Being a necessary metal, manganese is mostly obtained through dietary means. However, inhaling large concentrations of the metal can result in brain manganese build-up and manganism, a condition that resembles neurodegeneration. Manganese from food passes across the blood-brain barrier (BBB), but it is absorbed through the olfactory transport channel from inhalation, causing a build-up of the metal in the brain. In animal models, too much manganese impairs MnSOD, which leads to oxidative stress and pathophysiology of AD, including Aβ build-up and tau phosphorylation. Although the potential relationship with clinical AD has not yet been proven, manganese binds Aβ and elevated manganese exposure is connected with cognitive losses in human epidemiologic research. A specific type of neurodegenerative disease called manganism is linked to extremely high manganese exposure. The amounts of additional critical metals affect how much manganese has an impact on the nervous system. Mn might contribute to the development of AD. According to reports, the development of senile plaques is linked to deregulated Mn metabolism in AD patients as well as a malfunctioning Mn-SOD scavenger system.

Patients with neuropathology-confirmed AD have brains with decreased mitochondrial MnSOD activity. Additionally, it has been shown that Fe regulates the transport of Mn across the BBB and that a disturbed distribution of Fe has been linked to the pathophysiology of AD (Ward et al. 2015). The majority of intracellular Mn is stored in mitochondria, and an increase in Mn levels in this organelle can directly affect oxidative phosphorylation, limiting the activity of F1-ATPase and, as a result, cellular ATP generation (Hsu et al. 2018b). The excessive ROS are then produced by oxidative stress caused by high intra-mitochondrial Mn levels, which leads to mitochondrial malfunction. Its pro-oxidant capability is increased by the transition from Mn^{+2} to Mn^{+3}. Oxidative stress induced by Mn causes the mitochondrial transition to open.

2.12 Cadmium

Cadmium (Cd) is a proven human carcinogen and non-essential transition heavy metal. The main ways that people are exposed to cadmium are through their diet and cigarettes. Cadmium exposure through inhalation can pass past the olfactory bulb and blood-cerebrospinal fluid barrier and enter the brain. Cadmium induces oxidative stress, neuroinflammation, and neuronal death in animal models of the brain. By altering the BBB's permeability, causing AD to aggravate thereby creating tau neurofibrillary tangles as cadmium causes neurotoxicity. Cadmium may be linked to clinical AD and reduced cognitive performance specifically in human ageing research (Yokel 2000). Given the uncertainties surrounding cadmium transport to the brain, the pathophysiologic connection between environmental cadmium exposure and AD is, however, rather tenuous.

A metal like cadmium that is redox inactive indirectly causes oxidative stress. For sulfhydryl groups of thiols like glutathione and metallothionine, cadmium exhibits a strong attraction. The antioxidant defence system is interfered by either short-term high-level exposure or long-term persistent low-level exposure (Nandi et al. 2019). In neural cells as well as brain endothelial cells, cadmium causes oxidative stress. Glutathione detoxification is activated at low cadmium levels. Glutathione depletion occurs at larger concentrations with ongoing oxidative stress.

Another putative cadmium-AD route is poisoning of cholinergic neurons. Acetylcholinesterase is altered, and basal forebrain cholinergic neurons degenerate as a result of cadmium exposure, which accelerates cell death on cholinergic neurons. Cadmium impairs neurodevelopment and induces oxidative stress-dependent neuroinflammation. Direct effects on neuronal cells through oxidative stress, neuroinflammation, and apoptosis are well understood. By altering the BBB's permeability and combining with other neurotoxicants, cadmium may also cause neurotoxicity, aggregation of Aβ, and formation of tau neurofibrillary tangles (Lidsky 2014). Cadmium exposure causes pathogenic mechanisms that lead to cognitive dysfunction and AD pathogenesis.

2.13 Lead

Lead contamination is widespread as a result of recent and previous industrial usage. Lead enters the bloodstream through ingestion, inhalation, or cutaneous absorption before it can cross the blood-brain barrier (Deore et al. 2021). Lead is a strong neurotoxin that disrupts the brain in a generalised manner and leads to oxidative stress, mitochondrial damage, endoplasmic reticulum stress, excitotoxicity, altered homeostatic metal signalling, inflammation, and finally neuronal apoptosis (Du et al. 2017). Lead treatment results in memory problems and AD-related pathologies in animal models, including alterations in tau, APP, and Aβ.

Additionally, it disrupts Ca^{2+} homeostasis, prevents PKC115 from being phosphorylated, and lowers nitric oxide generation. Intelligence, executive functioning, memory, attention, processing speed, language, emotion, and motor and visuospatial skills are all negatively impacted by Pb exposure.

Microtubule-associated protein tau (MAPT): Transgenic mice treated with 0.2% lead acetate water during PNDs 1–20 had lead-related altered expression of MAPT and miR-34c, a miRNA that targets MAPT causing cytoskeleton stability impairment and neuronal dysfunction (Martinez-Finley et al. 2013). Rats of both sexes exposed to these conditions had increased tau protein in the forebrain and cerebellum and tau hyperphosphorylation. Mice and rats that were exposed to lead early in life had pathology that was related to Alzheimer's disease (Bihaqi and Zawia 2013). Increased tau mRNA, tau protein, its transcriptional regulators (Sp1 and Sp3), and site-specific tau hyperphosphorylation were all linked to early-life lead exposure.

2.14 Conclusion

In conclusion, metal-induced oxidative stress has been identified as a significant contributor to the development and advancement of Alzheimer's disease. Elevated levels of metals such as copper, aluminium, and iron have been found in the brains of individuals with Alzheimer's disease, and these metals can generate reactive oxygen species and lead to oxidative damage to cells and tissues. This damage can result in the accumulation of amyloid beta protein and tau tangles, which are key pathological features of Alzheimer's disease. Moreover, metal-induced oxidative stress can also lead to inflammation and cell death, which can further contribute to the deterioration of brain function in individuals with Alzheimer's disease. For example, studies have shown that individuals with Alzheimer's disease have higher levels of aluminium in their brains compared to healthy individuals. Additionally, researchers have found that increasing the levels of copper in the brains of mice leads to the accumulation of amyloid beta protein and the development of Alzheimer's disease-like symptoms. Similarly, studies have demonstrated that iron accumulation in the brain is associated with the development of Alzheimer's disease.

Targeting metal-induced oxidative stress may represent a promising strategy for the prevention and treatment of Alzheimer's disease. For example, studies have shown that chelating agents, which bind to metals and remove them from the body, can reduce oxidative stress and improve cognitive function in individuals with Alzheimer's disease. Additionally, dietary interventions, such as increasing the intake of antioxidants, may also help to mitigate metal-induced oxidative stress and prevent or delay the onset of Alzheimer's disease.

Acknowledgments The authors acknowledge the award of Research Fellowships and financial support by the Department of Pharmaceuticals, Ministry of Chemical and Fertilizer, Government of India. NIPER-R communication no. is 395.

Conflicts of Interest The authors declare no conflict of interest.

References

Acevedo K, Masaldan S, Opazo CM, Bush AI (2019) Redox active metals in neurodegenerative diseases. J Biol Inorg Chem 24(8):1141–1157. https://doi.org/10.1007/s00775-019-01731-9

Ackermann JA, Hofheinz K, Zaiss MM, Krönke G (2017) The double-edged role of 12/15-lipoxygenase during inflammation and immunity. Biochim Biophy Acta Mol Cell Biol Lipids 1862(4):371–381. https://doi.org/10.1016/j.bbalip.2016.07.014

Akbar H, Duan X, Saleem S, Davis AK, Zheng Y (2016) RhoA and Rac1 GTPases differentially regulate agonist-receptor mediated reactive oxygen species generation in platelets. PLoS One 11(9):e0163227. https://doi.org/10.1371/journal.pone.0163227

Angeli S, Barhydt T, Jacobs R, Killilea DW, Lithgow GJ, Andersen JK (2014) Manganese disturbs metal and protein homeostasis in Caenorhabditis elegans. Metallomics 6(10):1816–1823. https://doi.org/10.1039/c4mt00168k

Anonymous (2020) 2020 Alzheimer's disease facts and figures. Alzheimers Dement. https://doi.org/10.1002/alz.12068

Atwood CS, Obrenovich ME, Liu T, Chan H, Perry G, Smith MA, Martins RN (2003) Amyloid-beta: a chameleon walking in two worlds: a review of the trophic and toxic properties of amyloid-beta. Brain Res Brain Res Rev 43(1):1–16. https://doi.org/10.1016/s0165-0173(03)00174-7

Bader B, Nübling G, Mehle A, Nobile S, Kretzschmar H, Giese A (2011) Single particle analysis of tau oligomer formation induced by metal ions and organic solvents. Biochem Biophys Res Commun 411(1):190–196. https://doi.org/10.1016/j.bbrc.2011.06.135

Barnham KJ, McKinstry WJ, Multhaup G, Galatis D, Morton CJ, Curtain CC, Williamson NA, White AR, Hinds MG, Norton RS, Beyreuther K, Masters CL, Parker MW, Cappai R (2003) Structure of the Alzheimer's disease amyloid precursor protein copper binding domain. A regulator of neuronal copper homeostasis. J Biol Chem 278(19):17401–17407. https://doi.org/10.1074/jbc.M300629200

Barone E, Di Domenico F, Sultana R, Coccia R, Mancuso C, Perluigi M, Butterfield DA (2012) Heme oxygenase-1 posttranslational modifications in the brain of subjects with Alzheimer disease and mild cognitive impairment. Free Radic Biol Med 52(11–12):2292–2301. https://doi.org/10.1016/j.freeradbiomed.2012.03.020

Behl C (2005) Oxidative stress in Alzheimer's disease: implications for prevention and therapy. Subcell Biochem 38:65–78. https://doi.org/10.1007/0-387-23226-5_3

Bihaqi SW, Zawia NH (2013) Enhanced taupathy and AD-like pathology in aged primate brains decades after infantile exposure to lead (Pb). Neurotoxicology 39:95–101. https://doi.org/10.1016/j.neuro.2013.07.010

Blanc EM, Keller JN, Fernandez S, Mattson MP (1998) 4-hydroxynonenal, a lipid peroxidation product, impairs glutamate transport in cortical astrocytes. Glia 22(2):149–160. https://doi.org/10.1002/(sici)1098-1136(199802)22:2<149::aid-glia6>3.0.co;2-2

Bondy SC, Guo-Ross SX, Pien J (1998) Mechanisms underlying the aluminum-induced potentiation of the pro-oxidant properties of transition metals. Neurotoxicology 19(1):65–71

Bootman MD, Collins TJ, Peppiatt CM, Prothero LS, MacKenzie L, De Smet P, Travers M, Tovey SC, Seo JT, Berridge MJ, Ciccolini F, Lipp P (2001) Calcium signalling—an overview. Semin Cell Dev Biol 12(1):3–10. https://doi.org/10.1006/scdb.2000.0211

Carmel G, Mager EM, Binder LI, Kuret J (1996) The structural basis of monoclonal antibody Alz50's selectivity for Alzheimer's disease pathology. J Biol Chem 271(51):32789–32795. https://doi.org/10.1074/jbc.271.51.32789

Chen X, Guo C, Kong J (2012) Oxidative stress in neurodegenerative diseases. Neural Regen Res 7(5):376–385. https://doi.org/10.3969/j.issn.1673-5374.2012.05.009

Chen P, Parmalee N, Aschner M (2014) Genetic factors and manganese-induced neurotoxicity. Front Genet 5:265. https://doi.org/10.3389/fgene.2014.00265

Cho DH, Nakamura T, Fang J, Cieplak P, Godzik A, Gu Z, Lipton SA (2009) S-nitrosylation of Drp1 mediates beta-amyloid-related mitochondrial fission and neuronal injury. Science (New York, N.Y.) 324(5923):102–105. https://doi.org/10.1126/science.1171091

Cioffi F, Adam RHI, Broersen K (2019) Molecular mechanisms and genetics of oxidative stress in Alzheimer's disease. J Alzheimers Dis 72(4):981–1017. https://doi.org/10.3233/JAD-190863

Craven KM, Kochen WR, Hernandez CM, Flinn JM (2018) Zinc exacerbates tau pathology in a tau mouse model. J Alzheimers Dis 64(2):617–630. https://doi.org/10.3233/JAD-180151

Cristóvão JS, Santos R, Gomes CM (2016) Metals and neuronal metal binding proteins implicated in Alzheimer's disease. Oxidative Med Cell Longev 2016:9812178. https://doi.org/10.1155/2016/9812178

Csordás G, Renken C, Várnai P, Walter L, Weaver D, Buttle KF, Balla T, Mannella CA, Hajnóczky G (2006) Structural and functional features and significance of the physical linkage between ER and mitochondria. J Cell Biol 174(7):915–921. https://doi.org/10.1083/jcb.200604016

Cuajungco MP, Goldstein LE, Nunomura A, Smith MA, Lim JT, Atwood CS, Huang X, Farrag YW, Perry G, Bush AI (2000) Evidence that the beta-amyloid plaques of Alzheimer's disease represent the redox-silencing and entombment of abeta by zinc. J Biol Chem 275(26): 19439–19442. https://doi.org/10.1074/jbc.C000165200

Dalleau S, Baradat M, Guéraud F, Huc L (2013) Cell death and diseases related to oxidative stress: 4-hydroxynonenal (HNE) in the balance. Cell Death Differ 20(12):1615–1630. https://doi.org/10.1038/cdd.2013.138

de Godoy MA, Saraiva LM, de Carvalho LRP, Vasconcelos-Dos-Santos A, Beiral HJV, Ramos AB, Silva LRP, Leal RB, Monteiro VHS, Braga CV, de Araujo-Silva CA, Sinis LC, Bodart-Santos V, Kasai-Brunswick TH, Alcantara CL, Lima APCA, da Cunha-E Silva NL, Galina A, Vieyra A, De Felice FG et al (2018) Mesenchymal stem cells and cell-derived extracellular vesicles protect hippocampal neurons from oxidative stress and synapse damage induced by amyloid-β oligomers. J Biol Chem 293(6):1957–1975. https://doi.org/10.1074/jbc.M117.807180

Deore MS, Keerthana S, Naqvi S, Kumar A, Flora SJS (2021) Alpha-lipoic acid protects co-exposure to Lead and zinc oxide nanoparticles induced neuro, Immuno and male reproductive toxicity in rats. Front Pharmacol 12:626238. https://doi.org/10.3389/fphar.2021.626238

Dimitrov M, Alattia JR, Lemmin T, Lehal R, Fligier A, Houacine J, Hussain I, Radtke F, Dal Peraro M, Beher D, Fraering PC (2013) Alzheimer's disease mutations in APP but not γ-secretase modulators affect epsilon-cleavage-dependent AICD production. Nat Commun 4: 2246. https://doi.org/10.1038/ncomms3246

Du X, West MB, Cheng W, Ewert DL, Li W, Saunders D, Towner RA, Floyd RA, Kopke RD (2016) Ameliorative effects of antioxidants on the hippocampal accumulation of pathologic tau in a rat model of blast-induced traumatic brain injury. Oxidative Med Cell Longev 2016: 4159357. https://doi.org/10.1155/2016/4159357

Du K, Liu M, Pan Y, Zhong X, Wei M (2017) Association of Serum Manganese Levels with Alzheimer's disease and mild cognitive impairment: a systematic review and meta-analysis. Nutrients 9(3):231. https://doi.org/10.3390/nu9030231

Duchen MR (2000) Mitochondria and calcium: from cell signalling to cell death. J Physiol 529 Pt 1 (Pt 1):57–68. https://doi.org/10.1111/j.1469-7793.2000.00057.x

Fan L, Mao C, Hu X, Zhang S, Yang Z, Hu Z, Sun H, Fan Y, Dong Y, Yang J, Shi C, Xu Y (2020) New insights into the pathogenesis of Alzheimer's disease. Front Neurol 10:1312. https://doi.org/10.3389/fneur.2019.01312

Fang X, Yu SX, Lu Y, Bast RC Jr, Woodgett JR, Mills GB (2000) Phosphorylation and inactivation of glycogen synthase kinase 3 by protein kinase A. Proc Natl Acad Sci U S A 97(22): 11960–11965. https://doi.org/10.1073/pnas.220413597

Ferrer I, Gomez-Isla T, Puig B, Freixes M, Ribé E, Dalfó E, Avila J (2005) Current advances on different kinases involved in tau phosphorylation, and implications in Alzheimer's disease and tauopathies. Curr Alzheimer Res 2(1):3–18. https://doi.org/10.2174/1567205052772713

Fridovich I (1999) Fundamental aspects of reactive oxygen species, or what's the matter with oxygen? Ann N Y Acad Sci 893:13–18. https://doi.org/10.1111/j.1749-6632.1999.tb07814.x

Gammoh NZ, Rink L (2017) Zinc in infection and inflammation. Nutrients 9(6):624. https://doi.org/10.3390/nu9060624

Garai K, Sahoo B, Kaushalya SK, Desai R, Maiti S (2007) Zinc lowers amyloid-beta toxicity by selectively precipitating aggregation intermediates. Biochemistry 46(37):10655–10663. https://doi.org/10.1021/bi700798b

Garza-Lombó C, Posadas Y, Quintanar L, Gonsebatt ME, Franco R (2018) Neurotoxicity linked to dysfunctional metal ion homeostasis and xenobiotic metal exposure: redox signaling and oxidative stress. Antioxid Redox Signal 28(18):1669–1703. https://doi.org/10.1089/ars.2017.7272

Gaschler MM, Stockwell BR (2017) Lipid peroxidation in cell death. Biochem Biophys Res Commun 482(3):419–425. https://doi.org/10.1016/j.bbrc.2016.10.086

Gunter TE, Gerstner B, Lester T, Wojtovich AP, Malecki J, Swarts SG, Brookes PS, Gavin CE, Gunter KK (2010) An analysis of the effects of Mn2+ on oxidative phosphorylation in liver, brain, and heart mitochondria using state 3 oxidation rate assays. Toxicol Appl Pharmacol 249(1):65–75. https://doi.org/10.1016/j.taap.2010.08.018

Heston LL, White J (1978) Pedigrees of 30 families with Alzheimer disease: associations with defective organization of microfilaments and microtubules. Behav Genet 8(4):315–331. https://doi.org/10.1007/BF01067395

Hirai K, Aliev G, Nunomura A, Fujioka H, Russell RL, Atwood CS, Johnson AB, Kress Y, Vinters HV, Tabaton M, Shimohama S, Cash AD, Siedlak SL, Harris PL, Jones PK, Petersen RB, Perry G, Smith MA (2001) Mitochondrial abnormalities in Alzheimer's disease. J Neurosci 21(9):3017–3023. https://doi.org/10.1523/JNEUROSCI.21-09-03017.2001

Hsu TM, Noble EE, Reiner DJ, Liu CM, Suarez AN, Konanur VR, Hayes MR, Kanoski SE (2018a) Hippocampus ghrelin receptor signaling promotes socially-mediated learned food preference. Neuropharmacology 131:487–496. https://doi.org/10.1016/j.neuropharm.2017.11.039

Hsu HW, Bondy SC, Kitazawa M (2018b) Environmental and dietary exposure to copper and its cellular mechanisms linking to Alzheimer's disease. Toxicol Sci 163(2):338–345. https://doi.org/10.1093/toxsci/kfy025

Ishii T, Hayakawa H, Igawa T, Sekiguchi T, Sekiguchi M (2018) Specific binding of PCBP1 to heavily oxidized RNA to induce cell death. Proc Natl Acad Sci U S A 115(26):6715–6720. https://doi.org/10.1073/pnas.1806912115

Kawamata H, Manfredi G (2010) Import, maturation, and function of SOD1 and its copper chaperone CCS in the mitochondrial intermembrane space. Antioxid Redox Signal 13(9): 1375–1384. https://doi.org/10.1089/ars.2010.3212

Keller JN, Pang Z, Geddes JW, Begley JG, Germeyer A, Waeg G, Mattson MP (1997) Impairment of glucose and glutamate transport and induction of mitochondrial oxidative stress and dysfunction in synaptosomes by amyloid beta-peptide: role of the lipid peroxidation product

4-hydroxynonenal. J Neurochem 69(1):273–284. https://doi.org/10.1046/j.1471-4159.1997.69010273.x

Kern A, Mavrikaki M, Ullrich C, Albarran-Zeckler R, Brantley AF, Smith RG (2015) Hippocampal dopamine/DRD1 signaling dependent on the ghrelin receptor. Cell 163(5):1176–1190. https://doi.org/10.1016/j.cell.2015.10.062

Leigh GJ (1990) Nomenclature of inorganic chemistry. Recommendations. Blackwell Scientific Publications, Oxford

Lidsky TI (2014) Is the aluminum hypothesis dead? J Occup Environ Med 56(5 Suppl):S73–S79. https://doi.org/10.1097/JOM.0000000000000063

Liu J, Head E, Gharib AM, Yuan W, Ingersoll RT, Hagen TM, Cotman CW, Ames BN (2002) Memory loss in old rats is associated with brain mitochondrial decay and RNA/DNA oxidation: partial reversal by feeding acetyl-L-carnitine and/or R-alpha-lipoic acid. Proc Natl Acad Sci U S A 99(4):2356–2361. https://doi.org/10.1073/pnas.261709299

Lovell MA, Markesbery WR (2007) Oxidative DNA damage in mild cognitive impairment and late-stage Alzheimer's disease. Nucleic Acids Res 35(22):7497–7504. https://doi.org/10.1093/nar/gkm821

MacAllister RJ, Parry H, Kimoto M, Ogawa T, Russell RJ, Hodson H, Whitley GS, Vallance P (1996) Regulation of nitric oxide synthesis by dimethylarginine dimethylaminohydrolase. Br J Pharmacol 119(8):1533–1540. https://doi.org/10.1111/j.1476-5381.1996.tb16069.x

Mairet-Coello G, Courchet J, Pieraut S, Courchet V, Maximov A, Polleux F (2013) The CAMKK2-AMPK kinase pathway mediates the synaptotoxic effects of Aβ oligomers through tau phosphorylation. Neuron 78(1):94–108. https://doi.org/10.1016/j.neuron.2013.02.003

Malecki EA (2001) Manganese toxicity is associated with mitochondrial dysfunction and DNA fragmentation in rat primary striatal neurons. Brain Res Bull 55(2):225–228. https://doi.org/10.1016/s0361-9230(01)00456-7

Manisha WH, Rajak R, Jat D (2017) Oxidative stress and antioxidants: an overview. Int J Adv Res Rev 2(9):110–119. https://www.researchgate.net/publication/319468596

Mark RJ, Pang Z, Geddes JW, Uchida K, Mattson MP (1997) Amyloid beta-peptide impairs glucose transport in hippocampal and cortical neurons: involvement of membrane lipid peroxidation. J Neurosci 17(3):1046–1054. https://doi.org/10.1523/JNEUROSCI.17-03-01046.1997

Markesbery WR, Lovell MA (2006) DNA oxidation in Alzheimer's disease. Antioxid Redox Signal 8(11–12):2039–2045. https://doi.org/10.1089/ars.2006.8.2039

Martinez-Finley EJ, Gavin CE, Aschner M, Gunter TE (2013) Manganese neurotoxicity and the role of reactive oxygen species. Free Radic Biol Med 62:65–75. https://doi.org/10.1016/j.freeradbiomed.2013.01.032

Mattson MP (2004) Pathways towards and away from Alzheimer's disease. Nature 430(7000): 631–639. https://doi.org/10.1038/nature02621

Mezzaroba L, Alfieri DF, Colado Simão AN, Vissoci Reiche EM (2019) The role of zinc, copper, manganese and iron in neurodegenerative diseases. Neurotoxicology 74:230–241. https://doi.org/10.1016/j.neuro.2019.07.007

Mills E, Dong XP, Wang F, Xu H (2010) Mechanisms of brain iron transport: insight into neurodegeneration and CNS disorders. Future Med Chem 2(1):51–64. https://doi.org/10.4155/fmc.09.140

Moreira PI, Duarte AI, Santos MS, Rego AC, Oliveira CR (2009) An integrative view of the role of oxidative stress, mitochondria and insulin in Alzheimer's disease. J Alzheimers Dis 16(4): 741–761. https://doi.org/10.3233/JAD-2009-0972

Moulton PV, Yang W (2012) Air pollution, oxidative stress, and Alzheimer's disease. J Environ Public Health 2012:472751. https://doi.org/10.1155/2012/472751

Nandi A, Yan LJ, Jana CK, Das N (2019) Role of catalase in oxidative stress- and age-associated degenerative diseases. Oxidative Med Cell Longev 2019:9613090. https://doi.org/10.1155/2019/9613090

Naqvi S, Samim M, Abdin M, Ahmed FJ, Maitra A, Prashant C, Dinda AK (2010) Concentration-dependent toxicity of iron oxide nanoparticles mediated by increased oxidative stress. Int J Nanomedicine 5:983–989. https://doi.org/10.2147/IJN

Nordberg J, Arnér ES (2001) Reactive oxygen species, antioxidants, and the mammalian thioredoxin system. Free Radic Biol Med 31(11):1287–1312. https://doi.org/10.1016/s0891-5849(01)00724-9

Nunomura A, Perry G (2020) RNA and oxidative stress in Alzheimer's disease: focus on microRNAs. Oxidative Med Cell Longev 2020:2638130. https://doi.org/10.1155/2020/2638130

Pastorino L, Sun A, Lu PJ, Zhou XZ, Balastik M, Finn G, Wulf G, Lim J, Li SH, Li X, Xia W, Nicholson LK, Lu KP (2006) The prolyl isomerase Pin1 regulates amyloid precursor protein processing and amyloid-beta production. Nature 440(7083):528–534. https://doi.org/10.1038/nature04543

Pinton P, Pozzan T, Rizzuto R (1998) The Golgi apparatus is an inositol 1,4,5-trisphosphate-sensitive Ca2+ store, with functional properties distinct from those of the endoplasmic reticulum. EMBO J 17(18):5298–5308. https://doi.org/10.1093/emboj/17.18.5298

Praticò D (2005) Peripheral biomarkers of oxidative damage in Alzheimer's disease: the road ahead. Neurobiol Aging 26(5):581–583. https://doi.org/10.1016/j.neurobiolaging.2004.09.020

Readnower RD, Sauerbeck AD, Sullivan PG (2011) Mitochondria, amyloid β, and Alzheimer's disease. Int J Alzheimers Dis 2011:104545. https://doi.org/10.4061/2011/104545

Ribarič S (2018) Peptides as potential therapeutics for Alzheimer's disease. Molecules (Basel, Switzerland) 23(2):283. https://doi.org/10.3390/molecules23020283

Roos PM, Vesterberg O, Nordberg M (2006) Metals in motor neuron diseases. Exp Biol Med (Maywood, N.J.) 231(9):1481–1487. https://doi.org/10.1177/153537020623100906

Seminara RS, Jeet C, Biswas S, Kanwal B, Iftikhar W, Sakibuzzaman M, Rutkofsky IH (2018) The neurocognitive effects of ghrelin-induced signaling on the hippocampus: a promising approach to Alzheimer's disease. Cureus 10(9):e3285. https://doi.org/10.7759/cureus.3285

Seo YA, Li Y, Wessling-Resnick M (2013) Iron depletion increases manganese uptake and potentiates apoptosis through ER stress. Neurotoxicology 38:67–73. https://doi.org/10.1016/j.neuro.2013.06.002

Shelat PB, Chalimoniuk M, Wang JH, Strosznajder JB, Lee JC, Sun AY, Simonyi A, Sun GY (2008) Amyloid beta peptide and NMDA induce ROS from NADPH oxidase and AA release from cytosolic phospholipase A2 in cortical neurons. J Neurochem 106(1):45–55. https://doi.org/10.1111/j.1471-4159.2008.05347.x

Smith MA, Harris PL, Sayre LM, Perry G (1997) Iron accumulation in Alzheimer disease is a source of redox-generated free radicals. Proc Natl Acad Sci U S A 94(18):9866–9868. https://doi.org/10.1073/pnas.94.18.9866

Spiers JG, Cortina Chen HJ, Barry TL, Bourgognon JM, Steinert JR (2022) Redox stress and metal dys-homeostasis appear as hallmarks of early prion disease pathogenesis in mice. Free Radic Biol Med 192:182–190. https://doi.org/10.1016/j.freeradbiomed.2022.09.025

Strong MJ, Garruto RM, Joshi JG, Mundy WR, Shafer TJ (1996) Can the mechanisms of aluminum neurotoxicity be integrated into a unified scheme? J Toxicol Environ Health 48(6):599–613. https://doi.org/10.1080/009841096161096

Stutzmann GE, Caccamo A, LaFerla FM, Parker I (2004) Dysregulated IP3 signaling in cortical neurons of knock-in mice expressing an Alzheimer's-linked mutation in presenilin1 results in exaggerated Ca2+ signals and altered membrane excitability. J Neurosci 24(2):508–513. https://doi.org/10.1523/JNEUROSCI.4386-03.2004

Suh SW, Jensen KB, Jensen MS, Silva DS, Kesslak PJ, Danscher G, Frederickson CJ (2000) Histochemically-reactive zinc in amyloid plaques, angiopathy, and degenerating neurons of Alzheimer's diseased brains. Brain Res 852(2):274–278. https://doi.org/10.1016/s0006-8993(99)02096-x

Takeda A, Smith MA, Avilá J, Nunomura A, Siedlak SL, Zhu X, Perry G, Sayre LM (2000) In Alzheimer's disease, heme oxygenase is coincident with Alz50, an epitope of tau induced by

4-hydroxy-2-nonenal modification. J Neurochem 75(3):1234–1241. https://doi.org/10.1046/j.1471-4159.2000.0751234.x
Tamagno E, Guglielmotto M, Aragno M, Borghi R, Autelli R, Giliberto L, Muraca G, Danni O, Zhu X, Smith MA, Perry G, Jo DG, Mattson MP, Tabaton M (2008) Oxidative stress activates a positive feedback between the gamma- and beta-secretase cleavages of the beta-amyloid precursor protein. J Neurochem 104(3):683–695. https://doi.org/10.1111/j.1471-4159.2007.05072.x
Tanaka M, Jaruga P, Küpfer PA, Leumann CJ, Dizdaroglu M, Sonntag WE, Boon Chock P (2012) RNA oxidation catalyzed by cytochrome c leads to its depurination and cross-linking, which may facilitate cytochrome c release from mitochondria. Free Radic Biol Med 53(4):854–862. https://doi.org/10.1016/j.freeradbiomed.2012.05.044
Tong BC, Wu AJ, Li M, Cheung KH (2018) Calcium signaling in Alzheimer's disease & therapies. Biochim Biophys Acta Mol Cell Res 1865(11 Pt B):1745–1760. https://doi.org/10.1016/j.bbamcr.2018.07.018
Verma M, Vats A, Taneja V (2015) Toxic species in amyloid disorders: oligomers or mature fibrils. Ann Indian Acad Neurol 18(2):138–145. https://doi.org/10.4103/0972-2327.144284
Wang JX, Gao J, Ding SL, Wang K, Jiao JQ, Wang Y, Sun T, Zhou LY, Long B, Zhang XJ, Li Q, Liu JP, Feng C, Liu J, Gong Y, Zhou Z, Li PF (2015) Oxidative modification of miR-184 enables it to target Bcl-xL and Bcl-w. Mol Cell 59(1):50–61. https://doi.org/10.1016/j.molcel.2015.05.003
Wang L, Yin YL, Liu XZ, Shen P, Zheng YG, Lan XR, Lu CB, Wang JZ (2020) Current understanding of metal ions in the pathogenesis of Alzheimer's disease. Transl Neurodegener 9:10. https://doi.org/10.1186/s40035-020-00189-z
Ward RJ, Dexter DT, Crichton RR (2015) Ageing, neuroinflammation and neurodegeneration. Front Biosci (Schol Ed) 7(1):189–204. https://doi.org/10.2741/S433
Weingarten MD, Lockwood AH, Hwo SY, Kirschner MW (1975) A protein factor essential for microtubule assembly. Proc Natl Acad Sci U S A 72(5):1858–1862. https://doi.org/10.1073/pnas.72.5.1858
Wright RO, Baccarelli A (2007) Metals and neurotoxicology. J Nutr 137(12):2809–2813. https://doi.org/10.1093/jn/137.12.2809
Yang W, Omaye ST (2009) Air pollutants, oxidative stress and human health. Mutat Res 674(1–2):45–54. https://doi.org/10.1016/j.mrgentox.2008.10.005
Yokel RA (2000) The toxicology of aluminum in the brain: a review. Neurotoxicology 21(5):813–828
Zhang J, Cao R, Cai T, Aschner M, Zhao F, Yao T, Chen Y, Cao Z, Luo W, Chen J (2013) The role of autophagy dysregulation in manganese-induced dopaminergic neurodegeneration. Neurotox Res 24(4):478–490. https://doi.org/10.1007/s12640-013-9392-5

Neuroinflammation in Alzheimer's Disease 3

Santanu Kaity and Anoop Kumar

Abstract

Alzheimer's disease (AD) is one of the significant progressive disorders of the brain, which lead to the destruction of memory and cognitive power in elderly individuals. Mild-to-severe extent of neurodegeneration triggers AD in most cases. Patients with AD carry a higher chance of generation of dementia, memory loss, and gradual loss of basic intellect. According to a recent report, calcium dysfunction at the neuron level and microglia-mediated neuroinflammation are the leading causes of AD. The formation of amyloid plaques at the extracellular level and neurofibrillary tangles in the intracellular part of neurons is the most established AD hypothesis as explored by different research groups throughout the globe to date. However, the exact root cause of the disease pathogenesis is still unknown which has made the disease incurable. Multiple symptomatic treatment options exist for AD, but none of the existing and new drug candidates proved to have potent disease-curing activity. This chapter deals mainly with the role of neuroinflammation in the disease progression of AD. The effect of injury, immunity, infection, and aging on neuroinflammation is described in detail. Further, the chapter describes the molecular pathway of microglia and astrocyte activation and the subsequent release of different cytokines and chemokines, which trigger AD through neuroinflammation. The last part of the chapter describes the role of reactive oxygen species and infiltration of peripheral leukocytes in central nervous system (CNS) for disease progression. The exact

S. Kaity
Department of Pharmaceutics, National Institute of Pharmaceutical Education and Research (NIPER), Kolkata, West Bengal, India

A. Kumar (✉)
Department of Pharmacology, Delhi Pharmaceutical Sciences & Research University (DPSRU), New Delhi, India

A. Sharma, G. P. Modi (eds.), *Natural Product-Based Synthetic Drug Molecules in Alzheimer's Disease*, https://doi.org/10.1007/978-981-99-6038-5_3

cause of disease initiation and subsequent disease progression for AD is still a matter of extensive research. This chapter is intended to concisely depict a detail of till date reported molecular hypothesis for AD and the future perspectives, where we still have to shed light to make the disease manageable and curable in the near future.

Keywords

Alzheimer's disease · Neuroinflammation · β-Amyloid · Neurofibrillary tangles · Cytokines · Chemokines

Abbreviations

5-LOX	5-Lipoxygenase
A.D.	Anno Domini
AD	Alzheimer's disease
APOE	Apolipoprotein E
APP	Amyloid precursor protein
BACE-1	Beta-site amyloid precursor protein-cleaving enzyme 1
CLU	Clusterin
CNS	Central nervous system
COX-1	Cyclooxygenase
CR1	Complement receptor type 1
DS	Down syndrome
GWI	Gulf War illness
HPETES	5-Hydroxyeicosatetraenoic acid
IL	Interleukin
IRAK1	Interleukin-1 receptor-associated kinase 1
IRAK4	Interleukin-4 receptor-associated kinase 4
MAP Kinase	Mitogen-activated protein kinase
MyD88	Myeloid differentiation primary response 88
NFKB	Nuclear factor kappa B
NOX	NADPH oxidase
PGG2	Prostaglandin (PG) G2
PICLAM	Phosphatidylinositol-binding clathrin assembly
RNS	Reactive nitrogen species
ROS	Reactive oxygen species
TAK1	Transforming growth factor β-activated kinase 1
TBI	Traumatic brain injury
Term2	Triggering receptor expressed on myeloid cells
TLR	Toll-like receptor
TNF alpha	Tissue necrosis factor
τ-Protein	Tau protein
TRAF6	Tumor necrosis factor receptor-associated factor 6

USD	The United States dollar
β-Amyloid	Beta-amyloid

3.1 Introduction

Alzheimer's disease (AD) is the most common form of dementia, which leads to brain atrophy and loss of brain cells, which significantly disturbs the human cognitive power and leads to major behavioral changes. The major symptoms include loss of memory, loss of thinking ability, depression, apathy, loss of judgement ability, and many more metal complications, which make the daily life difficult for the affected individual (National Institute of Aging n.d.-a). The term "dementia" was coined near about 600 AD. It is a Latin word originated from "de" (loss), "ment" (mind), and "ia" (a state). Altogether, the word dementia means "a state out of mind" (Yang et al. 2016). Dementia can be classified into four major types, namely, frontotemporal dementia, Lewy body dementia, vascular dementia, and Alzheimer's disease (Geldmacher and Whitehouse 1996). The name "Alzheimer's disease" is after the name of a German doctor Alois Alzheimer who first time documented the symptoms of the disease. The term "Alzheimer's disease" was introduced in the year 1910 by the psychiatrist Emil Kraepelin, a colleague of Dr. Alzheimer (Cipriani et al. 2011). The major reasons behind initiation of AD may be related to genetic factors, environmental factors, and lifestyle. Age-related disease progression is very prominent in AD. The average age of appearance of symptoms of AD is identified as 65 years, and the number of affected persons increases gradually with age which reaches to almost 30% of people of average age group of 85 years and above. AD is classified into broadly two types, namely, early onset and late onset. Among the two types, the late-onset type is the most common type, and the clear cause is unknown till date. However, one genetic risk factor named apolipoprotein E *(APOE)* gene on chromosome 19 may increase the risk of disease initiation. In early-onset type of AD, symptoms appear in between 30 and 60 years of age group (National Institute of Aging n.d.-a).

Extensive research is still going on in different parts of the globe to identify the actual root cause of disease initiation and disease progression. Numerous hypotheses have been established as a probable molecular pathway; however, none of them have proved to explain all the disease-causing and -progressing aspects of AD. Among various established and most popular few hypotheses described as the major cause of AD are neuropathological changes due to formation of amyloid plaques in the form of beta-amyloid (Aβ) which deposits in the extracellular part and neurofibrillary tangles which deposit in the intracellular part and formed due to hyperphosphorylation of tau protein (Serrano-Pozo et al. 2011). Another hypothesis says that the formation of neuropil threads may be another reason for AD. The major difference between neurofibrillary tangles and neuropil threads is that tangles are fibrous inclusion components of intracytoplasmic part of either the cell body or the proximal part of the dendrites of the affected neuron. However, neuropil threads are a swollen filamentous structure containing dendrites. Neurites are the swollen filamentous material at the distal axon part (Savonenko et al. 2015). Dystrophic neuritis at the amyloid plaque region is another established hypothesis behind AD. It is established that Aβ promotes disruption of microtubules in dystrophic neuritis and further disrupts axonal transport. Further, the process accelerates the process of accumulation of amyloid precursor protein (APP) and the cleaving enzyme at the affected terminal and progressively worsens the condition by contributing to the Aβ formation (Sadleir et al. 2016). Neuronal loss and early synaptic loss are another proven fact behind the progression of AD. Microglial phagocytosis may be considered as a major reason behind synaptic loss (Subramanian et al. 2020).

Apart from the above-documented established multiple hypothesis, various cytokine- and chemokine-mediated neuroinflammation is considered one of the major factors behind disease progression. Neuroinflammation is the process of sensitizing the central nervous system (CNS) through the activation of innate immunity mechanism against external impacts like disease, infection, and injury (Spencer et al. 2012). Recent reports suggest that the process of disease initiation and progression of AD is not only associated with the neuronal factors, but few immunogenic responses are also responsible for the disease. An innate immune response is triggered by binding of misfolded and aggregated proteins with the microglial and astroglial pattern recognition receptor. The processes further relay the process of inflammatory mediator release and subsequent disease progression (Heneka et al. 2015). Multiple factors can be associated with the process of neuroinflammation. The cell-mediated neuroinflammation in AD is mainly triggered by astrocytes and microglia, which can produce a wide range of inflammatory components like chemokines, cytokines, prostanoids, cyclooxygenase, and reactive oxygen species (Heneka and O'Banion 2007; Calsolaro and Edison 2016).

An effort has been made in this chapter to give a detail of the role of neuroinflammation in the disease progression of AD. Role of various factors like injury, infection, immunity, aging, reactive oxygen species, and many other critical factors which cascade the process of neuroinflammation in AD is documented here. The proven molecular mechanisms behind the activation of till date explored self-defending mechanisms of CNS which are responsible for disease progression and severity are explained for better understanding. Apart from the above factors, the role of lifestyle, food habit, and obesity in triggering the process of neuroinflammation is also described in the subsequent sections of this chapter. A brief detail of various inflammatory markers and diagnosis scope, available treatment options, and future perspective is also documented.

3.2 Epidemiology

The global impact of AD is expected to increase continuously with the increment of average life expectancy and demographic aging. It is predicted that the developing countries will face a cozy burden because of AD (Zhang et al. 2021). As per the recent report by the World Health Organization (WHO), presently, more than 55 million people live with dementia and approximately ten million new cases are reported each year throughout the globe. Nearly about 60% of the total affected population are from low- and middle-income countries. AD, the most common form of dementia, may contribute to 60–70% of the total cases worldwide. A high portion of older people is increasing almost in all countries and so the disease burden. It is predicted that the cases of dementia may increase nearly about 78 million by the year 2030 and 139 million by 2050 (World Health Organisation n.d.). According to Alzheimer's disease facts and figures, 2020, the number of AD cases may increase from 5.8 million to 13.8 million by 2050 in America (Anonymous 2020). As per recent report, an estimated 6.5 million Americans aged ≥65 years were living with AD in 2022. About 1 in 9 people of ≥65 years has Alzheimer's dementia. The risk of disease appearance and progression increases with age. Chance of occurrence of AD is 5% in the age group of 65–74 years, 13.1% in the age group of 75–84 years, and 33.2% in the age group of 85 years and older (Rajan et al. 2021). There is some proven impact of sex on disease prevalence. In general, women lives longer than men. The age burden in case of women will be more than men. It is reported that among all AD-infected population, two-thirds are women. Almost, 12% of women and 9% of men are AD infected among the total population aged ≥65 years. It gives a clear indication that sex can impact the process of disease initiation and disease progression (Alzheimer's Disease Facts and Figures n.d.). It is even established that the geographic region and race are also having some crucial impact on disease progression. Older blacks have a higher tendency to develop AD than older whites in America (Power et al. 2021). The cost of the treatment for AD and other associated complication management is very high. The situation becomes worse because of the higher socio-economic burden than other noncommunicable diseases. In an international report, it is documented that the overall national caring cost for people having AD or any other type of dementia will reach nearly about 321 billion USD in America (Alzheimer's Disease Facts and Figures n.d.). This legacy may obviously not be economically possible for most of the third world and developing countries. Lack of specialist staff like doctors and nurses throughout the globe is also a major challenge to manage the affected population. A firm, structured policy should be undertaken among all countries throughout the globe to control the socio-economic complications of AD.

3.3 Neuroinflammation

Neuroinflammation is the spontaneous process of the CNS to sensitize the system regarding the external treats, which may be related to injury, infection, aging, and certain other factors. In this section, we will discuss about various factors which can trigger the process of neuroinflammation.

3.3.1 Etiology

The process of neuroinflammation may occur due to factors like trauma, infection, immunogenic process, redox iron, reactive oxidative species, and different oligomers of τ- and β-amyloid (Morales et al. 2014).

Brain injury: Traumatic brain injury (TBI) is identified as a potential initiator of AD. The initial inflammatory process activates the pathways to clear up the junk materials from the CNS by the phagocytic activity of the microglial cells. However, a prolonged process of inflammation process triggers the initiation process and gradual progression of AD. Cytokine-mediated inflammation mitigation process may increase the level of BACE-1 in certain cases, which activates APP production and subsequently increases Aβ level (Sastre et al. 2003a; Corrigan et al. 2011). The process may continue for several years, and gradually the impact of the activity of microglial phagocytosis process markedly reduces neuronal activities (Heneka et al. 2015).

Infection: Chronic infection or injury can trigger systemic inflammation and initiate the process of AD. Inflammation outside the CNS is referred to as systemic inflammation and has a role in initiating or promoting AD pathogenesis. Prolonged bacterial or viral infection has a deep link with activation and progression of multiple factors associated with AD. Infection with various gastric, enteric, oral, chlamydia, spirochetes, and different viral organisms, mainly herpes viruses, plays a role in triggering AD (Butler and Walker 2021). There are several hypotheses related to the role of multiple microorganisms in the AD progression. According to the reported literature, the process may work in three major ways where the pathogen can directly be the cause of disease initiation. According to the second hypothesis, the pathogen can act as a disease accelerator. The third hypothesis is the reverse mechanism one, where AD-infected persons are relatively more susceptible to infection by various chronic diseases caused by microorganisms (Butler and Walker 2021). In multiple studies, it has been reported that people affected with microbial infection and AD used to have higher levels of inflammatory cytokines, chemokines, and acute-phase protein in systemic circulation. Few meta-analyses showed elevated levels of IL-1ß, IL-6, and IL-18 in patients with AD (Lai et al. 2017). However, still it is not exactly proven whether infection and systemic inflammation induce AD or because of AD the affected individual becomes susceptible to infection and systemic inflammation.

Immunogenic process: Immunogenic responses play a very critical role in the initiation and progression of AD. Even autoimmune diseases like osteoarthritis have a positive impact on neuroinflammation and subsequent disease progression (Kyrkanides et al. 2011). Few cases of AD, which are mainly early onset in nature, are caused by gene mutation and transfer as heredity factor. This is a small proportion among all affected population. However, the late-onset type is most prevailing in nature and has a few genetic connections regarding disease progression. There are several genes which are not directly linked to Aβ production but have an impact on AD. Most of these AD-associated genes are involved in immune activation followed by neuroinflammation. Genes like Term2 and CD33 are strongly involved in neuroinflammation-associated AD. The exact mechanism of disease progression is yet to explore (Pimenova et al. 2018). A multiple number of genes like APOE, CLU, CR1, and PICLAM have been identified, which may have contribution to AD, and they act in a wide range of mole pathways. Among all those, a few have a very prominent impact on AD. Apolipoprotein E (APOE), which consists of three alleles (e2, e3, and e4), which are different from one another by a single amino acid, may increase the risk of AD by expression of APOE4 (Bu 2009).

Age: The process of aging is very natural, and a fall of all biological activities is seen after a certain age. The changes are very individual specific, and multiple factors like food habit, lifestyle, education, socio-economic condition, environment, and many more are directly related to the life span and geriatric condition. Age-related complications in AD are a very common and most prominent risk factor. Though age does not cause or initiate the disease, it is the most influential risk factor which may lead to the development of AD-related complications. Beyond 65 years of age, an increment of 5 years of age increases the risk of AD by almost double (National Institute of Aging n.d.-b). There are various established hypotheses to show the link between aging and AD. The theory of glycosylation is that there is an increased interaction between glucose and protein in neuron-affected AD (Baloyannis et al. 2019). Another hypothesis says that the somatic mutation, which leads to mutagenesis of various genes, leads to over-phosphorylation of tau protein, which leads to initiation and progression of AD. The increase of gene mutation and other metabolic errors increases the probability of occurrence of AD to manyfold (Fiford et al. 2018).

Environmental stressor: Stressors are defined as the factors which modify homeostasis and oblige the affected one to restore the homeostasis. There are multiple factors which can induce the process of homeostasis, and the factors may include heat, cold, chemicals, noise, sleep disturbance, social or domestic violence, stress, and many more which alone or in combination can create a stressor environment (O'Callaghan and Miller 2019). The stressor atmosphere further triggers the process of neuroinflammation with associated symptoms like anorexia, depression, social withdrawal, lethargy, and lack of cognitive intelligence. Prolonged exposure to stressor environment leads to a degenerative impact through neuroinflammation and can initiate the process of AD.

Reactive oxygen species (ROS) and reactive nitrogen species (RNS): Reactive oxygen species and reactive nitrogen species generate free radicals, which are responsible for triggering many diseases and aging (Kim et al. 2015). ROS and RNS intervene the redox signaling-mediated pathways of cell activities, namely cell growth and cell death. The antioxidant defense mechanism of the body which balances the ROS and RNS is hampered, and the buildup of oxidative stress induces various neurodegenerative diseases like AD. Transition metal-catalyzed oxidative stress formation with Aβ metal ligand can be a potential source of ROS formation in AD. Additionally, the dysfunctional mitochondria can play a significant role in increasing the ROS load in AD initiation and progression. Apart from that, ROS is generated as an inflammatory response from microglia where NADPH oxidase (NOX) plays a vital role in initiating the process. NOX is the enzyme which produces ROS through superoxide radical and hydrogen peroxide and finally damages the neuronal cell and triggers the disease progression (Ganguly et al. 2021).

Trisomy in Down syndrome (DS): Down syndrome is caused by triplication of chromosome 21, the place where amyloid precursor protein gene is encoded. In Down syndrome, Aβ starts to deposit at an early age, which gradually increases with time. Within 45 years of age, AD initiation and disease progression are observed through inflammation and neurofibrillary tangles (Wisniewski et al. 1985). Thus, DS increases the risk of AD in many instances. Young DS-affected population have higher neuroinflammatory profile in frontal cortex with microglial intermediate activation, rodlike microglia, and a higher rate of upregulation of major inflammatory cytokines (Lisi Flores-Aguilar et al. 2020). In elder patients also, there is an increase in inflammatory cytokines along with increase of dystrophic microglia. It is reported that about 50% DS-affected people of age above 60 years are having AD (National Down Syndrome Society n.d.).

Gulf War illness (GWI): It is a disorder which is characterized by fatigue, depression, muscle pain, cognitive impairment, and in few cases dermatological issues (Steele 2000). The symptoms are almost similar to the symptoms occurred by various toxic gas and material exposures during the 1991 Gulf War. The disease progression follows a similar pathway of neuroinflammation and neuroimmune response as was observed during the Gulf War. The mode of disease progression may not be explained on the basis of a single stressor; however, multiple factors like nerve agents, sarin, and organophosphorus compounds may have synergistically shown the impact on CNS in the form of neuroinflammation to initiate the AD and disease progression. The impact is still persistent, and it has been going on for the last 25 years (O'Callaghan and Miller 2019).

Family history: Sometimes, family history of AD in first-degree relatives increases the risk of AD. People with parent or sibling having AD are more likely and not necessarily can be affected by AD (Green et al. 2002). The major risk factor associated with APOE-e4 gene, however, is not a definite cause of AD. All of us are inherited with one or the other form of APOE genes, namely e2, e3, and e4. People who inherit a single copy of e4 gene from parents have a risk of AD, and the risk is increased if two copies of e4 gene are inherited from parents (Van Cauwenberghe et al. 2016). The APOE-e4 gene has a high neuroinflammatory effect on Aβ; additionally, it is responsible for tau-induced neuroinflammation. APOE-e4-activated neurodegeneration through tau-mediated mechanism is also another factor for AD initiation and subsequent disease progression (Shi et al. 2017).

Education attainment: Socio-economic condition and education attainment have an impact on AD. In a recent study, five different inflammatory markers, namely CRP, fibrinogen, IL-6, IL-1β, and TNF-α, were evaluated for a group of people with different socio-economic patterning and educational attainment. It was observed that CRP and fibrinogen level are influenced by the socio-economic patterning (Maurel et al. 2020). It is also a fact that social and educational engagement throughout life boosts brain health and keeps cognitive power healthier. It is reported that people with fewer years of education have a higher chance of losing cognitive powers at an early age than people who have studied for more years. A mentally nourishing and stimulating job where high level of intelligence is required keeps the brain more active. People with such intelligent engagement have very less chance of cognitive loss and AD (Grzywacz et al. 2016).

3.4 Mechanism of Neuroinflammation in AD

Various reports are available in the literature, which show involvement of neuroinflammation in AD. The complete mechanism is unclear so far; however, studies have shown that activation of microglia and astrocytes, excessive release of cytokines, etc. are involved (Akiyama et al. 2000; Ho et al. 2005; Sastre et al. 2006). The possible mechanism of neuroinflammation in AD is described below.

3.4.1 Microglial Activation

In the central nervous system (CNS), microglia are resident immune cells which are in inactive state in healthy individuals. M1 and M2 are two major types of microglia in which M1 is mainly responsible for the production of pro-inflammatory mediators, whereas M2 is responsible for anti-inflammatory mediators such as IL-10 and IL-13 (Tang and Le 2016; Yao and Zu 2020). These cells will become active in response to any threat such as infection and injury to the CNS. As mentioned in earlier sections, there is an accumulation of Aβ in AD patients; the accumulated Aβ results in the activation of microglia, which ultimately results in the phagocytosis of Aβ. However, after prolonged periods, microglia are no longer able to process Aβ.

Secretases are involved in the breakdown of amyloid precursor protein (APP), which is one of the integral membrane proteins for conversion into Aβ40 and Aβ42 (Evin et al. 2003; Sastre et al. 2003b). Normally, these breakdown products are sensed by toll-like receptors (TLRs) of microglia, which results in the activation of signaling pathways and is removed out from the body (Lehnardt 2010). However, accumulation of these products particularly Aβ42 within age activates microglia cells for a long period of time, which further activates myeloid differentiation primary response 88 (MyD88)-dependent (Xiang et al. 2015) and MyD88-independent pathways (Arroyo et al. 2011).

3.4.1.1 MyD88-Dependent Pathways

In MyD88-dependent pathways, the phosphorylation of IRAK1 in IRAK1 and IRAK4 complex results in binding of TRAF6. The binding of TRAF6 results in the activation of TAK1 kinase, which further activates NFKB and mitogen-activated protein (MAP) kinase pathways.

(a) **Activation of NFKB pathways**
 The NFKB pathways activate IKβ and phosphorylation of NFKB and IKβ complex, which results in the release of NFKB. The NFKB results in

transcription and synthesis and excessive release of TNFα and IL-1, which are the main inflammatory mediators in neuroinflammation (Fillit et al. 1991). NFKB is an important transcription factor associated with the promotion of inflammation through the upregulation of pro-inflammatory cytokines such as IL-1-beta, IL-6, IL-12, and tumor necrosis factor-alpha (TNF-α) and chemokines including IL-18 and adhesion molecules, e.g., matrix metalloproteinase (MMP) (Strauss et al. 1992).

(b) **Activation of MAPK pathways**
The mitogen-activated protein kinase (MAPK) pathways, triggering pro-inflammatory gene expression, result in neuroinflammation as shown in Fig. 3.1.

3.4.1.2 Activation of MyD88-Independent Pathways

Microglial cell activation could also result in the activation of MyD88-independent pathways. In MyD88-independent pathways, the phosphorylation of IRF 3/7 occurs, which results in the synthesis and release of inflammatory mediators. The excessive synthesis and release of these mediators result in the neuroinflammation as shown in Fig. 3.1. Further, the released inflammatory mediators also activate the neighboring astrocyte, which also results in the Aβ1–42 oligomer production.

3.4.2 Activation of Astrocytes

Astrocytes are the most abundant glial cells having a number of functions in the CNS. Various studies in literature have reported the activation of astrocytes by APP, amyloid aggregates, and fibrils. The activated astrocytes result in the stimulation of inflammatory cascades using various types of cytokines. The clinical studies have also shown astrogliosis in the postmortem brain tissue of AD patients.

Inflammatory cytokines and chemokines such as IL-1β and TNF-α also activate the astrocytes. Several cytokines (pro-inflammatory and anti-inflammatory) are identified in AD patients. Pro-inflammatory cytokines such as IL-1, TNF-α, IL-6, IL-8, and IFN-γ usually determine inflammation, whereas anti-inflammatory cytokines such as IL-1 receptor antagonist, IL-4, IL-6, IL-10, IL-11, and IL-13 control the pro-inflammatory molecules (Colombo and Farina 2016; Kwon and Koh 2020).

Simultaneously, the activated microglia and astrocytes also promote the deposition of Aβ. The released cytokines also upregulate enzymes (β-secretase and BACE-1), which are involved in the formation of Aβ (Venugopal et al. 2008).

3.4.3 Damage to Cell Membrane

The accumulation of Aβ42 does damage to cell membrane, which results in the activation of glial cells. The activated glial cells activate phospholipase A2, which does the breakdown of phospholipids (main component of cell membrane) into

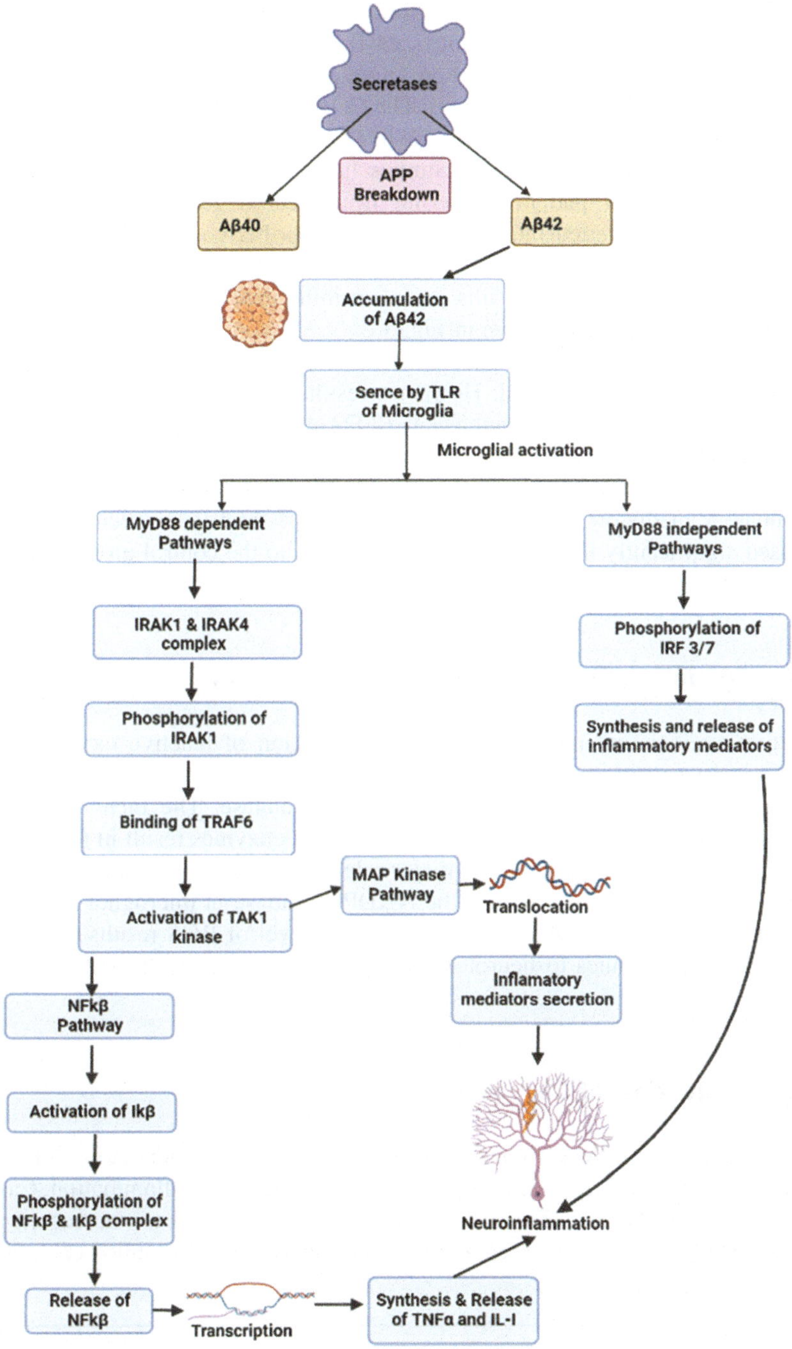

Fig. 3.1 Mechanism of neuroinflammation in AD through activation of MyD88-dependent and MyD88-independent pathways. APP amyloid precursor protein; TLR toll-like receptor; MyD88 myeloid differentiation primary response 88; IRAK1 interleukin-1 receptor-associated kinase 1; IRAK4 interleukin-4 receptor-associated kinase 4; TRAF6 tumor necrosis factor receptor-

arachidonic acid (AA). The AA is further converted into PGG2 by action of cyclooxygenases (COXs) and leukotrienes (LTs) by the action of lipoxygenases (LOXs). The prostaglandins (20-carbon saturated fatty acids) and leukotrienes also play a major role in neuroinflammation. The COX (COX1 and COX2) and LOX are normally expressed in brain, and their role in neurodegenerative diseases particularly COX2 has already been explored. Studies have shown increased expression of COX2 in neurons of patients suffering from neurodegenerative diseases. The PGG2 is further converted into PGH2, which is further converted into TxA2 and PGI2. The leukotrienes are further converted into LTA4 and LXB4. The excessive release of these cytokines results in neuroinflammation (Bazan et al. 2002; Hoozemans et al. 2002) as shown in Fig. 3.2.

Various studies have been conducted so far to check the level of cytokines in AD patients as compared to control. However, results of these studies are conflicting with each other. Recently, Anuradha et al. (2022) have also conducted an SLR and a meta-analysis of clinical studies to find out the association of inflammatory mediators with AD. The results of the meta-analysis have indicated that among all inflammatory mediators, the levels of IL-6, TGF-β1, and IL-1α were found to be increased significantly in AD patients as compared to the control group (Anuradha et al. 2022).

3.4.4 Oxidative Stress and Neuroinflammation

Oxidative stress is an imbalance between generation of reactive oxygen species (ROS), reactive nitrogen species (RNS), etc. and in the levels of antioxidant enzymes such as glutathione, superoxide dismutase, and catalase. The increased level of reactive species and decreased level of endogenous enzymes result in the oxidative stress. The NADPH oxidase plays a major role in the excessive production of ROS. The accumulated Aβ also activates the NADPH oxidase in microglial cells, which results in the production of ROS. The increased level of ROS results in oxidative stress, which contributes to neurotoxicity (Schilling and Eder 2011; Mhatre et al. 2004).

3.4.5 Activation of Caspases

Caspases are well-known proteases, which regulate cell death. However, a few caspases such as caspases 1, 3, 7, and 8 have also a role in the inflammation. Caspase 1 plays a role in the maturation of pro-inflammatory cytokines, whereas caspases 3, 7, and 8 play a role in the regulation of microglial cells (Burguillos et al. 2011).

Fig. 3.1 (continued) associated factor 6; TAK1 transforming growth factor β-activated kinase 1; MAP kinase mitogen-activated protein kinase; NFKB nuclear factor kappa B; IκB kinase; TNF alpha tissue necrosis factor alpha; IRFIFN regulatory factor 3

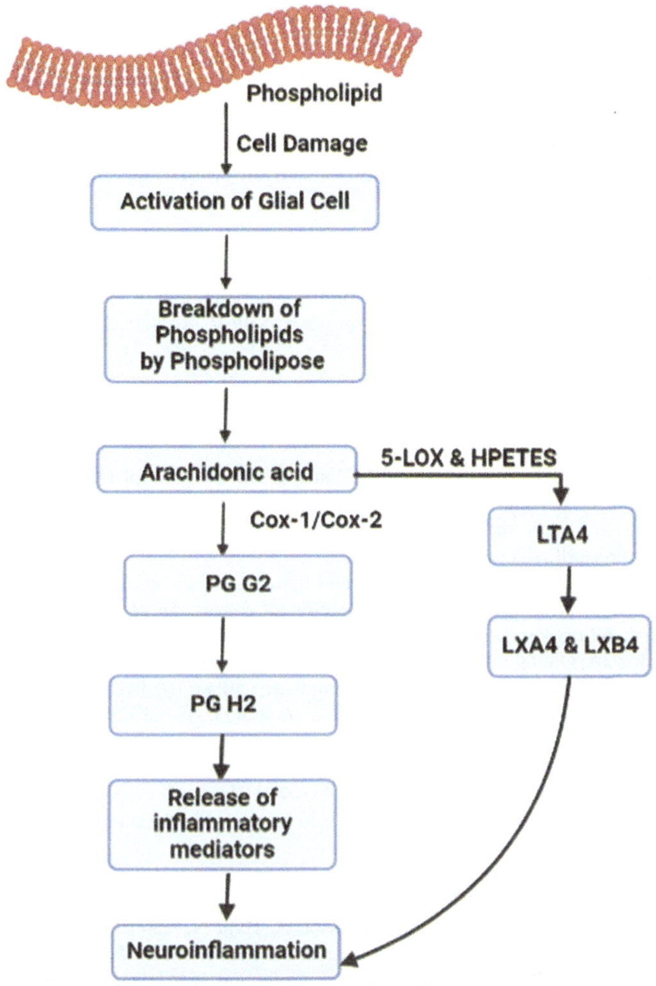

Fig. 3.2 Mechanism of neuroinflammation in AD through activation of arachidonic acid (AA) pathway. 5-LOX 5-lipoxygenase; HPETES 5-hydroxyeicosatetraenoic acid; COX-1 cyclooxygenase; PGG2 prostaglandin (PG) G2; leukotriene A4

3.4.6 Activation of Complement System

The role of complement system in neuroinflammation is also explored by researchers. Various complement factors such as C3a, C3b, and C5a play an important role in neuroinflammatory processes, particularly in the chemotaxis and phagocytic functions of microglial cells. Aβ is known to activate complement system in the alternative pathway (Shah et al. 2021; Bonifati and Kishore 2007).

3.5 Conclusion

Neuroinflammation plays a significant role in AD. Aggregated proteins, particularly Aβ42, sense by pattern recognition receptors of microglia and astroglia, which further activates the innate immune signaling pathways and results in the excessive release of various inflammatory mediators, which contribute to disease progression and severity.

Acknowledgments Author, Anoop Kumar, is thankful to the Vice Chancellor, Professor (Dr.) R.K. Goyal, for constant support and motivation for the completion of this work. Author, Santanu Kaity, is thankful to Prof. V. Ravichandiran, Director, NIPER, Kolkata, for motivation.

References

Akiyama H, Barger S, Barnum S, Bradt B, Bauer J, Cole GM et al (2000) Inflammation and Alzheimer's disease. Neurobiol Aging 21(3):383–421

Alzheimer's Disease Facts and Figures (n.d.). https://www.alz.org/media/documents/alzheimers-facts-and-figures.pdf. Accessed 12 Nov 2022

Anonymous (2020) Alzheimer's disease facts and figures. Alzheimers Dement 16:391–460

Anuradha U, Kumar A, Singh RK (2022) The clinical correlation of proinflammatory and anti-inflammatory biomarkers with Alzheimer disease: a meta-analysis. Neurol Sci 43(1):285–298

Arroyo DS, Soria JA, Gaviglio EA, Rodriguez-Galan MC, Iribarren P (2011) Toll-like receptors are key players in neurodegeneration. Int Immunopharmacol 11(10):1415–1421

Baloyannis S, Trevisan K, Cristina-Pereira R, Silva-Amaral D, Aversi-Ferreira TA (2019) Theories of aging and the prevalence of Alzheimer's disease. Biomed Res Int 2019:9171424

Bazan NG, Colangelo V, Lukiw WJ (2002) Prostaglandins and other lipid mediators in Alzheimer's disease. Prostaglandins Other Lipid Mediat 68:197–210

Bonifati DM, Kishore U (2007) Role of complement in neurodegeneration and neuroinflammation. Mol Immunol 44(5):999–1010

Bu G (2009) Apolipoprotein E and its receptors in Alzheimer's disease: pathways, pathogenesis and therapy. Nat Rev Neurosci 10:333–344

Burguillos MA, Deierborg T, Kavanagh E, Persson A, Hajji N, GarciaQuintanilla A et al (2011) Caspase signalling controls microglia activation and neurotoxicity. Nature 472:319–324

Butler L, Walker KA (2021) The role of chronic infection in Alzheimer's disease: instigators, co-conspirators, or bystanders? Curr Clin Microbiol Rep 8(4):199–212

Calsolaro V, Edison P (2016) Neuroinflammation in Alzheimer's disease: current evidence and future directions. Alzheimers Dement 12(6):719–732

Cipriani G, Dolciotti C, Picchi L, Bonuccelli U (2011) Alzheimer and his disease: a brief history. Neurol Sci 32(2):275–279

Colombo E, Farina C (2016) Astrocytes: key regulators of neuroinflammation. Trends Immunol 37(9):608–620

Corrigan F, Pham CL, Vink R, Blumbergs PC, Masters CL, van den Heuvel C, Cappai R (2011) The neuroprotective domains of the amyloid precursor protein, in traumatic brain injury, are located in the two growth factor domains. Brain Res 1378:137–143

Evin G, Zhu A, Holsinger RD, Masters CL, Li QX (2003) Proteolytic processing of the Alzheimer's disease amyloid precursor protein in brain and platelets. J Neurosci Res 74(3):386–392

Fiford CM, Ridgway GR, Cash DM et al (2018) Patterns of progressive atrophy vary with age in Alzheimer's disease patients. Neurobiol Aging 63:22–32

Fillit H, Ding WH, Buee L, Kalman J, Altstiel L, Lawlor B et al (1991) Elevated circulating tumor necrosis factor levels in Alzheimer's disease. Neurosci Lett 129:318–312

Ganguly U, Kaur U, Chakrabarti SS, Sharma P, Agrawal BK, Saso L, Chakrabarti S (2021) Oxidative stress, Neuroinflammation, and NADPH oxidase: implications in the pathogenesis and treatment of Alzheimer's disease. Oxid Med Cell Longev 2021:7086512. https://doi.org/10.1155/2021/7086512

Geldmacher DS, Whitehouse PJ (1996) Evaluation of dementia. N Engl J Med 335(5):330–336

Green RC, Cupples LA, Go R, Benke KS, Edeki T, Griffith PA et al (2002) Risk of dementia among white and African American relatives of patients with Alzheimer disease. JAMA 287(3): 329–336

Grzywacz JG, Segel-Karpas D, Lachman ME (2016) Workplace exposures and cognitive function during adulthood: evidence from National Survey of midlife development and the O*NET. J Occup Environ Med 58(6):535–541

Heneka MT, O'Banion MK (2007) Inflammatory processes in Alzheimer's disease. J Neuroimmunol 184(1–2):69–91

Heneka MT et al (2015) Lancet Neurol 14(4):388–405

Ho GJ, Drego R, Hakimian E, Masliah E (2005) Mechanisms of cell signaling and inflammation in Alzheimer's disease. Curr Drug Targets Inflamm Allergy 4(2):247–256

Hoozemans JJ, Veerhuis R, Janssen I, van Elk EJ, Rozemuller AJ, Eikelenboom P (2002) The role of cyclo-oxygenase 1 and 2 activity in prostaglandin E2 secretion by cultured human adult microglia: implications for Alzheimer's disease. Brain Res 951(2):218–226

Kim GH, Kim JE, Rhie SJ, Yoon S (2015) The role of oxidative stress in neurodegenerative diseases. Exp Neurobiol 24(4):325

Kwon HS, Koh SH (2020) Neuroinflammation in neurodegenerative disorders: the roles of microglia and astrocytes. Transl Neurodegener 9(1):1–12

Kyrkanides S, Tallents RH, Miller JH et al (2011) Osteoarthritis accelerates and exacerbates Alzheimer's disease pathology in mice. J Neuroinflammation 8:112

Lai KSP et al (2017) Peripheral inflammatory markers in Alzheimer's disease: a systematic review and meta-analysis of 175 studies. J Neurol Neurosurg Psychiatry 88(10):876–882

Lehnardt S (2010) Innate immunity and neuroinflammation in the CNS: the role of microglia in toll-like receptor-mediated neuronal injury. Glia 58(3):253–263

Lisi Flores-Aguilar M, Iulita F, Kovecses O, Torres MD, Levi SM, Zhang Y, Askenazi M, Wisniewski T, Jorge Busciglio A, Cuello C (2020) Evolution of neuroinflammation across the lifespan of individuals with down syndrome. Brain 143(12):3653–3671

Maurel M, Castagné R, Berger E, Bochud M, Chadeau-Hyam M, Fraga S, Gandini M, Hutri-Kähönen N, Jalkanen S, Kivimäki M, Marmot M, McCrory C, Preisig M, Raitakari O, Ricceri F, Salmi M, Steptoe A, Vineis P, Delpierre C, Kelly-Irving M (2020) Patterning of educational attainment across inflammatory markers: findings from a multi-cohort study. Brain Behav Immun 90:303–310

Mhatre M, Floyd RA, Hensley K (2004) Oxidative stress and neuroinflammation in Alzheimer's disease and amyotrophic lateral sclerosis: common links and potential therapeutic targets. J Alzheimers Dis 6(2):147–157

Morales I, Guzmán-Martínez L, Cerda-Troncoso C, Farías GA, Maccioni RB (2014) Neuroinflammation in the pathogenesis of Alzheimer's disease. A rational framework for the search of novel therapeutic approaches. Front Cell Neurosci 8:112

National Down Syndrome Society (n.d.) Alzheimer's disease and down syndrome. https://www.ndss.org/resources/alzheimers/. Accessed 18 Nov 2022

National Institute of Aging (n.d.-a). https://www.nia.nih.gov/health/what-are-signs-alzheimers-disease. Accessed 14 Sept 2022

National Institute of Aging (n.d.-b). https://www.nia.nih.gov/health/what-causes-alzheimers-disease#risk. Accessed 13 Nov 2022

O'Callaghan JP, Miller DB (2019) Neuroinflammation disorders exacerbated by environmental stressors. Metabolism 100:153951

Pimenova AA, Raj T, Goate AM (2018) Untangling genetic risk for Alzheimer's disease. Biol Psychiatry 83:300–310

Power MC, Bennett EE, Turner RW, Dowling NM, Ciarleglio A, Glymour MM, Gianattasio KZ (2021) Trends in relative incidence and prevalence of dementia across non-Hispanic black and white individuals in the United States, 2000-2016. JAMA Neurol 78(3):275–284

Rajan KB, Weuve J, Barnes LL, McAninch EA, Wilson RS, Evans DA (2021) Population estimate of people with clinical Alzheimer's disease and mild cognitive impairment in the United States (2020–2060). Alzheimers Dement 17(12):1966–1975

Sadleir KR, Kandalepas PC, Buggia-Prévot V, Nicholson DA, Thinakaran G, Vassar R (2016) Presynaptic dystrophic neurites surrounding amyloid plaques are sites of microtubule disruption, BACE1 elevation, and increased Aβ generation in Alzheimer's disease. Acta Neuropathol 132(2):235–256

Sastre M, Dewachter I, Landreth GE, Willson TM, Klockgether T, van Leuven F, Heneka MT (2003a) Nonsteroidal anti-inflammatory drugs and peroxisome proliferator-activated receptor-γ agonists modulate immunostimulated processing of amyloid precursor protein through regulation of β-secretase. J Neurosci 23(30):9796–9804

Sastre M, Dewachter I, Landreth GE, Willson TM, Klockgether T, van Leuven F et al (2003b) Nonsteroidal anti-inflammatory drugs and peroxisome proliferator-activated receptor-gamma agonists modulate immunostimulated processing of amyloid precursor protein through regulation of beta-secretase. J Neurosci 23:9796–9804

Sastre M, Klockgether T, Heneka MT (2006) Contribution of inflammatory processes to Alzheimer's disease: molecular mechanisms. Int J Dev Neurosci 24(2–3):167–176

Savonenko AV, Melnikova T, Li T, Price DL, Wong PC (2015) Chapter 21: Alzheimer disease. In: Neurobiology of brain disorders. Academic Press, pp 321–338

Schilling T, Eder C (2011) Amyloid-beta-induced reactive oxygen species production and priming are differentially regulated by ion channels in microglia. J Cell Physiol 226:3295–3302

Serrano-Pozo A, Frosch MP, Masliah E, Hyman BT (2011) Neuropathological alterations in Alzheimer disease. Cold Spring Harbor Perspect Med 1:a006189

Shah A, Kishore U, Shastri A (2021) Complement system in Alzheimer's disease. Int J Mol Sci 22(24):13647

Shi Y, Yamada K, Liddelow SA, Smith ST, Zhao L, Luo W et al (2017) ApoE4 markedly exacerbates tau-mediated neurodegeneration in a mouse model of tauopathy. Nature 549(7673):523–527

Spencer JP, Vafeiadou K, Williams RJ, Vauzour D (2012) Neuroinflammation: modulation by flavonoids and mechanisms of action. Mol Asp Med 33(1):83–97

Steele L (2000) Prevalence and patterns of gulf war illness in Kansas veterans: association of symptoms with characteristics of person, place, and time of military service. Am J Epidemiol 152(10):992–1002

Strauss S, Bauer J, Ganter U, Jonas U, Berger M, Volk B (1992) Detection of interleukin-6 and alpha 2-macroglobulin immunoreactivity in cortex and hippocampus of Alzheimer's disease patients. Lab Investig 66:223–230

Subramanian J, Savage JC, Tremblay MÈ (2020) Synaptic loss in Alzheimer's disease: mechanistic insights provided by two-photon in vivo imaging of transgenic mouse models. Front Cell Neurosci 14:592607

Tang Y, Le W (2016) Differential roles of M1 and M2 microglia in neurodegenerative diseases. Mol Neurobiol 53(2):1181–1194

Van Cauwenberghe C, Van Broeckhoven C, Sleegers K (2016) The genetic landscape of Alzheimer disease: clinical implications and perspectives. Genet Med 18(5):421–430

Venugopal C, Demos CM, Jagannatha Rao KS, Pappolla MA, Sambamurti K (2008) Beta-secretase: structure, function, and evolution. CNS Neurol Disord 7(3):278–294

Wisniewski KE, Wisniewski HM, Wen GY (1985) Occurrence of neuropathological changes and dementia of Alzheimer's disease in Down's syndrome. Ann Neurol 17(3):278–282

World Health Organisation (n.d.). https://www.who.int/news-room/fact-sheets/detail/dementia. Accessed 12 Nov 2022

Xiang W, Chao ZY, Feng DY (2015) Role of toll-like receptor/MYD88 signaling in neurodegenerative diseases. Rev Neurosci 26(4):407–414

Yang HD, Lee SB, Young LD (2016) History of Alzheimer's disease. Dement Neurocogn Disord 15(4):115–121

Yao K, Zu HB (2020) Microglial polarization: novel therapeutic mechanism against Alzheimer's disease. Inflammopharmacology 28(1):95–110

Zhang XX, Tian Y, Wang ZT, Ma YH, Tan L, Yu JT (2021) The epidemiology of Alzheimer's disease modifiable risk factors and prevention. J Prev Alzheimers Dis 8(3):313–321

Part II

Biomarkers and Diagnosis in Alzheimer's Disease

Biomarkers for Alzheimer's Disease

4

Mareechika Gaddam, Esther Rani Motamarri, and Abha Sharma

Abstract

Alzheimer's disease is a neurodegenerative disease, which affects intellectually is frequently observed in elderly generations and genetically linked individuals. The diagnosis of this disease is characterized by the presence of biomarkers such as amyloid beta ($A\beta_{1-42}$), phosphorylated tau (P-tau), and total tau (t-tau) protein causing NFTs' intercellularly interrupting synapse, which aids in transmitting the neuronal signals, thereby leading to memory loss and cognitive impairment failing to do basic tasks in daily life. Acquiring the diagnosis of this disease through the presence of biomarkers is categorized into two ways, invasive and noninvasive. The invasive way includes the collection of cerebrospinal fluid which is considered painful, but the advantage of this method is that it verifies the presence of the biomarkers accurately by taking into account its close association with the brain. This method is usually done at late stages for confirmation. The less painful or noninvasive ways include collecting samples from blood and ocular, olfactory, and oral fluids. These methods are in the preliminary stages of research that need further investigation to overcome methodological heterogeneity and discrepancy in accuracy and specificity.

Keywords

Alzheimer's disease · Cognitive impairment · Hippocampus atrophy · Synapse · Biomarkers · Invasive · Noninvasive · Amyloid beta ($A\beta_{1-42}$) · Total tau (t-tau) · Phosphorylated tau (P-tau) · Neurofibrillary tangles (NFTs) · Microsomal RNA · Autoantibodies · Isoprostanes · ELISA

M. Gaddam · E. R. Motamarri · A. Sharma (✉)
Department of Medicinal Chemistry, National Institute of Pharmaceutical Education and Research (NIPER)-Raebareli, Lucknow, India
e-mail: abha.sharma@niperraebareli.edu.in; abha.sharma@niperrbl.ac.in

A. Sharma, G. P. Modi (eds.), *Natural Product-Based Synthetic Drug Molecules in Alzheimer's Disease*, https://doi.org/10.1007/978-981-99-6038-5_4

Abbreviations

15-LOX	15-Lipoxygenase
AChE	Acetylcholine esterase
AD	Alzheimer's disease
ADNI	Alzheimer's disease neuroimaging initiative
AMPK	5′ Adenosine monophosphate-activated protein kinase
Apoε4	Apolipoprotein-E
APP	Amyloid precursor protein
Aβ	Beta-amyloid
BBB	Blood-brain barrier
BCSFB	Blood-cerebrospinal fluid barrier
CAA	Cerebral amyloid angiopathy
cdk5	Cyclin-dependent kinase 5
CDR	Clinical dementia rating
CHGA	Chromogranin A
CK 1	Casein kinase 1
CNS	Central nervous system
CSF	Cerebrospinal fluid
CSF-ISF exchange	Cerebrospinal fluid and interstitial fluid exchange
CT	Computerized tomography
DNA	Deoxyribonucleic acid
DS	Down syndrome
DYRK-1A	Dual-specificity tyrosine-phosphorylation-regulated kinase 1A
ELISA	Enzyme-linked immunosorbent assay
FDG	Fluoro-2-deoxy-D-glucose
FMRI	Functional magnetic resonance imaging
GSK-3	Glycogen synthase kinase-3
IL	Interleukin
IP-10	Interferon-induced protein 10
LC	Locus coeruleus
LM	Logical memory
LRP-1	Low-density lipoprotein receptor-related protein
Lys C	Lysozyme C
MALDI-MS	Matrix-assisted laser desorption/ionization-mass spectroscopy
MARKS	Microtubule affinity-regulating kinases
MCI	Mild cognitive impairment
MCP-1	Monocyte chemoattractant protein 1
miRNA	Microsomal RNA
MMSE	Mini-mental state examination
MRI	Magnetic resonance imaging
MS	Mass spectroscopy
Multiplex iTRAQ	

	Multiplexed isobaric tagging technology for relative quantitation
NCANP	Neurocan core protein
NFTs	Neurofibrillary tangles
NGF	Nerve growth factor
NMDA	N-methyl-D-aspartate
NPTX1	Neuronal pentraxin 1
NRXNs	Neurexins
NTP	Neuronal thread protein
OCT	Optical coherence tomography
PET	Positron-emission tomography
PGP-D	D-glycol protein
PiB	Pittsburgh compound B
PKA	Cyclic AMP-dependent protein kinase A
PP1	Protein phosphate-1
PP2A	Protein phosphate-2A
PP5	Protein phosphate-5
PS 2	Presenilin
P-tau	Phosphorylated tau
RNA	Ribonucleic acid
RNFL	Retinal nerve fiber layer
RT-PCR	Real-time reverse transcriptase-polymerase chain reaction
SCG2	Secretogranin 2
SELDI	Surface-enhanced laser desorption/ionization time-of-flight mass spectroscopy
SYN-2	Synaptic protein synapsin-2
t-tau	Total tau
VILIP-1	Visinin-like protein 1
WMS	Wechsler memory scale
β2M	β2-Macroglobulin

4.1 Introduction

Alzheimer's disease (AD) causes considerable loss of forebrain cholinergic neurons, which undoubtedly facilitates dementia-causing components that include frequent manifestation of cell death, senile plaques, and neurofibrillary tangles in the neocortex, hippocampus, and basal nucleus of Meynert. Abetment of elements for AD diagnosis is the reduction in cerebral cortex and hippocampal acetylcholine neurotransmitter marked as one of the most significant abnormal alterations, and specific genes that undergo mutation such as amyloid precursor protein (APP), apolipoprotein (Apoε4), and presenilin 2 (PS 2) are known today, leading to β-amyloid protein buildup in cortex and hippocampus that are key areas for cognitive function and memory. Furthermore, forebrain cholinergic neurons appear to be directly disrupted by excessive glutamate receptor stimulation, mainly the NMDA receptor (Wenk 2003). The main attributes to be considered for a fluid biomarker are disease specificity and accuracy, appropriate results according to the disease's level of infection, being noninvasive or unobtrusive concerning the retrieval of the fluid from the respective source, being convenient to employ on diverse individuals, and being effortless and cost-effective. The blood biomarkers satisfy most of these requirements, therefore counting it as a better source for knowing the disease advancement through the presence of biomarkers. Competently, the cerebrospinal fluid (CSF), although invasive due to the techniques involved in its collection, is of more importance considering the close association of CSF with the brain and also in furnishing the essential data regarding the chemical changes induced in the biological system happening in the brain during the preclinical stages of AD, while the other sources like blood, sweat, and tears give inefficient or inaccurate information about the disease progression. However, there are certain drawbacks when it comes to CSF because the mode of collection is invasive, extremely painful, and uncomfortable. On the other hand, if the mode of collection is blood, sweat, and tears, it is less painful and noninvasive, making these preferable over the former method (Ausó et al. 2020) and the detection of biomarkers of AD from these sources is characterized by the presence of $A\beta_{1-42}$ ($A\beta_{37}$, $A\beta_{36}$, $A\beta_{38}$, $A\beta_{40}$), total tau, phospho-tau-181 (single or multiplex), multiple phosphorylated tau epitopes, or isoform-effectuated aggregation entailing extreme harm to neurons that are acquired using the analytical techniques like ELISA, INNO-BIA AlzBio3 Luminex-based technology (Innogenetics), quantitative real-time RT-PCR, LC/electrospray ionization MS, SELDI, and MALDI (Humpel 2011).

4.2 Biomarkers of AD

A biomarker is typically defined by substances (unnatural compounds, particular enzymes, cells, proteins, hormones, or (DNA, RNA) genetic material) or imaging findings that are utilized as attributes to identify the existence of a specific physiological condition and may aid in clinical diagnosis. Furthermore, a huge increase in research from 1980 to the present shows that the use of biomarkers is increasing for aiding the diagnosis and prognosis of AD.

4.2.1 Biomarkers Based on AD Stages

Stage 1: It includes people who may have a functional impairment but not cognitive impairment, which may only be determined through neuropsychological techniques. There is growing evidence that specific biomarkers, such as the presence of amyloid imaging and a altered CSF $A\beta_{42}$ concentration, can indicate pathogenic alterations at an early preclinical period (Food and Administration

2018). The approval of AD treatment may be aided by early diagnosis based on biomarkers, which may have clinical advantages and result in better outcomes.

Stage 2: Although the patients have subtle cognitive alterations, and no functional deficits, they do not meet the requirements for dementia, which can be identified with the use of sensitive devices. The Food and Drug Administration (FDA) suggests that while diagnosing AD stage 2, sensitive neuropsychological tests and biomarker changes should be taken into consideration.

Stage 3: Pathophysiological biomarkers have been identified, and patients have begun to struggle with some measured everyday tasks. While the first two phases are preclinical, this stage of the disease is correlated with modest cognitive impairment.

Stages 4–6: Pathophysiological biomarkers are evident in the successive stages of mild, moderate, and severe Alzheimer's disease dementia, with decreased cognitive impairment (Omar and Preddy 2020).

4.3 Invasive Biomarkers

The chief origin of noradrenaline in the brain is the locus coeruleus (LC) that is rich in neuromelanin is thought to influence attention and memory. Tau neurofibrillary tangles (NFTs) are first noticed in LC during the asymptomatic stage of AD before being found in other cerebral regions such as the entorhinal cortex and the neocortex. Tau clumps appear in the LC before the normal neuronal loss as AD progresses. Studies utilizing unbiased stereology have shown neuronal loss that is concentrated in the rostral/middle region of the LC, progressing from 30% in the prodromal stage to 55% when dementia is diagnosed, as well as an average drop in LC volume of 8.4% for each Braak stage advancement. In terms of function, this neuronal loss has been linked to cognitive degeneration and diminishing noradrenaline levels in the brain and hippocampus (Ausó et al. 2020).

$A\beta_{42}$ is a 42-amino acid-long peptide associated with Aβ used as a biomarker for the diagnosis of AD. The concentration of $A\beta_{42}$ in CSF is negatively proportional to the number of Aβ plaques implying that the reducing levels of $A\beta_{42}$ and high concentration of Aβ plaques are observed in positron-emission tomography (PET) scans of CSF. The brain neuron contains tau protein in its axons. Escalated amounts of tau and phosphorylated tau are one of the main causes of AD (Gunes et al. 2022). As a result, if the patient responds well to treatment at the right time, the identification employed for in vivo imaging of early structural tissue modifications, like diminution in LC volume or metabolic changes, might aid in diagnosis and perhaps slows down the disease progression (Ausó et al. 2020).

4.3.1 CSF

The mean volume of CSF present in our brains' ventricles, cranial and spinal subarachnoid spaces is 150 mL of which 25 mL is present in ventricles and the rest 125 mL in subarachnoid spaces. One of the most commonly known functions of CSF is its hydromechanical protection. Apart from this, CSF is necessary for brain growth regulation or turnover of brains' ISF homeostasis and managing the function of neurons (Sakka et al. 2011). Obstruction in normal CSF circulation is thought to facilitate the cause of a wider range of advanced CNS pathologies like AD, one of the most commonly known neurodegenerative disorders (Simon and Iliff 2016).

Mild cognitive impairment (MCI) is an age-linked disorder that has obtained awareness among the scholarly community and common people. Most of the patients having MCI have inchoate AD, while others have benign MCI forms.

Neurodegeneration begins during the preclinical stage (2–3 decades prior to the clinical outset of the disease) due to an increase in tangle bundles at a certain threshold, and the first signs manifest (Blennow and Hampel 2003).

It is necessary to take into account proteins that are implicated in the abnormal breakdown of an APP as a peculiar biomarker for early diagnosis of AD. Specifically, the enzymes β-secretase 1 (BACE1) and presenilin 1 (PSEN1) are necessary for the breakdown of APP. PSEN1 and BACE1 activity levels were intensifying in the CSF of MCI individuals (García-Ayllón et al. 2013; Ewers et al. 2008). Furthermore, the Apoε4 gene has been linked to increased BACE1 (-β-secretase 1) expression (Ewers et al. 2008). It was found that BACE1 activity was not elevated in MCI patients with persistent MCI, but only in those whose cognitive dysfunction declined to lead to later phases of dementia [10], making BACE1 not a strong candidate for being Apoε4 noncarrier even though it appears to be of more importance and is a potential early-stage biomarker to identify changes in the amyloidogenic pathway in Apoε4 carriers (Ewers et al. 2008).

Neuroinflammation and synaptic dysfunction are additional early AD features to emphasize, which consequently become distinct markers of these processes that may play a significant role that is closely linked to cognitive decline (Calsolaro and Edison 2016). Secretogranin II (SCG2), and chromogranin A (CHGA) are some proteins that take part in vesicular transport. The concentration of these proteins was found to be considerably more prominent in the CSF of patients with MCI, particularly whose condition is getting worse as AD pathology advances when compared to the healthy individuals (Duits et al. 2018). One research found that higher CHGA levels in the CSF of healthy elderly individuals suggested a subsequent decline in $A\beta_{42}$ (Mattsson et al. 2013).

YKL-40 and visinin-like protein-1 are additional proteins that are vitally involved in inflammation (VILIP-1). In contrast to elderly people who are intellectually healthy, MCI and AD patients indicated a greater concentration of such compounds (Craig-Schapiro et al. 2010).

Interferon-induced protein-10 (IP-10) concentrations are elevated in the CSF of asymptomatic older generation who also showed increased concentrations of total tau and phospho-tau, which is another potential inflammatory marker (Bettcher et al. 2018).

Both MCI and AD patients continued to show increased concentrations of MCP-1, a low-molecular-weight cytokine associated with the inflammatory process (Dhiman et al. 2019).

Considering the existence of microRNAs (miRNAs), a large family of inherently occurring short noncoding RNAs that influence the amount of fully functional or mature miRNAs during the post transcriptional stage (Provost 2010), is also significant. The brain generates about 70% of the identified miRNAs, although some miRNA species are found in exosomes, making them promising prospects for use as biomarkers in clinical diagnosis. Only 40 of the 2000 human miRNAs that were discovered so far are highly expressed in the brain (Lukiw et al. 2013). The primary CSF biomarkers for clinical AD ($A\beta_{42}$, total tau, and phospho-tau) tend to remain consistent which is helpful for diagnosis but are insufficient for monitoring the course of the disease. CSF miRNAs have the benefit of targeting significant pathological AD genes despite being collected in an invasive way, which is far from optimal. A small family of genes may be affected by a single miRNA's ability to interfere with their expression. This is the situation regarding miRNA-125b (high in the case of AD), aiming at the cell cycle regulator CDKN2A and the enzymes 15-LOX and SYN-2. The possibility of using miRNAs as medicines in the subsequent years is more likely. Furthermore, the number of brain miRNAs was progressively reduced in areas of the brain that are susceptible to AD advancement (Mattsson et al. 2013), which may partly be explained by the dysregulation of particular miRNAs that may be associated with the etiopathogenesis of AD (Lukiw 2007) (Fig. 4.1).

Various biomarkers aid in detecting early morphological alterations of the brain observed in people already diagnosed with MCI comprising notable atrophy at different locations of the brain that include the LC or the hippocampus, and protein accumulation of amyloid plaques in extracellular Abeta plaques or intracellular tau prominences such as extracellular amyloid plaques or intracellular tau (accommodating NFTs). The cerebrospinal fluid of AD patients can be examined for indicators of degenerative processes linked to Alzheimer's disease (AD), such as synaptic impairments, neuroinflammation, oxidative stress, or neuronal death. A new and foreshadowing tool for early AD diagnosis is the detection of miRNAs (Ausó et al. 2020).

4.3.1.1 Diagnostic Markers

Diagnostic markers are very important when it comes to the effective treatment of AD by using acetylcholinesterase inhibitors as they facilitate the detection to know the progression of the disease. There are two types of diagnostic markers. They are state and stage markers.

4.3.1.2 State Markers

The severity of the disease's progress is mirrored by state markers. The total amount of tau protein is one illustration of a state marker representing the severity of neuronal deterioration and injury. There is a momentary rise in tau protein content in the CSF during acute diseases (e.g., ischemic stroke), and this rise is correlated with the extent of the infarct as determined by CT.

Additionally, disorders with severe neuronal degeneration, such as Creutzfeldt-Jakob disease, have the highest levels of increased tau protein concentrations in the

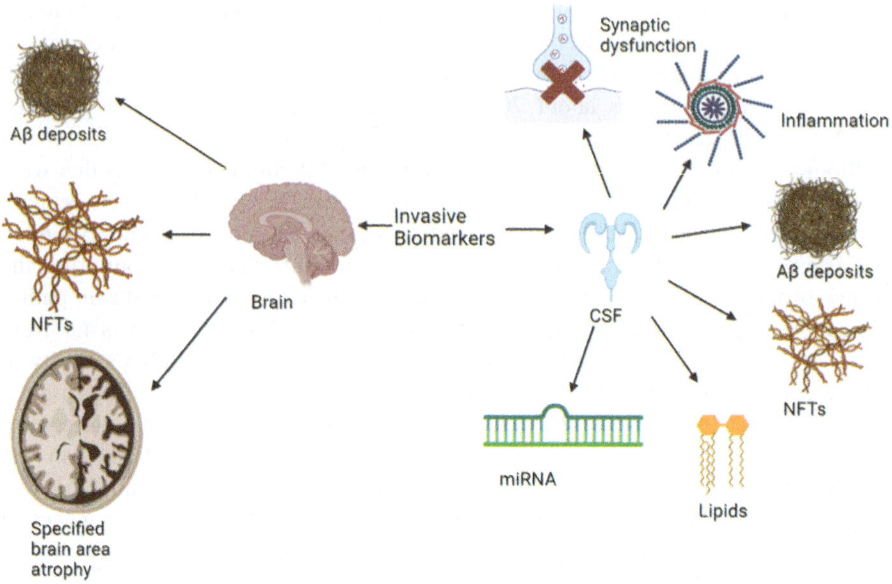

Fig. 4.1 Overview of various invasive indicators of AD

CSF, whereas gradual increase is observed in AD patients, but patients with depression have no or marginal neuronal degeneration and have a normal concentration. In patients with MCI, state markers, such as tau protein, are likely to have a greater predictive value than stage markers, like hippocampal shrinkage on CT or MRI.

4.3.1.3 Stage Markers

They calculate the degeneration process. Hippocampal atrophy as evaluated by CT or MRI is one illustration of a stage marker. Stage markers also include clinical rating measures, which quantify the intensity of cognitive decline. Regardless of the phase, hippocampal atrophy (9–15% reduction in hippocampus volume) is less severe than elevated tau protein concentrations (210–290% increase).

4.3.1.4 Biomarkers in CSF for AD

Because the CSF and the brain's extracellular space are directly associated, alterations in the brain's biochemistry always have an impact on the CSF. AD pathology only affects the brain, so CSF is a natural place to detect AD biomarkers. The proportions of total tau protein, $A\beta_{1-42}$ (the 42-amino acid version of Aβ), and phosphorylated tau protein in the CSF may potentially serve as biomarkers for these pathogenetic processes (Blennow and Hampel 2003).

4.3.1.4.1 $A\beta_{1-42}$

APP undergoes consecutive cleavages by β- and γ-secretases to form profibrillary $A\beta_{1-42}$, an essential constituent of senile plaques, chiefly composed of Aβ. $A\beta_{42}$, an extensively soluble protein, gets accumulated on the walls of leptomeningeal structure and CAA. Generally, Aβ originates in our brain due to synapsis. Considerably, it gets dispersed through being ingested by microglial cells and astrocytes flowing across the BBB aided by solute transporters as well as PGP (D-glycoprotein) and LRP-1 (low-density lipoprotein receptor-related protein) that are dispersed through the perivascular glymphatic pathway. CSF levels of amyloid is elevated in humans, when a person is awake where glymphatic pathway activity and Aβ clearance are considerably reduced. Then, as sleep sets in, interstitial Aβ levels decrease as perivascular CSF-ISF exchange becomes active and interstitial Aβ clearance is accelerated. These results suggest that the physiological clearance of Aβ from the brain is significantly influenced by perivascular CSF-ISF exchange along the glymphatic pathway. This physiological clearance of Aβ through CSF-ISF exchange is reduced due to alterations in BBB Aβ dispersion, leading to the accumulation of Aβ protein resulting in neurodegeneration (Simon and Iliff 2016).

4.3.1.4.2 Tau Protein

Tau protein exists in the human brain in six different variants and has a vast group of phosphorylation sites (Fig. 4.2). There are 85 possible sites for serine (S), threonine (T), and tyrosine (Y) phosphorylation in the phosphorylated tau protein. The proline-rich domain of tau, on each side of the microtubule-binding domain, contains a large number of phosphorylated tau residues (Blennow and Hampel 2003). In the case of neurodegenerative disorders, they are present in abnormal amounts inside neurons and neuronal axis as NFTs.

The six human central nervous system (CNS) tau variants are shown (Fig. 4.3). Exons 2 and 3 (E2 and E3) code for two distinct 28-amino acid sequences located close to the tau's N-terminal. Tau isoforms with 0 N are produced when E2 and E3 are absent, while tau isoforms with 1 N and 2 N are produced when E2 and E3 are present. The four incorrect-reoccurring microtubule-binding domains are represented by M1 through M4, with M2 being encoded by exon 10. When M2 is absent, 3R tau is produced, and when M2 is present, 4R tau is produced. The tau

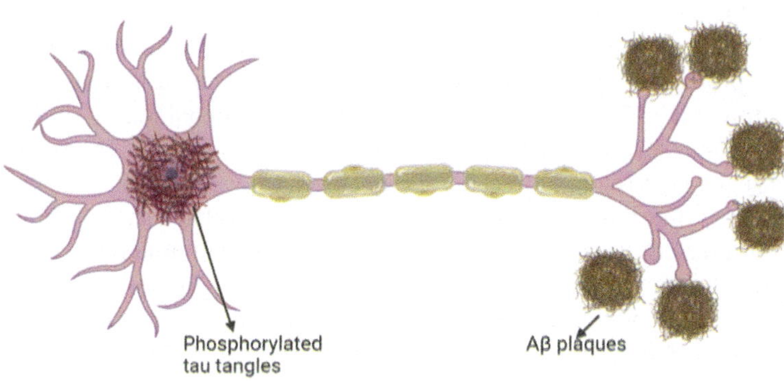

Fig. 4.2 Tau pathology through different phosphorylation sites

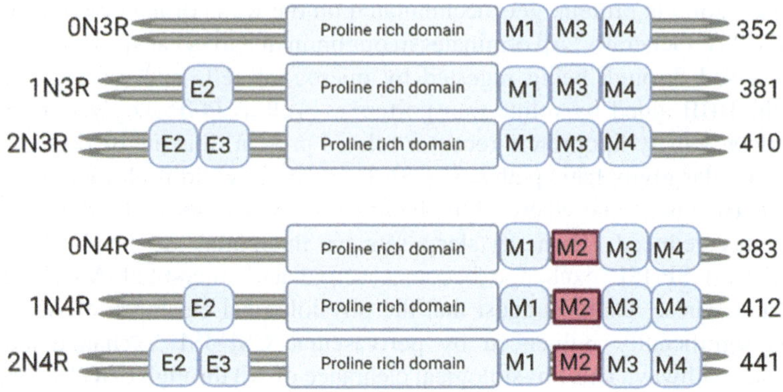

Fig. 4.3 Human tau genes and their six isoforms through phosphorylation and hyperphosphorylation

polypeptide's central proline-rich region is highlighted. As a result, alternative splicing results in tau proteins with 352–441 amino acids (Fig. 4.3).

There are about 45 phosphorylation sites known, most of which are in the proline-rich domain and the areas surrounding the microtubule-binding domain.

The concentration of tau protein as a biomarker for AD in CSF was first reported in 1993 for which the ELISA method was used along with monoclonal antibodies. Later, the ELISA methods depending on monoclonal antibodies were developed that helped in identifying ensemble forms of tau protein autonomous of their phosphorylation and development. Tau phosphorylation is controlled by certain kinases and phosphatases. Tau hyperphosphorylation in disease is thought to be the outcome of an imbalance in tau kinase and phosphatase activity. GSK-3, cdk5, and AMPK are examples of proline-directed tau kinases. CK1, MARKs, PKA, and DYRK-1A are examples of non-proline-directed kinases. Tyrosine kinases include Syk, Fyn, and Abl. And also, several phosphatases include PP1, PP2A, and PP5 and dephosphorylated tau (Noble et al. 2013).

4.3.1.4.3 Phosphorylated Tau Protein

Phosphorylated tau protein is the latest biomarker for AD to be identified. For several tau protein phosphorylated homologs, such as threonine 181 and 231, threonine 181, threonine 231 and serine 235, threonine 199, threonine 231, and threonine 396 and 404, several techniques have been developed. The amount of phosphorylated tau protein in the CSF likely correlates with the level of phosphorylation of tau in the brain. But the amount of phosphorylated tau protein remains unchanged after an acute stroke with regard to the overall amount of tau protein. Moreover, despite a marginal increase in total tau protein concentrations, phosphorylated tau protein concentrations do not rise in Creutzfeldt-Jakob disease.

These results demonstrate that the concentration of phosphorylated tau protein in the CSF also depicts the phosphorylation state of tau and, consequently, the formation of tangles in AD, rather than just serving as a signal for neuronal damage like total tau protein (**Blennow and Hampel** 2003).

4.3.2 Association of Biological Biomarkers in CSF: The Diagnosis of AD in Different Cases

4.3.2.1 Role of Plasma Amyloid Proteins in Patients whose Diagnosis is Hereditary

At times, elevation in Aβ plasma concentration is noticed in patients with AD, especially those who are related and have pre-symptomatic carriers of mutations, making AD inheritable and genetically involved. It is observed in patients whose diagnosis is not hereditary and is not seen in spasmodic cases.

4.3.2.2 Role of Plasma Amyloid Proteins in Down Syndrome Patients who are Diagnosed with AD

Down syndrome (DS) patients ensure neuropathological alterations that are in commensuration with AD probably as a result of APP gene triplication and overexpression.

The rise in $A\beta_{42}$ and $A\beta_{40}$ plasma levels in DS patients is presumed as a primary cause in the facilitation of amyloidosis and the advancement of dementia.

A survey was done which included 30 individuals with DS where 21 of them did not develop dementia, but the eight patients developed. The cognitive and daily task performance of all 30 of them was evaluated along with blood samples every year for 5 years. During this delve, eight of them were diagnosed with dementia. The most unusual thing observed was that there were not any fluctuations in $A\beta_{40}$ and $A\beta_{42}$ plasma concentration compared to a healthy person. To straighten up, $A\beta_{42}$ and $A\beta_{40}$ are not convincing enough that they facilitate the advancement of AD (Frey et al. 2005).

4.3.2.3 Role of Sex Differentiation Linked to AD Biomarkers' Prominence and Intellectual Depreciation

There is a pretentious rife of AD in women because of the disease's commonness and acuteness. It has been revealed in earlier autopsy research that women affected with AD are more vulnerable to clinical implications of AD. By investigating whether sex influences the existing correlation between CSF biomarker levels and consequences of brain aging, Mary E.I. Koran et al. expanded this research work broader. In this investigation, individuals with normal intellect ($n = 348$), MCI ($n = 565$), and AD ($n = 185$) were chosen in addition to the people selected from ADNI, employing multivariate degeneration models. They obtained an analysis that impacts baseline levels of CSF biomarker $A\beta_{42}$ and an ensemble of tau on cross-sectional, extensive brain aging as an aftermath. They also procured a substantial interlinkage betwixt sex and $A\beta_{42}$ on the progression of hippocampal atrophy ($p = 0.002$), extensive memory loss ($p = 0.017$), and cognitive impairment ($p = 0.025$) analogously; they also noticed a substantial interaction betwixt sex and overall tau protein on extensive hippocampal atrophy ($p = 0.008$) and cognitive decline ($p = 0.034$). Sexual diversity is shown to be prominent in people with MCI whose literacy rate is low.

4.3.2.3.1 Influence of Sex on $A\beta_{42}$

After the initial assessment, participants were monitored for an average of 2.5 years (range: 0–9 years). Initial memory performance (t (1198) = 1.58, $p = 0.11$), executive function performance (t (1198) = 0.29, $p = 0.77$), hippocampus volume (t (1082) = 0.93, $p = 0.35$, and t (1082) = 0.12, $p = 0.90$, respectively), or sex did not influence $A\beta_{42}$. Throughout cumulative comparisons, there existed a correlation between sex and $A\beta_{42}$ where three of the five manifested brains aging during evaluation included cognitive performance and left hippocampal shrinkage (t (2739) = 3.07, $p = 0.002$, and t (4154) = 2.39, $p = 0.02$). In all cases, the female sex was linked to deteriorating outcomes when there was a low $A\beta_{42}$ level. Low CSF $A\beta_{42}$ is a sign of higher brain $A\beta_{42}$. Concerning right hippocampus atrophy, sex did not interact with $A\beta_{42}$(t (2739) = 1.54, $p = 0.12$).

4.3.2.3.2 Influence of Sex on Tau

After an initial assessment, sex is not associated with tau concerning cognitive decline (t (1198) = 0.51, $p = 0.61$), right hippocampus volume (t (1082) = 0.49, $p = 0.62$) or left hippocampal volume (t (1082) = 0.44, $p = 0.66$), and memory decline (t (1198) = 0.22, $p = 0.83$). Throughout cumulative comparisons, the association of tau with sex was observed which included left hippocampus shrinkage and cognitive decline (t(2739)=−2.61, p=0.009) and (t (4132) =−2.11, $p = 0.03$, respectively). When there was an elevated amount of total tau, the female sex was linked to deteriorating executive function and hippocampus atrophy. But sex did not interfere with the ensemble tau level associated with overall cognitive performance (t (4154) = 1.20, $p = 0.23$) or right hippocampus atrophy (t (2739) = 0.56, $p = 0.58$) (Koran et al. 2017).

4.3.2.4 Role of CSF Abundance and Clearance Associated with the Prominence of AD Biomarkers and Other Factors

The BBB Aβ efflux pathways in the geriatric brain are hampered, which may lead to amyloid aggregation and later neuronal death. This dysfunction causes the downregulation of PGP and LRP-1 that aid in CSF production. However, as age progresses, the expression of the choroid plexus (CP) Aβ efflux transporter escalates indicating that the CP and BCSFB could serve a contributing role in amyloid clearance from the ventricular CSF and concomitant ISF. With senescence, AD symptoms show deposition of amyloid within the CP and reduced CSF secretion by the CPs; therefore, the effectiveness of amyloid clearance with the bulk reabsorption of CSF via arachnoid villi and along cranial nerve sheaths minimizes the rate of CSF turnover.

The brains of elderly people also exhibit disrupted glymphatic pathway function, which includes altered perivascular CSF-ISF exchange and impeded clearance of interstitial amyloid β. Uncertainty exists regarding the possibility that decreased CSF secretion by the CP is contributing to the perivascular CSF flow through the aging brain slowing.

Glymphatic pathway function was slowed when CSF secretion was experimentally disrupted by providing carbonic anhydrase inhibitor acetazolamide, which infers that lowered CSF secretion at the CP could be a contributor to disrupted lymphatic pathway function in the aging brain. But perivascular AQP4 that takes part in CSF production whose localization was also substantially diminished in the aging brain was attributed to a delayed perivascular CSF inflow into and through the brain parenchyma. It has not yet been formally ascertained whether perivascular AQP4 location loss stimulates amyloid aggregation in the aging brain. Nevertheless, the ability of Aβ deposits that lead to AQP4 mislocalization, be it in the form of senile plaques or CAA, indicates the existence of a growing pathogenic cycle, with reactive astrogliosis obstructing glymphatic pathway function as well as Aβ clearance, leading to amyloid deposition and furthermore neuroinflammation, resulting in disruption of the glymphatic pathway (Simon and Iliff 2016).

4.4 Noninvasive Biomarkers

There is a need for the development of early noninvasive diagnostic techniques for AD; even though the established neuroimaging techniques show promising results, they are expensive and cause radiation. Collection of CSF fluid needs lumbar puncture, so the diagnosis/analysis of peripheral fluids is a good alternative which includes blood, urine, and ocular, oral, and olfactory fluids (Gleerup et al. 2019).

4.4.1 Oral

Studies suggest that saliva is one of the potential sources of a noninvasive biomarker for AD diagnosis; biomarkers could be produced directly in salivary glands/diffused from blood. Degeneration of neurons or nerve cells is observed in AD; it is also observed that this causes damage to nerve cells related to autonomic nervous system (ANS) and blood. ANS regulates the production of saliva via three major salivary glands (parotid, sublingual, submandibular); thus, there will be alteration in the production and composition of saliva. Various studies have identified and determined $A\beta_{40}$, $A\beta_{42}$, P-tau, and t-tau in the saliva of patients diagnosed with AD (Tvarijonaviciute et al. 2020).

4.4.1.1 Beta Amyloid

The pathological hallmarks of AD include Aβ accumulation and tau hyperphosphorylation. Pareja et al. reported a statistical increase of $A\beta_{42}$ in the saliva of people with mild AD and no visible change in people with severe AD (Bermejo-Pareja et al. 2010). Later, studies by Sabbagh et al. reported a 2.45-fold rise in levels of $A\beta_{42}$ at all stages of AD (mild to moderate) (Liang and Lu 2019).

4.4.1.2 Tau

AD is marked by tau protein hyperphosphorylation and aggregation; t-tau (total tau) represents the intensity of neuronal damage, and P-tau represents the phosphorylated state of tau, which has been assessed in various studies. Origin of salivary tau remains unknown; there are numerous possibilities which include the possibility of salivary tau protein release from nerves that innervate salivary glands as they are proximal to CNS via cranial nerves. Another possibility includes the secretion of tau by acinal epithelial cells of salivary glands. The P-tau/t-tau ratio was shown to be significantly higher in AD participants compared to cognitively healthy older subjects, particularly at the S_{396} phosphorylation site.

4.4.1.3 Acetylcholinesterase (AChE)

Acetylcholine (ACh) cholinergic neurotransmitter is metabolized to its subsequent components, i.e., choline and acetic acid, by AChE. Cholinergic neurons, which are essential for memory and learning, are damaged at an early stage of AD. Because of differences in sample methodologies and potential nonlinear changes in enzyme activity correlated with progression of the disease, data on changes in AchE activity linked with AD in peripheral bodily fluids have been inconsistent (Tvarijonaviciute et al. 2020). Balchitiari et al. state that despite the fact that salivary AChE activity in persons with AD was lower than in the control group, there were no significant changes (Bakhtiari et al. 2017).

4.4.1.4 Lactoferrin

According to a few studies and evidence reported, bacterial/viral infections in the brain may cause or exacerbate the pathophysiology of AD. Because of its antibacterial protein composition, saliva constitutes one of the body's first lines of defense. Lactoferrin is an Aβ-binding glycoprotein, a foremost antimicrobial peptide in saliva. Chero et al. reported decreased levels of saliva lactoferrin. Decreased levels of saliva lactoferrin increase the risk by greater than 77% of converting into amnestic MCI and AD dementia (Carro et al. 2017). The accuracy of salivary lactoferrin diagnosis was found to be greater than that obtained from CSF biomarkers (T-tau, CSF Aβ) (Liang and Lu 2019).

4.4.1.5 Others

Reports have proposed a relation between high salivary sugar levels and progress of AD. Therefore, salivary sugar levels (trehalose) may serve as a potential biomarker of AD (Lukiw 2007) (Fig. 4.4).

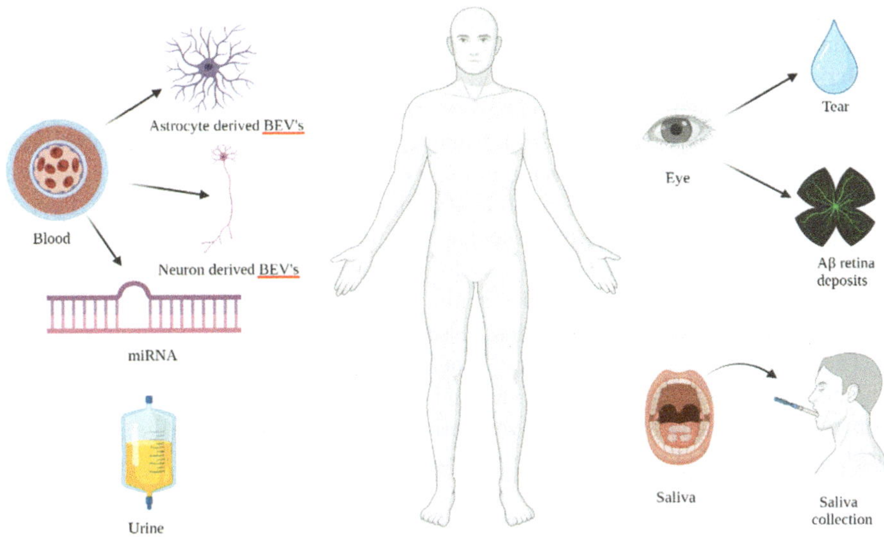

Fig. 4.4 An illustration of noninvasive biomarkers: eyes, saliva, urine, and blood

4.4.2 Eye

4.4.2.1 Retinal Biomarkers

4.4.2.1.1 Retinal Nerve Fiber Layer (RNFL) and Optic Nerve

Hinton et al. presented the first histopathological evidence of retinal abnormalities in AD, i.e., widespread ganglion cell losses, retinal nerve fiber layer (RNFL) thinning, and optic nerve degeneration in comparison to healthy controls (Hinton et al. 1986).

Blanks et al. (1996) provide the basis for the usage of retinal imaging as a potential biomarker for AD. In vivo assessments of the nerve fiber layer and the optic nerve are often done using scanning laser ophthalmoscopy (SLO) and optic coherence tomography (OCT), with SLO being preferable because it uses a confocal scanning laser system that enables accurate depth resolution. SLO was used by Kergoat et al. (2001) and Danesh-Meyer et al. (2006) for identifying RNFL thinning in retina's peripapillary and macula areas (Lim et al. 2016).

Berisha et al. observed RNFL thinning in AD; these assessments are supported by various studies in patients with MCI and AD, which showed that ganglion cell complex (including ganglion cell bodies, their dendrites, and RNFL) at macula is a significantly better indicator than RNFL thickness. In addition, this method is used in order to differ AD and MCI from healthy controls (Berisha et al. 2007).

4.4.2.1.2 Retinal Blood Flow and Vasculature

Researchers have theorized about the possible value of evaluating the blood volume of the eye as a biomarker of AD due to the resemblances between the blood vessels of the brain and the eye. Common vascular issues linked with AD include decreased blood vessel density, decreased blood flow, vasoconstriction, and inadequate Aβ clearance.

Berisha et al.'s (2007) study showed reduced venous blood flow and substantial narrowing of retinal veins in comparison to healthy controls by using Doppler imaging device (Berisha et al. 2007). Cheung et al. (2014), using fundus photography and automated vessel segmentation software, showed (1) a decrease in retinal vessel caliber and (2) a decrease in fractal dimension (a measurement of the overall branching structure of the retinal vascular tree). (3) There is no change in the angle subtended between both daughter vessels at each bifurcation, which is the vessel branching angle. (4) An increase in vascular tortuosity, or the average percentage of curliness/"non-straightness" of retinal vessels, is also seen.

Thus, vascular measures may be beneficial for screening, but the question is whether age, atherosclerosis, and vascular related problems can be distinguished from AD-related alteration; moreover, retinal vascular parameters that can distinguish MCI from AD are yet unclear (Cheung et al. 2015).

4.4.2.1.3 Intraretinal Tau and Amyloid Deposition

Ratnayaka et al.'s studies confirmed that retinal ganglion cells express APP in the inner nuclear layer of the retina and retinal pigment epithelium; these retinal ganglion cells also express an enzyme which is capable of producing Aβ, i.e., β-secretase (Ratnayaka et al. 2015). AD deposits (tagging): first successful demonstration of noninvasive in vivo visualization of curcumin-bound fluorescent Aβ in the bitransgenic mice retina, which possesses the mutated genes that regulate presenilin-1 (PS-1) and APP.

The CSIRO in Australia presented extensive preliminary results in humans using curcumin contrast in the patients with proven brain deposits using (Pittsburg compound B) PiB-PET imaging.

However, other studies have not observed the same results, suggesting that P-tau could be a potential biomarker compared to Aβ; presence of P-tau is demonstrated using SLO imaging of FBS [(trans, trans)-1-fluoro-2,5-bis-(3-hydroxycarbonyl-4-hydroxy) styryl benzene] bound tau in the retina of P_{310S} human tau mouse line.

Despite being hopeful, HO et al. (2014) were unable to locate Aβ- or P-tau deposits in the retina of AD donors. The authors conclude that AD hallmarks do not accumulate in the eye like that of the brain, despite the possibility that using different assays, such as limited retinal cross sections instead of whole mounts and paraffin-embedded sections rather than frozen tissues, may reduce the sensitivity of these proteins' identification (Ho et al. 2013).

AD deposits (visualizing): cross polarizers are newly developed imaging technology that can discriminate between controls and ex vivo human volunteer retinas with AD (Campbell et al. 2015; Hamel et al. 2016). By using the fibrillary arrangement of Aβ, this technique produces distinct modifications to birefringence that may be quantified using Mueller matrix polarimetry (Campbell et al. 2015; Hamel et al. 2016).

4.4.2.1.4 Choroidal Thickness

The choroid, which is placed between the retina and the outermost layer of the eye, is specifically responsible for the blood flow to the retina. Modern OCT devices include enhanced depth imaging (EDI) capabilities that use longer wavelengths in the 1060 nm range (Wong et al. 2011), allowing for deeper penetration into the eye's layers and viewing of the choroid (Wong et al. 2011).

According to Gharbiya et al. (2014), studies of AD participants ($n = 21$) had choroid that was 30% thinner than that of healthy controls ($n = 21$) directly below and within 1 mm of the fovea, and RNFL thinning was not observed; this indicates that choroidal thinning in AD occurred before RNFL alterations (Gharbiya et al. 2014). The major challenge is that choroidal thinning endures substantial diurnal variations (Kinoshita et al., 2016). Moreover, choroidal thinning is a symptom of several other conditions, including uveitis, chronic obstructive pulmonary disease, myopia, and aging (Barteselli et al., 2012) (Kinoshita et al. 2017; Shah et al. 2012).

4.4.2.2 Nonretinal Biomarker

4.4.2.2.1 Pupillary Reactions

The dilator muscles and iris sphincter work in harmony to balance forces that affect pupil size and pupillary response to light. The former is innervated by parasympathetic cholinergic receptors emerging from the Edinger-Westphal nucleus, while the other is innervated by postganglionic sympathetic noradrenaline receptors emerging from the superior cervical ganglion.

Together, these findings demonstrate that high-speed video pupillography pupil flash responses, which directly evaluate the integrity of the cholinergic system in the nervous system, may be a promising biomarker. It is unclear if using AD drugs such as cholinesterase inhibitors can minimize this outcome. Indeed, further research is needed to determine the dependability and reproducibility of these growing technologies (Lim et al. 2016).

4.4.2.2.2 Crystalline Lens

The lens is a transparent biconvex structure in most land vertebrate eyes; the crystalline lens, together with the cornea, aqueous and vitreous humors, is responsible for focusing incident light from the outside world towards the retina.

The crystalline lens is primarily composed of crystalline proteins in higher concentration, giving the ability to examine protein aggregation, which contributes to cataract formation (van Wijngaarden et al. 2017). Studies have identified Aβ deposits in crystalline lenses of rats, rabbits, monkeys, and transgenic mice. The first report of an AD-specific cataract in humans was made by Goldstein et al. (2003), who found deposits in the cytoplasm of the equatorial supranuclear/deep cortical lens fiber cells (Goldstein et al. 2003). This AD cataract does not impede vision and is difficult to detect during a regular eye examination until full dilatation is achieved; studies have failed to detect Aβ in human lenses using either confocal Raman spectroscopy or standard immune histochemistry methods (Lim et al. 2016). Kerbage et al. (2013) performed two small clinical studies on people with AD;

researchers used in vivo imaging of lens. The researchers used a fluorescent ligand (aftobetin hydrochloride) with Aβ affinity and exhibited ocular penetration after topical application of an ointment (Kerbage et al. 2013).

Using in vivo laser fluorescence spectroscopy, the researchers were able to identify supranuclear amyloid in the lenses of the majority of AD patients with 95% specificity and 85% sensitivity in a group of 20 AD and 20 healthy control subjects (Kerbage et al. 2015).

The clinical prediction was more accurate using this approach than AβPET imaging (florbetapir F18). If this strategy is duplicated in a bigger study with a diverse range of people with variable disease severity, it may show promise as a biomarker for AD. The disadvantage of this strategy is that it will be ineffective in those who have had cataract surgery (van Wijngaarden et al. 2017).

4.4.2.2.3 Eye Movements

Due to suboptimal eye movements, AD patients are known to have difficulty in reading; they demonstrate decreased eye movement velocity, failure to concentrate on a target, and failure to track a moving target; these errors are observed due to damage to cortex and brain stem neural generators. Further studies are needed to establish a distinct AD signature, eye movements, as a possible biomarker (Lim et al. 2016).

4.4.2.3 Potential Future Biomarkers

4.4.2.3.1 Ocular Fluid Biomarkers

Since the 1990s, proteomics has been utilized for extensive protein analysis in both health and diseased conditions, including AD. Aqueous humor samples were subjected to proteomic analysis for presence of the AD-related peptides. Tear fluid, which is simple to obtain, has recently been demonstrated to be viable clinically (Kalló et al. 2016) and has demonstrated promise in the diagnosis of neurological illnesses like glaucoma (Kalló et al. 2016; Tezel 2013).

miRNA has attracted a lot of interest as a subclass of noncoding RNA regulatory molecules that may regulate gene expression at the post transcriptional stage by interacting with the untranslated region of mRNAs that they target. Numerous potential miRNAs have already been shown to play a part in controlling the production of amyloid, including hsa-miR-106, -153, -101, -29, and -107 (Kumar et al. 2013). The viability of miRNA molecules obtained from tears as a fluid biomarker is further investigated (Kumar et al. 2013).

4.4.2.3.2 Corneal Nerve Imaging

The three layers of corneal nerves are found between the Bowman's layer and the basal epithelium (subbasal nerve plexus), the Bowman's layer and the anterior stroma (subepithelial nerve plexus), and the stroma (stromal nerve plexus) (Lim et al. 2016).

The nerve growth factor plays an important role in the growth and maintenance of neuronal cells in the cornea, central nervous system, and peripheral nervous system.

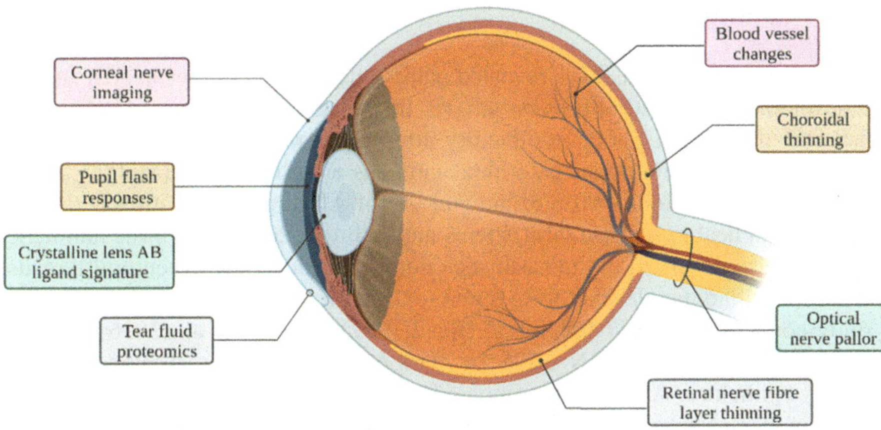

Fig. 4.5 Ocular biomarkers of AD

Cholinergic neurodegeneration in AD has been connected to impaired retrograde nerve growth factor (NGF) transport (Rocco et al. 2018). The viability of the corneal nerve may serve as an AD biomarker. Given the direct imaging capabilities of corneal nerves, a functional measure of axonal transport may be more effective in identifying AD. The various ocular parts shown in Fig. 4.5 can serve as a source of AD biomarkers.

4.4.3 Urine

Urine is now a popular source of illness indicators due to its noninvasive nature. Its composition comprises cellular components, metabolic substances, and proteins derived from glomerular filtration. Urogenital tract secretion/renal tube secretion reflects the pathophysiological and metabolic condition of an individual (Bălaşa et al. 2020). Studies have reported the presence of protein biomarkers in urine, which include (AD7c-NTP) AD-associated NTP; neural thread protein (NTP) levels are increased in CSF of people with AD, and thus, it has been considered as a possible AD biomarker, detected by using monoclonal antibody assay. Studies by L. Ma et al. showed a rise in urinary levels of NTP in patients with MCI and AD. As the specificity is unclear, further research is needed to distinguish AD from other types of dementia (Ma et al. 2015).

Oxidative stress is one of the important factors in AD; research into the molecules associated with this brain damage may provide possible early AD indicators in urine (Bălaşa et al. 2020). Isoprostane 8,12-iso-iPF (2-alpha)-VI, a free amino acid product of lipid peroxidation, was shown to significantly increase in several studies. This finding may imply a pathological transition from MCI to AD dementia (García-Blanco et al. 2018; Praticò et al. 2002).

4.4.4 Olfactory

Postmortem investigations have revealed pathological abnormalities in AD, such as the development of neurofibrillary tangles in the anterior olfactory nucleus and olfactory bulb regions crucial in olfactory information processing. Braak staging is used for early AD pathology in entorhinal and trans-entorhinal areas (Murphy 2019). Based on it, studies found 90% of tau and 9% of Aβ in patients with AD (loss of smell) than in healthy individuals (Attems and Jellinger 2006). The basic hallmarks of AD are also observed in the nasal secretions, neuritic plaques, neuropil threads, and neurofibrillary tangles that were found to be in olfactory bulbs of AD patients in a study conducted in 1987 (Ohm and Braak 1987). Moon J et al. (2016) reported increased levels of microRNA-206 by performing an intranasal biopsy of the olfactory epithelia in AD patients compared to healthy controls. Preliminary research on oral, ocular, olfactory, and urine biofluid potential noninvasive biomarkers is still ongoing, which requires additional study to address methodological variability and difference in accuracy and specificity (Moon et al. 2016).

4.4.5 Blood

Compared to other bodily fluids like CSF, blood is the most easily accessible biological sample. It offers affordable clinical diagnostic or screening procedures and is even practical for achieving repeatable outcomes in clinical trials. Biofluid blood has been established as a biomarker in the diagnosis and study of cancer and cardiovascular disease; as a result, it may function as a crucial component in the early detection of AD (O'Bryant et al. 2017). Aβ is considered as a potential blood biomarker due to its ability to pass through the BBB and its presence in the prodromic stage of AD (Omar and Preddy 2020).

4.4.5.1 Aβ

The results on $A\beta_{42}$ and $A\beta_{40}$ levels in the plasma of AD patients throughout the first decade of the 2000s were inconsistent and contrary. Mayeux et al. discovered elevated plasma $A\beta_{42}$ levels, however not plasma $A\beta_{40}$ levels, in AD patients. When compared to people with low plasma $A\beta_{42}$, the chance of developing AD was more than twice as high in individuals with high plasma $A\beta_{42}$. In further research, van Oijen et al. observed that high plasma $A\beta_{40}$ concentration is linked to an improved risk of dementia (Preische et al. 2019). However, other studies discovered that a lesser plasma $A\beta_{42}/A\beta_{40}$ ratio is linked to higher cognitive decline among older adults. This discrepancy results from the clinical evaluation stage or the combination of different kinds of dementia.

Trials based on immunological precipitation-mass spectrometry-based assays were conducted, which demonstrated a considerable decrease in both $A\beta_{40}$ and $A\beta_{42}$ concentrations in plasma, to obtain reliability and precision in the analysis of biomarkers (Nakamura et al. 2018; Ovod et al. 2017). A novel ELISA approach was used to examine in vivo levels of Aβ oligomers vs. Aβ monomers in both plasma and brain tissues of individuals with sporadic and familial AD, which revealed a strict association between the Aβ oligomers and amount of $A\beta_{42}$; however, the concentration of plasma Aβ is not clear yet.

4.4.5.2 Tau

Studies have indicated that p-tau is a late-stage biomarker of AD (Omar and Preddy 2020), and because t-tau levels are so low in MCI and/or AD collated to CSF, it is essentially undetectable in these conditions. Increased t-tau levels are seen in the plasma of AD patients compared to healthy controls and MCI using an ultrasensitive approach created by Zetlerberg et al.; however, it was unable to differentiate among patients with stable MCI and MCI that acquired AD (Zetterberg et al. 2013). Tatebe et al. developed a novel ultrasensitive immune assay for P-tau 181 quantification, which showed increased levels of p-tau 181 in patients with AD (Tatebe et al. 2017).

4.4.5.3 Isoprostanes

Increased plasma levels of isoprostanes (F2-IsoP) are observed in patients with AD (Praticò et al. 2000); it is considered as an unspecific biomarker for AD due to increased levels of isoprostanes in other diseases associated with oxidative stress (Montine et al. 2002).

4.4.5.4 Post translational Modification (PTM)-Plasma

PTMs are generally modifications (non-coded DNA) to the structure/composition of proteins that generate neo-epitopes (novel and distinct protein components); modifications include isomerization, methylation, acetylation, nitrosylation, etc. (Karsdal et al. 2010). Few neo-epitopes are discovered and utilized as biochemical markers to track disease severity. $A\beta_{1-40}$, $A\beta_{1-42}$, and phosphorylated tau serve as post translational modified protein species; however, they are not established as a biomarker of AD in the blood largely as a result of technical reasons (Thambisetty and Lovestone 2010; Mayeux and Schupf 2011). Pathologically, AD is an interesting disease for PTM-based biomarker development due to the formation of Aβ, P-tau, and other proteins linked to neuronal pathologies such as DNA binding/ protein 43 and alpha-synuclein.

4.4.5.5 Autoantibodies

Autoantibodies are established to be present in AD, but it is unclear whether they are more likely to be defensive or destructive in terms of pathology. However, given that many of them are present in both blood and CSF, their potential as biomarkers for AD is of significant notice (Colasanti et al. 2010).

Autoantibodies against Aβ have drawn a lot of interest, and there are signs that they may have diagnostic value (Gustaw-Rothenberg et al. 2010). They exist in both forms: antigen-bound and free antibodies. It is still unclear how beneficial measuring these autoantibodies is for AD diagnosis and/or prognosis, even though the instruments to do so have been developed (Maftei et al. 2012). Serum autoantibody profiling was utilized in a study by Nagele and colleagues, who discovered that autoantibodies are prevalent in serum from early to advanced stages of AD. In a panel of ten autoantibodies, they also found remarkable specificity and sensitivity (90%) for AD diagnosis. It is yet unclear, nevertheless, whether these features are more prominent in higher AD populations and are unique to AD compared to other dementias. Hence, before using this strategy, there is still a lot of validation to be done (Nagele et al. 2011).

4.4.5.6 MicroRNA

As compared to autoantibodies, here is a strong confirmation that changes in microRNA levels are connected to various aspects of AD pathology (Geekiyanage et al. 2012); as a result, efforts are being made to measure changes in the concentration of specific microRNAs in blood as AD biomarkers. The theory that blood microRNAs might act as biomarkers for AD and/or MCI has some value, although there are currently only a few studies employing a limited number of individuals. Compared to AD, MCI had greater variation in microRNA levels. The lack of overlap between two studies and the study's comparatively small sample sizes point to the need for additional research to fully comprehend the utility of these blood-based AD biomarkers (Sheinerman et al. 2012).

4.4.5.7 Blood-Derived Genetic Markers

It is well known that Apoε4 alleles increase the risk of AD; they are considered as genetic biomarkers for early-onset types of AD. Epigenetic studies on the Apoε4 allele have suggested that this gene may have a role in AD. More significantly, similar modifications were seen in corresponding blood lymphocytes as well as brain tissues (Wang et al. 2008). Another study found that in comparison to healthy control subjects, AD patients expressed hypermethylated promoter regions of the Apoε4 gene (Kaddurah-Daouk et al. 2011). The fact that this study only examined brain samples emphasizes the necessity for additional analysis of epigenetic modifications in blood-derived samples along with larger patient populations (Patel et al. 2011). However, these results indicate that abnormal gene methylation may have a major effect on AD, and if it can be found in blood-derived cells like lymphocytes, it might potentially be an AD biomarker (Henriksen et al. 2014).

4.4.5.8 Transcriptome

Several separate organizations have described the possible utility of blood-based gene expression profiling in the diagnosis of brain diseases (Sharp et al. 2006; Burczynski and Dorner 2006; Gladkevich et al. 2004). The transcriptome, or collection of messenger RNAs in a specific cell or tissue type, represents essential information for determining the final expression of the proteome that is used to develop a blood-based signature to distinguish AD patients from asymptomatic control subjects (Henriksen et al. 2014).

Two transcriptome-based strategies, one employing a 96-gene set and the other a 136-gene strategy, have been reported. Both methods produced useful diagnostic results, by comparing AD patients with control subjects who were of similar age. The study by Rye and colleagues, which showed a connection to CSF biomarker levels, emphasized the potential of these strategies (Rye et al. 2011). Interestingly, the study by Booij and colleagues reported a distinction between AD and patients with Parkinson's (Booij et al. 2011). These investigations may aid in the development of novel ideas about the AD pathophysiology and may explicate various physiological phenomena, such as inflammatory processes, involved in the onset of the disease. Even so, there are still critical data gaps for example, validation in a large clinical group as the AD Neuroimaging Initiative or the Australian Imaging Biomarkers and Lifestyle research (Fehlbaum-Beurdeley et al. 2012).

4.4.5.9 Markers of Inflammation

Amyloid deposition in the AD brain triggers several reactive inflammatory responses, including astrocytosis, microgliosis, an increase in proinflammatory cytokines, acute-phase reactions, and complement activation (Weiner and Selkoe 2002). Because most of these proteins have difficulty crossing the BBB, it is difficult to determine if the buildup of cytokines and acute-phase reactants in brain are similarly reflected in serum or plasma. Alternatively, AD may be associated with a more widespread immune dysregulation that is detectable in plasma. The measurement of immune mediators in AD serum or plasma is a topic of significant debate in the literature. While various cytokines such as IL-12, and interferon remain unchanged, inflammatory substances such as C-reactive protein, interleukin (IL)-1, tumor necrosis factor, IL-6, IL-6 receptor complex, 1-antichymotrypsin, and transforming growth factor show variable changes among studies (Teunissen et al. 2002).

For instance, IL-6 has been widely studied in AD. A receptor complex made up of the IL-6 receptor subunit and glycoprotein 130 mediates the effects of IL-6, a cytokine involved in inflammation, acute-phase reactions, and cell proliferation. Some studies have found elevated IL-6 levels in AD plasma and serum (Panossian et al. 2003; Kalman et al. 1997; Licastro et al. 2000; Maes et al. 1999; Singh and Guthikonda 1997; Tarkowski et al. 1999), while other cohorts did not find this to be the case (Angelis et al. 1998; Blum-Degen et al. 1995; Chao et al. 1994; van Duijn et al. 1990). Different plasma collection techniques, assay methodologies, assay sensitivity, limited sample numbers, varied patient demographics, impact of disease severity, age, and comorbid inflammatory condition are few confounding issues in this research. To determine if IL-6 can be an effective biomarker for AD, more studies are needed (Lönneborg 2008).

4.4.6 Biomarkers Through Noninvasive Diagnostic Methods

4.4.6.1 Cognitive Biomarkers

The current noninvasive AD diagnostic criteria are based on a person's medical history, cognitive and neuropsychological state, and clinical rating score obtained from the Mini-Mental State Examination (MMSE), the Wechsler Memory Scale (WMS), Clinical Dementia Rating (CDR), and Logical Memory (LM) test (Lopez et al. 2011). The MMSE (0–30) score classifies a person as having mild dementia if their score falls between 20 and 24; moderate dementia if their score falls between 13 and 20; and severe dementia if their score falls below 12 (Folstein et al. 1975).

4.4.6.2 Imaging Biomarkers

A second line of diagnostic criteria for the diagnosis of AD has been imaging of the brain, including positron-emission tomography (PET), magnetic resonance imaging (MRI), and functional magnetic resonance imaging (fMRI) (Omar and Preddy 2020). Current guidelines follow structural imaging in a scientific setting, such as magnetic resonance imaging (MRI) or computed tomography (CT), which is primarily necessary for the examination of patients presenting with a cognitive/dementia disease (Sheikh-Bahaei et al. 2017). MRI is utilized as a structural and functional imaging tool to examine the visualization of AD-related cortical atrophy and alterations in brain connectivity (Scheltens et al. 1992).

The functional connectivity of the brain was also shown by fMRI investigations, including abnormalities in the hippocampus (Wang et al. 2006). Positron-emission tomography (PET) is a sophisticated imaging technique that detects Aβ-related metabolic activity and plaque deposition in AD by labeling substances with transient positron-emitting radionuclides (Kadir et al. 2012). PiB and fluoro-2-deoxy-D-glucose (FDG), two extensively used PET tracers, show good sensitivity and specificity in imaging biomarkers of plaque-associated amyloid progression in humans, particularly in the initial phases (Cohen and Klunk 2014).

4.5 Conclusion

$A\beta_{42}$, P-tau, and t-tau have already been identified as CSF biomarkers for AD. They are well known for their great diagnostic specificity. Several more candidates in noninvasive biological fluids are currently being studied for their potential therapeutic utility in early AD diagnosis and prognosis. So far, a growing number of molecules have been found, with NFL showing the most promise in both CSF and blood, lactoferrin in saliva, and $A\beta_{42}$ and p-tau in plasma.

However, most current AD fluidic biomarkers are obtained through a single study, or there is a substantial inconsistency in outcomes obtained from different studies, including pre-analytical sample process standardization and analytical method validation. Other criteria such as our age, presence of multiple medical conditions, and disease diagnosis/stage must also be thoroughly considered.

These restrictions are slightly relieved by advanced imaging techniques that enable the detection of structural and functional biomarkers associated with AD offering information that is simple to understand and highly accurate for identifying the disease stage of AD. In this context, recently developed biosensors are showing promise as rapid, affordable, and straightforward alternatives for diagnosing AD, even in its early stages.

These techniques are powerful analytical methods with elevated sensitivity and selectivity, which can be used to detect AD biomarkers in physiological fluids such as blood, urine, and tears.

Combining more than one of the aforementioned indicators, particularly fluidic molecular examination and imaging studies, is now commonly accepted as the best strategy to accurately diagnose AD and disease stage progression. Depending on the case being assessed, several biomarkers and procedures may be combined.

References

Angelis P, Scharf S, Mander A, Vajda F, Christophidis N (1998) Serum interleukin-6 and interleukin-6 soluble receptor in Alzheimer's disease. Neurosci Lett 244(2):106–108

Attems J, Jellinger KA (2006) Olfactory tau pathology in Alzheimer disease and mild cognitive impairment. Clin Neuropathol 25(6):265

Ausó E, Gómez-Vicente V, Esquiva G (2020) Biomarkers for Alzheimer's disease early diagnosis. J Pers Med 10(3):114

Bakhtiari S, Moghadam NB, Ehsani M, Mortazavi H, Sabour S, Bakhshi M (2017) Can salivary acetylcholinesterase be a diagnostic biomarker for Alzheimer? J Clin Diagn Res 11(1):ZC58

Bălaşa AF, Chircov C, Grumezescu AM (2020) Body fluid biomarkers for Alzheimer's disease—an up-to-date overview. Biomedicine 8(10):421

Berisha F, Feke GT, Trempe CL, McMeel JW, Schepens CL (2007) Retinal abnormalities in early Alzheimer's disease. Invest Ophthalmol Vis Sci 48(5):2285–2289

Bermejo-Pareja F, Antequera D, Vargas T, Molina JA, Carro E (2010) Saliva levels of Abeta1-42 as potential biomarker of Alzheimer's disease: a pilot study. BMC Neurol 10(1):1–7

Bettcher BM, Johnson SC, Fitch R, Casaletto KB, Heffernan KS, Asthana S et al (2018) Cerebrospinal fluid and plasma levels of inflammation differentially relate to CNS markers of Alzheimer's disease pathology and neuronal damage. J Alzheimers Dis 62(1):385–397

Blennow K, Hampel H (2003) CSF markers for incipient Alzheimer's disease. Lancet Neurol 2(10): 605–613

Blum-Degen D, Müller T, Kuhn W, Gerlach M, Przuntek H, Riederer P (1995) Interleukin-1 beta and interleukin-6 are elevated in the cerebrospinal fluid of Alzheimer's and de novo Parkinson's disease patients. Neurosci Lett 202(1–2):17–20

Booij BB, Lindahl T, Wetterberg P, Skaane NV, Sæbø S, Feten G et al (2011) A gene expression pattern in blood for the early detection of Alzheimer's disease. J Alzheimers Dis 23(1):109–119

Burczynski ME, Dorner AJ (2006) Transcriptional profiling of peripheral blood cells in clinical pharmacogenomic studies. Pharmacogenomics 7:187–202

Calsolaro V, Edison P (2016) Neuroinflammation in Alzheimer's disease: current evidence and future directions. Alzheimers Dement 12(6):719–732

Campbell MC, De Vries D, Emptage L, Cookson C, Kisilak M, Bueno JM, Avila FJ (2015) Polarization properties of amyloid beta in the retina of the eye as a biomarker of Alzheimer's disease. In: Bio-optics: design and application. Optica Publishing Group, p BM3A-4

Carro E, Bartolomé F, Bermejo-Pareja F, Villarejo-Galende A, Molina JA, Ortiz P et al (2017) Early diagnosis of mild cognitive impairment and Alzheimer's disease based on salivary lactoferrin. Alzheimer's Dement 8:131–138

Chao CC, Ala TA, Hu S, Crossley KB, Sherman RE, Peterson PK, Frey WH 2nd (1994) Serum cytokine levels in patients with Alzheimer's disease. Clin Diagn Lab Immunol 1(4):433

Cheung CYL, Ong YT, Hilal S, Ikram MK, Low S, Ong YL et al (2015) Retinal ganglion cell analysis using high-definition optical coherence tomography in patients with mild cognitive impairment and Alzheimer's disease. J Alzheimers Dis 45(1):45–56

Cohen AD, Klunk WE (2014) Early detection of Alzheimer's disease using PiB and FDG PET. Neurobiol Dis 72:117–122

Colasanti T, Barbati C, Rosano G, Malorni W, Ortona E (2010) Autoantibodies in patients with Alzheimer's disease: pathogenetic role and potential use as biomarkers of disease progression. Autoimmun Rev 9(12):807–811

Craig-Schapiro R, Perrin RJ, Roe CM, Xiong C, Carter D, Cairns NJ et al (2010) YKL-40: a novel prognostic fluid biomarker for preclinical Alzheimer's disease. Biol Psychiatry 68(10):903–912

Dhiman K, Blennow K, Zetterberg H, Martins RN, Gupta VB (2019) Cerebrospinal fluid biomarkers for understanding multiple aspects of Alzheimer's disease pathogenesis. Cell Mol Life Sci 76:1833–1863

Duits FH, Brinkmalm G, Teunissen CE, Brinkmalm A, Scheltens P, Van der Flier WM et al (2018) Synaptic proteins in CSF as potential novel biomarkers for prognosis in prodromal Alzheimer's disease. Alzheimers Res Ther 10(1):1–9

Ewers M, Zhong Z, Bürger K, Wallin A, Blennow K, Teipel SJ et al (2008) Increased CSF-BACE 1 activity is associated with ApoE-ε4 genotype in subjects with mild cognitive impairment and Alzheimer's disease. Brain 131(5):1252–1258

Fehlbaum-Beurdeley P, Sol O, Désiré L, Touchon J, Dantoine T, Vercelletto M et al (2012) Validation of AclarusDx™, a blood-based transcriptomic signature for the diagnosis of Alzheimer's disease. J Alzheimers Dis 32(1):169–181

Folstein MF, Folstein SE, McHugh PR (1975) "Mini-mental state": a practical method for grading the cognitive state of patients for the clinician. J Psychiatr Res 12(3):189–198

Food U, Administration D (2018) Early Alzheimer's disease: developing drugs for treatment–guidance for industry. https://www.fda.gov/downloads/Drugs/GuidanceComplianceRegulatoryInformation/Guidances/UCM596728.pdf. Accessed 10 Mar 2023

Frey H, Mattila K, Korolainen M et al (2005) Problems associated with biological markers of Alzheimer's disease. Neurochem Res 30:1501–1510

García-Ayllón MS, Campanari ML, Brinkmalm G, Rábano A, Alom J, Saura CA et al (2013) CSF Presenilin-1 complexes are increased in Alzheimer's disease. Acta Neuropathol Commun 1(1): 1–13

García-Blanco A, Peña-Bautista C, Oger C, Vigor C, Galano JM, Durand T et al (2018) Reliable determination of new lipid peroxidation compounds as potential early Alzheimer disease biomarkers. Talanta 184:193–201

Geekiyanage H, Jicha GA, Nelson PT, Chan C (2012) Blood serum miRNA: non-invasive biomarkers for Alzheimer's disease. Exp Neurol 235(2):491–496

Gharbiya M, Trebbastoni A, Parisi F, Manganiello S, Cruciani F, D'Antonio F et al (2014) Choroidal thinning as a new finding in Alzheimer's disease: evidence from enhanced depth imaging spectral domain optical coherence tomography. J Alzheimers Dis 40(4):907–917

Gladkevich A, Kauffman HF, Korf J (2004) Lymphocytes as a neural probe: potential for studying psychiatric disorders. Prog Neuro-Psychopharmacol Biol Psychiatry 28(3):559–576

Gleerup HS, Hasselbalch SG, Simonsen AH (2019) Biomarkers for Alzheimer's disease in saliva: a systematic review. Dis Markers 2019:1

Goldstein LE, Muffat JA, Cherny RA, Moir RD, Ericsson MH, Huang X et al (2003) Cytosolic β-amyloid deposition and supranuclear cataracts in lenses from people with Alzheimer's disease. Lancet 361(9365):1258–1265

Gunes S, Aizawa Y, Sugashi T, Sugimoto M, Rodrigues PP (2022) Biomarkers for Alzheimer's disease in the current state: a narrative review. Int J Mol Sci 23(9):4962

Gustaw-Rothenberg KA, Siedlak SL, Bonda DJ, Lerner A, Tabaton M, Perry G, Smith MA (2010) Dissociated amyloid-β antibody levels as a serum biomarker for the progression of Alzheimer's disease: a population-based study. Exp Gerontol 45(1):47–52

Hamel MT, Emptage L, DeVries D, Oliveros C, Chow T, Shah N et al (2016) Polarization properties of amyloid deposits in the retinas of an animal model of Alzheimer's disease differ in those with and without cognitive impairment. Invest Ophthalmol Vis Sci 57(12):2216–2216

Henriksen K, O'Bryant SE, Hampel H, Trojanowski JQ, Montine TJ, Jeromin A et al (2014) The future of blood-based biomarkers for Alzheimer's disease. Alzheimers Dement 10(1):115–131

Hinton DR, Sadun AA, Blanks JC, Miller CA (1986) Optic-nerve degeneration in Alzheimer's disease. N Engl J Med 315(8):485–487

Ho M, Liu DT, Chan VC, Lam DS (2013) Choroidal thickness measurement in myopic eyes by enhanced depth optical coherence tomography. Ophthalmology 120(9):1909–1914

Humpel C (2011) Identifying and validating biomarkers for Alzheimer's disease. Trends Biotechnol 29(1):26–32

Kaddurah-Daouk R, Rozen S, Matson W, Han X, Hulette CM, Burke JR et al (2011) Metabolomic changes in autopsy-confirmed Alzheimer's disease. Alzheimers Dement 7(3):309–317

Kadir A, Almkvist O, Forsberg A, Wall A, Engler H, Långström B, Nordberg A (2012) Dynamic changes in PET amyloid and FDG imaging at different stages of Alzheimer's disease. Neurobiol Aging 33(1):198–1e1

Kalló G, Emri M, Varga Z, Ujhelyi B, Tőzsér J, Csutak A, Csősz É (2016) Changes in the chemical barrier composition of tears in Alzheimer's disease reveal potential tear diagnostic biomarkers. PLoS One 11(6):e0158000

Kalman J, Juhasz A, Laird G, Dickens P, Jardanhazy T, Rimanoczy A et al (1997) Serum interleukin-6 levels correlate with the severity of dementia in down syndrome and in Alzheimer's disease. Acta Neurol Scand 96(4):236–240

Karsdal MA, Henriksen K, Leeming DJ, Woodworth T, Vassiliadis E, Bay-Jensen AC (2010) Novel combinations of post-translational modification (PTM) neo-epitopes provide tissue-specific biochemical markers—are they the cause or the consequence of the disease? Clin Biochem 43(10–11):793–804

Kerbage C, Sadowsky CH, Jennings D, Cagle GD, Hartung PD (2013) Alzheimer's disease diagnosis by detecting exogenous fluorescent signal of ligand bound to Beta amyloid in the lens of human eye: an exploratory study. Front Neurol 4:62

Kerbage C, Sadowsky CH, Tariot PN, Agronin M, Alva G, Turner FD et al (2015) Detection of amyloid β signature in the lens and its correlation in the brain to aid in the diagnosis of Alzheimer's disease. Am J Alzheimers Dis Other Dement 30(8):738–745

Kinoshita T, Mitamura Y, Shinomiya K, Egawa M, Iwata A, Fujihara A et al (2017) Diurnal variations in luminal and stromal areas of choroid in normal eyes. Br J Ophthalmol 101(3): 360–364

Koran MEI, Wagener M, Hohman TJ, Alzheimer's Neuroimaging Initiative (2017) Sex differences in the association between AD biomarkers and cognitive decline. Brain Imaging Behav 11(1): 205–213

Kumar P, Dezso Z, MacKenzie C, Oestreicher J, Agoulnik S, Byrne M et al (2013) Circulating miRNA biomarkers for Alzheimer's disease. PLoS One 8(7):e69807

Liang D, Lu H (2019) Salivary biological biomarkers for Alzheimer's disease. Arch Oral Biol 105: 5–12

Licastro F, Pedrini S, Caputo L, Annoni G, Davis LJ, Ferri C et al (2000) Increased plasma levels of interleukin-1, interleukin-6 and α-1-antichymotrypsin in patients with Alzheimer's disease: peripheral inflammation or signals from the brain? J Neuroimmunol 103(1):97–102

Lim JK, Li QX, He Z, Vingrys AJ, Wong VH, Currier N et al (2016) The eye as a biomarker for Alzheimer's disease. Front Neurosci 10:536

Lönneborg A (2008) Biomarkers for Alzheimer disease in cerebrospinal fluid, urine, and blood. Mol Diagn Ther 12:307–320

Lopez OL, McDade E, Riverol M, Becker JT (2011) Evolution of the diagnostic criteria for degenerative and cognitive disorders. Curr Opin Neurol 24(6):532–541

Lukiw WJ (2007) Micro-RNA speciation in fetal, adult and Alzheimer's disease hippocampus. Neuroreport 18(3):297–300

Lukiw WJ, Andreeva TV, Grigorenko AP, Rogaev EI (2013) Studying micro RNA function and dysfunction in Alzheimer's disease. Front Genet 3:327

Ma L, Chen J, Wang R, Han Y, Zhang J, Dong W et al (2015) The level of Alzheimer-associated neuronal thread protein in urine may be an important biomarker of mild cognitive impairment. J Clin Neurosci 22(4):649–652

Maes M, DeVos N, Wauters A, Demedts P, Maurits V, Neels H et al (1999) Inflammatory markers in younger vs elderly normal volunteers and in patients with Alzheimer's disease. J Psychiatr Res 33(5):397–405

Maftei M, Thurm F, Leirer VM, von Arnim CA, Elbert T, Przybylski M et al (2012) Antigen-bound and free β-amyloid autoantibodies in serum of healthy adults. PLoS One 7(9):e44516

Mattsson N, Insel P, Nosheny R, Zetterberg H, Trojanowski JQ, Shaw LM et al (2013) CSF protein biomarkers predicting longitudinal reduction of CSF β-amyloid42 in cognitively healthy elders. Transl Psychiatry 3(8):e293–e293

Mayeux R, Schupf N (2011) Blood-based biomarkers for Alzheimer's disease: plasma Aβ40 and Aβ42, and genetic variants. Neurobiol Aging 32:S10–S19

Montine TJ, Quinn JF, Milatovic D, Silbert LC, Dang T, Sanchez S et al (2002) Peripheral F2-isoprostanes and F4-neuroprostanes are not increased in Alzheimer's disease. Ann Neurol 52(2):175–179

Moon J, Lee ST, Kong IG, Byun JI, Sunwoo JS, Shin JW et al (2016) Early diagnosis of Alzheimer's disease from elevated olfactory mucosal miR-206 level. Sci Rep 6(1):1–9

Murphy C (2019) Olfactory and other sensory impairments in Alzheimer disease. Nat Rev Neurol 15(1):11–24

Nagele E, Han M, DeMarshall C, Belinka B, Nagele R (2011) Diagnosis of Alzheimer's disease based on disease-specific autoantibody profiles in human sera. PLoS One 6(8):e23112

Nakamura A, Kaneko N, Villemagne VL, Kato T, Doecke J, Doré V et al (2018) High performance plasma amyloid-β biomarkers for Alzheimer's disease. Nature 554(7691):249–254

Noble W, Hanger DP, Miller CC, Lovestone S (2013) The importance of tau phosphorylation for neurodegenerative diseases. Front Neurol 4:83

O'Bryant SE, Mielke MM, Rissman RA, Lista S, Vanderstichele H, Zetterberg H et al (2017) Blood-based biomarkers in Alzheimer disease: current state of the science and a novel collaborative paradigm for advancing from discovery to clinic. Alzheimers Dement 13(1):45–58

Ohm TG, Braak H (1987) Olfactory bulb changes in Alzheimer's disease. Acta Neuropathol 73(4): 365–369

Omar SH, Preddy J (2020) Advantages and pitfalls in fluid biomarkers for diagnosis of Alzheimer's disease. J Pers Med 10(3):63

Ovod V, Ramsey KN, Mawuenyega KG, Bollinger JG, Hicks T, Schneider T et al (2017) Amyloid β concentrations and stable isotope labeling kinetics of human plasma specific to central nervous system amyloidosis. Alzheimers Dement 13(8):841–849

Panossian LA, Porter VR, Valenzuela HF, Zhu X, Reback E, Masterman D et al (2003) Telomere shortening in T cells correlates with Alzheimer's disease status. Neurobiol Aging 24(1):77–84

Patel S, Shah RJ, Coleman P, Sabbagh M (2011) Potential peripheral biomarkers for the diagnosis of Alzheimer's disease. Int J Alzheimers Dis 2011:572495

Praticò D, Clark CM, Lee VMY, Trojanowski JQ, Rokach J, FitzGerald GA (2000) Increased 8, 12-iso-iPF2α-VI in Alzheimer's disease: correlation of a noninvasive index of lipid peroxidation with disease severity. Ann Neurol 48(5):809–812

Praticò D, Clark CM, Liun F, Lee VYM, Trojanowski JQ (2002) Increase of brain oxidative stress in mild cognitive impairment: a possible predictor of Alzheimer disease. Arch Neurol 59(6): 972–976

Preische O, Schultz SA, Apel A, Kuhle J, Kaeser SA, Barro C et al (2019) Serum neurofilament dynamics predicts neurodegeneration and clinical progression in presymptomatic Alzheimer's disease. Nat Med 25(2):277–283

Provost P (2010) Interpretation and applicability of microRNA data to the context of Alzheimer's and age-related diseases. Aging (Albany NY) 2(3):166

Ratnayaka JA, Serpell LC, Lotery AJ (2015) Dementia of the eye: the role of amyloid beta in retinal degeneration. Eye 29(8):1013–1026

Rocco ML, Soligo M, Manni L, Aloe L (2018) Nerve growth factor: early studies and recent clinical trials. Curr Neuropharmacol 16(10):1455–1465

Rye P, Booij BB, Grave G, Lindahl T, Kristiansen L, Andersen HM et al (2011) A novel blood test for the early detection of Alzheimer's disease. J Alzheimers Dis 23(1):121–129

Sakka L, Coll G, Chazal J (2011) Anatomy and physiology of cerebrospinal fluid. Eur Ann Otorhinolaryngol Head Neck Dis 128(6):309–316

Scheltens P, Leys D, Barkhof F, Huglo D, Weinstein HC, Vermersch P et al (1992) Atrophy of medial temporal lobes on MRI in "probable" Alzheimer's disease and normal ageing: diagnostic value and neuropsychological correlates. J Neurol Neurosurg Psychiatry 55(10):967–972

Shah RD, Randleman JB, Grossniklaus HE (2012) Spontaneous corneal clearing after Descemet's stripping without endothelial replacement. Ophthalmology 119(2):256–260

Sharp FR, Xu H, Lit L, Walker W, Apperson M, Gilbert DL et al (2006) The future of genomic profiling of neurological diseases using blood. Arch Neurol 63(11):1529–1536

Sheikh-Bahaei N, Sajjadi SA, Manavaki R, Gillard JH (2017) Imaging biomarkers in Alzheimer's disease: a practical guide for clinicians. J Alzheimers Disease Rep 1(1):71–88

Sheinerman KS, Tsivinsky VG, Crawford F, Mullan MJ, Abdullah L, Umansky SR (2012) Plasma microRNA biomarkers for detection of mild cognitive impairment. Aging 4(9):590–605

Simon MJ, Iliff JJ (2016) Regulation of cerebrospinal fluid (CSF) flow in neurodegenerative, neurovascular and neuroinflammatory disease. Biochim Biophys Acta 1862(3):442–451

Singh VK, Guthikonda P (1997) Circulating cytokines in Alzheimer's disease. J Psychiatr Res 31(6):657–660

Tarkowski E, Blennow K, Wallin A, Tarkowski A (1999) Intracerebral production of tumor necrosis factor-[alpha], a local neuroprotective agent, in Alzheimer disease and vascular dementia. J Clin Immunol 19(4):223

Tatebe H, Kasai T, Ohmichi T, Kishi Y, Kakeya T, Waragai M et al (2017) Quantification of plasma phosphorylated tau to use as a biomarker for brain Alzheimer pathology: pilot case-control studies including patients with Alzheimer's disease and down syndrome. Mol Neurodegener 12: 1–11

Teunissen CE, De Vente J, Steinbusch HWM, De Bruijn C (2002) Biochemical markers related to Alzheimer's dementia in serum and cerebrospinal fluid. Neurobiol Aging 23(4):485–508

Tezel G (2013) A proteomics view of the molecular mechanisms and biomarkers of glaucomatous neurodegeneration. Prog Retin Eye Res 35:18–43

Thambisetty M, Lovestone S (2010) Blood-based biomarkers of Alzheimer's disease: challenging but feasible. Biomark Med 4(1):65–79

Tvarijonaviciute A, Zamora C, Ceron JJ, Bravo-Cantero AF, Pardo-Marin L, Valverde S, Lopez-Jornet P (2020) Salivary biomarkers in Alzheimer's disease. Clin Oral Investig 24:3437–3444

van Duijn CM, Hofman A, Nagelkerken L (1990) Serum levels of interleukin-6 are not elevated in patients with Alzheimer's disease. Neurosci Lett 108(3):350–354

van Wijngaarden P, Hadoux X, Alwan M, Keel S, Dirani M (2017) Emerging ocular biomarkers of Alzheimer disease. Clin Exp Ophthalmol 45(1):54–61

Wang L, Zang Y, He Y, Liang M, Zhang X, Tian L et al (2006) Changes in hippocampal connectivity in the early stages of Alzheimer's disease: evidence from resting state fMRI. NeuroImage 31(2):496–504

Wang SC, Oelze B, Schumacher A (2008) Age-specific epigenetic drift in late-onset Alzheimer's disease. PLoS One 3(7):e2698

Weiner HL, Selkoe DJ (2002) Inflammation and therapeutic vaccination in CNS diseases. Nature 420(6917):879–884

Wenk GL (2003) Neuropathologic changes in Alzheimer's disease. J Clin Psychiatry 64(Suppl 9): 7–10

Wong IY, Koizumi H, Lai WW (2011) Enhanced depth imaging optical coherence tomography. Ophthalmic Surg Lasers Imaging 42 Suppl:S75–S84

Zetterberg H, Andreasson U, Hansson O, Wu G, Sankaranarayanan S, Andersson ME et al (2008) Elevated cerebrospinal fluid BACE1 activity in incipient Alzheimer disease. Arch Neurol 65(8): 1102–1107

Zetterberg H, Wilson D, Andreasson U, Minthon L, Blennow K, Randall J, Hansson O (2013) Plasma tau levels in Alzheimer's disease. Alzheimers Res Ther 5(2):1–3

Fluorescent Organic Molecules as Diagnostic and Theranostic Tools for Alzheimer's Disease

5

Kondakamarla Imran, Yashasvi Anant Dhamapurkar, and Abha Sharma

Abstract

Diagnosis of Alzheimer's disease (AD) is crucial especially for the complete comprehension of pathology and clinical therapy of the disorder, as well as in declining the severity of the diseased condition. There are two major biomarkers associated with AD. One is an accumulation of insoluble amyloid beta, which later on gets transformed to amyloid plaques in extracellular space, and the other one is an aggregation of hyperphosphorylated tau, which has the capability to form neurofibrillary tangles in the intracellular space of neurons. Massive research efforts and resources are currently being dedicated towards the development of tools and methodologies for the identification and measurement of biomarkers indicative of molecular alterations in the brain and imminent neurodegeneration. The latest literature is showing a great interest towards the invention of a single chemical agent to deliver diagnosis and therapy simultaneously, which is entitled as "theranostics". A theranostic agent is one that has both therapeutic and diagnostic qualities. Unlike the typical therapeutic method, in which two distinct drugs serve as deputies to achieve two different objectives, theranostic agents combine these two goals in a single agent to achieve them. Theranostics, due to their great potentiality, helps in avoiding the possible variations in biodistribution pattern that exist when we use separate diagnostics and therapeutic agents. In the framework of theranostics, the goal is so to create the capacity to simultaneously monitor the impaired tissue, release kinetics, and therapeutic treatment efficacy. In recent years, several naturally derived compounds have been discovered, which are well proven in their efficiency to

K. Imran · Y. A. Dhamapurkar · A. Sharma (✉)
Department of Medicinal Chemistry, National Institute of Pharmaceutical Education and Research, Raebareli, Lucknow, Uttar Pradesh, India
e-mail: abha.sharma@niperraebareli.edu.in; abha.sharma@niperrbl.ac.in

A. Sharma, G. P. Modi (eds.), *Natural Product-Based Synthetic Drug Molecules in Alzheimer's Disease*, https://doi.org/10.1007/978-981-99-6038-5_5

treat various neurological disorders like AD and have showed defensive mechanisms against neuroinflammatory reactions and neurotoxic injuries. This concept of theranostics is now emerging as a new trend for the possibility of searching for new treatments for AD.

Keywords

Alzheimer's disease (AD) · Amyloid beta · Neurofibrillary tangles · Theranostics · Natural products

5.1 Introduction

Alzheimer's disease (AD) is one of the major neurological disorders having complex pathological features. The most properly researched pathological biomarkers of AD, amyloid beta (Aβ) and neurofibrillary tangles (NFTs) in cerebral tissues, are employed to diagnose this disease (Breijyeh and Karaman 2020). This disease majorly affects older populations globally, and it is the leading cause of persistent brain injury that cannot be repaired, resulting in altered functionality and behaviour. β-Amyloidogenesis is caused by the self-assemblance of peptide forms of Aβ, which results in the formation of soluble misfolded forms of Aβ (oligomers and dimers that are much more harmful to the neurological tissues); they ultimately transform into insoluble fibrils with a rich β-sheet structure. Tauopathies, on the other hand, refer to the anomalous self-assemblance of tau proteins into NFTs. It is believed that the development of Aβ and tau aggregation greatly contributes to AD pathogenesis, resulting in vascular damage and persistent memory deficits in AD patients (Iqbal et al. 2010). The condition known as amyloidopathy causes Aβ species to accumulate from outside neurons in the form of plaques on the fibrillary nerve, which might interfere with neuronal signal transmission at the synaptic site and contribute to neuronal cell death. Whilst also NFTs destabilize microtubules and interrupt key processes such as axonal transport, neuronal signal transmission, and also transport of other important nutrients within neurons, all of this may lead to neuronal cell death (Wang and Mandelkow 2016). There is still research being done into early AD pathogenesis variables, but they are linked to a variety of factors such as age, gender (particularly, females are more susceptible to this disease) (Vina and Lloret 2010), pathogenic factors such as cholinergic neuron depletion which reduces acetylcholine (ACh) levels (Singh et al. 2020), a genetic disorder such as Down's syndrome (Lott and Dierssen 2010), inflammatory stimuli, increased levels of reactive oxygen species (ROS) (Manoharan et al. 2016), and in addition to this metal ion dyshomeostasis such as Cu, Zn, and Fe (Eckroat et al. 2013; Akiyama et al. 2000; Sharma et al. 2019). AD starts to begin 10–30 years before the clinical symptoms appear in the disease. So, as a result, patients are unaware of the disease's early progression (Beason-Held et al. 2013; Perl 2010; Bulic et al. 2010; Chintamaneni and Bhaskar 2012). Professionals find it difficult to make a clinical diagnosis of AD, especially in the early stages. A thorough and critical evaluation of the patient's neurobiological and neuropsychological evidence is necessary to differentiate AD from many other types of dementia, like Parkinson's disease (PD), cerebrovascular associated dementia (CVAD), Lewy body dementia (LBD), and frontotemporal lobar degeneration (FTLD). Besides continued efforts by AD researchers, "theranostics" is a promising approach as this combines diagnosis and treatment in a single small molecule (Schneider et al. 2007; Viswanathan et al. 2009; Barker et al. 2002).

5.2 Hypothesis of AD

AD is a very complex disease which involves various factors. The precise pathogenesis of this disease is still unknown due to the complex nature of the human brain and the lack of research tools. Various hypotheses have been developed about AD, such as those involving the amyloid cascade hypothesis, tau hypothesis, cholinergic hypothesis, oxidative stress, inflammation, etc. (Liu et al. 2019).

5.2.1 Amyloid Hypothesis

The amyloid hypothesis is also called an Aβ hypothesis or amyloid cascade hypothesis, proposed by John Hardy and David Allsop in 1991. This hypothesis states that the imbalance between the production and clearance of hazardous Aβ peptides leads to impairment and subsequent neuronal cell death (Selkoe 1991). Aβ production occurs via two biochemical pathways, namely non-amyloidogenic and amyloidogenic pathways (Fig. 5.1). In the non-amyloidogenic pathway, amyloid precursor protein (APP) is cleaved by α-secretase followed by γ-secretase, resulting in a normal functional protein. However, in the amyloidogenic pathway, this protein is first cleaved by β-secretase, and then γ-secretase results in the formation of amyloid species, mainly Aβ_{40} and Aβ_{42}, which are the main constituents of the Aβ plaque. In aged persons, the metabolic ability for the degradation of Aβ is decreased; due to this, accumulation of Aβ will happen (Gu and Guo 2013). An increase in levels of this major Aβ species (Aβ_{40} and Aβ_{42}) causes the formation of amyloid fibrils, which get accumulated and further develop into senile plaques, causing neurotoxicity and tau pathology induction leading to neurodegeneration and consequently neuronal cell death (DeTure and Dickson 2019). The APP genes which are present on chromosome 21 and mutations in some genes like PSEN1 and PSEN2 affect catabolism and anabolism of Aβ, resulting in the accumulation of Aβ and also the rapid development of neurodegeneration (Vetrivel et al. 2006). These pathogenic APP mutations have been detected near the cleavage sites of β- or γ-secretase which are involved in the increased formation of Aβ or alteration occurs in the production of Aβ (De Jonghe et al. 2001). Patients with Down's syndrome with trisomy 21 display the symptoms like AD at the age of about 40 years, which was thought to be involved due to the increase in the amount of APP to 1.5 times the normal amount. Furthermore, APP locus duplication resulted in autosomal dominant early-

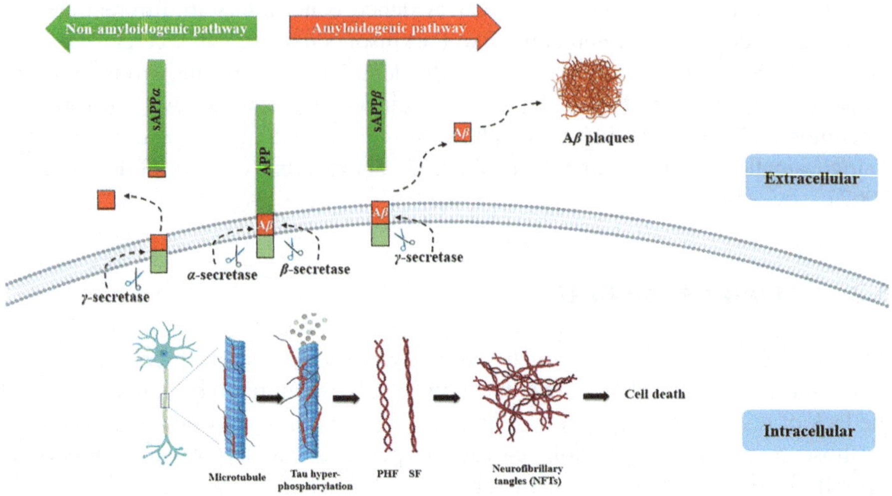

Fig. 5.1 Schematic representation of amyloid and tau pathology

onset AD and also the deposition of a massive amount of Aβ peptides. Besides this, mutations in PSEN1 and PSEN2 (major components of γ-secretase) cause other familial AD. These APP and presenilin mutations are strongly linked to the Aβ production and also amyloid fibril formation, which are primary pathogenic causes of AD (Kametani and Hasegawa 2018).

5.2.2 Tau Hypothesis

Neurofibrillary tangles are one of the major hallmarks of AD, mainly composed of tau protein. A change in the conformation of the tau protein causes its detachment from microtubules in pathogenic conditions. The conformational condition of tau proteins, as well as the equilibrium of the two main regulatory proteolytic enzymes, tau kinases and phosphatases, governs the fate of the microtubule-associated tau protein in neurobiology (Rodríguez-Martín et al. 2013). Tau protein hyperphosphorylation causes self-assembly, which leads to the formation of prefibrillar oligomeric aggregates. Aggregation progresses result in the development of paired helical filaments (PHFs) and, eventually, high-compact-structured neurofibrillary tangles (Fig. 5.1). The microtubule-associated tau protein (MATP) gene, found on chromosome 17q21.31, encodes the tau protein in humans, which consists of 16 exons. Based on the alternative mRNA splicing of exons 2, 3, and 10, six tau protein isoforms are recognized and differentiated (Goedert and Jakes 1990). The 2N4R isoform of human tau is the longest one (tau441), which is divided into four functional domains based on the nature of amino acids or their interactions with microtubules (Sergeant et al. 2008; Fitzpatrick et al. 2017). Exons 9–12 encode the repeating units R1–R4, which are separated from one another by a spacer region

(13 or 14 residues), which enables microtubule binding and also promotes their assembly. Each repeat unit is made up of a strand of 18 amino acid residues that are highly conserved (Brandt and Lee 1993). In the region ^{275}VQINKKLDLS^{285}, Lys280, Ser356, and Ser262 play a critical role in the tau-microtubule interaction (Cohen et al. 2011; Biernat et al. 1993; Pradeepkiran and Reddy 2019). PHF6 (VQIVYK) and PHF6* (VQIINK), two of the six amino acid residue segments, have been identified as being responsible for tau PHF aggregation in AD. Some amino acid residues such as serine residues, Ser285, Ser 289, Ser 293, and Ser 305, and tyrosine residue at 310 (Tyr310), play important roles in the phosphorylation of tau protein. Therefore, finding and validating serine-targeted site-specific kinase-dependent inhibitors are probably the most promising therapeutic strategy for creating potent new treatments for pTau-associated abnormalities (Pradeepkiran and Reddy 2019).

5.3 Diagnosis of AD

Diagnosis of disease by determining the presence of biomarkers in biological fluids is a very critical step for prescribing medicines to the patients. Plenty of literature evidence states that the Aβ proteins and tau proteins are the major biomarkers despite several pathological hallmarks of AD. In the primary stage of disease, diagnostic agents play an important role in the early detection, progression, and prevention of the disease. The assessment of the patient's cognitive and behavioural states, both of which are indicators of final stages of neurotoxicity, is the basis for the current clinical diagnosis of AD. The development of methods and procedures for the identification and assessment of biomarkers, which are the primary indications of molecular alterations in the brain and neurodegeneration, is currently the focus of significant research efforts and funding. Various techniques like magnetic resonance imaging (MRI) (Wolz et al. 2011), positron-emission tomography (PET) (Vlassenko et al. 2012), and single-photon emission computed tomography (SPECT) (Maya et al. 2009) are currently used in clinical trials. However, these sophisticated techniques such as PET, MRI, and SPECT are not only difficult to use but also very expensive for the majority of the population (Shimojo et al. 2015). As compared to the other techniques, optical imaging techniques offer good results in the context of the application. Also, this technique has various advantages such as being rapid, inexpensive, reliable, and easy for handling. Recently available techniques like multi-photon excitation techniques require both clinicians and scientists to capture the brain images (in vivo) in animal models (Svoboda and Yasuda 2006). However, biomarkers of AD that are found within the brain and biofluids can be easily detected and seen with the right resolution using optical imaging techniques. In addition, these optical imaging methods are economical and affordable for the majority of the population (Xu et al. 2016; Ono and Saji 2015). Furthermore, colorimetric probes allowed scientists to detect the Aβ aggregates with the naked eye (Rajasekhar et al. 2016). With the recent advances in optical imaging techniques and modern spectroscopy, this technique is no longer limited to in vitro bioimaging of biomarkers and detection in different assays, but it has developed into a trustworthy in vivo imaging device. For instance, multi-photon excitation fluorescence spectroscopy can be helpful as a diagnostic tool for otherwise incurable clinical manifestations seen with AD progression. Early detection of AD biomarkers like Aβ plaques and NFTs using this technique can be beneficial (Svoboda and Yasuda 2006). Functional biosensors and fluorescent proteins have been reported to be able to track the physiological condition of an animal's brain by adjusting the sensor's biophysical characteristics in response to the living creature. Because of this, optical detection and imaging instruments are undoubtedly effective methods for the early identification of AD (Shimojo et al. 2015). Due to its numerous benefits and possible application in enhanced therapy and diagnostics in the near future, fluorescent molecular probes are currently receiving a great deal of attention from the research community. When creating optimal fluorescent probes for the detection and imaging of protein or peptide (tau or Aβ) aggregation species in the brain, certain design principles must be kept in mind (Arora et al. 2020):

1. Extreme alloform selectivity for a protein or peptide.
2. Straightforward synthesis.
3. A significant Stokes shift in the probe's emission profile is required to prevent self-absorption.
4. Activate the fluorescence response.
5. Increased quantum yield.
6. Appreciable binding constant.
7. The probe's emission wavelength must be larger than 450 nm (>450 nm), typically in the red and near-infrared (NIR) area, in order to minimize background fluorescence.
8. Minimal photobleaching.
9. Metabolic stability.
10. Quick clearance from the brain.
11. Blood-brain barrier (BBB) penetration of significant magnitude.
12. Nontoxicity and biocompatibility.

5.3.1 A Summary of Various Diagnostic Imaging Methods and Their Clinical Relevance

Various imaging techniques (Fig. 5.2) have been studied for use as part of diagnostic tools for the diagnosis of disease. These are summarized in Table 5.1.

5.4 Fluorescent Molecules as Diagnostic Tools in AD

Fluorescence diagnostics has recently emerged as an appealing and promising choice for diagnosing and researching the progress of neurodegenerative diseases due to its non-invasive, speedy, delicate, simple, economical, real-time, and

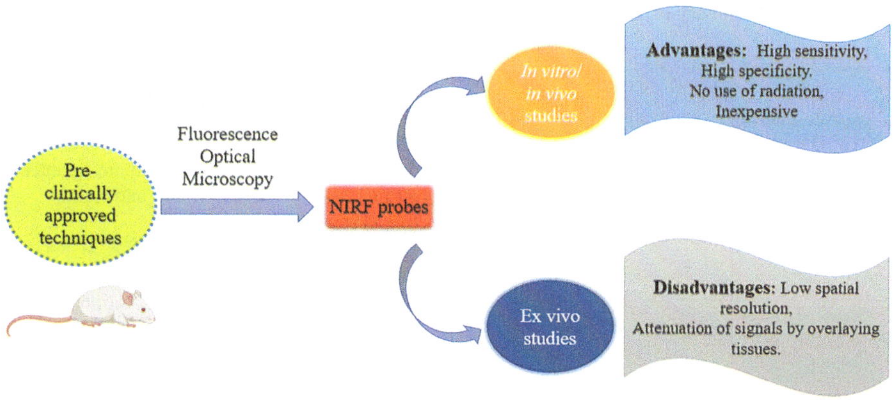

Fig. 5.2 Schematic representation of imaging techniques

increased resolution advantages (Ntziachristos 2006). Fluorescence has extensive applications (quantitative measurements, mapping, detecting, and theranostic purposes) as well as the probes should be varied according to the application (Fig. 5.3). Several fluorescence systems can break down or remove amyloid fibrils in contrast to their detecting function, which could help with the management of AD. The probes must be customized for each circumstance since the hurdles that fluorescence imaging must overcome in each are widely diverse. For instance, near-infrared (NIR) biosensors are desperately required for tissue implementation given that the blood-brain barrier (BBB) is still not a restriction for in vitro tests; the sensor for in vivo studies must fulfil several specifications, like washout kinetic parameters and relatively low biological toxic effects; there exists a considerable increase of albumin in the serum, which would be particularly susceptible to interact with the testing results and demands fluorophores with high selectivity (Licha and Olbrich 2005).

5.4.1 Fluorescent Probes for AD Biomarker Detection

Several fluorescent scaffolds are being modified for enhancing their optical and imaging properties for the development as diagnostic tools and are briefed in the subsections.

5.4.1.1 Fluorogenic and Standard Optical Probes

Commercially accessible dyes such as thioflavin-T and Congo red have been extensively utilized to identify tau fibrils and Aβ aggregates in vitro. These dyes are still not the best for deep tissue detection due to their low emission wavelengths (500–550 nm), which are frequently shorter than that of the autofluorescence of bioactive substances (Peng et al. 2019). One of the most thoroughly researched structural analogues of ThT, PET tracer ^{11}C-labelled Pittsburgh compound-B

Table 5.1 A summary of various diagnostic imaging methods and their clinical relevance

Techniques	Traditional characters
	Radioligand or nuclear imaging techniques
Positron-emission tomography (Johnson et al. 2013; Harada et al. 2015; Collier et al. 2017; Jie et al. 2021)	• Radiolabelling with a specific target site, i.e. receptor, transporter, and enzyme, a wide variety of readily available synthetic ligands, fast metabolism measurement, and rapid clinical translation • Nature of probes—^{11}C, ^{18}F, ^{124}I, ^{68}Ga, and ^{64}Cu • Tauvid (PET radiotracer)—approved by the FDA in 2020, useful in clinical tau pathology detection in AD • Amyloid protein-PET—^{18}F-flutemetamol, ^{18}F-florbetaben—demonstrate the presence of brain amyloid in mild cognitive impairment • Tau protein-PET—^{18}F-MK-6240, ^{18}F-AV-1451—excellent selectivity and noise ratio for neurofibrillary tangles
Magnetic resonance imaging (MRI) (Johnson et al. 2012; Hojjati et al. 2019; Coimbra et al. 2006)	• Cerebrospinal fluid, white matter, and grey matter visualization • Type of probes—^{1}H, ^{19}F, protamine sulphate, ^{23}Na poly-L-lysine • Structural MRI is used to measure cortical thickness as well as hippocampus volume, atrophy rate, and medial temporal lobe atrophy • The modification of the neural network in the AD brain was investigated using resting-state functional MRI (rs-fMRI) • Functional magnetic resonance imaging (fMRI) is helpful in assessing the health of the brain's neuroanatomical network in the early stages of Alzheimer's disease • Perfusion MRI—useful in the detection of areas with lower cerebral blood flow and volume
Single-photon-emission computed tomography (SPECT) (Hammoud et al. 2007)	• Nature of probes—^{99}mTc, ^{123}I, ^{125}I, ^{111}In • Regional brain perfusion • Target-specific radiolabelling • Physical half-life that is longer • There is no conceivable limit to the spatial resolution • Fast clinical translation
	Optical imaging techniques
Fluorescence molecular tomography (Zhang et al. 2015)	• Type of probes—artificial NIR probes, fluorescent proteins • Reliable and inexpensive • Real-time detection • Wide availability

(continued)

Table 5.1 (continued)

Techniques	Traditional characters
	• Potential for detecting Aβ plaques in vivo, among other available imaging technologies
Bioluminescence imaging (BLI) (Mezzanotte et al. 2017)	• Source of luminescence: luciferin, coelenterazine • It involves the use of luciferase enzyme to generate light • Luciferase enzyme is genetically modified to target any biological cell • Helpful for tracking biological processes quantitatively both in vitro and in vivo
	Other technologies
Optical coherence tomography (OCT) (Cunha et al. 2016)	• Non-invasive imaging technique • Helpful for determining neuronal fundus integrity • OCT makes it possible to evaluate RGC layer deterioration in an indirect manner • It makes it easier to determine how much the retinal nerve fibre layer's peripapillaries have shrunk (RNFL) • Offers clear cross-sectional pictures of the retinal structure
Magnetic encephalography (MEG) or electroencephalography (EEG) (Pineda-Pardo et al. 2014)	• High temporal resolution is provided by MEG and EEG for the non-invasive investigation of minute fluctuations in brain activity • They facilitate the capturing of the neuritic magnetic field, leading to a more precise neuronal index score than haemodynamic methods
Regional metabolic cerebral blood flow biomarkers (Graff et al. 2022)	• Nature of probes: ^{18}F-FDG-PET • To investigate the localized glucose metabolism • Comparatively improved sensitivity or specificity for separating AD from healthy subjects
Two-photon excited fluorescence laser scanning microscopy (TPLSM) (Theer et al. 2003)	• TPLSM is a versatile method that provides clinically useful information about dynamic analysis and cell signalling in the brain • This method enables real-time recording of existing cellular activity by experts

(^{11}C-PIB), is being used worldwide for the clear and specific diagnosis of fibrillar amyloid plaques in AD. This is very useful for distinguishing non-amyloid-related dementia conditions, such as frontotemporal lobar degeneration (FTLD). The very first US Food and Drug Administration-approved PET tracer radioisotope to measure brain amyloid deposition is 18F-AV-45 (florbetapir). Fascinatingly, the greater

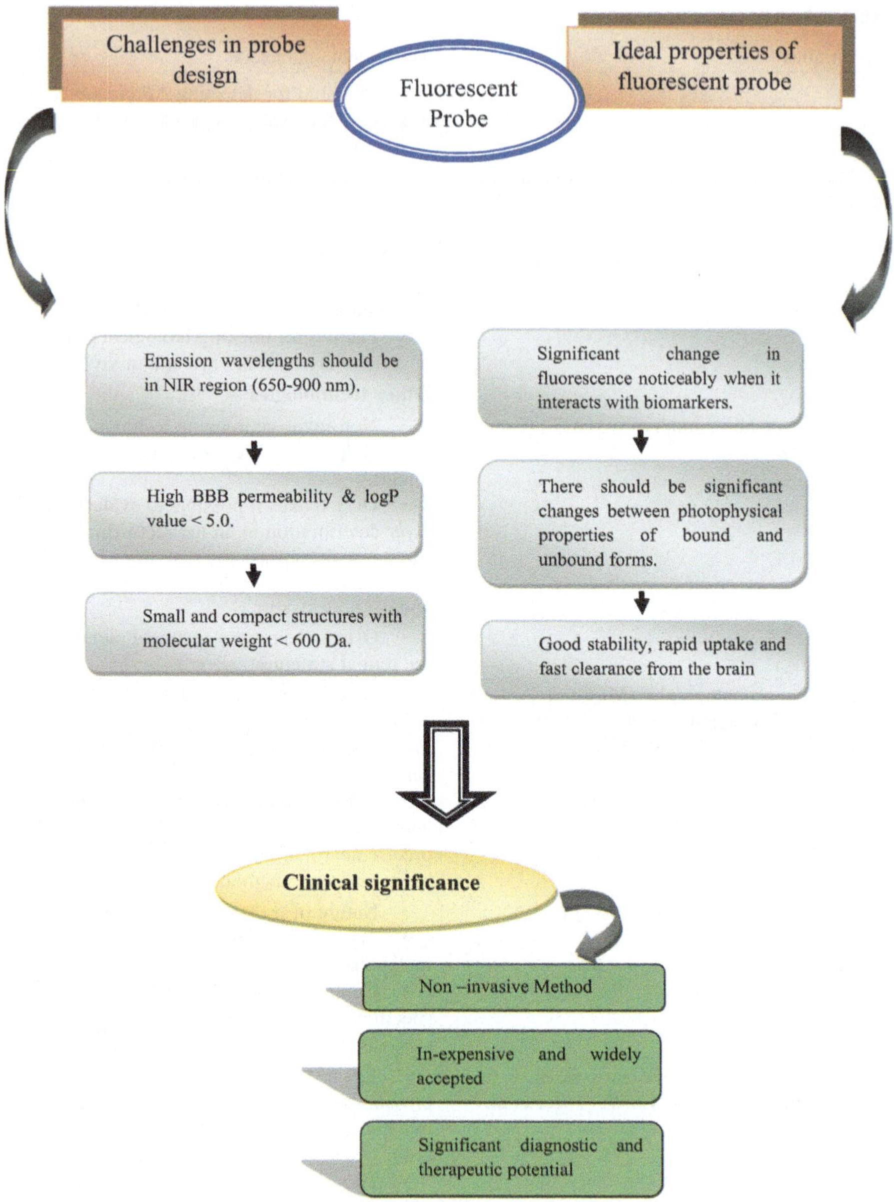

Fig. 5.3 Ideal characteristic features of fluorescent probes in AD

marker level in the brain areas with amyloid deposits is made possible by the fluorine-18 analogue **3** (Fig. 5.4) due to the presence of larger half-life (about 5.48 times) compared to ^{11}C-PIB (Wong et al. 2010). Cost-effective alternatives could be a preferable option for early AD diagnosis considering the high cost, restricted

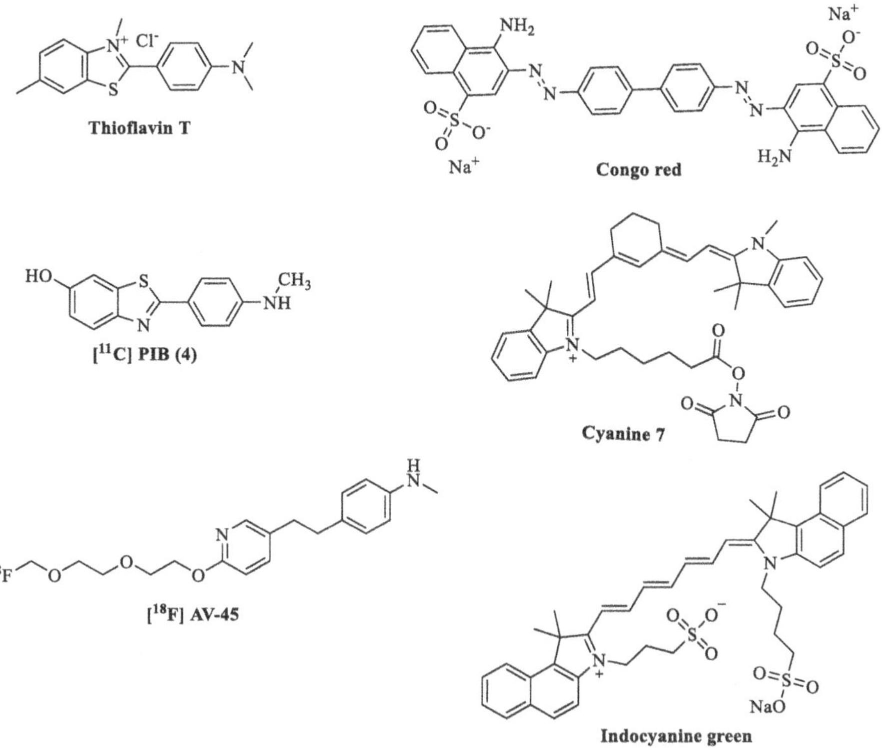

Fig. 5.4 Chemical structures of commercially available fluorescent dyes

practical application, requirement of radioactivity substances, and long-lasting collecting data method in some techniques like PET and SPECT techniques. Numerous different NIR fluorophores (Fig. 5.4), like cyanine dyes (Cy7), indocyanine green dyes (ICG), Alexa fluor dyes, and SRfluor dyes, display restricted BBB permeation and higher affinity for bioactive proteins other than the genetic disorder peptides in the brain. These NIR probes have a greater molecular weight and intrinsic charge transfer (ICT) (Cui et al. 2014).

5.4.2 Classification of Fluorescent Probes for Amyloid (Aβ) Species and Tau Species Based on Fluorophore Structure

5.4.2.1 Thioflavin Derivatives for Aβ Peptides

The invention of fluorescent probes for AD has provided a significant amount of focus to the benzothiazole ring containing acceptor and donor subunits linked by a free C-C rotational bond. The two most often employed fluorescent probes in in vitro studies of AD are thioflavin-T and thioflavin-S. The cationic charge on the benzothiazole ring in these structures limits brain penetration, and their characteristic

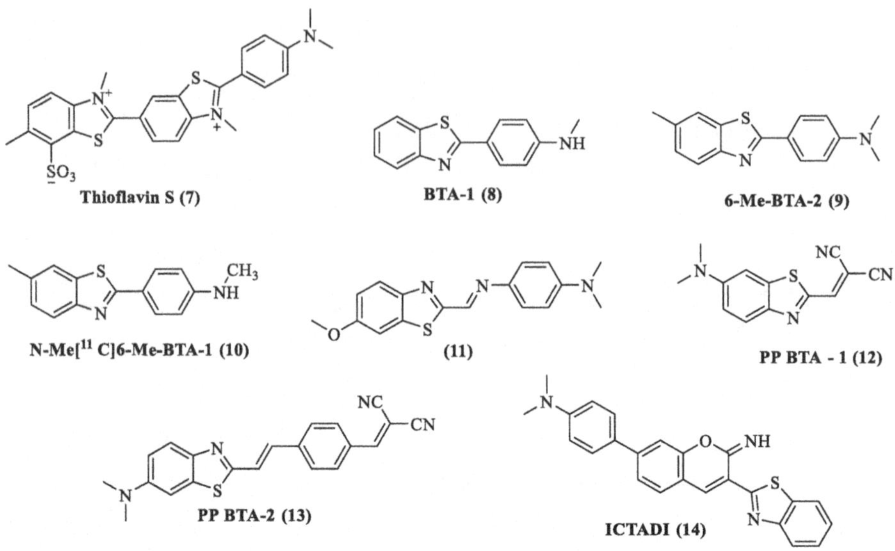

Fig. 5.5 Chemical structures of thioflavin-T derivatives for Aβ peptides

short emission limits their clinical relevance in AD diagnosis. Figure 5.5 represents various examples of thioflavin-derived fluorescent probes having an affinity towards Aβ peptides. The probes 8, 9, and 10 are uncharged thioflavin-T derivatives that have an affinity to bind with Aβ protein and readily enter the human brain. These probes bind to A$\beta_{(1-40)}$ and stained both tangles and plaque forms in the post-mortem AD brain (Klunk et al. 2001).

New Schiff-based benzothiazole molecules with strong affinity for Aβ plaques in AD brain were produced and tested by Gan et al. In this series, compound (probe 11) with an N,N-dimethyl amino group shows high binding affinity in in vitro fluorescent staining of brain sections of post-mortem AD patients (Gan et al. 2013). Similar to this, benzothiazole-based probes were designed and created using a push-pull mechanism, PP-BTA 1 and PP-BTA2. These two probes succeeded in staining of Aβ plaques in brain slices of transgenic mice as well as in human AD patients (Xu et al. 2016). The differentiation of Aβ subspecies, i.e. A$\beta_{(1-40)}$ and A$\beta_{(1-42)}$, is considered a big task for small fluorescent probes even if there are only two amino acids' differentiation present between them. So, in order to overcome this problem, probe 14 was synthesized which showed that it succeeded in differentiating both A$\beta_{(1-40)}$ and A$\beta_{(1-42)}$ in solutions, plasma, and brain tissues. These probes might be a crucial part of an AD diagnostic method (Yang et al. 2020).

5.4.2.2 Thioflavin Derivatives for Tau Proteins

For the purpose of identifying tau aggregation and preventing neurodegeneration, clinical and preclinical feature analysis will be beneficial. Tau aggregation plays a key role in neurotoxicity as a hallmark to assess the extent of severity linked to the different stages of degenerative AD. This study describes the discovery of

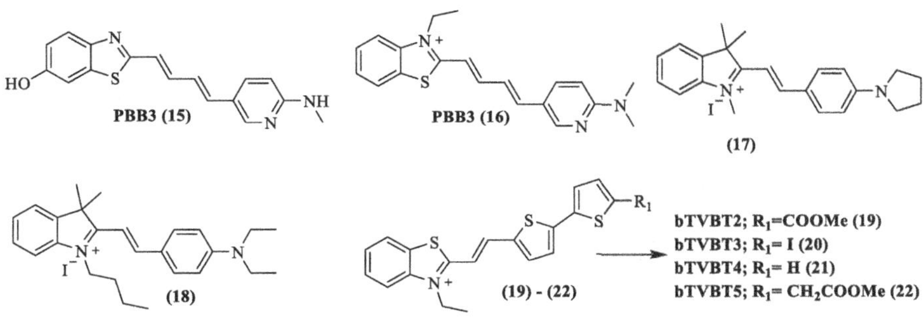

Fig. 5.6 Chemical structures of thioflavin-T derivatives for aggregated tau proteins

fluorescent probes that demonstrate a significant role in tau aggregation in AD and new knowledge regarding tau selectivity. A series of compounds containing phenyl/pyridinyl-butadienyl-benzothiazoles/benzo-thiazoliums (PBBs) for staining various tau species in the brains of AD patients and animal studies of these disorders have been studied. Probes 15 and 16 (Fig. 5.6) are shown to have outstanding detection capacity and adequate selectivity for detecting tau species in living beings (Maruyama et al. 2013). Similarly, two (probes 17 and 18) water-soluble 2-styrylindolium probes for fluorescent imaging of NFTs in AD have been developed. The staining experiments confirm that these probes may show selectivity to NFTs even in the presence of amyloid beta species with low or no toxicity in various cell line studies like liver hepatocellular carcinoma cells (Gu et al. 2012). In order to distinguish senile plaques and NFTs spectrally, Nilsson et al. devised a series of oligothiophene hybrids called bithiophene-vinyl-benzothiazole/benzimidazole. The experimental studies demonstrated that these ligands could be key for determining new therapeutic techniques to monitor the progression of AD (Shirani et al. 2017).

5.4.2.3 Curcumin Derivatives for Aβ Peptides

The chemical structure of curcumin contains two o-methoxy phenols, which are joined symmetrically by an unsaturated diketone linker, which is responsible for its keto-enol tautomerism. As a result, curcumin demonstrates a variety of remarkable photophysical and photochemical characteristics. Curcumin's maximum absorption range in various organic solvents is in between 408 and 430 nm, whereas its highest emission range is 460–560 nm (Priyadarsini 2009). These results indicate that curcumin may be among the most promising precursors for the development of theranostic chromophores for medical diagnosis and treatment administration in AD. To get over drawbacks including a lower emission wavelength, a BBB with restricted permeability, and other issues like curcumin-related fast metabolism sensitivity, Ran's group created curcumin analogues (CRANAD 1 and CRANAD 2) using a new difluoroboronate ring (Fig. 5.7). CRANAD-2 on binding with Aβ aggregates shows various changes like 70-fold increase in fluorescence intensity and a significant increase in quantum yield (QY) (Ran et al. 2009).

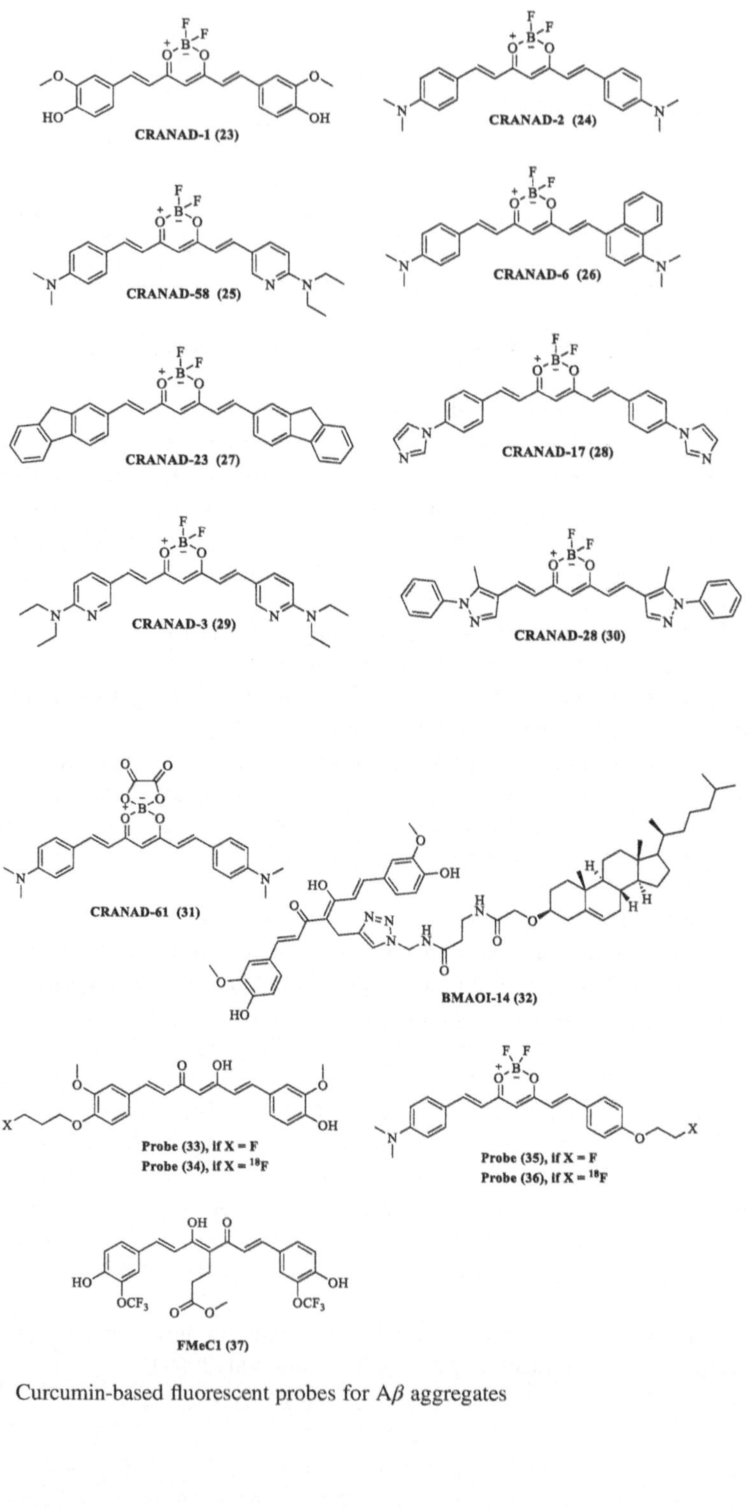

Fig. 5.7 Curcumin-based fluorescent probes for Aβ aggregates

CRANAD-58 was designed by Zhang et al., which shows significant fluorescent property changes on binding with both soluble and insoluble forms of Aβ species in vitro. Probes 26 and 27 share a common structural feature of probe 24, with little changes at phenyl ring moiety. Later, an imidazole ring containing monodentate curcumin scaffold CRANAD 17 was developed that could inhibit copper-induced Aβ_{42} cross linking (Zhang et al. 2013). CRANAD-3 was developed, which can detect both soluble and insoluble forms of Aβ species. According to in vivo imaging, CRANAD-3 can recognize the initial phases of molecular pathology because its NIRF signal was 2.29 times higher in 4-month-old transgenic AD (APP/PS1) mice than it was in age-matched wild-type mice (Zhang et al. 2015).

CRANAD-28 is a bifunctional curcumin analogue, which was designed to overcome the low QY associated with the curcumin analogues. Its in vivo two-photon imaging shows that it is BBB permeable and could stain plaques and cerebral amyloid angiopathies (Zhang et al. 2014). CRANAD-61 has the potential to detect active amyloid plaques that are surrounded by high ROS species (Yang et al. 2017). Liu et al. created a bivalent multifunctional Aβ oligomerization inhibitor (BMAOI) that has cell membrane lipid raft on one side for treating AD and curcumin on the other for detecting amyloid peptides (Liu et al. 2012). Probe 33 is a fluoropropyl-substituted curcumin analogue that exhibits high binding affinity for Aβ species, whereas its ^{18}F-labelled radioactive derivative probe 34 failed to get better in vivo Aβ imaging results (Ryu et al. 2006). Probes 35 and 36 exhibited favourable BBB permeability and were able to distinguish Aβ plaques in transgenic mouse brain sections (Kim et al. 2019). Similar to this, Yanagiswa et al. examined ^{19}F-containing curcumin derivative (FMeC1), which exists in both keto and enol forms, out of which enol form has selectivity towards Aβ aggregates while the keto form does not. These studies examine the diagnostic value of molecular NIR probes that can be used with more flexible equipment like PET/SPECT and MRI (Yanagisawa et al. 2011).

5.4.2.4 Curcumin Derivatives for Tau Proteins

Schmidt and his colleagues synthesized bis (aryl vinyl) pyrazines, pyrimidines, and pyridazines as imaging agents for tau fibrils, of which pyrimidine derivatives show binding affinity for both NFT and Aβ. In vitro studies suggested that probes 38 and 39 exhibit high affinity for tau aggregates than Aβ species (Boländer et al. 2012). Similarly, a new derivative based on curcumin with various aromatic substituents was developed, out of which probe 40 having a dimethylamino-2,6-dimethoxy phenyl moiety shows greater differences in fluorescent properties after binding with tau fibrils in live cells (Park et al. 2015). The 3,5-dimethoxy-N,N-dimethylaniline-4-yl moiety, a previously found core scaffold, and the distinctive donor-Π-acceptor architecture of multiple NIRF probes were combined to create probe 41 (Seo et al. 2016). Probe 42 is a difluoroboron diketone derivative which shows a turn-on fluorescence response for tau fibrils. Additionally, the mechanism for tau specificity was examined, and the results show that the molecular rotor-like characteristic feature of the probe allows for the specific identification of tau aggregates to emit high fluorescence qualities (Park et al. 2017) (Fig. 5.8).

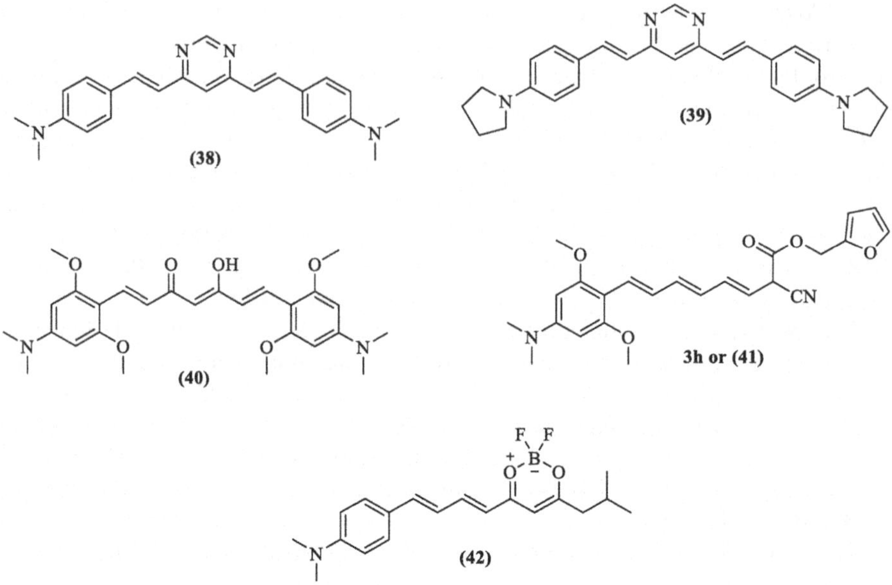

Fig. 5.8 Curcumin-based derivatives for aggregated tau proteins

5.4.2.5 BODIPY Derivatives for Amyloid Aggregates

In recent years, fluorophores with boron dipyrromethene (BODIPY) have drawn a lot of attention as a small-molecule fluorescent probe for imaging. As a fluorescent scaffold, BODIPY has exceptional photophysical characteristics, including strong photostability, high fluorescence QY, high extinction coefficient, and narrow emission bandwidth. These characteristics make them a good approach for designing of fluorescent probes (Lee et al. 2009).

Ono and his group developed a BODIPY-based labelled radioligand probe 43 and used in SPECT for the first time. The BODIPY derivative displayed a strong affinity for the $A\beta_{(1-42)}$ aggregates in binding tests conducted in vitro. In a mouse brain, probe 43 vividly stained $A\beta$ plaques, demonstrating its affinity for $A\beta$ aggregates in vitro. These results imply that additional structural modifications to the BODIPY backbone may be used in dual SPECT/fluorescent probes that might be used to image $A\beta$ plaques (Ono et al. 2010). Later, to increase its biological affinity, they created another fluorescent probe (BAP-1) 43 by switching the thiophene phenyl chain for p-dimethyl amino phenyl. The probe showed better binding for synthetic $A\beta$ aggregates. Ex vivo fluorescence labelling of Tg2576 mouse brain slices following BAP-1 injection revealed selective $A\beta$ plaque binding with negligible nonspecific binding (Ono et al. 2012). Watanabe and his colleagues created a series of new probes (BAPs) after being inspired by probe 44. The developed structures are thiophenyl-containing analogue of BAP-2, BAP-3 contains furanyl, BAP-4 contains phenyl thiophenyl, and BAP-5 contains the phenylfuranyl moiety. These probes are designed on the basis of such a plan that BODIPY fluorophore with a dimethylaminoheteroaryl group could enhance the photophysical activity of the probes. These molecules extensively marked Aβ plaques in the brain area, which was consistent with their in vitro affinity for Aβ aggregation (Watanabe et al. 2013).

Using information from earlier findings on BODIPY derivatives, Sozmen and his team created a number of new BODIPY probes that are conjugated to styryl (49–53). These findings showed considerably good staining properties in brain sections from Tg2576 mice (Sozmen et al. 2014). Ren et al. synthesized another BODIPY derivative probe 54; the developed probe showed better results in fluorescence properties on binding with soluble and insoluble Aβ species. Additionally, it effectively stained the brains of 6-month-old APPswe/PSEN 1De9 Tg and wild-type mice, according to in vivo experiments (Yang et al. 2019a) (Fig. 5.9).

5.4.2.6 BODIPY Derivatives for Tau Proteins

Ojida and his team developed a new fluorescent probe (probe 55, Fig. 5.10) containing two nuclear Zn (II) complexes for the identification of NFTs. This probe is made of two Zn(II)-2,2-dipicolylamine complexes that serve as an active site for phosphorylated amino acid units and a fluorescent BODIPY unit. A hippocampal tissue cut from an AD patient was used for histological imaging, and it was discovered that it fluorescently visualizes NFT deposits and distinguishes between NFTs and senile plaques (Ojida et al. 2009).

A newly found effective candidate for the real-time detection of in vitro tau protein fibrillization was found using a quinoxaline-functionalized styryl-BODIPY derivative probe 56. In vitro studies suggested that the compound is non-toxic to living cells and it has good cell permeability (Mani et al. 2017). Another probe, 4,4-difluoro-4-bora-3a,4a-diaza-s-indacene BD-tau (probe 57), was synthesized, which shows increased fluorescence intensities on binding to both aggregated and pre-aggregated tau proteins (Lim et al. 2017). Verwilst and his group reported probes 58 and 59, which show a high selectivity for self-assembled microtubule-associated tau protein both in solution and cell-based experiments. Additionally, these fluorophores might be used as the foundation for the creation of therapeutically applicable sensors, such as those dependent on PET imaging (Verwilst et al. 2017).

5.4.2.7 Donor-Π-Acceptor-Based Probes (DANIR) for Amyloid Species

DANIR cyanine derivatives are the top option for designing the fluorescent probes because of their sensitivity and affinity for the protein of interest as well as the favourable spectrophotometric characteristics that can be attained with the right donor and acceptor moiety linked through conjugated pi-electron linear alkyl chain backbone. In 2013, Mengchao and his team developed a family of NIRF imaging probes with a donor-acceptor architecture that is coupled by an electron chain for selectivity in Aβ plaques. The synthesized probes with a longer conjugated Π system show maximum emission, which comes under the category of NIRF probes. The DANIR-2c upon interaction with the A$\beta_{(1-42)}$ aggregates shows better fluorescence intensity (Cui et al. 2014). Similarly, Fu et al. synthesized a class of DANIR probes, which contains p-dimethylaniline donor groups and various acceptor groups like Meldrum acid (MAAD-3), dimedone (DMDAD-3), methyl cyanoacetate (MCADD-

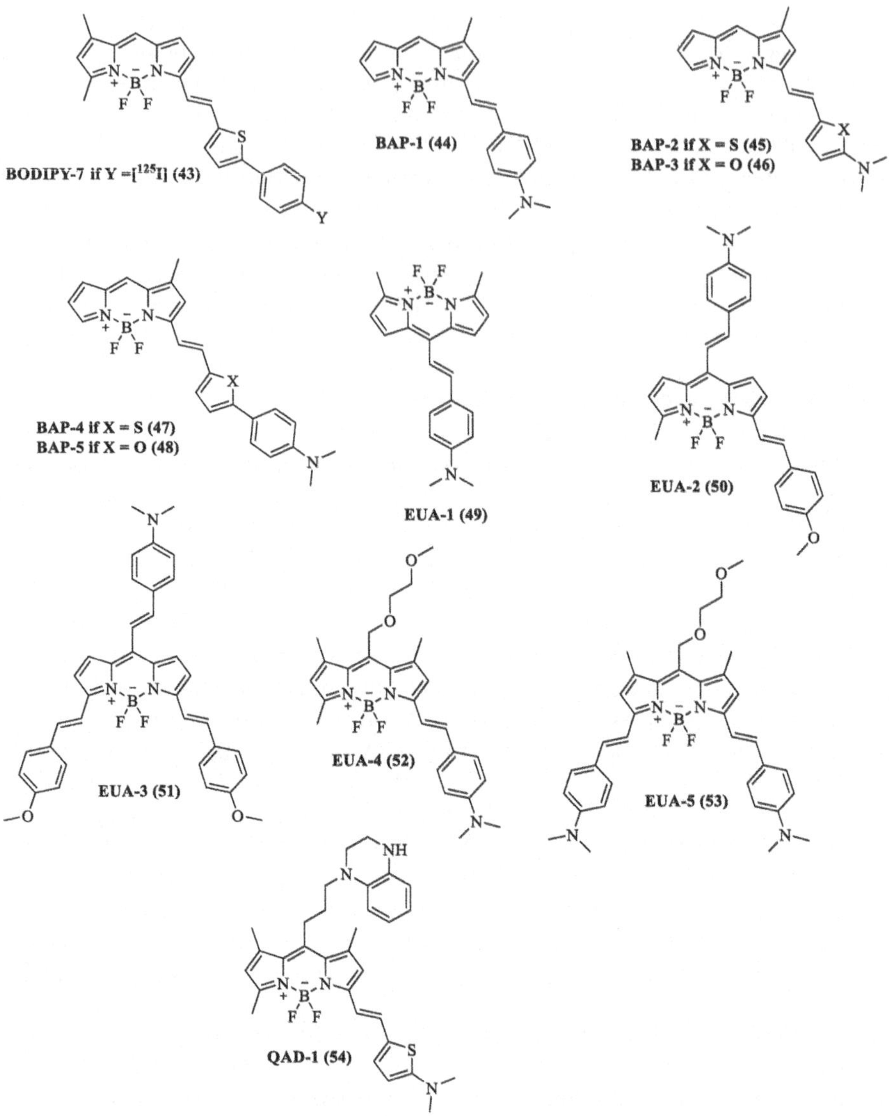

Fig. 5.9 Chemical structures of BODIPY derivatives for Aβ aggregates

3), and dimethylmalonate (DMMAD-3), which are interlinked through polyene spacers. These four probes with the longest conjugated double-bond architecture showed good binding affinity to A$\beta_{(1-42)}$ aggregates (Fu et al. 2014). DANIR-based fluorescent probe (BBTO-3) containing planar barbituric acid derivatives as stronger acceptor groups is developed in order to get greater affinity and longer emission wavelength. The BBTOM-3 was inert towards photochemical isomerization and shows good results upon interaction with the Aβ aggregates (Zhou et al. 2015).

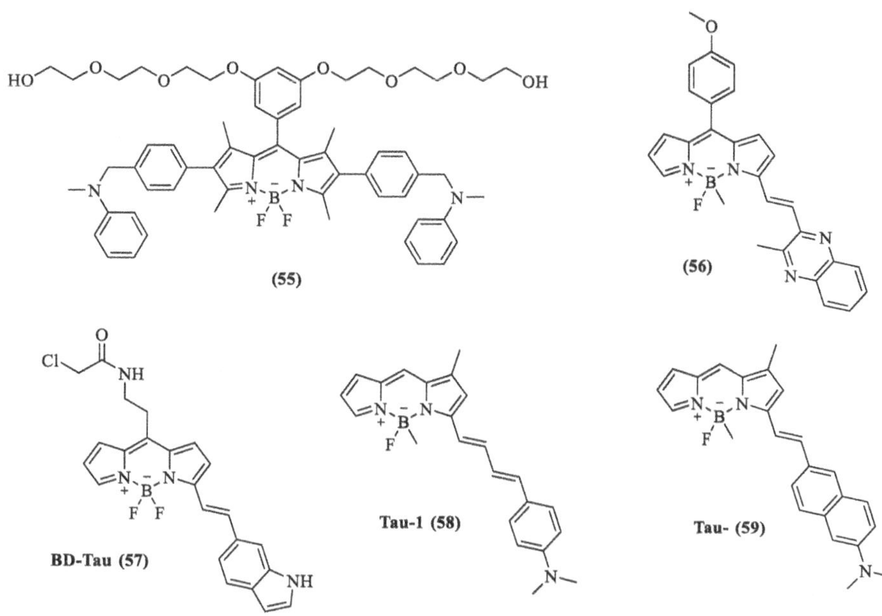

Fig. 5.10 Chemical structures of BODIPY derivatives for tau aggregates

In order to improve the DANIR probes' weak brain pharmacokinetic properties and boost their polarity, Fu et al. created DANIR-8c, which has hydroxyethyl-substituted naphthalene as the electron donor moiety and a malononitrile group as the electron acceptor. DANIR 8c shows excellent fluorescence features with emission above 670 nm upon binding to Aβ aggregates and also its high sensitivity towards the Aβ species (Fu et al. 2016). Similarly, Petric and his group explored the involvement of hydrophobic naphthalene moiety in DDNP/FDDNP for bioimaging of plaques in AD. In addition, FDDNP as a PET probe was also studied for in vivo analysis of NFTs in AD patients (Agdeppa et al. 2003).

Fu and his team developed dimethylaminonaphthyl derivatives as a donor group in DANIR-3b and DANIR-3c in a push-pull design with cyanoacetonitrile as an acceptor group. These two probes show higher QY compared to the DANIR 2c and good emission properties upon binding with the amyloid aggregates (Fig. 5.11). The selectivity and binding capability for Aβ plaques were further suggested by their histological imaging in 14–19-month-old female Tg mice models (Fu et al. 2015).

5.4.2.8 Naphthalene Derivatives for Aβ Aggregates and Tau Fibrils

Later in the year 2012, a series of aminonaphthalenyl 2-cyanoacrylate-based probes (ANCA series) were synthesized which can differentiate different forms of amyloid species in the brain. The differentiation ability of these probes is because of the stabilization of both ground state and excited state as a function of the polarity of their environment. This shorter emission wavelength of these probes below the NIR range indicates the later invention of structural modification to design a desirable

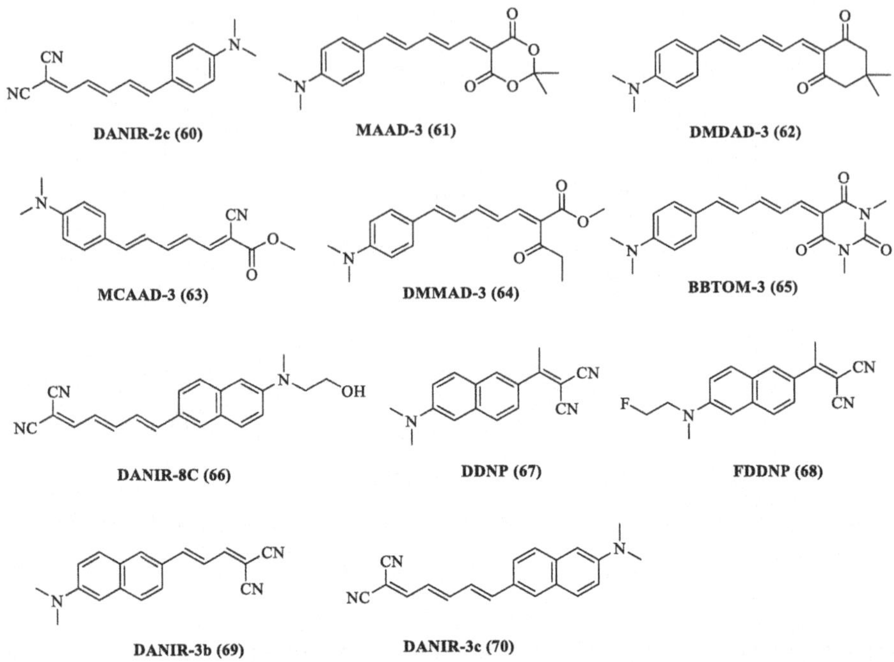

Fig. 5.11 Chemical structures of DANIR derivatives for Aβ aggregates and tau fibrils

ANCA-NIR probe (Cao et al. 2012). Similarly, in the year 2016, a novel spiropyran-based fluorescent probe (AN-SP) that shows high binding affinity and specific nature against Aβ oligomers both in vitro and in vivo was synthesized. Bioimaging studies proved that AN-SP can penetrate BBB and specifically detect Aβ oligomers in brain tissues of transgenic mice (Fig. 5.12) (Lv et al. 2016).

2018 saw the creation and synthesis of several novel amino-aryl cyanoacrylate probes by Cao and colleagues. Their design approach is heavily reliant on the superior specificity and fluorescence imaging of ANCA probes for amyloid plaques associated with AD. The amyloidogenic proteins in AD were successfully stained by these probes (Cao et al. 2018).

5.4.2.9 Quinoxaline-Naphthalene Derivatives for Aβ Protein and Tau Fibrils

The enhanced fluorescent properties of the quinoxaline-based probes are due to the presence of the two nitrogen atoms, which have the hydrogen-bonding property. In the year 2013, a series of pyrazine and quinoxaline probes (QNNs) with high affinity and selectivity for tau aggregates were synthesized. These probes show better features like fluorescent turn-on effects and favourable brain kinetics. In vivo fluorescence imaging results show that these probes could differentiate tau transgenic mice from wild-type controls with specific labelling of cerebral tau tangles (Zhu et al. 2018). Zhou et al. later developed another quinoxaline ring-containing

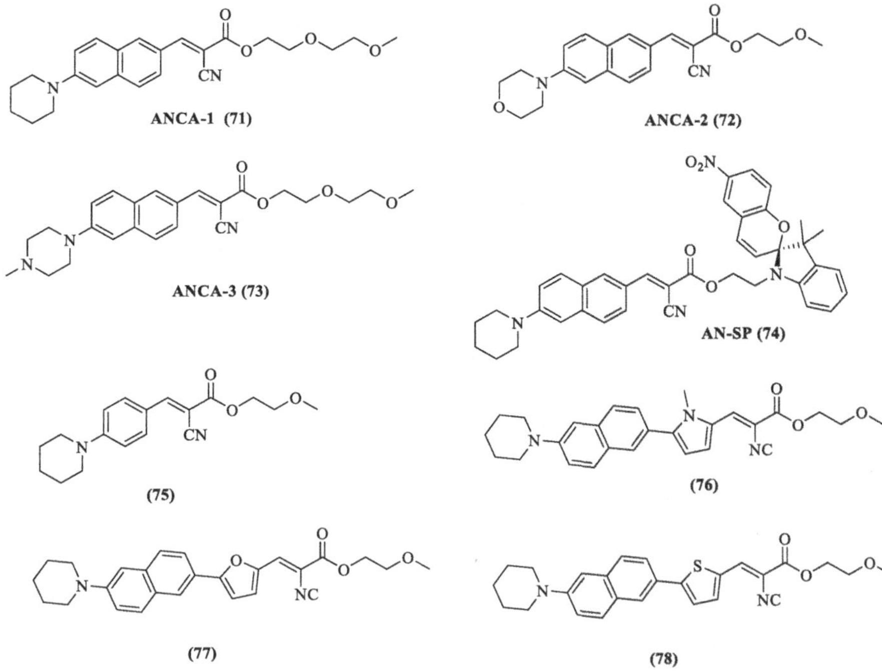

Fig. 5.12 Chemical structures of naphthalene-containing probes for A*β* proteins and tau fibrils

probe QN-18 (Fig. 5.13), by replacing the naphthalene core with a quinoxaline ring in the DANIR-3b structural moiety. The data from in vitro and in vivo studies showed that this probe has high selectivity and binding ability for A*β* aggregates relative to tau proteins (Zhou et al. 2019).

5.4.2.10 Novel Benzoselenazole Derivatives for A*β* Plaques

A planar benzoselenazole-based NIRF fluorescent probe with high amyloid aggregation binding affinity was created by Sundaram and his colleagues (Fig. 5.14). The ex vivo data shows that it has a remarkable capacity to stain A*β* plaques. Inspired from the above results, the same group later developed its F-radiolabelled ^{18}F-florbetaben (probe 82), which has better pharmacokinetic characteristic features and evolved as a probe that has clinical capability in in vivo imaging of AD (Sundaram et al. 2016).

5.4.2.11 Cyanine-Based Derivatives for Amyloid Peptide and Tau Proteins

The first ratiometric cyanine-based probe, CyDPA2 (Fig. 5.15), was created in 2013 by Bai and his team. It is made up of two zinc dipicolylamine structures that are linked to the parent probe IR820. The photophysical studies suggested that this probe can bind specifically to tau protein (Kim et al. 2013). In 2019, Yag et al. made the hemocyanin derivative ZT-1 (Fig. 5.15) by substituting the pivaloylacetonitrile

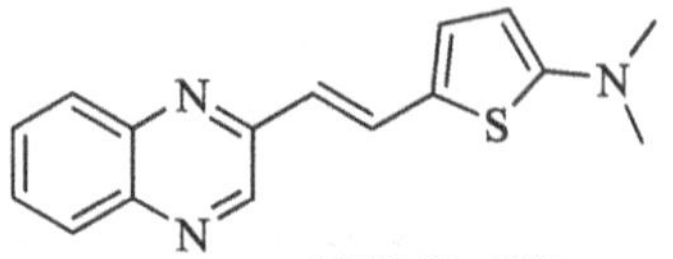

QNN-3b (79)

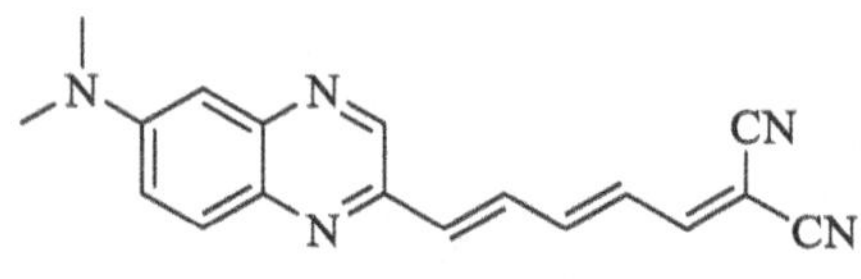

QN-18 (80)

Fig. 5.13 Chemical structures of quinoxaline ring-containing probes for Aβ proteins and tau fibrils

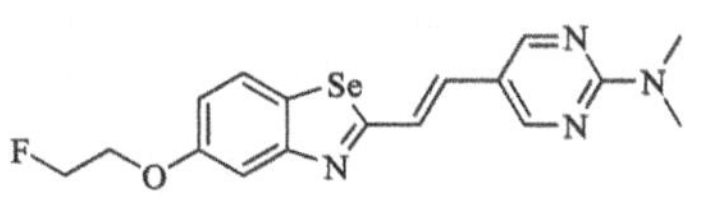

81

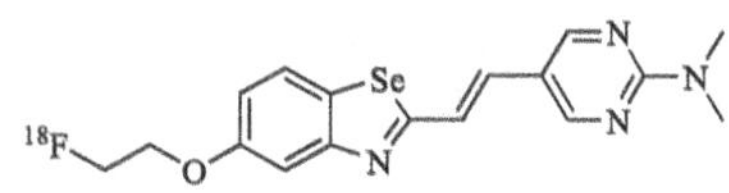

82

Fig. 5.14 Chemical structures of benzoselenazole-containing probe for Aβ proteins and tau fibrils

group for the cationic portion of IR-780. In vivo fluorescence imaging data shows that the probe is successful in detecting amyloid plaques in pre-clinical stages. All these results encouraged the designing of cyanine-based probes that can detect amyloid and tau proteins in the AD brain tissues distinctly (Yang et al. 2019b).

5.4.2.12 Thiophene Derivatives for Aβ Aggregates

Thiophenes are the best basic components to use as the foundation for the synthesis of conjugated π systems because of their enormous and alluring structural variation, which enables tuning of the electronic properties over a broad range. Their exceptional chemical and physical properties and high polarizing power of sulphur atoms in thiophene rings stabilize the conjugated chain and result in outstanding charge-transport properties. These properties make thiophene molecules useful for designing new fluorescent probes for various applications (Cheng et al. 2012).

Swager and his team created numerous D-A-based new thiophene-containing NIRF probes, including NIAD-4, NIAD-11, and NIAD-16, which are effective in amyloid imaging, based on the aforementioned features (Fig. 5.16). NIAD-4 shows better results like 400-fold enhancement in fluorescence properties and high QY, and much better binding affinity for aggregates of Aβ. NIAD-11 exhibits increased fluorescence characters on binding with the Aβ aggregates. NIAD-16 shows a red shift upon staining vascular amyloid plaques (Nesterov et al. 2005; Raymond et al. 2008).

Recently, Fu and his group developed a thiophene-π extension core-based fluorescent probe QM-FN-SO_3 to extend its emission wavelength towards NIR region and to increase its BBB permeability. This probe makes a breakthrough in in vivo findings of Aβ plaques with greater affinity and acts as an excellent approach to some of the commercially available probes such as ThT or ThS (Fu et al. 2019).

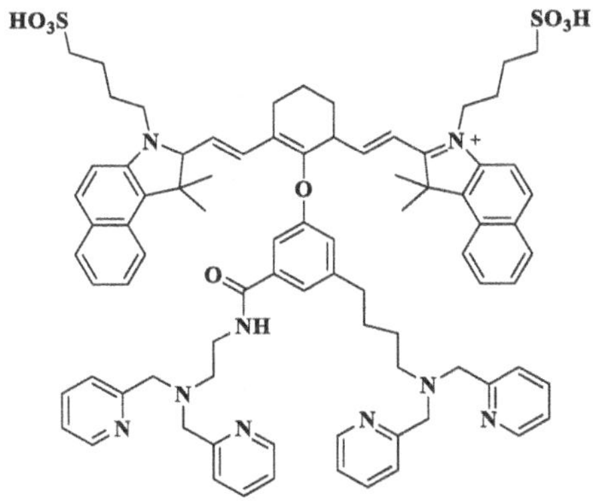

CyDPA2 (83)

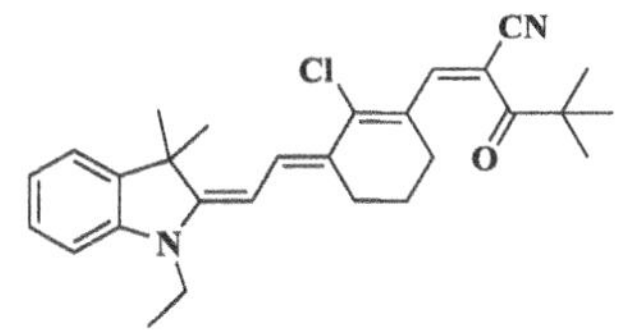

ZT1 (84)

Fig. 5.15 Chemical structures of cyanine derivatives for Aβ peptides and tau proteins

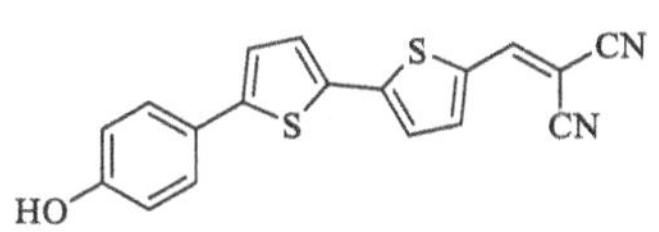

NIAD-4 (85)

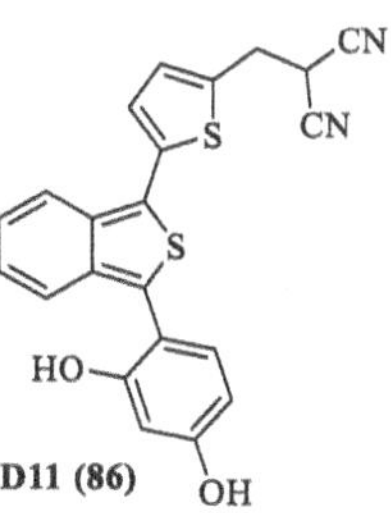

NIAD11 (86)

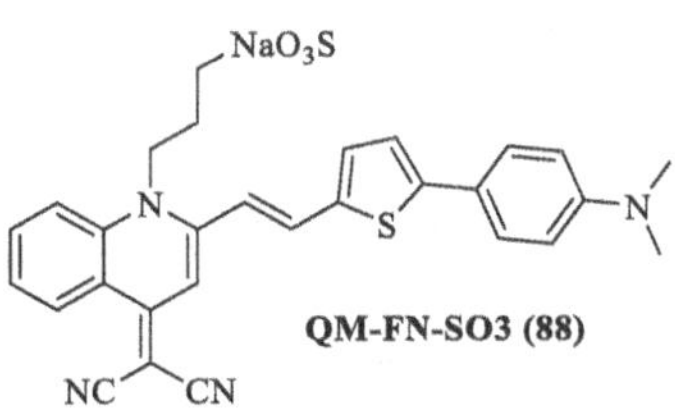

QM-FN-SO3 (88)

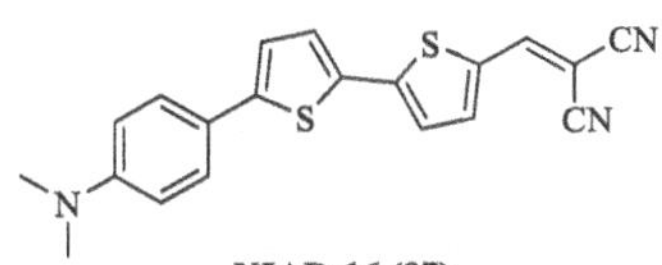

NIAD-16 (87)

Fig. 5.16 Chemical structures of thiophene derivatives for *Aβ* proteins

5.4.2.13 Oligothiophene-Containing Probes for Aβ and Tau Fibrils

Nilsson et al. reported luminescent-conjugated oligothiophene derivatives (LCOs), out of which the pentamer formyl thiophene acetic acid (pFTTA) shows better fluorescence properties. pFTTA is able to spectrally differentiate Aβ deposits and NFTs (Simon et al. 2014). Another probe KTAA is analogous to the above probe that contains acetyl group in place of acetic acid in the previous probe. This probe shows greater fluorescence change in emission wavelength on binding with neuritic plaques and tau fibrils (Åslund et al. 2009).

Klingstedt and his team described the degree of spectral resolution as a function of a number of thiophene rings from tetrathiophene (q-FTAA), pentathiophene (p-FTTA), hexathiophene (hx-FTTA), and (h-FTAA) heptathiophene, as well as

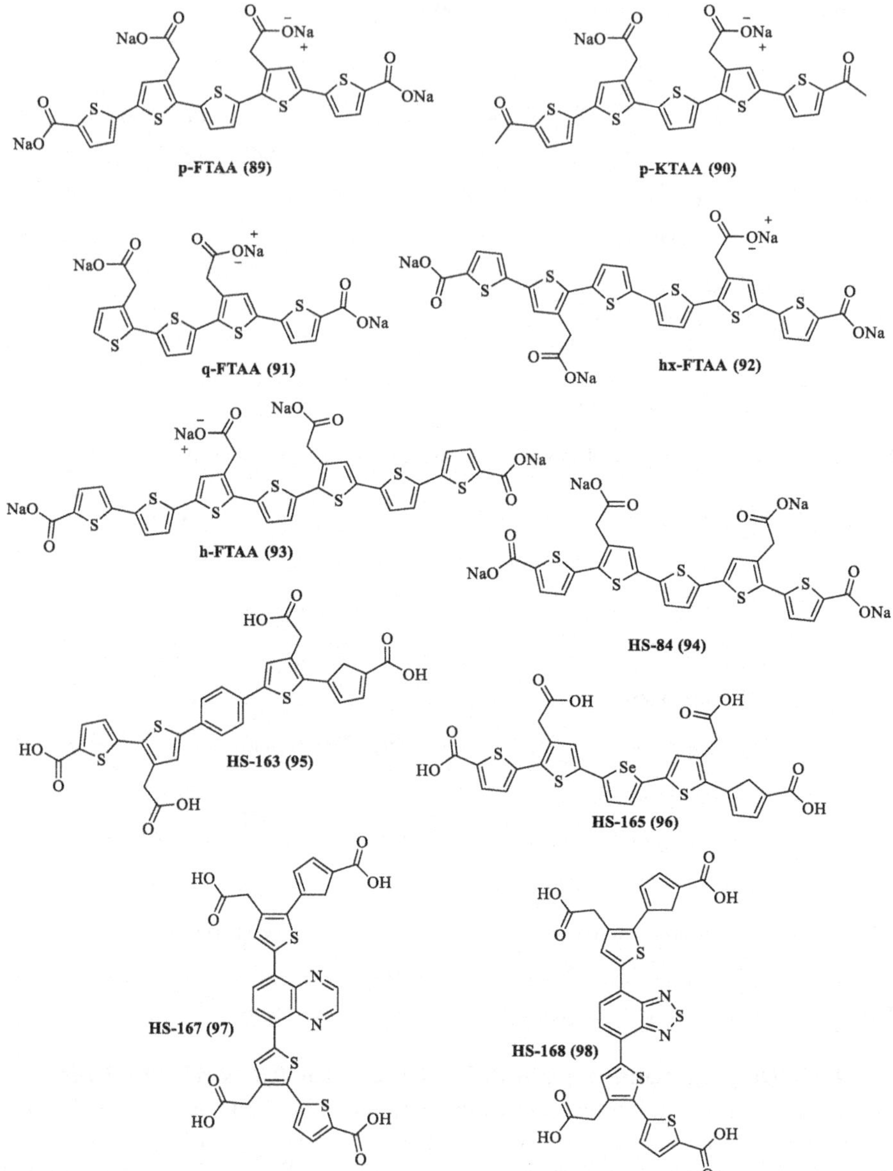

Fig. 5.17 Chemical structures of oligothiophene derivatives for Aβ protein and tau fibrils

the significance of terminal carboxylic acid in spectral differentiation of amylopathies. All these probes have the ability to image neurotic plaques and aggregates of tau protein (Fig. 5.17) (Klingstedt et al. 2011; Verwilst et al. 2018).

Shirani and colleagues reported a series of anionic pentameric oligothiophene fluorescent probes HS-163, HS-165, HS-167, and HS-169. The photophysical studies showed that these derivatives on binding with AD hall markers exhibit two peaks in absorption wavelengths. These probes emerged as the potential ones to stain vascular amyloid deposits and NFTs. In conclusion, the different optical properties of these probes for amyloid species and NFTs encourage their use in fluorescence imaging of the AD biomarkers (Shirani et al. 2015).

5.5 Natural Products as Theranostics in AD

The last 10 years have seen the development of a new strategy for the management of AD, i.e. the theranostic strategy: a single compound having the capability of diagnosis and therapeutic treatment. A number of fluorescent probes for the precise labelling, identification, and imaging of Aβ plaques both in vitro and in vivo are in use. At first, these probes are designed for imaging various biomarkers of AD. Later, these are evolved as theranostics, which might be useful for both imaging and inhibition of amyloid and tau protein aggregation (Cantore 2019).

Nowadays, several natural compound derivatives have gained attention in the management of AD as theranostic agents such as curcumin derivatives. In the year 2012, Li and Wong reported cyanine fluorescent probes which have the potential to bind with the Aβ_{1-40} peptides, plaques, and aggregate forms, resulting in an increase in fluorescence properties, which will make them a direct imaging agent in AD. These probes are non-toxic to the neural cell lines; to be more important, one of these probes named SLOH shows strong inhibition activity on the Aβ fibrillogenesis. The neuroprotective activity was also demonstrated, which shows that it could pass BBB and binds with the Aβ plaques, which make it count as a novel fluorescent probe that acts as a theranostic agent in AD (Yang et al. 2012).

Bolognesi and his team developed a styryl quinoline fluorophore (G8) as a diagnostic and therapeutic small molecule in the year 2013. This molecule offers several advantages like low molecular weight in comparison to the previously reported molecules, a small molecule which can be easily modified further in order to increase its fluorescence response and amyloid staining abilities. In addition to these, it offers anti-fibrillar activity which was proved by in vitro and in vivo studies, with low toxicity (Staderini et al. 2013). In 2014, Moore, Ran, and colleagues developed a new approach to inhibit Aβ aggregation by neutralizing the positive charges present on the amino acids of the Aβs. They reported that PiB-C, a conjugation product of PiB and crown ether, will prominently decrease the aggregation in vitro. It also proved as a bioimaging agent, which could label Aβ plaques in AD mice. Introduction of the crown ether group will be helpful not only for the neutralization purpose but also for increasing the hydrophobicity of the Aβ surface. PiB-C could be considered as a theranostic agent, because it has the potential to act as a diagnostic as well as a therapeutic agent (Tian et al. 2014).

Similarly, another theranostic molecule, the curcumin-based CRANAD-28, was also developed by the same group in 2014. At first, they developed the curcumin derivative as a diagnostic agent that could detect the soluble and insoluble forms of the Aβ. To overcome the low QY drawbacks of the curcumin analogues, they planned that replacing the phenyl rings with the pyrazoles will be helpful for the increment of fluorescence properties. By considering the above facts, they designed and synthesized the CRANAD-28. This molecule has good tissue penetration capability, because of its higher emission wavelengths. CRANAD-28 not only has imaging potential but also inhibits the cross linking of $A\beta_{(42)}$ induced by metal ions and natural conditions (Zhang et al. 2014). These features make it a promising theranostic agent.

Wang and his team, in 2015, reported a dual functional fluorescent molecule (TBT) that can bind and disassemble metal-linked $A\beta$ aggregates and also image the amyloid species in the brain of the AD mice. The binding behaviour of TBT as a metal chelator for Zn and Cu was demonstrated by the fluorescence titration methods. The binding affinity for aggregate forms of $A\beta$ was also confirmed by using a ThT fluorescence competition assay. Considering the amyloid specificity, neuroprotectivity, and efficient BBB permeability, it offers a great advantage for designing the probe for the chelation therapy and significant potential as a theranostic tool in AD (Yang et al. 2016). In 2017, Dao and his colleagues synthesized a series of new theranostic probes derived from phenothiazine moiety with the donor-acceptor architecture for the detection of $A\beta$ plaques. Most of these molecules exhibit their emission wavelength in the range of NIR. In in vitro studies and binding experiments, they showed good affinity for the amyloid aggregates. Compounds 4a1 and 5a1 show high affinity for Aβ species and also good fluorescence imaging of the $A\beta$ plaques in AD brain models. These molecules offer a great advantage of the promising profile of amyloid species aggregation together with good stability in mouse serum and a low toxic profile to human neuronal cells (Dao et al. 2017).

Ma et al. designed and developed a new benzothiazole agent (BPB) by combining the two beta-sheet targeting groups into one molecule to act by both detecting and inhibiting the aggregates of amyloid species. BPB can selectively measure the soluble $A\beta$ oligomer level and can differentiate the aggregate forms. Animal study results also showed that it can cross BBB and can stain the $A\beta$ plaques in the brains of the AD. Aside from its diagnostic value, BPB prevents the synthesis of ROS, safeguards neurons from neurotoxicity, and breaks down aggregates. $A\beta$ aggregates are formed by metal ions in the brain homogenates of APP transgenic mice. So, BPB could be a potential diagnostic as well as therapeutic agent for AD (Ma et al. 2020). Yang and co-workers developed a multifunctional fluorescent probe Cur-N-BF2 with aggregation-induced emission AIE characteristics for the detection of the $A\beta$ fibrils and plaques and inhibition of $A\beta$ fibrillation. Investigation reports suggested that there is a significant increase in fluorescence property and also selectivity of probe towards the $A\beta$ plaques and disaggregation capability towards $A\beta$ fibrils. The good neuronal cell protection ability also makes it a promising theranostic agent in the study of AD (Yang et al. 2019c).

Wong and his team have reported a number of uniquely charged NIR theranostic probes based on A-PI-D, including DMA-SLOH, DBA-SLOH, and DPA-SLM. The maximum binding affinity and selectivity, improved stability in bovine serum, good lipophilic character, low toxicity, and BBB permeability are all displayed by

DBA-SLOH with a moderate alkyl chain length. This substance inhibits Aβ aggregation and performs in vivo NIR imaging of Aβ species. This approach raises a fresh thought for creating theranostic probes with improved BBB permeability. These findings support the DBA-more SLOH's optimistic function in the diagnosis of AD in animal models and medicines for AD treatment (Li et al. 2016).

Similarly, in the year 2021, DBAN-SLM, DBAN-SLOH, and DBAN-OSLM (Fig. 5.18) turned-on fluorescent probes containing a 2-naphthylamine structure with dibutyl groups were introduced into the probe structure to increase its lipophilic character and BBB permeability. These probes on binding with the different Aβ species exhibit an increase in the emission wavelength. All these probes are non-toxic and importantly show an effective inhibition on Aβ self-aggregation and better neuronal cell protection activity against the Aβ-induced toxicity. DBAN-SLM is a desirable probe with real-time imaging of the Aβ oligomer in Tg-AD mice model (Wang et al. 2021).

5.5.1 Nanotechnology-Based Approach

The amyloid cascade hypothesis is most often utilized to discuss the pathophysiological features of Alzheimer's disease (AD). The mechanism states that Aβ aggregation is the primary cause of chronic neurotoxicity in AD. However, medications developed to target this component have not shown much effect in the treatment of AD. Despite a critical need for treatment and substantial interdisciplinary team investigation in the pharmaceutical and biological fields, no viable curative drug has yet been developed, and pharmacotherapy for AD remains symptomatic. Until now, most anti-AD drugs used in clinical trials and clinics have focused on the brain's central nervous system (CNS). Unfortunately, these medications did not cross or were not present in pharmacologically significant concentrations in the CNS due to their physicochemical nature and the blood-brain barrier (BBB), a protective barrier. The BBB represents the most crucial consideration in the creation of new CNS medications. Pharmaceuticals are often made up of tiny compounds that do not penetrate the BBB (Potschka 2010). Various attempts have been undertaken over the last two decades to address this critical issue by developing various strategies that enhance medication passage through the BBB.

The BBB is now being penetrated by a variety of strategies, including nanotechnology, which has shown positive results in delivering medications to the desired CNS area. Techniques based on nanotechnology have made it possible to distribute imaging agents in addition to drugs. Smartly constructed nanoparticles (NPs) can be used in AD diagnosis in addition to medicine. Nanotechnology-based tactics and approaches have gotten a lot of attention because they have a good chance of overcoming the constraints of BBB passage. Nanotechnology-based theranostics combine highly sensitive molecular identification with precise medicine targeting, which can enhance AD treatment by delaying memory loss and extending life. Both organic and non-organic nanoparticles were successfully investigated as theranostics against AD. Quantum dots (QD), dendrimers, fullerenes, and carbon nanotubes are

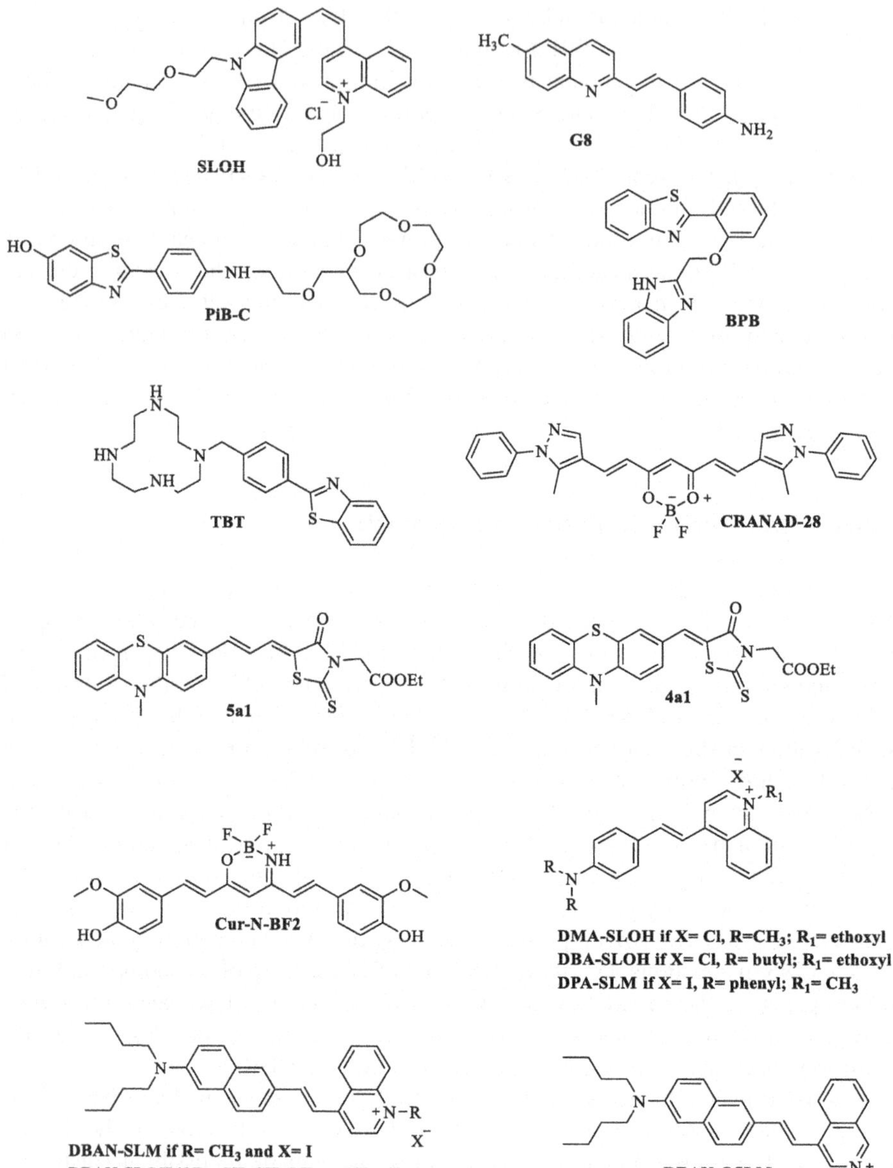

Fig. 5.18 Chemical structures of various fluorescent probes as theranostics in AD

examples of organic and inorganic nanomaterials that can be utilized to make theranostic nanocarriers. Other examples include polymeric nanoparticles, nanocapsules, nanospheres, micelles, emulsions, and liposomes (CNT) (Wong et al. 2012). A blueprint for theranostic nanocarriers that can cross the BBB and have both therapeutic and diagnostic capabilities has been developed over the past 10 years as a result of multidisciplinary nanotechnology research (Gauthier and Molinuevo 2013).

5.5.1.1 Nanotechnology in Alzheimer's Disease Diagnosis

The potential for an early diagnosis of AD is the current focus in the production of NPs for neuroimaging and molecular recognition of biomarkers. Very effective signal transduction is possible with nanotechnology, which could help with AD early detection. The physiological, chemical, and/or biological characteristics of specifically made nanoparticles are the main foundation for this potential use of nanotechnology in imaging and detection (Nazem and Mansoori 2011).

5.5.1.1.1 In Vitro Diagnostics

Nanoparticle Conjugates

Protein biomarkers can be found using DNA-NP complexes even in very low molar concentrations (even as low as 10–18 moles per liter). This detection method makes use of gold NPs linked to antibodies for certain proteins. This approach has yielded promising results in in vitro AD diagnosis (Georganopoulou et al. 2005).

Localized Surface Plasmon Resonance-Based Nanosensor (LSPR)

Surface plasmon resonance (SPR), a resonating and cumulative oscillation of outer electrons in solid material, occurs when exposed to incoming light. In all directions, photon emission and absorption have a similar frequency. Certain inorganic NPs have a characteristic SPR phenomenon that makes them useful for imaging and theranostic uses in a range of disorders, including AD. Several metallic NP-based ultra-sensitive and economical methods for identifying AD biomarkers like Aβ-derived diffusible ligands (ADDL) have recently been developed and are based on the SPR effect (Svenson 2013).

Scanning Tunnelling Microscopy (STM) and Two-Photon Rayleigh Spectroscopy

Specific antibody fragment immobilization on GNPs is a component of STM. When STM passes GNPs, the frequency of the generated pulse, which is shown on the device's scanner tip, determines the tunnelling pattern that results from a change in substrate or antigen concentration. With this method, Aβ can be found at concentrations as low as 10 fg/mL (Kang et al. 2009). Recently, the changed signal from GNP scattering was seen using two-photon Rayleigh spectroscopy to accurately evaluate tau protein. This method is rapid, focused, and extremely sensitive, able to detect concentrations as low as 1 pg/mL (Neely et al. 2009).

5.5.1.1.2 In Vivo Nanodiagnostics for Alzheimer's Disease

Magnetic Resonance Imaging (MRI)

In recent years, a great deal of research has been done on magnetic or iron oxide nanoparticles (IONPs) as MRI contrast agents (Viola et al. 2015). An antibody-superparamagnetic iron oxide nanoparticle (SPION) combination has been created by Sillerud et al. for in vivo MR imaging of Aβ plaques in transgenic animal models of AD. Their results showed that newly developed SPIONs considerably enhanced the MRI contrast in mouse brains with Aβ plaque (Sillerud et al. 2013).

Optical Imaging

In vivo imaging of molecular biomarkers in various disorders, including AD, is currently being studied using optical imaging (OI) using a specific near-infrared (NIR) fluorescent dye. The most common prerequisites for an AD diagnostic test are the imaging probe's ability to cross the BBB and its specificity to the target of AD-related biomarkers (Nesterov et al. 2005).

5.5.1.2 Nanotechnology in Alzheimer's Disease Therapy

With improvements in AD cell biology, pathology, and neurophysiology, as well as the upcoming smart NP design capable of crossing the blood-brain barrier and focusing on the desired site into the CNS, nanomaterial approaches focused on neuroprotective and neurodegenerative techniques in AD can greatly benefit. Over the past 10 years, a number of NPs have been investigated to increase the bioavailability and therapeutic effectiveness of AD medicines.

5.5.1.2.1 Nanogels, Fullerene, and Nanoceria

Because of their improved drug encapsulation, stability, and capacity to be targeted depending on many factors like pH, temperature, and ionic strength, nanogels are appealing drug delivery techniques. Ikeda et al. created novel monomeric peptide-integrated pullulan (CHP) nanogels containing cholesterol to explore the possibility for therapeutic targeting. Colloidal NPs effectively decreased the aggregation of Aβ, reducing its toxicity to PC12 cells (Ikeda et al. 2006).

Another important NP from the carbon family known as fullerene (C60) has been investigated for use in AD therapy as a neuroprotective antioxidant and free source of antioxidant (Dugan et al. 2001).

The antioxidant activity of cerium oxide (CeO2) nanoparticles, commonly known as nanoceria, has been demonstrated to have a neuroprotective effect on in vitro models of AD. The generated oxygen vacancies in cerium's two redox states, Ce2+ and Ce4+, make them powerful antioxidants (D'Angelo et al. 2009).

5.5.1.2.2 Dendrimers

Nanoscale monodispersed polymers called dendrimers have a set number of spatially ordered peripheral functional groups, a regular and highly branched 3D architecture (Stiriba et al. 2002). Usually, a series of reactions are carried out in iterative succession to prepare them. Experts in pharmacology and biomedicine have recently become interested in these NPs. It makes sense to target the Aβ peptide by functionalizing the nearby functional groups with the desired ligands. The potential of dendrimer NPs to deconstruct existing amyloid aggregates and their inhibitory influence on Aβ aggregation were both demonstrated by Chafekar et al. after they created KLVFF-functionalized dendrimer NPs (Chafekar et al. 2007).

5.5.1.2.3 Gold Nanoparticles

Gold nanoparticles (GNPs) have undergone extensive study for their theranostic and drug delivery capabilities in a range of medicinal applications, including AD treatment. By producing local thermal energy, GNPs dissolved Aβ-peptide clumps and prevented additional Aβ-peptide aggregation when utilized in mild microwave fields by Kogan et al. In order to destroy an Aβ fibril bond (10–20 J binding energy per non-covalent link) every microsecond without damaging covalent bonds, which are two times stronger, GNPs release heat energy at a rate of 10–14 J/s (Kogan et al. 2006).

5.5.1.2.4 Diamondoid and Its Derivatives

Several antibacterial and antiviral drugs that are either on the market or in various stages of development are derived from diamondoids and their derivatives (Mansoori 2008). Memantine, a commercially available diamondoid chemical with FDA approval, is used to treat moderate-to-severe AD. Memantine is a neuroprotective medication that prevents excitotoxicity, overexcitation of glutamate's membrane receptors, or excitatory neurotransmitter's excitotoxicity, which kills neurons. Excitotoxicity kills neurons by overactivating glutamate receptors of the N-methyl-d-aspartate (NMDA) type. Memantine has little impact on normal activity but lowers excessive NMDA receptor activity.

5.5.1.2.5 Quantum Dots

Quantum dots (QDs) are yet another sort of brilliant dye produced by nanotechnology. The remarkable fluorescence properties of these nanoscale semiconductor crystals include low photobleaching, good stability, a broad absorption spectrum, and a very restricted size-dependent adjustable emission spectrum. QDs are now able to recognize and track a variety of physiological and pathological molecular events across time in comparison to conventional fluorescent dyes. Because there are various biomarkers associated with the pathology of AD, and because their presence in combination with one another may help rule out other possible diagnoses, this simultaneous multiple labelling characteristic is especially important in the diagnosis of AD (Jaiswal et al. 2003).

However, the in vivo uses of the semiconductor materials currently used in QDs, such as cadmium selenide and cadmium sulphate, remain unknown. Of course, several studies have found that encapsulating QDs in polymers (such as phospholipids) or coating them with polyethylene glycol (PEG) reduces their toxicity (Dubertret et al. 2002). Tokurako and his group created a nanoprobe for amyloid aggregation and oligomerization utilizing PEG-coated QDs as Aβ_{42} labels. They investigated A$\beta_{(42)}$ oligomerization behaviour in solution and on live cells, as well as microglia ingestion of A$\beta_{(42)}$ monomers and oligomers. However, in order to apply this diagnostic technology to monitor A$\beta_{(42)}$ biochemical behaviour in vivo, extra attention should be paid, in addition to QD safety considerations, to the successful passage of these QD-Aβ nanoprobes via the BBB (Tokuraku et al. 2009).

5.6 Natural compounds in Clinical Trials for AD

Despite major studies on this disease, just a few medicines capable of delaying disease development are currently available. Till today, there has been no report of one such molecular probe that has both diagnostic and therapeutic applications carried forward in clinical trials to counter-attack AD. Several substances with pharmacological effects derived from plants, animals, and microbes have been shown in recent years to have positive effects for the treatment of AD, targeting various pathogenic processes. Natural compounds were classified into two categories: bioactive compounds and natural extracts. A bioactive chemical is a medicinal molecule in this context, whereas a natural extract is a mixture of many molecules. Several bioactive compounds include vitamins, bryostatin, melatonin, resveratrol, curcumin, etc., whereas natural extracts include *Ginkgo biloba*, saffron, papaya, etc. that are in clinical trial stage. As a result, a variety of natural substances may be important in the prevention of AD and have been shown to be effective in numerous preclinical and clinical trials (Andrade et al. 2019).

5.7 Conclusion

Early diagnosis of the AD is vital in understanding the pathophysiology as well as to delay the progress of the diseased condition. In this chapter, we mainly focused on both diagnosis and therapeutic approaches to the AD. Here in, we highlighted the importance of fluorescent probes that are derived from various naturally inspired molecules such as curcumin and some others in diagnostic as well as therapeutic applications. Thereafter, we mentioned the various nanotechnology approaches towards AD diagnosis and treatment. Till today, there has been no such evidence that a single fluorescent probe that has theranostic potentiality is carried forward to the clinical study stage in drug development. This chapter provides knowledge about the importance of fluorescent probes in understanding the pathophysiological conditions as well as the disease state.

References

Agdeppa ED, Kepe V, Liu J, Small GW, Huang SC, Petrič A et al (2003) 2-Dialkylamino-6-acylmalononitrile substituted naphthalenes (DDNP analogs): novel diagnostic and therapeutic tools in Alzheimer's disease. Mol Imaging Biol 5(6):404–417

Akiyama H, Barger S, Barnum S, Bradt B, Bauer J, Cole GM et al (2000) Inflammation and Alzheimer's disease. Neurobiol Aging 21(3):383–421

Andrade S, Ramalho MJ, Loureiro JA, Pereira MDC (2019) Natural compounds for Alzheimer's disease therapy: a systematic review of preclinical and clinical studies. Int J Mol Sci 20(9):2313

Arora H, Ramesh M, Rajasekhar K, Govindaraju T (2020) Molecular tools to detect alloforms of Aβ and tau: implications for multiplexing and multimodal diagnosis of Alzheimer's disease. Bull Chem Soc Jpn 93(4):507–546

Åslund A, Sigurdson CJ, Klingstedt T, Grathwohl S, Bolmont T, Dickstein DL et al (2009) Novel pentameric thiophene derivatives for in vitro and in vivo optical imaging of a plethora of protein aggregates in cerebral amyloidoses. ACS Chem Biol 4(8):673–684

Barker WW, Luis CA, Kashuba A, Luis M, Harwood DG, Loewenstein D et al (2002) Relative frequencies of Alzheimer disease, Lewy body, vascular and frontotemporal dementia, and hippocampal sclerosis in the State of Florida Brain Bank. Alzheimer Dis Assoc Disord 16(4): 203–212

Beason-Held LL, Goh JO, An Y, Kraut MA, O'Brien RJ, Ferrucci L, Resnick SM (2013) Changes in brain function occur years before the onset of cognitive impairment. J Neurosci 33(46): 18008–18014

Biernat J, Gustke N, Drewes G, Mandelkow E (1993) Phosphorylation of Ser262 strongly reduces binding of tau to microtubules: distinction between PHF-like immunoreactivity and microtubule binding. Neuron 11(1):153–163

Boländer A, Kieser D, Voss C, Bauer S, Schön C, Burgold S et al (2012) Bis (arylvinyl) pyrazines, -pyrimidines, and -pyridazines as imaging agents for tau fibrils and β-amyloid plaques in Alzheimer's disease models. J Med Chem 55(21):9170–9180

Brandt R, Lee G (1993) Functional organization of microtubule-associated protein tau. Identification of regions which affect microtubule growth, nucleation, and bundle formation in vitro. J Biol Chem 268(5):3414–3419

Breijyeh Z, Karaman R (2020) Comprehensive review on Alzheimer's disease: causes and treatment. Molecules 25(24):5789

Bulic B, Pickhardt M, Mandelkow EM, Mandelkow E (2010) Tau protein and tau aggregation inhibitors. Neuropharmacology 59(4–5):276–289

Cantore M (2019) New perspective in Alzheimer's disease-theranostic strategy. Biomed J Sci Tech Res 15(1):11108–11111

Cao K, Farahi M, Dakanali M, Chang WM, Sigurdson CJ, Theodorakis EA, Yang J (2012) Aminonaphthalene 2-cyanoacrylate (ANCA) probes fluorescently discriminate between amyloid-β and prion plaques in brain. J Am Chem Soc 134(42):17338–17341

Cao KJ, Elbel KM, Cifelli JL, Cirera J, Sigurdson CJ, Paesani F et al (2018) Solvation-guided design of fluorescent probes for discrimination of amyloids. Sci Rep 8(1):6950

Chafekar SM, Malda H, Merkx M, Meijer EW, Viertl D, Lashuel HA et al (2007) Branched KLVFF tetramers strongly potentiate inhibition of β-amyloid aggregation. Chembiochem 8(15): 1857–1864

Cheng X, Li S, Jia H, Zhong A, Zhong C, Feng J et al (2012) Fluorescent and colorimetric probes for mercury (II): tunable structures of electron donor and π-conjugated bridge. Chem A Eur J 18(6):1691–1699

Chintamaneni M, Bhaskar M (2012) Biomarkers in Alzheimer's disease: a review. ISRN Pharmacol 2012:984786

Cohen TJ, Guo JL, Hurtado DE, Kwong LK, Mills IP, Trojanowski JQ, Lee VM (2011) The acetylation of tau inhibits its function and promotes pathological tau aggregation. Nat Commun 2(1):252

Coimbra A, Williams DS, Hostetler ED (2006) The role of MRI and PET/SPECT in Alzheimer's disease. Curr Top Med Chem 6(6):629–647

Collier TL, Yokell DL, Livni E, Rice PA, Celen S, Serdons K et al (2017) cGMP production of the radiopharmaceutical [18F] MK-6240 for PET imaging of human neurofibrillary tangles. J Label Compd Radiopharm 60(5):263–269

Cui M, Ono M, Watanabe H, Kimura H, Liu B, Saji H (2014) Smart near-infrared fluorescence probes with donor–acceptor structure for in vivo detection of β-amyloid deposits. J Am Chem Soc 136(9):3388–3394

Cunha LP, Almeida ALM, Costa-Cunha LVF, Costa CF, Monteiro ML (2016) The role of optical coherence tomography in Alzheimer's disease. Int J Retina Vitreous 2(1):1–11

D'Angelo B, Santucci S, Benedetti E, Di Loreto S, Phani RA, Falone S et al (2009) Cerium oxide nanoparticles trigger neuronal survival in a human Alzheimer disease model by modulating BDNF pathway. Curr Nanosci 5(2):167–176

Dao P, Ye F, Liu Y, Du ZY, Zhang K, Dong CZ et al (2017) Development of phenothiazine-based theranostic compounds that act both as inhibitors of β-amyloid aggregation and as imaging probes for amyloid plaques in Alzheimer's disease. ACS Chem Neurosci 8(4):798–806

De Jonghe C, Esselens C, Kumar-Singh S, Craessaerts K, Serneels S, Checler F et al (2001) Pathogenic APP mutations near the γ-secretase cleavage site differentially affect Aβ secretion and APP C-terminal fragment stability. Hum Mol Genet 10(16):1665–1671

DeTure MA, Dickson DW (2019) The neuropathological diagnosis of Alzheimer's disease. Mol Neurodegener 14(1):32

Dubertret B, Skourides P, Norris DJ, Noireaux V, Brivanlou AH, Libchaber A (2002) In vivo imaging of quantum dots encapsulated in phospholipid micelles. Science 298(5599):1759–1762

Dugan LL, Lovett EG, Quick KL, Lotharius J, Lin TT, O'Malley KL (2001) Fullerene-based antioxidants and neurodegenerative disorders. Parkinsonism Relat Disord 7(3):243–246

Eckroat TJ, Mayhoub AS, Garneau-Tsodikova S (2013) Amyloid-β probes: review of structure–activity and brain-kinetics relationships. Beilstein J Org Chem 9(1):1012–1044

Fitzpatrick AW, Falcon B, He S, Murzin AG, Murshudov G, Garringer HJ et al (2017) Cryo-EM structures of tau filaments from Alzheimer's disease. Nature 547(7662):185–190

Fu H, Cui M, Tu P, Pan Z, Liu B (2014) Evaluation of molecules based on the electron donor–acceptor architecture as near-infrared β-amyloidal-targeting probes. Chem Commun 50(80): 11875–11878

Fu H, Cui M, Zhao L, Tu P, Zhou K, Dai J, Liu B (2015) Highly sensitive near-infrared fluorophores for in vivo detection of amyloid-β plaques in Alzheimer's disease. J Med Chem 58(17):6972–6983

Fu H, Tu P, Zhao L, Dai J, Liu B, Cui M (2016) Amyloid-β deposits target efficient near-infrared fluorescent probes: synthesis, in vitro evaluation, and in vivo imaging. Anal Chem 88(3): 1944–1950

Fu W, Yan C, Guo Z, Zhang J, Zhang H, Tian H, Zhu WH (2019) Rational design of near-infrared aggregation-induced-emission-active probes: in situ mapping of amyloid-β plaques with ultrasensitivity and high-fidelity. J Am Chem Soc 141(7):3171–3177

Gan C, Zhou L, Zhao Z, Wang H (2013) Benzothiazole Schiff-bases as potential imaging agents for β-amyloid plaques in Alzheimer's disease. Med Chem Res 22:4069–4074

Gauthier S, Molinuevo JL (2013) Benefits of combined cholinesterase inhibitor and memantine treatment in moderate–severe Alzheimer's disease. Alzheimers Dement 9(3):326–331

Georganopoulou DG, Chang L, Nam JM, Thaxton CS, Mufson EJ, Klein WL, Mirkin CA (2005) Nanoparticle-based detection in cerebral spinal fluid of a soluble pathogenic biomarker for Alzheimer's disease. Proc Natl Acad Sci 102(7):2273–2276

Goedert M, Jakes R (1990) Expression of separate isoforms of human tau protein: correlation with the tau pattern in brain and effects on tubulin polymerization. EMBO J 9(13):4225–4230

Graff BJ, Harrison SL, Payne SJ, El-Bouri WK (2022) Regional cerebral blood flow changes in healthy ageing and Alzheimer's disease: a narrative review. Cerebrovasc Dis:1–10

Gu L, Guo Z (2013) Alzheimer's Aβ42 and Aβ40 peptides form interlaced amyloid fibrils. J Neurochem 126(3):305–311

Gu J, Anumala UR, Monte FL, Kramer T, von Haußen RH, Hölzer J et al (2012) 2-Styrylindolium based fluorescent probes visualize neurofibrillary tangles in Alzheimer's disease. Bioorg Med Chem Lett 22(24):7667–7671

Hammoud DA, Hoffman JM, Pomper MG (2007) Molecular neuroimaging: from conventional to emerging techniques. Radiology 245(1):21–42

Harada R, Okamura N, Furumoto S, Furukawa K, Ishiki A, Tomita N et al (2015) [18 F] THK-5117 PET for assessing neurofibrillary pathology in Alzheimer's disease. Eur J Nucl Med Mol Imaging 42:1052–1061

Hojjati SH, Ebrahimzadeh A, Babajani-Feremi A (2019) Identification of the early stage of Alzheimer's disease using structural MRI and resting-state fMRI. Front Neurol 10:904

Ikeda K, Okada T, Sawada SI, Akiyoshi K, Matsuzaki K (2006) Inhibition of the formation of amyloid β-protein fibrils using biocompatible nanogels as artificial chaperones. FEBS Lett 580(28–29):6587–6595

Iqbal K, Liu F, Gong CX, Grundke-Iqbal I (2010) Tau in Alzheimer disease and related tauopathies. Curr Alzheimer Res 7(8):656–664

Jaiswal JK, Mattoussi H, Mauro JM, Simon SM (2003) Long-term multiple color imaging of live cells using quantum dot bioconjugates. Nat Biotechnol 21(1):47–51

Jie CV, Treyer V, Schibli R, Mu L (2021) Tauvid™: the first FDA-approved PET tracer for imaging tau pathology in Alzheimer's disease. Pharmaceuticals 14(2):110

Johnson KA, Fox NC, Sperling RA, Klunk WE (2012) Brain imaging in Alzheimer disease. Cold Spring Harb Perspect Med 2(4):a006213

Johnson KA, Minoshima S, Bohnen NI, Donohoe KJ, Foster NL, Herscovitch P et al (2013) Appropriate use criteria for amyloid PET: a report of the amyloid imaging task force, the Society of Nuclear Medicine and Molecular Imaging, and the Alzheimer's association. Alzheimers Dement 9(1):E1–E16

Kametani F, Hasegawa M (2018) Reconsideration of amyloid hypothesis and tau hypothesis in Alzheimer's disease. Front Neurosci 12:25

Kang DY, Lee JH, Oh BK, Choi JW (2009) Ultra-sensitive immunosensor for β-amyloid (1–42) using scanning tunneling microscopy-based electrical detection. Biosens Bioelectron 24(5): 1431–1436

Kim HY, Sengupta U, Shao P, Guerrero-Muñoz MJ, Kayed R, Bai M (2013) Alzheimer's disease imaging with a novel tau targeted near infrared ratiometric probe. Am J Nucl Med Mol Imaging 3(2):102

Kim H, Im YH, Ahn J, Yang J, Choi JY, Lee KH et al (2019) Synthesis and in vivo characterization of 18F-labeled difluoroboron-curcumin derivative for β-amyloid plaque imaging. Sci Rep 9(1): 6747

Klingstedt T, Åslund A, Simon RA, Johansson LB, Mason JJ, Nyström S et al (2011) Synthesis of a library of oligothiophenes and their utilization as fluorescent ligands for spectral assignment of protein aggregates. Org Biomol Chem 9(24):8356–8370

Klunk WE, Wang Y, Huang GF, Debnath ML, Holt DP, Mathis CA (2001) Uncharged thioflavin-T derivatives bind to amyloid-beta protein with high affinity and readily enter the brain. Life Sci 69(13):1471–1484

Kogan MJ, Bastus NG, Amigo R, Grillo-Bosch D, Araya E, Turiel A et al (2006) Nanoparticle-mediated local and remote manipulation of protein aggregation. Nano Lett 6(1):110–115

Lee JS, Kang NY, Kim YK, Samanta A, Feng S, Kim HK et al (2009) Synthesis of a BODIPY library and its application to the development of live cell glucagon imaging probe. J Am Chem Soc 131(29):10077–10082

Li Y, Xu D, Ho SL, Li HW, Yang R, Wong MS (2016) A theranostic agent for in vivo near-infrared imaging of β-amyloid species and inhibition of β-amyloid aggregation. Biomaterials 94:84–92

Licha K, Olbrich C (2005) Optical imaging in drug discovery and diagnostic applications. Adv Drug Deliv Rev 57(8):1087–1108

Lim S, Haque MM, Su D, Kim D, Lee JS, Chang YT, Kim YK (2017) Development of a BODIPY-based fluorescent probe for imaging pathological tau aggregates in live cells. Chem Commun 53(10):1607–1610

Liu K, Guo TL, Chojnacki J, Lee HG, Wang X, Siedlak SL et al (2012) Bivalent ligand containing curcumin and cholesterol as a fluorescence probe for Aβ plaques in Alzheimer's disease. ACS Chem Neurosci 3(2):141–146

Liu PP, Xie Y, Meng XY, Kang JS (2019) History and progress of hypotheses and clinical trials for Alzheimer's disease. Signal Transduct Target Ther 4(1):29

Lott IT, Dierssen M (2010) Cognitive deficits and associated neurological complications in individuals with Down's syndrome. Lancet Neurol 9(6):623–633

Lv G, Sun A, Wei P, Zhang N, Lan H, Yi T (2016) A spiropyran-based fluorescent probe for the specific detection of β-amyloid peptide oligomers in Alzheimer's disease. Chem Commun 52(57):8865–8868

Ma X, Wang Y, Hua J, Xu C, Yang T, Yuan J et al (2020) A β-sheet-targeted theranostic agent for diagnosing and preventing aggregation of pathogenic peptides in Alzheimer's disease. SCIENCE CHINA Chem 63:73–82

Mani V, Krishnakumar VG, Gupta S, Mori S, Gupta I (2017) Synthesis and characterization of styryl-BODIPY derivatives for monitoring in vitro tau aggregation. Sensors Actuators B Chem 244:673–683

Manoharan S, Guillemin GJ, Abiramasundari RS, Essa MM, Akbar M, Akbar MD (2016) The role of reactive oxygen species in the pathogenesis of Alzheimer's disease, Parkinson's disease, and Huntington's disease: a mini review. Oxidative Med Cell Longev 2016:1

Mansoori GA (2008) Diamondoid molecules. Adv Chem Phys 136:207–258

Maruyama M, Shimada H, Suhara T, Shinotoh H, Ji B, Maeda J et al (2013) Imaging of tau pathology in a tauopathy mouse model and in Alzheimer patients compared to normal controls. Neuron 79(6):1094–1108

Maya Y, Ono M, Watanabe H, Haratake M, Saji H, Nakayama M (2009) Novel radioiodinated aurones as probes for SPECT imaging of β-amyloid plaques in the brain. Bioconjug Chem 20(1):95–101

Mezzanotte L, van't Root M, Karatas H, Goun EA, Löwik CW (2017) In vivo molecular bioluminescence imaging: new tools and applications. Trends Biotechnol 35(7):640–652

Nazem A, Mansoori GA (2011) Nanotechnology for Alzheimer's disease detection and treatment. Insciences J 1(4):169–193

Neely A, Perry C, Varisli B, Singh AK, Arbneshi T, Senapati D et al (2009) Ultrasensitive and highly selective detection of Alzheimer's disease biomarker using two-photon Rayleigh scattering properties of gold nanoparticle. ACS Nano 3(9):2834–2840

Nesterov EE, Skoch J, Hyman BT, Klunk WE, Bacskai BJ, Swager TM (2005) In vivo optical imaging of amyloid aggregates in brain: design of fluorescent markers. Angew Chem 117(34): 5588–5592

Ntziachristos V (2006) Fluorescence molecular imaging. Annu Rev Biomed Eng 8:1–33

Ojida A, Sakamoto T, Inoue MA, Fujishima SH, Lippens G, Hamachi I (2009) Fluorescent BODIPY-based Zn (II) complex as a molecular probe for selective detection of neurofibrillary tangles in the brains of Alzheimer's disease patients. J Am Chem Soc 131(18):6543–6548

Ono M, Saji H (2015) Recent advances in molecular imaging probes for β-amyloid plaques. Med Chem Comm 6(3):391–402

Ono M, Ishikawa M, Kimura H, Hayashi S, Matsumura K, Watanabe H et al (2010) Development of dual functional SPECT/fluorescent probes for imaging cerebral β-amyloid plaques. Bioorg Med Chem Lett 20(13):3885–3888

Ono M, Watanabe H, Kimura H, Saji H (2012) BODIPY-based molecular probe for imaging of cerebral β-amyloid plaques. ACS Chem Neurosci 3(4):319–324

Park KS, Seo Y, Kim MK, Kim K, Kim YK, Choo H, Chong Y (2015) A curcumin-based molecular probe for near-infrared fluorescence imaging of tau fibrils in Alzheimer's disease. Org Biomol Chem 13(46):11194–11199

Park KS, Kim MK, Seo Y, Ha T, Yoo K, Hyeon SJ et al (2017) A difluoroboron β-diketonate probe shows "turn-on" near-infrared fluorescence specific for tau fibrils. ACS Chem Neurosci 8(10): 2124–2131

Peng C, Wang X, Li Y, Li HW, Wong MS (2019) Versatile fluorescent probes for near-infrared imaging of amyloid-β species in Alzheimer's disease mouse model. J Mater Chem B 7(12): 1986–1995

Perl DP (2010) Neuropathology of Alzheimer's disease. Mount Sinai J Med 77(1):32–42

Pineda-Pardo JA, Bruña R, Woolrich M, Marcos A, Nobre AC, Maestú F, Vidaurre D (2014) Guiding functional connectivity estimation by structural connectivity in MEG: an application to discrimination of conditions of mild cognitive impairment. NeuroImage 101:765–777

Potschka H (2010) Targeting the brain–surmounting or bypassing the blood–brain barrier. Drug Deliv:411–431

Pradeepkiran JA, Reddy PH (2019) Structure based design and molecular docking studies for phosphorylated tau inhibitors in Alzheimer's disease. Cell 8(3):260

Priyadarsini KI (2009) Photophysics, photochemistry and photobiology of curcumin: studies from organic solutions, bio-mimetics and living cells. J Photochem Photobiol C: Photochem Rev 10(2):81–95

Rajasekhar K, Narayanaswamy N, Murugan NA, Kuang G, Ågren H, Govindaraju T (2016) A high affinity red fluorescence and colorimetric probe for amyloid β aggregates. Sci Rep 6(1):1–10

Ran C, Xu X, Raymond SB, Ferrara BJ, Neal K, Bacskai BJ et al (2009) Design, synthesis, and testing of difluoroboron-derivatized curcumins as near-infrared probes for in vivo detection of amyloid-β deposits. J Am Chem Soc 131(42):15257–15261

Raymond SB, Skoch J, Hills ID, Nesterov EE, Swager TM, Bacskai BJ (2008) Smart optical probes for near-infrared fluorescence imaging of Alzheimer's disease pathology. Eur J Nucl Med Mol Imaging 35:93–98

Rodríguez-Martín T, Cuchillo-Ibáñez I, Noble W, Nyenya F, Anderton BH, Hanger DP (2013) Tau phosphorylation affects its axonal transport and degradation. Neurobiol Aging 34(9): 2146–2157

Ryu EK, Choe YS, Lee KH, Choi Y, Kim BT (2006) Curcumin and dehydrozingerone derivatives: synthesis, radiolabeling, and evaluation for β-amyloid plaque imaging. J Med Chem 49(20): 6111–6119

Schneider JA, Arvanitakis Z, Bang W, Bennett DA (2007) Mixed brain pathologies account for most dementia cases in community-dwelling older persons. Neurology 69(24):2197–2204

Selkoe DJ (1991) The molecular pathology of Alzheimer's disease. Neuron 6(4):487–498

Seo Y, Park KS, Ha T, Kim MK, Hwang YJ, Lee J et al (2016) A smart near-infrared fluorescence probe for selective detection of tau fibrils in Alzheimer's disease. ACS Chem Neurosci 7(11): 1474–1481

Sergeant N, Bretteville A, Hamdane M, Caillet-Boudin ML, Grognet P, Bombois S et al (2008) Biochemistry of tau in Alzheimer's disease and related neurological disorders. Expert Rev Proteomics 5(2):207–224

Sharma P, Tripathi A, Tripathi PN, Prajapati SK, Seth A, Tripathi MK et al (2019) Design and development of multitarget-directed N-Benzylpiperidine analogs as potential candidates for the treatment of Alzheimer's disease. Eur J Med Chem 167:510–524

Shimojo M, Higuchi M, Suhara T, Sahara N (2015) Imaging multimodalities for dissecting Alzheimer's disease: advanced technologies of positron emission tomography and fluorescence imaging. Front Neurosci 9:482

Shirani H, Linares M, Sigurdson CJ, Lindgren M, Norman P, Nilsson KPR (2015) A palette of fluorescent thiophene-based ligands for the identification of protein aggregates. Chem A Eur J 21(43):15133–15137

Shirani H, Appelqvist H, Bäck M, Klingstedt T, Cairns NJ, Nilsson KPR (2017) Synthesis of Thiophene-based optical ligands that selectively detect tau pathology in Alzheimer's disease. Chem A Eur J 23(67):17127–17135

Sillerud LO, Solberg NO, Chamberlain R, Orlando RA, Heidrich JE, Brown DC et al (2013) SPION-enhanced magnetic resonance imaging of Alzheimer's disease plaques in AβPP/PS-1 transgenic mouse brain. J Alzheimers Dis 34(2):349–365

Simon RA, Shirani H, Åslund KA, Bäck M, Haroutunian V, Gandy S, Nilsson KPR (2014) Pentameric thiophene-based ligands that spectrally discriminate amyloid-β and tau aggregates display distinct Solvatochromism and viscosity-induced spectral shifts. Chem A Eur J 20(39): 12537–12543

Singh YP, Tej GNVC, Pandey A, Priya K, Pandey P, Shankar G et al (2020) Design, synthesis and biological evaluation of novel naturally-inspired multifunctional molecules for the management of Alzheimer's disease. Eur J Med Chem 198:112257

Sozmen F, Kolemen S, Kumada HO, Ono M, Saji H, Akkaya EU (2014) Designing BODIPY-based probes for fluorescence imaging of β-amyloid plaques. RSC Adv 4(92):51032–51037

Staderini M, Aulić S, Bartolini M, Tran HNA, González-Ruiz V, Pérez DI et al (2013) A fluorescent styrylquinoline with combined therapeutic and diagnostic activities against Alzheimer's and prion diseases. ACS Med Chem Lett 4(2):225–229

Stiriba SE, Frey H, Haag R (2002) Dendritic polymers in biomedical applications: from potential to clinical use in diagnostics and therapy. Angew Chem Int Ed 41(8):1329–1334

Sundaram GSM, Dhavale DD, Prior JL, Yan P, Cirrito J, Rath NP et al (2016) Fluselenamyl: a novel benzoselenazole derivative for PET detection of amyloid plaques (Aβ) in Alzheimer's disease. Sci Rep 6(1):1–12

Svenson S (2013) Theranostics: are we there yet? Mol Pharm 10(3):848–856

Svoboda K, Yasuda R (2006) Principles of two-photon excitation microscopy and its applications to neuroscience. Neuron 50(6):823–839

Theer P, Hasan MT, Denk W (2003) Two-photon imaging to a depth of 1000 μm in living brains by use of a Ti:Al2O3 regenerative amplifier. Opt Lett 28(12):1022–1024

Tian Y, Zhang X, Li Y, Shoup TM, Teng X, Elmaleh DR et al (2014) Crown ethers attenuate aggregation of amyloid beta of Alzheimer's disease. Chem Commun 50(99):15792–15795

Tokuraku K, Marquardt M, Ikezu T (2009) Real-time imaging and quantification of amyloid-β peptide aggregates by novel quantum-dot nanoprobes. PLoS One 4(12):e8492

Verwilst P, Kim HR, Seo J, Sohn NW, Cha SY, Kim Y et al (2017) Rational design of in vivo tau tangle-selective near-infrared fluorophores: expanding the BODIPY universe. J Am Chem Soc 139(38):13393–13403

Verwilst P, Kim HS, Kim S, Kang C, Kim JS (2018) Shedding light on tau protein aggregation: the progress in developing highly selective fluorophores. Chem Soc Rev 47(7):2249–2265

Vetrivel KS, Zhang YW, Xu H, Thinakaran G (2006) Pathological and physiological functions of presenilins. Mol Neurodegener 1:1–12

Vina J, Lloret A (2010) Why women have more Alzheimer's disease than men: gender and mitochondrial toxicity of amyloid-β peptide. J Alzheimers Dis 20(s2):S527–S533

Viola KL, Sbarboro J, Sureka R, De M, Bicca MA, Wang J et al (2015) Towards non-invasive diagnostic imaging of early-stage Alzheimer's disease. Nat Nanotechnol 10(1):91–98

Viswanathan A, Rocca WA, Tzourio C (2009) Vascular risk factors and dementia: how to move forward? Neurology 72(4):368–374

Vlassenko AG, Benzinger TL, Morris JC (2012) PET amyloid-beta imaging in preclinical Alzheimer's disease. Biochim Biophys Acta 1822(3):370–379

Wang Y, Mandelkow E (2016) Tau in physiology and pathology. Nat Rev Neurosci 17(1):22–35

Wang X, Wang C, Chan HN, Ashok I, Krishnamoorthi SK, Li M et al (2021) Amyloid-β oligomer targeted theranostic probes for in vivo NIR imaging and inhibition of self-aggregation and amyloid-β induced ROS generation. Talanta 224:121830

Watanabe H, Ono M, Matsumura K, Yoshimura M, Kimura H, Saji H (2013) Molecular imaging of β-amyloid plaques with near-infrared boron dipyrromethane (BODIPY)-based fluorescent probes. Mol Imaging 12(5):7290–2013

Wolz R, Julkunen V, Koikkalainen J, Niskanen E, Zhang DP, Rueckert D et al (2011) Multi-method analysis of MRI images in early diagnostics of Alzheimer's disease. PLoS One 6(10):e25446

Wong DF, Rosenberg PB, Zhou Y, Kumar A, Raymont V, Ravert HT et al (2010) In vivo imaging of amyloid deposition in Alzheimer disease using the radioligand 18F-AV-45 (florbetapir F 18). J Nucl Med 51(6):913–920

Wong HL, Wu XY, Bendayan R (2012) Nanotechnological advances for the delivery of CNS therapeutics. Adv Drug Deliv Rev 64(7):686–700

Xu MM, Ren WM, Tang XC, Hu YH, Zhang HY (2016) Advances in development of fluorescent probes for detecting amyloid-β aggregates. Acta Pharmacol Sin 37(6):719–730

Yanagisawa D, Amatsubo T, Morikawa S, Taguchi H, Urushitani M, Shirai N et al (2011) In vivo detection of amyloid β deposition using 19F magnetic resonance imaging with a 19F-containing curcumin derivative in a mouse model of Alzheimer's disease. Neuroscience 184:120–127

Yang W, Wong Y, Ng OT, Bai LP, Kwong DW, Ke Y et al (2012) Inhibition of beta-amyloid peptide aggregation by multifunctional carbazole-based fluorophores. Angew Chem 124(8): 1840–1846

Yang T, Wang X, Zhang C, Ma X, Wang K, Wang Y et al (2016) Specific self-monitoring of metal-associated amyloid-β peptide disaggregation by a fluorescent chelator. Chem Commun 52(11): 2245–2248

Yang J, Zhang X, Yuan P, Yang J, Xu Y, Grutzendler J et al (2017) Oxalate-curcumin–based probe for micro-and macroimaging of reactive oxygen species in Alzheimer's disease. Proc Natl Acad Sci 114(47):12384–12389

Yang J, Zeng F, Ge Y, Peng K, Li X, Li Y, Xu Y (2019a) Development of near-infrared fluorescent probes for use in Alzheimer's disease diagnosis. Bioconjug Chem 31(1):2–15

Yang HL, Fang SQ, Tang YW, Wang C, Luo H, Qu LL et al (2019b) A hemicyanine derivative for near-infrared imaging of β-amyloid plaques in Alzheimer's disease. Eur J Med Chem 179:736–743

Yang Y, Li S, Zhang Q, Kuang Y, Qin A, Gao M et al (2019c) An AIE-active theranostic probe for light-up detection of Aβ aggregates and protection of neuronal cells. J Mater Chem B 7(15): 2434–2441

Yang J, Zhu B, Yin W, Han Z, Zheng C, Wang P, Ran C (2020) Differentiating Aβ40 and Aβ42 in amyloid plaques with a small molecule fluorescence probe. Chem Sci 11(20):5238–5245

Zhang X, Tian Y, Li Z, Tian X, Sun H, Liu H et al (2013) Design and synthesis of curcumin analogues for in vivo fluorescence imaging and inhibiting copper-induced cross-linking of amyloid beta species in Alzheimer's disease. J Am Chem Soc 135(44):16397–16409

Zhang X, Tian Y, Yuan P, Li Y, Yaseen MA, Grutzendler J et al (2014) A bifunctional curcumin analogue for two-photon imaging and inhibiting crosslinking of amyloid beta in Alzheimer's disease. Chem Commun 50(78):11550–11553

Zhang X, Tian Y, Zhang C, Tian X, Ross AW, Moir RD et al (2015) Near-infrared fluorescence molecular imaging of amyloid beta species and monitoring therapy in animal models of Alzheimer's disease. Proc Natl Acad Sci 112(31):9734–9739

Zhou K, Fu H, Feng L, Cui M, Dai J, Liu B (2015) The synthesis and evaluation of near-infrared probes with barbituric acid acceptors for in vivo detection of amyloid plaques. Chem Commun 51(58):11665–11668

Zhou K, Yuan C, Dai B, Wang K, Chen Y, Ma D et al (2019) Environment-sensitive near-infrared probe for fluorescent discrimination of Aβ and tau fibrils in AD brain. J Med Chem 62(14): 6694–6704

Zhu B, Zhang T, Jiang Q, Li Y, Fu Y, Dai J et al (2018) Synthesis and evaluation of pyrazine and quinoxaline fluorophores for in vivo detection of cerebral tau tangles in Alzheimer's models. Chem Commun 54(82):11558–11561

Nanostructure-Based Molecules as Diagnostic and Theranostic Tools in Alzheimer's Disease

6

Teeja Suthar and Keerti Jain

Abstract

Alzheimer's disease (AD), a neurodegenerative disorder characterized by progressive, continuous, and irreversible degeneration of structure and function of neurons, is a major cognitive dysfunction amongst elderly population. Drugs with different pharmacology have been developed and approved for the treatment of AD, but these treatments are not curative and merely provide symptomatic relief. Successful treatment of AD requires drug targeting to the central nervous system in effective therapeutic concentrations. Blood-brain barrier (BBB) is a major hurdle in transporting drugs to the brain. Nanostructure-based materials can cross the BBB by virtue of their very small size. Further, surface modification of these materials could aid in targeting the drugs to specific cells/organs, and to achieve higher drug concentrations at the required site. Nanostructure-based imaging modalities, therapeutic agents, and theranostics provide highly sensitive detection at molecular levels, optimal drug targeting, and combined effect. Recent advancements in the field of nanotechnology have enabled the combined delivery of diagnostic as well as therapeutic agents to remarkably improve the outcomes of AD treatment. In this chapter, we have discussed a brief overview of AD, recent advancements in the detection of neurotransmitters and applications of nanostructure-based molecules in the early detection of AD, targeted nanomedicines, and theranostic nanomaterials in the management of AD. We have also discussed the toxicity concerns with nanostructure-based materials and future perspectives of theranostics as personalized medicine.

T. Suthar · K. Jain (✉)
Department of Pharmaceutics, National Institute of Pharmaceutical Education and Research (NIPER)-Raebareli, Lucknow, India
e-mail: keertijain.02@niperraebareli.edu.in

A. Sharma, G. P. Modi (eds.), *Natural Product-Based Synthetic Drug Molecules in Alzheimer's Disease*, https://doi.org/10.1007/978-981-99-6038-5_6

Keywords

Nanotechnology · Alzheimer's disease · Diagnosis · Theranostic · Targeting · Neurotransmitters

Abbreviations

ACh	Acetylcholine
AChE	Acetylcholine esterase
AD	Alzheimer's disease
APP	Amyloid precursor protein
Aβ	Amyloid β
BBB	Blood-brain barrier
CHG	Glutaraldehyde cross-linked, chitosan, and hyaluronic acid co-assembled nanoparticles
CNS	Central nervous system
CSF	Cerebrospinal fluid
DSPE-PEG	1,2-Dioleoyl-*sn*-glycero-3-phosphoethanolamine-*n*-[poly(ethylene glycol)]
ELISA	Enzyme-linked immunosorbent assay
FF	Phenylalanine-phenylalanine dipeptide
LSPR	Localized surface plasmon resonance
MRI	Magnetic resonance imaging
NFTs	Neurofibrillary tangles
NIR	Near infrared
POC	Point-of-care
PSEN	Presenilin
SERS	Surface-enhanced Raman scattering
SPIONs	Superparamagnetic iron oxide nanoparticles

6.1 Introduction

Alzheimer's disease (AD) is named after Alois Alzheimer, who described the first case in a 55-year-old female patient (Auguste Deter) in 1906 (Amiri et al. 2013). It is a neurodegenerative disorder characterized by continuous and irreversible degeneration of structure and function of neurons in the central nervous system (CNS) and peripheral nervous system. The main pathological hallmarks of AD are accumulation of amyloid β (Aβ) plaques outside the cells, formation of neurofibrillary tangles (NFTs) made up of hyperphosphorylated tau protein inside the cells, atrophy of cerebral cortex, and degeneration of cholinergic neurons of basal forebrain. It is associated with various cognitive impairments and is the foremost cause of dementia. The number of people living with dementia worldwide is currently projected at

50 million, and it is predicted to reach 82 million in 2030 and about 152 million by the year 2050 (Dementia n.d.; Dementia statistics | Alzheimer's Disease International (ADI) n.d.). The occurrence rate is higher in the elderly population than young adults owing to reduced or altered hormone secretions, increased oxidative stress, and neuronal inflammation (Alzheimer Association 2019; Apostolova 2016).

To date, several hypotheses have been proposed to describe the pathophysiology of AD including amyloid cascade hypothesis, tau hypothesis, cholinergic hypothesis, and oxidative stress hypothesis. The most widely accepted hypothesis to explain the pathophysiology of AD is the amyloid cascade hypothesis stating that Aβ is the foremost contributor in the progression of AD (Suthar and Navneet 2021). However, the drugs aimed at targeting Aβ have not garnered much clinical success. Despite the advancements as well as vast research in pharmaceutical and biomedical field and an imperative medical need, no effective curative therapies have been developed for AD. The clinical treatment recommended for these disorders merely provides a symptomatic relief (Breijyeh and Karaman 2020). CNS is the major target site of action for the drugs used to treat AD. Most of the drugs fail to reach the CNS in sufficient concentrations either due to their physicochemical properties or the presence of protective barrier called the blood-brain barrier (BBB). The BBB is the major barrier between the blood and the brain acting as the gatekeeper of the brain for extraneous materials (Lalatsa and Butt 2018; Gauro et al. 2021; Suthar et al. 2022). Various efforts have been made by researchers all over the world in the past few decades to design drug delivery systems that can easily traverse across the BBB.

Nanotechnology and nanostructure-based molecules have shown promising outcomes in delivering drugs to the CNS in effective therapeutic concentrations. Nanotechnology confers suitable attributes to the drugs, so that they can easily traverse across the BBB (Suthar et al. 2023; Ojha et al. 2021). Nanostructure-based materials have shown promises for delivery of imaging/probing as well as therapeutic agents (Jain et al. 2021; Mehra et al. 2016). Further, nanostructure-based molecules can be utilized to effectively deliver the diagnostic, therapeutic, or theranostic agents to the brain. Theranostics are nano-sized agents that offer highly sensitive molecular detection and efficient drug targeting by serving as both diagnostic and therapeutic agents, simultaneously (Jain and Jain 2023; Ahmad et al. 2022; Chauhan et al. 2022; Bajwa et al. 2015). In this chapter, we have discussed various biosensors and biomarkers in AD, advancements in the detection of neurotransmitters, and applications of nanostructure-based molecules in diagnosis and theranostics of AD.

6.2 Biosensors in Alzheimer's Disease

A biosensor is an analytical tool that combines a biological element with a physicochemical detector. Research in the field of biosensors have been ongoing since over two decades, and their range of application includes medical diagnostics, food, sports, and environment monitoring (Nemane et al. 2019). They present great potential in providing simple, rapid, and cost-effective analyses of many diseases

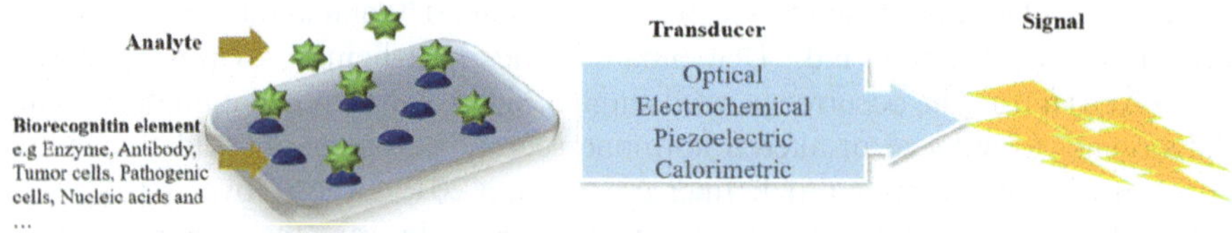

Fig. 6.1 Schematic representation of a biosensor. (*Reproduced with permission from* (Ameri et al. 2020))

by various point-of-care (POC) biosensors, which provide high degree of compliance to the patients by performing in situ analysis with a limited number of resources.

A typical biosensor is composed of two parts that are connected to one another, i.e., the biorecognition layer and the transducer. Figure 6.1 depicts the diagrammatic presentation of a typical biosensor. Biorecognition layer is generally made up of biomaterials that interact with the analyte of interest such as enzymes, antigens, or antibodies. The transducer converts the biological response into a quantifiable signal. Electrochemical, optical, and piezoelectric approaches are some of the widely used techniques for signal transduction. Biosensors are generally classified on the basis of either the biorecognition layer used in their construction or the method used for their signal transduction (Mohanty and Koucianos 2006).

Biosensors, especially POC biosensors, have shown remarkable potential in the improvement of AD diagnosis. Biosensor-based diagnostic tests are usually very simple, requiring only a minute quantity of sample and almost no hands-on time for the analysis. Further, biosensors can provide complex fluid analyses owing to their high selectivity and specificity towards the analyte of interest. The cost of biosensor-based diagnosis of AD is much lower than the neuroimaging techniques. Further, they are less invasive than other methods of detection as they possess the ability to detect biomarkers in other bodily fluids than the cerebrospinal fluid (CSF). The method of detection of CSF biomarkers requires lumbar puncture, which is very painful and noncompliant to the patients. Also, repeated sampling of CSF affects the level of CSF biomarkers, making them inadequate for monitoring disease progression. Neuroimaging is a very costly diagnosis, which is only advantageous in diagnosing the advanced state of the disease when symptoms are prominent and brain structure has changed. On the other hand, biosensors can provide early-stage diagnosis, since they rely on the detection of biomarkers (Brazaca et al. 2020; Shui et al. 2018). In the following subsections, we will be discussing the biosensors for the diagnosis of AD according to their recognition layer and transduction techniques.

6.2.1 Based on Transduction Mechanism

A transducer converts the interaction of the bioreceptor with the detector into an easily measurable and quantifiable signal. Selection of a suitable transducer greatly impacts the sensitivity as well as the limit of detection of the biosensor. Also, the compatibility of the transducer with the bioreceptor is a prerequisite in the development of a biosensor. Some of the major types of transduction mechanisms include electrochemical, optical, electronic, and piezoelectric biosensors (Naresh and Lee 2021). Electrochemical and optical transducers are the main transducers used in the detection of biomarkers of AD.

6.2.1.1 Electrochemical Biosensors

Electrochemical transduction-based biosensors have been in existence for a long time, but it continues to be a very active area of research. The last decade has seen a remarkable expansion in the use of electrochemical biosensors in areas such as food, environment, forensics, and medicine. Electrochemical biosensors have made it possible to diagnose a vast number of diseases and ailments in almost no time. They provide a highly specific, sensitive, and rapid detection of the analyte of interest. They can convert the chemical signals into quantifiable signals using the potentiometric, amperometric, conductometric, and impedimetric transducers. These methods work by monitoring the changes in various parameters like potential, current, conductivity, or impedance. These changes usually occur as a result of reduction or oxidation of electroactive species caused due to the type or concentration of the analyte of interest (Eltzov et al. 2011).

6.2.1.2 Optical Biosensors

Optical biosensors consist of a biorecognition layer and an optical transducer system, which produces a signal as a function of concentration of analyte. Optical transducers detect the changes in the optical properties like absorption, refraction index, fluorescence, and wavelength as a result of any biological reaction. They provide real-time detection of analyte of interest with a very high specificity and selectivity. Optical biosensors including fluorescence-based, colorimetric, localized surface plasmon resonance (LSPR), and surface-enhanced Raman scattering (SERS) detectors have shown promising results in quantifying the major biomarkers of AD, i.e., amyloid β and tau protein for early-stage detection of AD (Shui et al. 2018). Fluorescence-based biosensors are the most explored optical biosensors as they offer real-time fluorescence detection with very high sensitivity. Colorimetric biosensors are used for POC detections as they detect the analyte via color changes that are easily observable via naked eye. LSPR biosensors detect the changes in the refractive index upon biological interaction as a function of concentration of the analyte. SERS-based biosensors can also detect various forms of Aβ in the diagnosis of AD. Other optical biosensors to detect AD biomarkers include scanning tunneling microscopy and near-infrared (NIR) fluorescent imaging (Phan et al. 2021).

6.2.2 Based on Biorecognition Element

Biorecognition element is made up of biological materials that function by interacting with the analyte of interest (e.g., enzymes, antigens, or antibodies). A biorecognition layer allows the measurement of the analyte of interest in many complex samples due to the high specificity and selectivity of the biorecognition element. The most frequently used biorecognition elements in AD diagnosis include enzymes, antigens, antibodies, cells, and genetic material (Carneiro et al. 2020). Antibodies, enzymes, and genetic material are discussed briefly in the following subsections.

6.2.2.1 Antibodies

The high binding affinity of antibodies to their respective antigens makes them very useful as a bioanalytical device. Antigen-antibody interact via non-covalent interactions, and specific arrangements are required between them for binding to occur. The application of antibodies ranges from basic science to most advanced and applied sciences. The high specificity, selectivity, sensitivity, and homogeneity of antibodies make them an ideal recognition element in the development of biosensors. The easy availability, regeneration possibility, and requirement of low sample volumes are some of the major attributes that make antibodies an excellent recognition element. The major constraints with antibodies include their physical and thermal instabilities and complex in vitro and in vivo models (Kaushik et al. 2016a; Carneiro et al. 2017).

6.2.2.2 Enzymes

Enzymes are another widely used biorecognition element that acts as biological catalysts for chemical reactions. They offer great specificity to their substrates, and they are generally more selective towards the analyte of interest. They act as biorecognition element by causing changes in color, pH, or temperature, which can be detected by various transducers. Acetylcholine esterase (AChE) is one of the most used biorecognition enzymes in the diagnosis of AD (Brazaca et al. 2020).

6.2.2.3 Genetic Material

It is a well-established fact that genetic mutations are the major cause of various life-threatening diseases. Use of DNA or RNA as a recognition layer provides easy and quick detection of various diseases that are related to genetic mutations and polymorphisms. The detection of diseases occurs on the basis of hybridization between the sample and single-stranded capture sequence. AD, early-onset type as well as late-onset type, can be diagnosed by analyzing the genome of patients. Single-gene mutations in chromosomes 1, 14, and 21 lead to abnormalities in amyloid precursor protein, presenilin 1 (PSEN-1), and presenilin 2 (PSEN-1), respectively, leading to early-onset-type AD. Polymorphism of ApoE is the main causal factor for late-onset-type AD, which can be detected by analyzing the genome of patients (Hardy 2010).

6.3 Biomarkers for Detection of Alzheimer's Disease

Biomarkers are a group of clinical signs that accurately define the medical state of the patient, unlike the symptoms described and perceived by patients. National Institutes of Health Biomarkers Definitions Working Group defined biomarkers as "a characteristic that is objectively measured and evaluated as an indicator of pharmacologic responses, or normal or pathogenic biological processes, for a therapeutic intervention." Biomarkers are important to understand any physiological or pathological processes in the body, for developing newer diagnostic or therapeutic strategies, as well as to evaluate the pharmacological effects of any therapy. Biological markers could be measured inside or outside the body, and they may indicate prevalence of any disease in the body in its early stages (Aronson and Ferner 2017). Many reports suggest that biomarkers such as biological molecules and some genetic mutations can be utilized to evaluate overall pathophysiological conditions of aged patients to diagnose AD. The development of AD is slow and gradual, with cognitive dysfunction being the first symptom which starts appearing almost after a decade. This makes it difficult to identify the pathology of AD based on clinical symptoms alone. Biomarkers offer an early-stage diagnosis of AD and guide in decision-making process for the management of AD (Mantzavinos and Alexiou 2017; Wattamwar and Mathuranath 2010). An ideal biomarker should fit in the criteria decided by scientists worldwide as stated in Table 6.1. In this chapter, we have discussed CSF, plasma, and genetic mutation-based biomarkers in brief for detection of AD.

6.3.1 CSF Biomarkers

CSF biomarkers provide an accurate diagnosis of various neurological disorders including AD. Though the protein content in CSF is much lower than in blood, CSF is still a reliable source for developing biomarkers in AD. CSF reflects the exact pathological changes as it is in intimate contact with the extracellular space in the brain. Many reports suggested elevated levels of total tau (T-tau) and phosphorylated tau (P-tau) and reduced $A\beta_{1-42}$ in CSF analysis of AD patients. These biomarkers show a very high diagnostic accuracy with 85–90% specificity (Shui et al. 2018; Wattamwar and Mathuranath 2010; Blennow and Zetterberg 2009). The diagnostic method utilizing CSF biomarkers involves lumbar puncture, which is an invasive

Table 6.1 Criteria for an ideal biomarker

Should be highly specific and sensitive (with specificity and sensitivity of >80%)
Should reflect the aging of the brain
Should be able to describe the pathophysiological processes in brain
Should be able to reflect the pharmacological changes in patient
Should detect AD in its early stages (e.g., in mild cognitive enhancements)
Should be reliable, reproducible, noninvasive, and cost-effective

and very painful procedure. Further, enzyme-linked immunosorbent assay (ELISA) is the commonly employed assay for CSF analysis, which requires a very lengthy technique and costly reagents (Kaushik et al. 2016b). ApoE4 is another causal factor for AD, which can be utilized as a CSF biomarker for the diagnosis of AD (Genin et al. 2011).

6.3.2 Plasma Biomarkers

The high-cost, invasive nature and patient noncompliance associated with diagnosis of AD using CSF biomarkers make them difficult to be incorporated in the routine clinical practice. Thus, the focus of the scientists has shifted towards lesser invasive methods such as the use of plasma biomarkers. Plasma-based biomarkers are making their way towards routine screening of dementia, as blood collection is relatively noninvasive and inexpensive to perform (Wattamwar and Mathuranath 2010). Advancements in AD research have led to the identification of several biomarkers of AD in plasma, serum, and blood. The plasma screening of AD patients revealed two categories of biomarkers. The first one is a set of molecules with altered levels in plasma samples that collectively show a pathological fingerprint. Clusterin, fetuin B, pancreatic prohormone, and prostate-specific antigen complexed to α1-antichymotrypsin are a set of proteins used for the detection of AD. The second type includes biomarkers discretely related to AD like Aβ, tau, and ApoE (Sattlecker et al. 2014). These molecules are also the pathological hallmarks of AD, as their blood levels are deregulated, and hence, they can be used for early-stage AD detection. The levels of $A\beta_{42}$ and $A\beta_{40}$ are well-known biomarkers of AD, and their CSF and plasma levels increase with age but decrease with cognitive impairments due to accumulation of Aβ. Araclon Biotech Ltd. developed ELISA for colorimetric measurement of $A\beta_{1-40}/A\beta_{1-42}$ ratio in patients having mild cognitive impairments (Pesini et al. 2012). Although these molecules are promising biomarkers for AD diagnosis, further research is still required to bring them into routine clinical practice.

6.3.3 Genetic Mutations

The genetics of AD has not been completely understood owing to its complex nature and heterogeneity due to mutations and polymorphisms in multiple genes along with other nongenetic factors. Genetics play a crucial part in the development of early-onset AD, while late-onset AD is usually a result of genetic, ecological, and lifestyle factors (Alzheimer's Disease Genetics Fact Sheet I National Institute on Aging n.d.). The genetic mutations in AD are typically congenital and present high possibility of early-onset familial AD. The genetic mutations leading to early-onset familial AD are usually transmitted in an autosomal-dominant fashion; that is, they are less prevalent but have relatively high penetrance, while late-onset AD is related with polymorphisms which have high prevalence and low penetrance (Brazaca et al.

2020; Bertram et al. 2010). Early-onset familial AD generally occurs via single-gene mutations on chromosomes 21, 14, and 1, leading to abnormalities in amyloid precursor protein (APP), PSEN-1, and PSEN-2, respectively. Mutations in APP account for less than 5% of all early-onset familial AD cases. Mutations in PSEN-1 and PSEN-2 lead to an increase in the levels of Aβ plaques (Cruts et al. 1996). Late-onset AD usually occurs as a result of polymorphisms in ApoE gene located on chromosome 19. ApoE gene has three alleles namely, ε2, ε3, and ε4. Polymorphisms in ε4 allele are responsible for reduced amyloid clearance and increased plaque buildup leading to an increased risk of late-onset AD (Coon et al. 2007). Thus, in conclusion, genetic mutations and polymorphisms associated with familial AD cases present a viable biomarker for the detection of AD.

6.4 The Application of Nanostructure-Based Molecules in the Diagnosis of Alzheimer's Disease

Early diagnosis of AD before the occurrence of irreversible brain damage or complete mental decline gives hope in the treatment of AD. The AD biomarkers are conventionally detected either by determining the total tau/Aβ concentrations in CSF/plasma or by targeting the suspected pathogenic biomarkers (e.g., phosphorylated tau or cleaved tau) (Blennow and Zetterberg 2018). The former approach does not provide very convincing results as such biomarkers may be present in both healthy and AD patients, whereas the latter approach gives more conclusive results, but their quantification by conventional ELISA or western blotting may not be possible as they are present in very low concentrations (Fluri 2010).

Nowadays, the research in the field of AD diagnosis is directed towards the development of nanostructure-based molecules for imaging and molecular detection of AD biomarkers. Nanostructure-based molecules could give a very potent signal transduction, aiding in the early diagnosis of AD. These smartly designed molecules provide detection on the basis of their physical, chemical, or biological properties. These probes should have high specificity and sensitivity, moderate lipophilicity, high in vivo stability, and ability to cross the BBB nondestructively (Mpambani et al. 2018). The use of nanostructure-based molecular detection probes for brain imaging using magnetic resonance imaging (MRI) or positron-emission tomography could serve as a promising approach to detect the AD biomarkers like $A\beta_{42}$ and tau protein (Bilal et al. 2020).

MRI is a widely used diagnostic technique which relies on the relaxation time of protons excited with an external magnetic field. It detects the changes in signals caused by rotational orientation of protons of biological fluids. The most widely used magnetic nanoparticles are iron oxides, manganese oxide, and rare earth fluoride nanoparticles. The development of AD is usually very slow and gradual, which makes it difficult to diagnose AD till its symptoms start to appear. It is a well-known fact that oxidative stress and inflammation occur way before the appearance of diagnosable symptoms. He et al. developed a multifunctional anti-Aβ antibody 4G8-conjugated, brain-targeted, ROS-activatable MnO_2 nanoparticle-based theranostic nanoconstruct system for enhanced MRI signals in affected areas. The theranostic nanoconstructs showed a 1.51–2.24-fold enhancement in T1-weighted MRI signals in the CSF proportional to the cerebral ROS levels with 88.9% sensitivity and 100% specificity on intravenous administration in AD mouse model. These nanoconstructs could allow for early-stage detection of AD prior to cognitive deficits (He et al. 2020).

Nanostructure-based molecules have shown the ability to traverse across the BBB and improve the MRI signals and thus aid in early-stage diagnosis of AD. Yin et al. synthesized anti-amyloid antibody-grafted meso-2,3-dimercaptosuccinic acid-coated magnetite nanoparticles for MRI signal enhancement in the diagnosis of AD. The colony-forming assay demonstrated the biocompatibility of immunomagnetic nanoparticles in C6 cells. Their results revealed that the relaxivities increased with an increase in the density of immunomagnetic nanoparticles, indicating their applicability as excellent negative contrast agents for the diagnosis of AD (Yin et al. 2015).

It is an established fact that detection of Aβ plaques in its early stages leads to rapid detection of AD and can support more conclusive and effective therapeutic interventions. Mpambani et al. studied the ability of hybrid nanomaterials containing gadolinium fluoride-based paramagnetic nanoparticles surrounded by luminescent-conjugated oligothiophenes to detect Aβ aggregates using MRI/fluorescence dual-targeting approach. Transmission electron microscopy (TEM) and fluorescence microscopy results revealed the ability of these contrast agents to selectively bind to the Aβ deposits. The results of in vivo experiments also revealed that these contrast agents could diffuse through the BBB and selectively stain Aβ fibrils. The hybrid contrast media was able to provide dual-MRI/fluorescent imaging for detection of AD biomarkers (Mpambani et al. 2018).

These reports suggest that nanostructure-based molecules or nanocarriers can act as excellent contrast agents in the diagnosis of various diseases, particularly AD. In the next section of this chapter, we have discussed the applications of nanostructure-based molecules in AD theranostics for simultaneous diagnosis and treatment of AD.

6.5 The Application of Nanostructure-Based Molecules in Alzheimer's Disease Theranostics

Conventionally, the diagnosis and treatment of AD are carried out separately, which makes it difficult and time-consuming to efficiently monitor the progress of the treatment. A simultaneous or combined diagnosis and treatment of AD may prove to be a more efficient, personalized, and patient-friendly approach. The term "theranostic" was coined by John Funkhouser to describe the materials that combine the diagnostic and therapeutic modalities (Tripathi et al. 2021). Theranostics are molecular level agents that offer highly sensitive detection and efficient drug targeting by serving as both diagnostic and therapeutic agents, simultaneously. Theranostics also provides sufficient pharmacokinetic information of the targeted drug throughout the treatment and thus offers a huge number of aided advantages over concrete diagnosis or treatment (Ahmad et al. 2022; Jain and Zhong 2022). Various nanostructure-based molecules can be used as carriers for delivering imaging modalities and therapeutic agents to the target sites. Targeting ligands can be anchored on the surface of these molecules to guide them to the target site. Theranostic nanocarriers or nanotheranostics can be formulated from various organic and inorganic nanomaterials including polymeric nanoparticles, dendrimers, quantum dots, and lipidic nanoparticles (Aulić et al. 2013).

Ruan et al. developed nanotheranostic platform loaded with curcumin and superparamagnetic iron oxide nanoparticles (SPIONs) encapsulated by diblock 1,2-dioleoyl-*sn*-glycero-3-phosphoethanolamine-*n*-[poly(ethylene glycol)] (DSPE-PEG) surface modified with CRT (CRTIGPSVC) and QSH (QSHYRHISPAQV) peptides. The developed nanotheranostic platform demonstrated high peptide-targeted BBB permeation and significant reduction of Aβ plaques in APP/PSEN1 transgenic mice as observed by MRI. Further, there was a significant enhancement in the spatial learning and memory due to brain-derived neurotrophic factor-induced neuroprotection (Ruan et al. 2022).

Wang et al. formulated glutaraldehyde cross-linked, chitosan and hyaluronic acid co-assembled nanoparticles (CHG nanoparticles) as a nanotheranostic module for imaging Aβ fibrils and inhibiting Aβ aggregation. The developed nanoparticles demonstrated a high binding affinity towards Aβ oligomers, whereas they showed weak emission for the Aβ monomers. The results of binding affinity studies revealed that theranostic nanoparticles specifically bind to negatively charged Aβ aggregates due to the electrostatic and hydrophobic interactions. The ThT fluorescence assay method was used to detect the aggregation process, and the results revealed that ThT fluorescence intensity reduced on increasing the concentration of CHG nanoparticles and the nanoparticles over 360 μg/mL concentration completely inhibited the Aβ aggregation. In vivo assays with *Caenorhabditis elegans* CL2006 showed the potential of CHG nanoparticles for imaging Aβ fibrils and inhibiting the Aβ aggregation. Figure 6.2 shows the in vivo fluorescence images of *C. elegans* stained with ThT and CHG nanoparticles (Wang et al. 2021).

Sharma et al. formulated self-fluorescent dopamine tryptophan nanoparticles for dual imaging and aggregation inhibition of Aβ polypeptides for AD theranostic applications. The anti-aggregation potential of the nanotheranostics was evaluated on in vitro amyloid aggregation model, phenylalanine-phenylalanine (FF) dipeptide, and the results revealed that the nanoparticles showed positive anti-aggregation behavior towards both FF dipeptide and preformed Aβ fibrils as observed by electron microscopy and circular dichroism spectroscopy. The confocal fluorescence images depicted red and green fluorescence in the cytoplasm of nanoparticle-treated cells, which confirmed successful internalization and accumulation of self-fluorescent nanoparticles (Fig. 6.3). Further, the theranostic nanoparticles showed neuroprotective effects in SH-SY5Y cells against FF fibril-induced toxicity (Fig. 6.3d). It also attenuated the Aβ accumulation and reduced the cognitive deficits in intracerebroventricular streptozotocin (ICV-STZ)-induced rat model. In conclusion, the developed nanotheranostics served as excellent bioimaging agents and reduced cognitive impairments (Sharma et al. 2020).

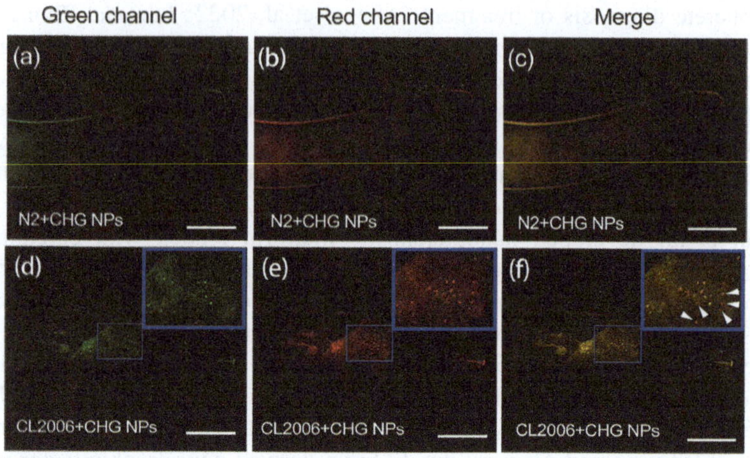

Fig. 6.2 In vivo fluorescence imaging of *C. elegans* strain of (**a**–**c**) N2 and (**d**–**f**) CL2006. Nematodes were stained with 10 μM ThT (green channel) and 30 μg/mL CHG NPs (red channel). (*Adapted with permission from* (Wang et al. 2021))

Wang et al. synthesized Aβ oligomer-specific surface-functionalized, Aβ oligomer-selective cyanine dye (F-SLOH)-modified Gd-based mesoporous silica nanoparticles for dual-NIR/MRI contrast agent for detection of Aβ fibrils. The nanoparticle probe showed high NIR and MRI sensitivity upon binding with Aβ fibrils. They studied the potential of nanoparticle probe for in vivo Aβ imaging and targeting in three different mice groups aged 7, 9, and 11 months, and the results showed age-dependent modifications in fluorescence intensities and higher Aβ content in older mice. It effectively inhibited Aβ aggregation and fibrillation and showed neuroprotection against Aβ-induced toxicity and ROS generation in APP/PSEN1 transgenic mice model (Wang et al. 2020).

In another study, Cai et al. developed a nanotheranostic platform consisting of ultrasmall SPIONs coupled to a phenothiazine-based NIR fluorescent dye. In vivo NIR fluorescence and MR imaging studies in APP/PSEN1 transgenic mice revealed the ability of the developed nanotheranostic to perform simultaneous in vivo NIR fluorescence and MRI of Aβ plaques in the mouse brain. Further, the nanotheranostics was able to significantly enhance the cell viability which was reduced by $A\beta_{1-42}$ in SH-SY5Y human neuroblastoma cells (Fig. 6.4b) and, thus, could prevent the AD-associated apoptosis. In addition, they also studied the inhibitory effects of nanotheranostics on Aβ aggregation and disaggregation of preformed Aβ fibrils, and results revealed that ultrasmall SPIONs could disaggregate the Aβ fibrils at much lower concentrations than phenothiazine-based small molecules (Fig. 6.4a) (Cai et al. 2020). Various theranostic approaches involving nanostructure-based molecules are enlisted in Table 6.2.

Theranostics has emerged as a very appealing approach for simultaneous diagnosis and treatment of various life-threatening diseases, particularly neurodegenerative disorders as they offer highly specific, sensitive, noninvasive, and reliable detection

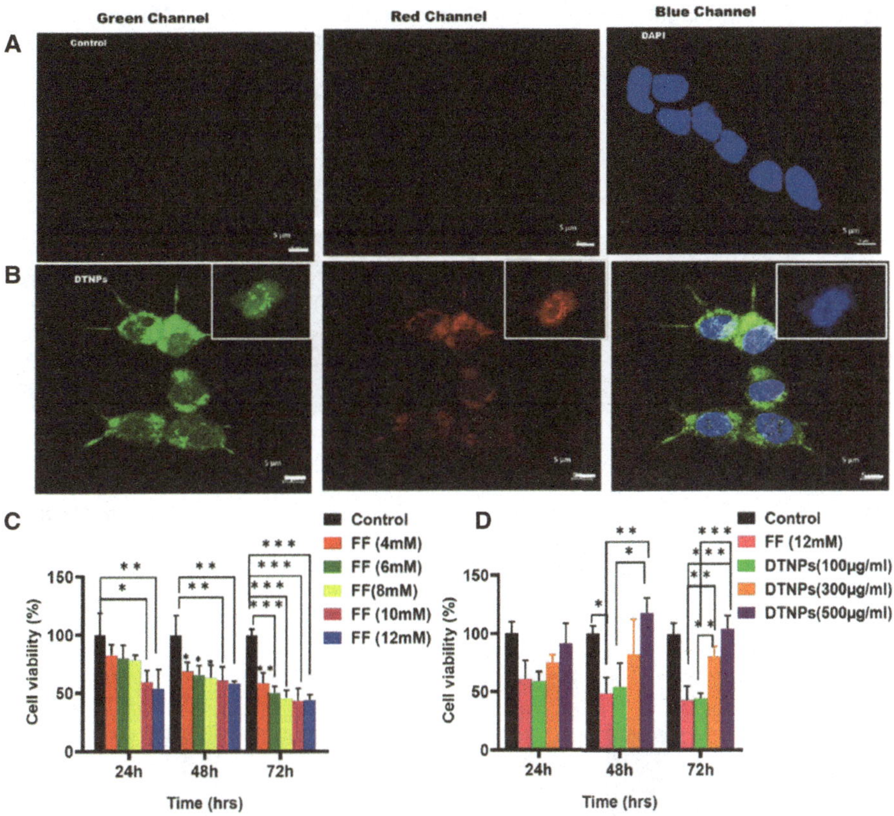

Fig. 6.3 Confocal fluorescence images depicting cellular uptake of DTNPs in SH-SY5Y cells. (**a**) DAPI-stained SH-SY5Y cells imaged in green, red, and blue channel without DTNPs; (**b**) DAPI-stained SH-SY5Y cells treated with DTNPs imaged in green, red, and blue channel; (**c**) cytotoxicity of FF towards SH-SY5Y cells examined by MTT assay; and (**d**) protective effects of DTNPs against FF-induced cytotoxicity in SH-SY5Y cells. (*Adapted with permission from* (Sharma et al. 2020))

in the early stages of the disease. The aforementioned studies suggest the potential of nanostructure-based molecules in the theranostics of AD. The next section of this chapter deals with the toxicity concerns associated with the use of nanotechnology in AD.

6.6 Toxicity Concerns with Use of Nanotechnology in Alzheimer's Disease

Nanotechnology-based carriers have been widely utilized in drug delivery applications, as they allow for accurate site-specific delivery of drugs and reduce the associated adverse effects. Engineered nanomaterials have garnered a lot of attention owing to their nano-size and relatively high surface area.

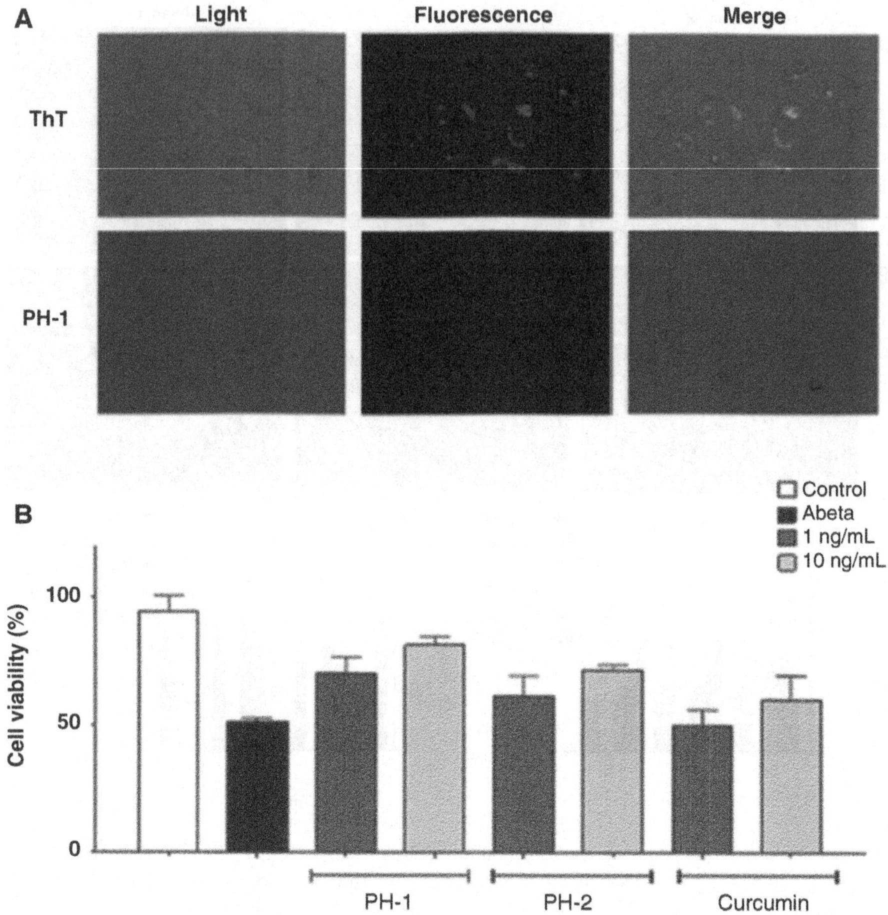

Fig. 6.4 Inhibition of intracellular self-aggregation of Aβ in SH-SY5Y cells on treatment with Aβ monomer and PH-1 or ThT for 2 days (**a**), and effect of PH-1, PH-2, or curcumin against the toxicity induced by $A\beta_{1-42}$ in SH-SY5Y cells (**b**). (*Reproduced with permission from* (Cai et al. 2020)). [PH-1, PH-2: DPA-PEGylated USPIONs labeled with phenothiazine-based small compounds]

Nanostructure-based molecules including diagnostic and therapeutic nanomodalities have been considered in the interventions for the diagnosis and treatment of AD. The small size of these materials facilitates their uptake and transport into and across the cells. However, very limited data is available on the toxicity profile and fate of these nanomolecules in the biological systems (Khan and Shanker 2015; Jain et al. 2010; Jain 2018). They can illicit an immune response either through direct handling or exposure, so care should be taken in their handling and usage. Their biological characterization should be performed with utmost care and diligence to ensure their safety and eliminate the associated health hazards. Various in vitro and in vivo studies for testing immune responses have shown that many nanostructures,

Table 6.2 Various nanostructure-based molecules explored for AD theranostics

S. no.	Type of nanocarrier	Imaging agent	Therapeutic agent	In vitro/in vivo model	Outcome	Ref.
1.	DSPE-PEG SPIONs	SPIONs	Curcumin	HT22 cells/APP/PSEN1 double-transgenic mice	– Demonstrated high peptide-targeted BBB penetration – Highly sensitive detection of Aβ plaques – Improved learning due to brain-derived neurotrophic factor-induced neuroprotection	Ruan et al. (2022)
2.	Glutaraldehyde cross-linked chitosan-hyaluronic acid nanoparticles (CHG NPs)	CHG NPs	CHG NPs	SH-SY5Y cells/ *Caenorhabditis elegans* CL2006 nematodes	– The nanoparticles demonstrated weak emission with the Aβ monomers and high emission and sensitivity for Aβ oligomers – In vitro experiments and in vivo assays with *C. elegans* CL2006 demonstrated the potential of nanoparticles for imaging Aβ fibrils and inhibiting Aβ deposits	Wang et al. (2021)
3.	Dopamine tryptophan-nanocomposites	Self-fluorescent nanoparticles	Dopamine, tryptophan	SH-SY5Y cells/rat model with ICV-STZ-induced memory deficits	– Effectively inhibited the aggregation of preformed FF fibrils and Aβ-polypeptide aggregates – Served as an efficient bioimaging agent and improved cognitive deficits simultaneously	Sharma et al. (2020)
4.	F-SLOH@SiO_2-modified Gd-based NPs	Aβ oligomer-selective cyanine dye, F-SLOH	F-SLOH-modified Gd^{3+}-based NPs	SH-SY5Y cells/9-month-old APP/PSEN1 double-transgenic mice	– Demonstrated high NIR and MR imaging sensitivity upon binding with Aβ species – The nanoparticles efficiently inhibited Aβ aggregation and fibrillation and showed	Wang et al. (2020)

(continued)

Table 6.2 (continued)

S. no.	Type of nanocarrier	Imaging agent	Therapeutic agent	In vitro/in vivo model	Outcome	Ref.
					neuroprotection against Aβ-induced toxicity and ROS generation	
5.	Ultrasmall SPIONs	Phenothiazine-based NIR fluorescent dye	Ultrasmall SPIONs	SH-SY5Y cells/9-month-old APP/PSEN1 double-transgenic mice	– Simultaneous in vivo NIR fluorescence and MRI of Aβ plaques in the AD transgenic mouse brain – Prevent Aβ aggregation and disaggregation of preformed Aβ fibrils – Significant enhancement of cell viability reduced by $A\beta_{1-42}$ in SH-SY5Y cells	Cai et al. (2020)
6.	Functionalized multiwalled carbon nanotubes	Pittsburgh compound B	Carbon nanotubes	Female C57BL/6 mice	– Significantly improved brain accumulation of conjugates in comparison to free metal complexes – Functionalized carbon nanotubes could effectively cross the BBB	Costa et al. (2018)

including metal and metal oxide-based nanoparticles, exhibit pro-inflammatory effects. The ability of these nanoparticles to exhibit immune response highly depends on their size, shape, and surface chemistry (Kononenko et al. 2015). In various studies involving metal and metal oxide nanoparticles, an abnormally high concentration of metals was found in the Aβ plaques. It can be attributed to the fact that metals have a tendency to increase the rate of Aβ aggregation, thus promoting AD progression (Liu et al. 2006). Thus, nanotechnology and nanotoxicology can be considered as the two sides of the same coin, as the same nano-size which provides benefits may also lead to undesired effects. Although most of the current toxicity protocols are sufficiently able to identify the hazards and toxic effects associated with nanostructure-based molecules, new testing methods should be researched to further establish the safety of these nanostructures (Carro et al. 2019).

6.7 Conclusion

With the rapid progression and advancements in nanotechnology, nanostructure-based materials have shown promising outcomes in the management of various diseases due to their excellent physicochemical properties. Nanoscale size and tunable surface of nanostructure-based materials make them promising diagnostic as well as therapeutic agents for the management of AD. Nanostructure-based materials can cross the BBB and simultaneously deliver the therapeutic agent at the target site. Some recent reports suggest that nanostructure-based materials can effectively slow down the progression of AD by various mechanisms including, by acting as drug carriers, facilitating brain targeting, inhibiting aggregation of Aβ, facilitating degradation of Aβ, abolishing tau aggregates, reducing tau hyperphosphorylation, and relieving the oxidative stress. These nanomaterials need to be explored further to establish their in vivo safety. Surface modifications and other such methods can be utilized to improve the target specificity and reduce off-target accumulation of therapeutic agents. Nanostructure-based materials are emerging as potential theranostics, but further research is required to understand the role and mechanisms of these materials during the diagnosis and treatment of AD.

Acknowledgements The authors are grateful to the Department of Pharmaceuticals, Ministry of Chemicals and Fertilizers, Government of India, for providing facilities for writing this chapter. The NIPER-R communication number for this chapter is NIPER-R/Communication/400.

Conflict of Interest The authors report no conflict of interest related to this chapter.

References

Ahmad J, Rizwanullah M, Suthar T, Albarqi HA, Ahmad MZ, Vuddanda PR et al (2022) Receptor-targeted surface-engineered nanomaterials for breast cancer imaging and theranostic

applications. Crit Rev Ther Drug Carrier Syst 39:1–44. https://doi.org/10.1615/CRITREVTHERDRUGCARRIERSYST.2022040686
Alzheimer Association (2019) Early signs and symptoms of Alzheimer's. Alzheimers Dement:1–88
Alzheimer's Disease Genetics Fact Sheet | National Institute on Aging (n.d.). https://www.nia.nih.gov/health/alzheimers-disease-genetics-fact-sheet. Accessed 15 Dec 2022
Amiri H, Saeidi K, Borhani P, Manafirad A, Ghavami M, Zerbi V (2013) Alzheimer's disease: pathophysiology and applications of magnetic nanoparticles as MRI theranostic agents. ACS Chem Neurosci 4(11):1417–1429. https://doi.org/10.1021/cn4001582
Ameri M, Shabaninejad Z, Movahedpour A, Sahebkar A, Mohammadi S, Hosseindoost S et al (2020) Biosensors for detection of tau protein as an Alzheimer's disease marker. Int J Biol Macromol 162:1100–1108. https://doi.org/10.1016/J.IJBIOMAC.2020.06.239
Apostolova LG (2016) Alzheimer disease. CONTIN Lifelong Learn Neurol 22:419–434. https://doi.org/10.1212/CON.0000000000000307
Aronson JK, Ferner RE. Biomarkers—a general review. Curr Protoc Pharmacol 2017;76: 9.23.1–9.23.17. doi:https://doi.org/10.1002/CPPH.19
Aulić S, Bolognesi ML, Legname G (2013) Small-molecule theranostic probes: a promising future in neurodegenerative diseases. Int J Cell Biol 2013. https://doi.org/10.1155/2013/150952
Bajwa N, Kumar Mehra N, Jain K, Kumar JN (2015) Targeted anticancer drug delivery through anthracycline antibiotic bearing functionalized quantum dots. Artif Cells Nanomed Biotechnol 44:1774–1782. https://doi.org/10.3109/21691401.2015.1102740
Bertram L, Lill CM, Tanzi RE (2010) The genetics of Alzheimer disease: back to the future. Neuron 68:270–281. https://doi.org/10.1016/J.NEURON.2010.10.013
Bilal M, Barani M, Sabir F, Rahdar A, Kyzas GZ (2020) Nanomaterials for the treatment and diagnosis of Alzheimer's disease: an overview. NanoImpact 20:100251. https://doi.org/10.1016/J.IMPACT.2020.100251
Blennow K, Zetterberg H (2009) Cerebrospinal fluid biomarkers for Alzheimer's disease. J Alzheimers Dis 18:413–417. https://doi.org/10.3233/JAD-2009-1177
Blennow K, Zetterberg H (2018) Biomarkers for Alzheimer's disease: current status and prospects for the future. J Intern Med 284:643–663. https://doi.org/10.1111/JOIM.12816
Brazaca LC, Sampaio I, Zucolotto V, Janegitz BC (2020) Applications of biosensors in Alzheimer's disease diagnosis. Talanta 210:120644. https://doi.org/10.1016/j.talanta.2019.120644
Breijyeh Z, Karaman R (2020) Comprehensive review on Alzheimer's disease: causes and treatment. Molecules 25:25. https://doi.org/10.3390/molecules25245789
Cai J, Dao P, Chen H, Yan L, Li YL, Zhang W et al (2020) Ultrasmall superparamagnetic iron oxide nanoparticles-bound NIR dyes: novel theranostic agents for Alzheimer's disease. Dyes Pigments 173:107968. https://doi.org/10.1016/J.DYEPIG.2019.107968
Carneiro P, Loureiro J, Delerue-Matos C, Morais S, do Carmo Pereira M (2017) Alzheimer's disease: development of a sensitive label-free electrochemical immunosensor for detection of amyloid beta peptide. Sens Actuators B Chem 239:157–165. https://doi.org/10.1016/J.SNB.2016.07.181
Carneiro P, Morais S, do Carmo Pereira M (2020) Biosensors on the road to early diagnostic and surveillance of Alzheimer's disease. Talanta 211:120700. https://doi.org/10.1016/J.TALANTA.2019.120700
Carro CE, Pilozzi AR, Huang X (2019) Nanoneurotoxicity and potential nanotheranostics for Alzheimer's disease. EC Pharmacol Toxicol 7:1
Chauhan S, Naqvi S, Jain K (2022) Dendrimers and its theranostic applications in infectious diseases. In: Nanotheranostics for treatment and diagnosis of infectious diseases, pp 199–228. https://doi.org/10.1016/B978-0-323-91201-3.00004-9
Coon KD, Myers AJ, Craig DW, Webster JA, Pearson J, Lince DH et al (2007) A high-density whole-genome association study reveals that APOE is the major susceptibility gene for sporadic late-onset Alzheimer's disease. J Clin Psychiatry 68:613–618. https://doi.org/10.4088/JCP.V68N0419

Costa PM, Wang JTW, Morfin JF, Khanum T, To W, Sosabowski J et al (2018) Functionalised carbon nanotubes enhance brain delivery of amyloid-targeting Pittsburgh compound B (PiB)-derived ligands. Nano 2:168–183. https://doi.org/10.7150/NTNO.23125

Cruts M, Hendriks L, van Broeckhoven C (1996) The presenilin genes: a new gene family involved in Alzheimer disease pathology. Hum Mol Genet 5:1449–1455. https://doi.org/10.1093/HMG/5.SUPPLEMENT_1.1449

Dementia (n.d.). https://www.who.int/news-room/fact-sheets/detail/dementia. Accessed 29 Dec 2020

Dementia statistics | Alzheimer's Disease International (ADI) (n.d.). https://www.alzint.org/about/dementia-facts-figures/dementia-statistics/. Accessed 29 Dec 2020

Eltzov E, Cosnier S, Marks RS (2011) Biosensors based on combined optical and electrochemical transduction for molecular diagnostics. Expert Rev Mol Diagn 11:533–546. https://doi.org/10.1586/ERM.11.38

Fluri F (2010) Clinical nanomedicine: nanomedical approaches in Alzheimer's disease. Eur J Nanomed 3:7–12. https://doi.org/10.1515/EJNM.2010.3.1.7

Gauro R, Nandave M, Jain VK, Jain K (2021) Advances in dendrimer-mediated targeted drug delivery to the brain. J Nanopart Res 23:1–20. https://doi.org/10.1007/S11051-021-05175-8/METRICS

Genin E, Hannequin D, Wallon D, Sleegers K, Hiltunen M, Combarros O et al (2011) APOE and Alzheimer disease: a major gene with semi-dominant inheritance. Mol Psychiatry 16:903–907. https://doi.org/10.1038/MP.2011.52

Hardy J (2010) New insights into the genetics of Alzheimer's disease. Ann Med 28:255–258. https://doi.org/10.3109/07853899609033127

He C, Ahmed T, Abbasi AZ, Li LY, Foltz WD, Cai P et al (2020) Multifunctional bioreactive-nanoconstructs for sensitive and accurate MRI of cerebrospinal fluid pathology and intervention of Alzheimer's disease. Nano Today 35:100965. https://doi.org/10.1016/J.NANTOD.2020.100965

Jain K (2018) Nanohybrids of dendrimers and carbon nanotubes: a benefaction or forfeit in drug delivery? Nanosci Nanotech Asia 9:21–29. https://doi.org/10.2174/2210681208666171204163622

Jain K, Zhong J (2022) Theranostic applications of nanomaterials. Curr Pharm Des 28:77–77. https://doi.org/10.2174/138161282802211223150153

Jain K, Kesharwani P, Gupta U, Jain NK (2010) Dendrimer toxicity: let's meet the challenge. Int J Pharm 394:122–142. https://doi.org/10.1016/J.IJPHARM.2010.04.027

Jain K, Shukla R, Yadav A, Ujjwal RR, Flora SJS (2021) 3D Printing in Development of Nanomedicines. Nanomaterials 11:420. https://doi.org/10.3390/NANO11020420

Jain K, Jain NK (2023) Multifunctional And Targeted Theranostic Nanomedicines: Formulation, Design And Applications. Springer Nature, ISBN: 978–981–99-0540-9; 1–417

Kaushik A, Shah P, Vabbina PK, Jayant RD, Tiwari S, Vashist A et al (2016a) A label-free electrochemical immunosensor for beta-amyloid detection. Anal Methods 8:6115–6120. https://doi.org/10.1039/C6AY01910B

Kaushik A, Jayant RD, Tiwari S, Vashist A, Nair M (2016b) Nano-biosensors to detect beta-amyloid for Alzheimer's disease management. Biosens Bioelectron 80:273–287. https://doi.org/10.1016/J.BIOS.2016.01.065

Khan HA, Shanker R (2015) Toxicity of nanomaterials. Biomed Res Int 2015:2–4

Kononenko V, Narat M, Drobne D (2015) Nanoparticle interaction with the immune system. Arh Hig Rada Toksikol 66:97–108. https://doi.org/10.1515/AIHT-2015-66-2582

Lalatsa A, Butt AM (2018) Physiology of the blood-brain barrier and mechanisms of transport across the BBB. Elsevier. https://doi.org/10.1016/B978-0-12-812218-1.00003-8

Liu G, Huang W, Moir RD, Vanderburg CR, Lai B, Peng Z et al (2006) Metal exposure and Alzheimer's pathogenesis. J Struct Biol 155:45–51. https://doi.org/10.1016/J.JSB.2005.12.011

Mantzavinos V, Alexiou A (2017) Biomarkers for Alzheimer's disease diagnosis. Curr Alzheimer Res 14:1149–1154. https://doi.org/10.2174/1567205014666170203125942

Mehra NK, Jain K, Jain NK (2016) Multifunctional carbon nanotubes in cancer therapy and imaging. Nanobiomater Med Imaging:421–453. https://doi.org/10.1016/B978-0-323-41736-5.00014-5

Mohanty SP, Koucianos E (2006) Biosensors: a tutorial review. IEEE Potentials 25:35–40. https://doi.org/10.1109/MP.2006.1649009

Mpambani F, Åslund AKO, Lerouge F, Reitan N, Huuse EM, Widerøe M et al (2018) Two-photon fluorescence and magnetic resonance specific imaging of a β amyloid using hybrid nano-GdF 3 contrast media. ACS Appl Bio Mater 1(2):462–472. https://doi.org/10.1021/acsabm.8b00191

Naresh V, Lee N (2021) A review on biosensors and recent development of nanostructured materials-enabled biosensors. Sensors 21:1109. https://doi.org/10.3390/S21041109

Nemane ST, Gholve SB, Bhusnure OG, Mule ST, Ingle P, v. (2019) Biosensors: an emerging technology in pharmaceutical industry. J Drug Deliv Therapeutics 9:643–647. https://doi.org/10.22270/JDDT.V9I4.3034

Ojha B, Jain VK, Mehra NK, Jain K (2021) Nanotechnology: introduction and basic concepts. Dendrim Nanomed:1–17. https://doi.org/10.1201/9781003029915-1

Pesini P, Pérez-Grijalba V, Monleón I, Boada M, Tárraga L, Martínez-Lage P et al (2012) Reliable measurements of the β-amyloid pool in blood could help in the early diagnosis of AD. Int J Alzheimers Dis 2012:1. https://doi.org/10.1155/2012/604141

Phan LMT, Hoang TX, Vo TAT, Pham HL, Le HTN, Chinnadayyala SR et al (2021) Nanomaterial-based optical and electrochemical biosensors for amyloid beta and tau: potential for early diagnosis of Alzheimer's disease. Expert Rev Mol Diagn 21:175–193. https://doi.org/10.1080/14737159.2021.1887732

Ruan Y, Xiong Y, Fang W, Yu Q, Mai Y, Cao Z et al (2022) Highly sensitive curcumin-conjugated nanotheranostic platform for detecting amyloid-beta plaques by magnetic resonance imaging and reversing cognitive deficits of Alzheimer's disease via NLRP3-inhibition. J Nanobiotech 20:20. https://doi.org/10.1186/S12951-022-01524-4

Sattlecker M, Kiddle SJ, Newhouse S, Proitsi P, Nelson S, Williams S et al (2014) Alzheimer's disease biomarker discovery using SOMAscan multiplexed protein technology. Alzheimers Dement 10:724–734. https://doi.org/10.1016/J.JALZ.2013.09.016

Sharma M, Tiwari V, Shukla S, Panda JJ (2020) Fluorescent dopamine-tryptophan nanocomposites as dual-imaging and antiaggregation agents: new generation of amyloid Theranostics with trimeric effects. ACS Appl Mater Interfaces 12:44180–44194. https://doi.org/10.1021/ACSAMI.0C13223

Shui B, Tao D, Florea A, Cheng J, Zhao Q, Gu Y et al (2018) Biosensors for Alzheimer's disease biomarker detection: a review. Biochimie 147:13–24. https://doi.org/10.1016/J.BIOCHI.2017.12.015

Suthar T, Navneet JK (2021) Nutraceuticals against neurodegeneration: understanding the mechanistic pathways. In: Lokhande JN, Pathak Y (eds) Nutraceuticals for aging and anti-aging: basic understanding and clinical evidence, 1st edn. CRC Press. https://doi.org/10.1201/9781003110866-7

Suthar T, Jain VK, Popli H, Jain K (2022) Nanoemulsions as effective carriers for targeting brain tumors. In: Kumar L, Pathak YY (eds) Nanocarriers for drug-targeting brain tumors. Elsevier, pp 347–363. https://doi.org/10.1016/B978-0-323-90773-6.00008-7

Suthar T, Patel P, Singh P, Datusalia AK, Yadav AK, Jain K (2023) Hesperidin microemulsion: Formulation optimization, characterization, and in vitro evaluation. J Drug Deliv Sci Tech 80 (104166):1–12. https://doi.org/10.1016/j.jddst.2023.104166

Tripathi P, Shukla P, Bieberich E (2021) Theranostic applications of nanomaterials in Alzheimer's disease: a multifunctional approach. Curr Pharm Des 28:116–132. https://doi.org/10.2174/1381612827666211122153946

Wang C, Wang X, Chan HN, Liu G, Wang Z, Li HW et al (2020) Amyloid-β oligomer-targeted gadolinium-based NIR/MR dual-modal theranostic nanoprobe for Alzheimer's disease. Adv Funct Mater 30:1909529. https://doi.org/10.1002/ADFM.201909529

Wang W, Liu M, Gao W, Sun Y, Dong X (2021) Coassembled chitosan-hyaluronic acid nanoparticles as a Theranostic agent targeting Alzheimer's β-amyloid. ACS Appl Mater Interfaces 13:55879–55889. https://doi.org/10.1021/ACSAMI.1C17267

Wattamwar PR, Mathuranath PS (2010) An overview of biomarkers in Alzheimer's disease. Ann Indian Acad Neurol 13:S116. https://doi.org/10.4103/0972-2327.74256

Yin Z, Yu T, Xu Y (2015) Preparation of amyloid immuno-nanoparticles as potential MRI contrast agents for Alzheimer's disease diagnosis. J Nanosci Nanotechnol 15:6429–6434. https://doi.org/10.1166/JNN.2015.11296

Part III

Drug Derivatives: A Strategy for New Multifunctional Drug Development for Alzheimer's Disease

7 Approved Cholinesterase Inhibitor-Based Derivatives: Synthesis and Their Biological Evaluation

Mange Ram Yadav, Prashant R. Murumkar, Karan Joshi, Rahul Barot, and Rasana Yadav

Abstract

Discovery of therapeutics for multifactorial diseases and disorders involving multiple pathophysiological pathways is still a challenge for the researchers. Alzheimer's disease (AD) is one such disease that involves multiple largely ambiguous pathophysiological pathways, which makes it a challenging thrust area of research around the globe. The current approved drugs for the management of the disease include rivastigmine, donepezil, galantamine (all AChE inhibitors), and memantine (NMDA receptor antagonist). Furthermore, bulk of the research on AD focuses on multi-target directed ligand (MTDL) approach which involves combining together two or more distinct pharmacophores to obtain potent MTDLs. In this chapter, we have discussed synthesis and biological evaluation of the most active novel moieties developed on the basis of the existing anti-Alzheimer's drugs reported in the last 5 years, along with the in silico overview of their interactions with the biological targets.

Keywords

Alzheimer's disease · Multifactorial disease · AChE inhibitor · BuChE inhibitor NMDA blocker · Pharmacophore · Multi-target directed ligand

M. R. Yadav (✉)
Research and Development Cell, Parul University, Vadodara, Gujarat, India

P. R. Murumkar · K. Joshi · R. Barot · R. Yadav
Faculty of Pharmacy, Kalabhavan Campus, The Maharaja Sayajirao University of Baroda, Vadodara, Gujarat, India

A. Sharma, G. P. Modi (eds.), *Natural Product-Based Synthetic Drug Molecules in Alzheimer's Disease*, https://doi.org/10.1007/978-981-99-6038-5_7

7.1 Introduction

Alzheimer's disease (AD) is considered to be a nonreversible chronic neurological disease currently. It has affected almost 55 million people till date and would affect around 78 million people worldwide by 2030 (https://www.Alzint.Org/Resource/World-Alzheimer-Report-2021/). Serious efforts have been made by the scientific community at various levels for the development of newer anti-AD drugs all over the world. Common symptoms of AD include memory loss and challenges in solving day-to-day problems and planning of work, along with decreased judgment and mood swings. Major causes listed for AD include neuron death, diminishing levels of neurotransmitter acetylcholine (ACh), metal dyshomeostasis, oxidative stress, formation of amyloid plaques, and tau protein disruption in the brain (Ashraf et al. 2019; Ge et al. 2022). Recently, activation of choline acetyltransferase (CHAT) activators (Kumar et al. 2019; Darreh-Shori et al. 2019) has been considered as an emerging approach for alleviating the problem, wherein acetylcholine synthesis is increased, thereby increasing the concentration of this neurotransmitter in the brain. A combination of drugs having the ability to increase the synthesis of acetylcholine (ACh) through CHAT activation along with inhibition of AChE and BuChE could be a significant breakthrough to treat the AD (Leng and Edison 2021).

Since AD is a multifactorial disease, there is no singular treatment for the disease (Yadav et al. 2023a). Use of multi-target directed ligands (MTDLs) as drugs is the most commonly advocated approach for the management of AD (Yadav et al. 2018). As far as MTDLs for AChE inhibitors are concerned, they consist of two or more pharmacophores having the potential to bind to the peripheral active site (PAS) and catalytic anionic site (CAS) of AChE fused together to form unique molecules having the potential to exhibit potent effect in inhibiting the breakdown of the neurotransmitter ACh, thereby increasing ACh levels in the brains of the AD patients (Shidore et al. 2016; Barmade et al. 2017). Novel compounds have been synthesized which act on multiple targets, like inhibition of AChE and BuChE enzymes, inhibition of Aβ aggregation, preventing tau hyper-phosphorylation, having metal chelating ability, and ability to scavenge free radicals.

7.2 Current Treatment Available for AD Patients

Currently available treatment for AD includes four major drugs (**1–4**). Out of them, three drugs donepezil (**1**), rivastigmine (**2**), and galantamine (**3**) (natural product) are AChE inhibitors, whereas memantine (**4**) is an NMDA receptor blocker. Tacrine (**5**), the first anti-AD drug introduced in the market in 1993, is another AChE inhibitor which had to be withdrawn from the market in 2013 due to its hepatotoxic side effect (Gupta 2014). A new drug in the form of monoclonal antibody Aduhelm was approved by the FDA in 2021 as a β-amyloid aggregation inhibitor, but the drug does not have much presence in the market. Figure 7.1 shows a brief timeline of approved drugs for AD from 1993 till 2021 (Zhang et al. 2022). Thus, discovering new drugs to combat AD remains a vital necessity for the whole world, as there is no

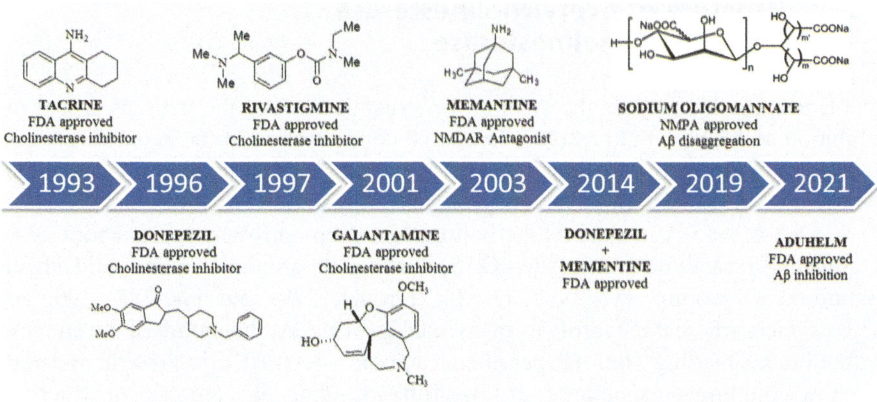

Fig. 7.1 Chronology of approved anti-Alzheimer's drugs

specific drug which can be used to terminate or reverse the progression of AD. Therefore, research for the discovery of newer molecules remains a priority as the number of dementia patients is increasing with every passing day.

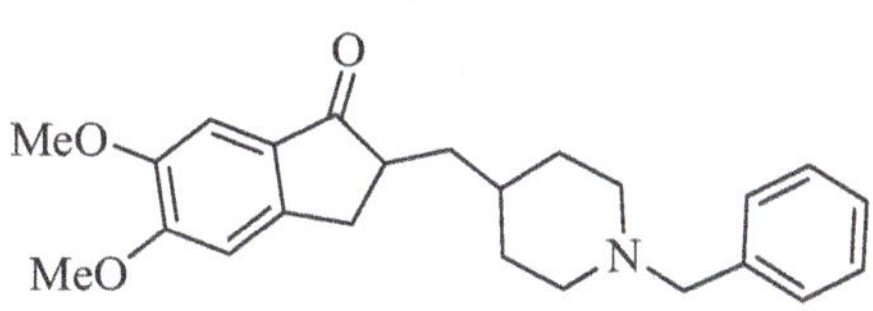

(1)

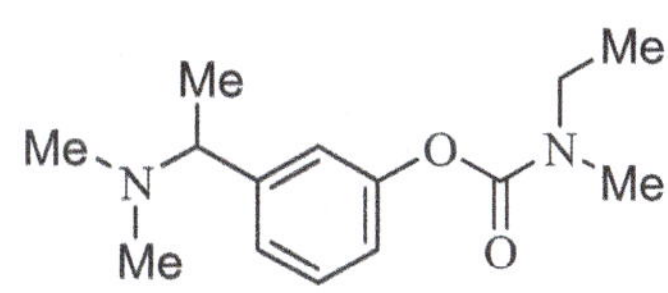

(2)

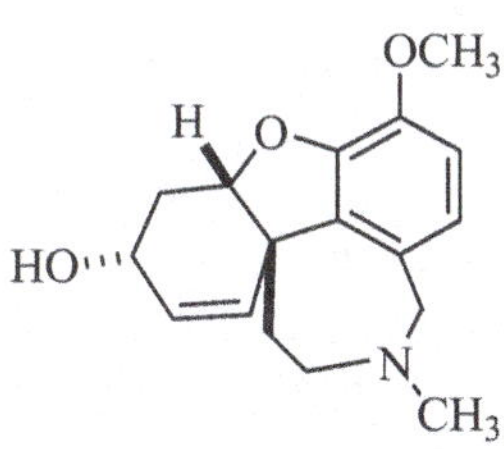

(3)

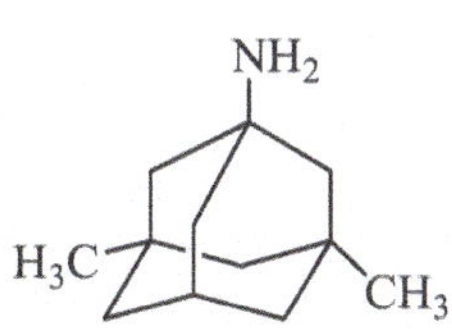

(4)

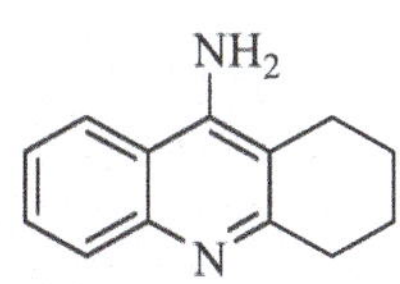

(5)

Before designing of any new derivative, in silico interactions produced by the standard drugs with acetylcholinesterase and butyrylcholinesterase enzymes would give enough insight to assess the stability of the protein-ligand complexes. Before considering the molecular interactions of the ligands with the enzymes, knowledge of the structures and active sites of the enzymes AChE and BuChE becomes important to see the correlation of the groups, which would be interacting with the enzymes' active sites.

7.3 Structure of Acetylcholinesterase and Butyrylcholinesterase

AChE enzyme hydrolyzes the ACh at the synaptic cleft of the neurons where the inhibition of AChE would result in increased concentration of the neurotransmitter, therefore compensating its deficit, which is a characteristic feature in AD patients (Martorana et al. 2010). The enzyme has ellipsoidal structure and is about 45 Å × 60 Å × 65 Å in size. At the bottom of a deep-narrow gorge of about 20 Å, it contains a catalytic active site (CAS) having 14 aromatic amino acid chains composed of mainly Tyr-70, 121, 334, Trp 84, 279, and Phe-288, 290, and 330 residues, where the hydrolysis of ACh takes place. At the mouth of the enzyme, an additional binding site, the peripheral anionic site (PAS), is present. Between these two binding sites, an array of low-affinity binding sites are present which are spread throughout the gorge that facilitates the binding of the ligands to the CAS site of the enzyme (Harel et al. 1993; Galdeano et al. 2012).

Butyrylcholinesterase (BuChE) is another hydrolyzing enzyme responsible for the hydrolysis of ACh having about 65% homology with the amino acid sequence of AChE. The gorge volume of the BuChE enzyme is about 200 $Å^3$ larger than the cavity of AChE, as it has a deficit of 6 aromatic amino acid residues out of the 14 amino acids present in AChE; therefore, BuChE is supposed to be less stereoselective for the substrate structure, for the same reason as given above (Nachon et al. 2003; Saxena et al. 1997, 1999; Johnson et al. 2003; Szegletes et al. 1999; Lane and He 2009). Moreover, the regulation of ACh hydrolysis in BuChE is done at the peripheral site; hence, it is the main binding site for the substrates, and therefore its binding with higher substrate concentration results in inhibitory activity, while lower concentration may result in accelerated rate of hydrolysis (Johnson et al. 2003; Szegletes et al. 1999). The AChE gorge comprises 14 aromatic residues, which make a substantial portion of the gorge, while the catalytic site at the bottom contains π-electron cloud of Trp84 along with Phe330, which assists in the binding of ACh through π-cation interactions to the enzyme through its quaternary ammonium center. Trp279 is additionally present at the entrance of the peripheral site (PAS), which is a site for binding of the substrate at the entry point. The active site of BuChE is located deep in the narrow gorge at the bottom of the enzyme, into which ACh diffuses and gets hydrolyzed. AChE is the major hydrolyzing enzyme in case of normal brain, and BuChE plays a supportive role, whereas BuChE is the main hydrolyzing enzyme for ACh in the peripheral nervous system. As the disease progresses, the brain AChE level and the ratio of AChE/BuChE gradually decrease in the AD patients, and the role of BuChE becomes more prominent as a vital therapeutic target to treat the AD patients (Lane and He 2009; Giacobini 2004). It is suggested that BuChE and AChE dual inhibition could broaden the scope of treatment by synergizing the effectiveness of medication and offer supplementary neuroprotection (Venneri et al. 2005).

The interaction of standard drugs like donepezil and rivastigmine with the enzyme AChE is important in order to understand the vital interactions that are responsible for inhibition of the enzyme activity. These interactions would be helpful to have a broader idea for designing new molecules as AChE inhibitors.

7.4 Interactions of Acetylcholine with Acetylcholinesterase

The neurotransmitter acetylcholine is responsible for the functioning of cholinergic or parasympathetic nervous system. Its concentration in the system is controlled by the enzyme acetylcholinesterase, which is responsible for hydrolyzing the neurotransmitter to choline and acetate. Thus, it is important to discuss the interactions between ACh and AChE (Fig. 7.2) to understand their binding pattern.

The neurotransmitter contains two fragments, an ester component and a quaternary ammonium ion component on either side of the ethylene chain. The carbon of the carbonyl group of the ester binds to the enzyme's catalytic triad through the OH of Ser200, while the quaternary ammonium ion binds to the anionic sub-site which consists of Glu199, Trp86, and Phe330 in the gorge. The PAS site does not have any interaction with acetylcholine due to its small structure. It is imperative to understand the interactions of the choline component with the enzyme to get an idea about the active binding sites of the enzyme along with important amino acids taking part in the interaction (Martins et al. 2021).

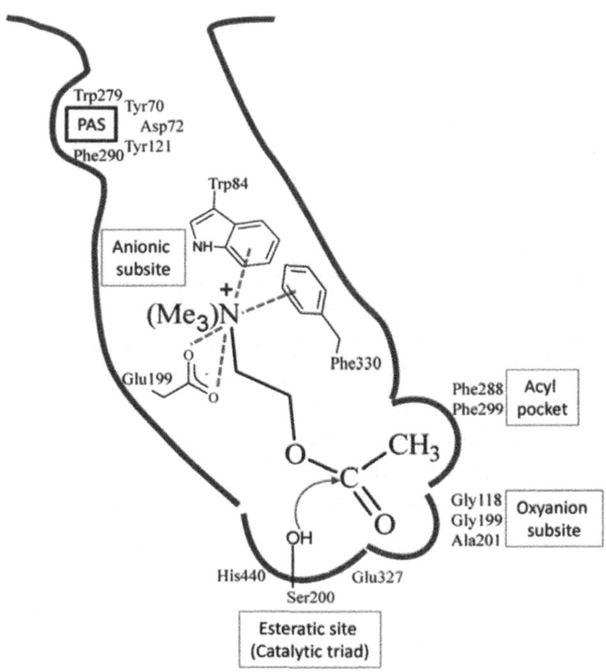

Fig. 7.2 Acetylcholine docked in the binding site of AChE enzyme gorge (Agatonovic-Kustrin et al. 2018)

7.5 Interaction of Rivastigmine with Cholinesterase

The structure of rivastigmine consists of two distinct side arms, a tertiary amine and a carbamate moiety at positions 1 and 3 of the central phenyl ring. The carbamate part consists of two fragments, an aliphatic carbamate group and an aromatic residue to which the carbamate is attached. The carbamate group binds to the CAS site of the AChE receptor by interacting with the hydroxyl groups of Ser200 and Ser203, whereas the aryl part of the structure interacts with Trp84 of the PAS site (Lin et al. 2005; Bar-On et al. 2002). It is hypothesized that the CAS is also responsible for hydrolysis of the carbamate group, but the rate of hydrolysis is quite slow. Therefore, rivastigmine along with other carbamate-containing derivatives is considered to be a pseudo-irreversible inhibitor (Mishra et al. 2021; Martins et al. 2021).

7.5.1 Designing of Rivastigmine-Based New Molecules

On the basis of rivastigmine structure, a plethora of new molecules have been designed, synthesized, and evaluated, which were found to exhibit potent anticholinesterase activity. The following section describes the best of the compounds exhibiting cholinesterase inhibitory activity as reported in literature in recent years along with their synthesis.

It is observed that drugs in general containing amide bond bind to a large number of receptors and enzymes eliciting biological responses. This observation underlines the importance of amide- and carbamate-based scaffolds in drug molecules and makes such scaffolds a dominant choice for the designing of new drugs. Based on fragment reassembly and in silico modeling approaches, Jiang et al. (2021) reported cannabidiol (CBD)-carbamate hybrids as selective BuChE inhibitors. CBD is able to insert itself into the BuChE binding groove, leading to multiple molecular interactions between Trp332 and the phenyl ring via π-π interactions, and hydrophobic interactions among Trp82 and His438, and cyclohexene ring, along with formation of two hydrogen bonds between Asp70 and -OH group, and with the Trp332 fragment. In the carbamate group containing drugs, the carbamate group can be considered as the main interacting group with the enzymes/receptors in the modern-day pharmacotherapy for AD as cholinesterase inhibitors. Based on the in silico studies, Jiang et al. (2021) designed and synthesized novel compounds among which compound **6** exhibited a highly selective and potent BuChE inhibitory activity having an IC_{50} value of 5.3 nM and SI of >4000. Additionally, compound **6** showed good BBB-penetrating ability, antioxidant activity, and pseudo-irreversible inhibition of BuChE with IC_{50} of 13 nM and k_2 value of 0.26 min^{-1}, offering neuroprotection, thus behaving like a good drug entity.

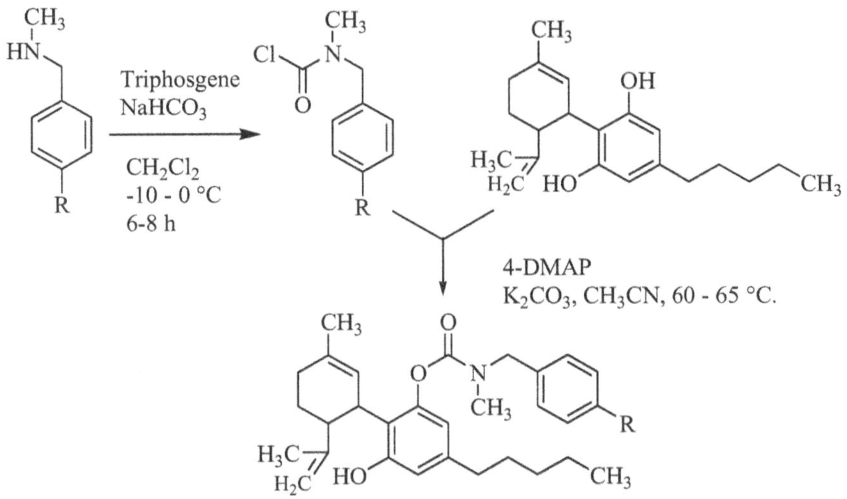

(**6**) R = Cl
(**7**) R = Br

Scheme 7.1 Synthesis of compounds **6** and **7**

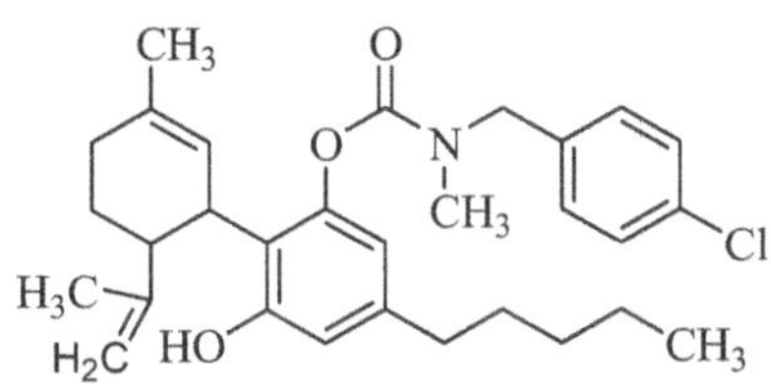

(6)

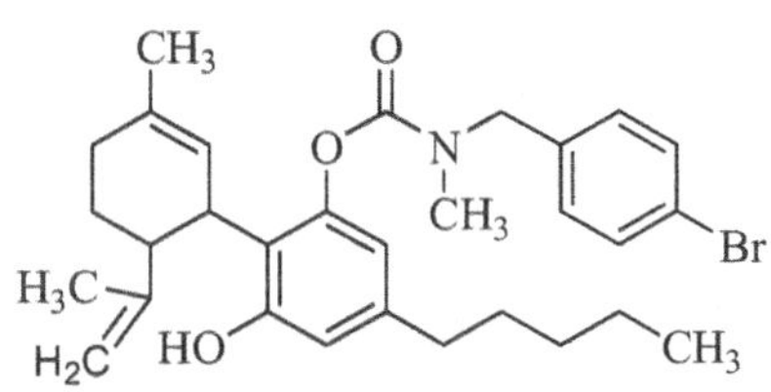

(7)

Compound **7** exhibited inhibition of $_{eq}$BuChE in nanomolar range having an IC_{50} value of 7.3 nM and selectivity index of >4000. Both the compounds **6** and **7** showed much potent hBuChE inhibition (IC_{50} values of 1.77 and 2.15 μM, respectively) in comparison to rivastigmine.

For the synthesis of compounds **6** and **7**, the substituted amines were converted to the respective carbamic chlorides by treating them with triphosgene in cold conditions. The respective carbamic acid chlorides were condensed with the cannabinoid derivative in the presence of 4-(dimethylamino)pyridine (DMAP) as base and acetonitrile as solvent at 65 °C to obtain the desired derivatives (**6**, **7**) (Scheme 7.1).

To study the effect of carbamate binding to BuChE, Scheiner et al. (2021) and Hoffmann et al. (2019) carried out creative modification of carbamate moiety introducing diverse heterocycles and alkyl spacers and performed systematic evaluation of the resulting compounds, including binding affinity to the enzyme and carbamate binding kinetics. A group of compounds having alkyl spacers (2–10 methylenes) between different heterocycles (e.g., tetrahydroquinazoline,

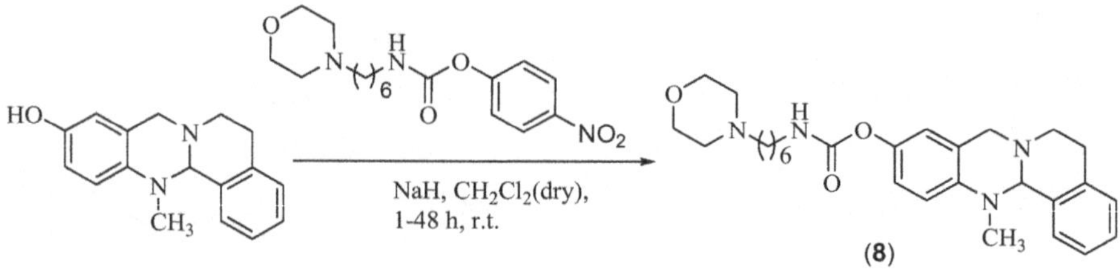

(8)

Scheme 7.2 Synthesis of compound **8**

benzimidazole, piperidine, and morpholine) and the carbamate residue were synthesized as per Scheme 7.2, and the resulting compounds were evaluated for their potential to inhibit the enzyme. Compound **8** was found to be the most potent compound among the series on the basis of enzyme inhibition and kinetic studies of the synthesized inhibitors, with BuChE inhibitory activity (IC_{50} value of 49.3 nM). The compounds also displayed neuroprotective activity against glutamate treatment and improved specifically the Aβ-induced memory impairments.

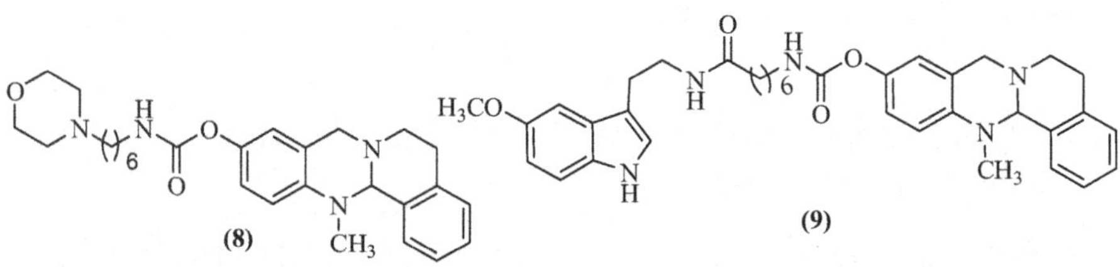

(8) (9)

Compound **8** was synthesized by nucleophilic acyl substitution reaction using the tetrahydroquinazoline derivative and 4-nitrophenylcarbamate-containing morpholine derivative in the presence of pyrophoric sodium hydride (NaH) to offer the desired product (**8**) (Scheme 7.2).

Multifactorial pathogenesis of AD involving multiple features such as metal dyshomeostasis, neuroinflammation, and oxidative stress is considered to be the causative factor for the development of AD. Scheiner et al. (2021) and Hoffmann et al. (2019) incorporated antioxidant moieties such as melatonin, ferulic acid, Trolox, and cinnamic acid into the carbamate-bearing compounds and developed some potent MTDLs. Among the synthesized compounds, compound **9** having melatonin and tetrahydroquinazoline moieties joined through carbamate linkage offered an IC_{50} value of 102 nM with additional neuroprotective and antioxidant activities.

For the synthesis of compound **9**, a suitably substituted indole derivative was reacted with 4-nitrophenyl chloroformate to offer a substituted carbamate derivative, which was further subjected to nucleophilic substitution reaction with tetrahydroquinazoline derivative in the presence of sodium hydride to offer the desired compound **9** (Scheme 7.3).

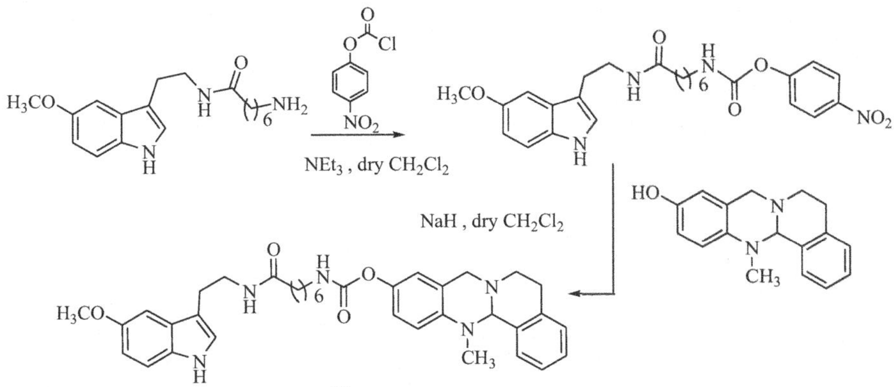

(9)

Scheme 7.3 Synthesis of compound **9**

Serotonin (5-HT), an endogenous neurotransmitter, plays numerous roles in physiological processes like memory and learning and in many pathological conditions. It was observed that the concentration of serotonin receptors (5-HTRs) got reduced in the cortex region in AD patients to prevent the decreasing levels of ACh, as a means of circumventing the changes caused by the pathophysiology of the disease. Substantial proof appeared that activation of 5-HTRs could significantly reduce cholinergic transmission, and ACh production could be stimulated by inhibition of 5-HTRs, indicating that we could probably combat AD by blocking 5-HTRs and inhibiting AChE. Based on this observation, Toublet et al. (2021) reported pleiotropic prodrugs as dual-butyrylcholinesterase inhibitors and 5-HT_6 receptor antagonists. Among the reported series, compound **10** behaves as a covalent inhibitor of BuChE (IC_{50} = 0.97 μM). The synthesis of compound **10** was carried out using Scheme 7.4 employing 6-fluoro-*1H*-indole and converting it into an aldehyde using Vilsmeier-Haack aldehyde synthesis in the presence of $POCl_3$ and DMF. The aldehyde is then reacted with nitromethane in the presence of ammonium acetate under reflux to produce an alkene, which is further reduced to an alkyl amine. The amine portion of indole derivative is reacted with the carbamate-containing aldehyde to produce a Schiff base, which is reduced to an amine in situ, to offer the desired compound **10** (Scheme 7.4).

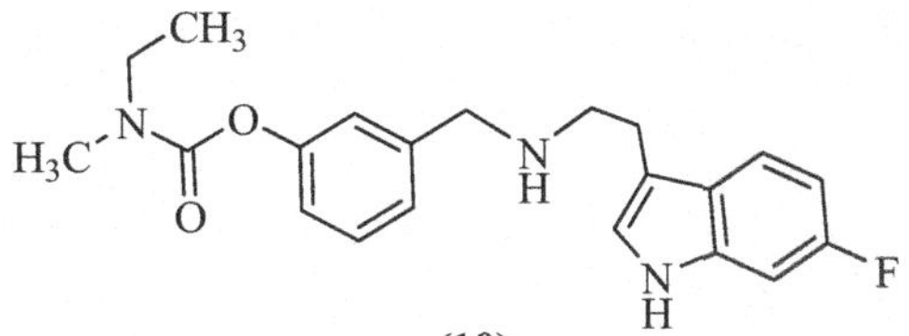

(10)

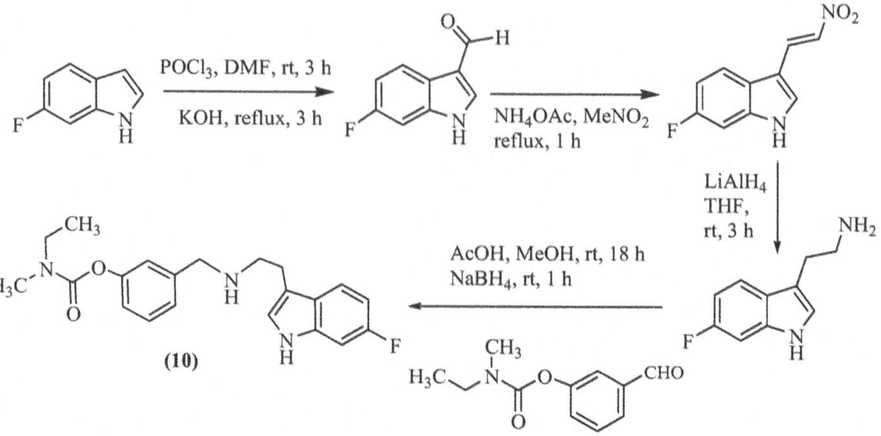

Scheme 7.4 Synthesis of compound **10**

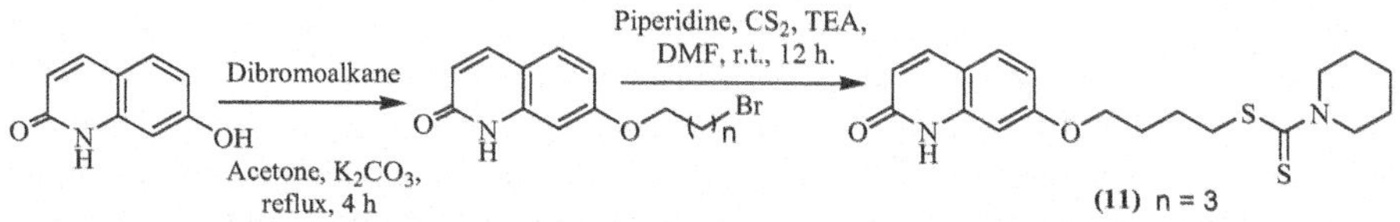

Scheme 7.5 Synthesis of compound **11**

Recently, Fu et al. (2020) reported analogs of quinolinones having a dithiocarbamate group. In vitro evaluation showed that compound **11** inhibited self-induced Aβ aggregation and acted as a mixed type of cholinesterase inhibitor. It was observed to be an effective inhibitor at 25 μM concentration of self-induced Aβ aggregation (30.67%) and the best inhibitor at 100 μM of Aβ aggregation induced by AChE (29.02%). Compound **11** acted as a potent inhibitor of both *hAChE* (IC_{50} value of 0.16 μM) and *eeAChE* (IC_{50} value of 0.22 μM). Kinetic studies and molecular modeling studies showed that the compound **11** caused mixed type of inhibition with the capability to bind simultaneously to both the catalytic active site (CAS) and the peripheral anionic site (PAS) of AChE. Compound **11** was synthesized as per Scheme 7.5 wherein the hydroxyquinoline derivative was reacted with dibromoalkane to produce an ether which was further reacted with appropriate secondary amine in the presence of carbon disulfide in DMF to offer compound **11**.

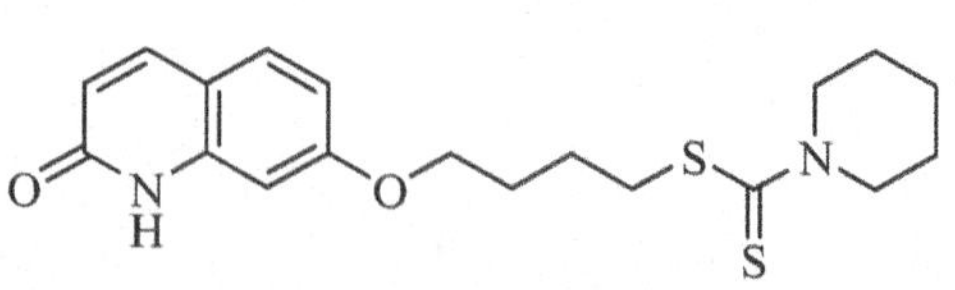

(11)

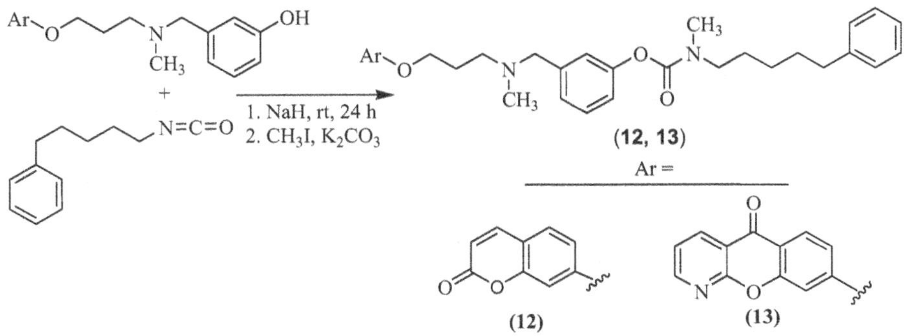

Scheme 7.6 Synthesis of compounds **12** and **13**

Coumarin, a well-known naturally occurring product, is reported to have excellent AChE inhibition potency. The molecule has infused strong interest among the researchers for developing novel agents against AD using this scaffold. Rampa et al. (2012) reported carbamate-based coumarin derivatives as anti-AD agents. Biological evaluation of the prepared derivatives showed that compounds **12** and **13** exhibited inhibitory activity in the nanomolar range against AChE, BChE, and fatty acid amide hydrolase (FAAH). This was a totally new and first-of-its-kind approach adopted by the researchers to develop novel entities exhibiting anti-AD effects, wherein the designed compounds possessed dual-FAAH-cholinesterase inhibitory activities. Compound **12** showed IC_{50} value of 74.9 nM against *h*AChE and 1.57 nM against *h*BuChE. Compound **13** showed IC_{50} value of 89.5 nM against *h*AChE and 1.71 nM against *h*BuChE. Both the compounds **12** and **13** were synthesized as per Scheme 7.6, wherein substituted phenol derivatives for the respective compounds were reacted with 5-isocyanatopentylbenzene in the presence of sodium hydride to offer compounds **12** and **13**, respectively.

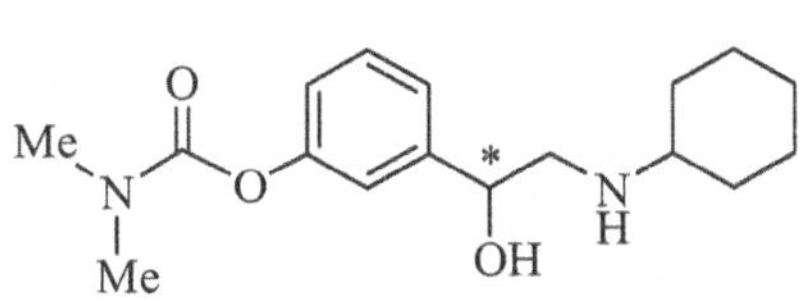

14

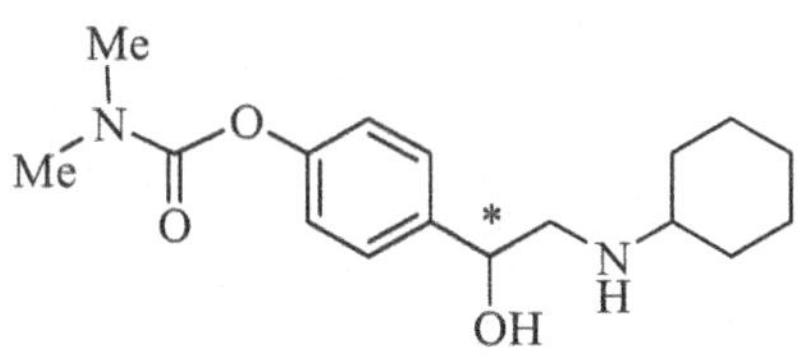

15

Terbutaline, a well-known β_2-adrenergic agonist, is used as an anti-asthmatic drug. Bambuterol (BMB) is its prodrug, which gets hydrolyzed by ChE to terbutaline, the active drug. BMB exhibited weak inhibitory activity (IC_{50} value of 3×10^{-5} M) against AChE but strong inhibitory activity against BuChE (IC_{50} value of 3×10^{-9} M). Based on these observations, Wu et al. (2020) designed and synthesized some carbamate derivatives as AChE and BuChE inhibitors. Among

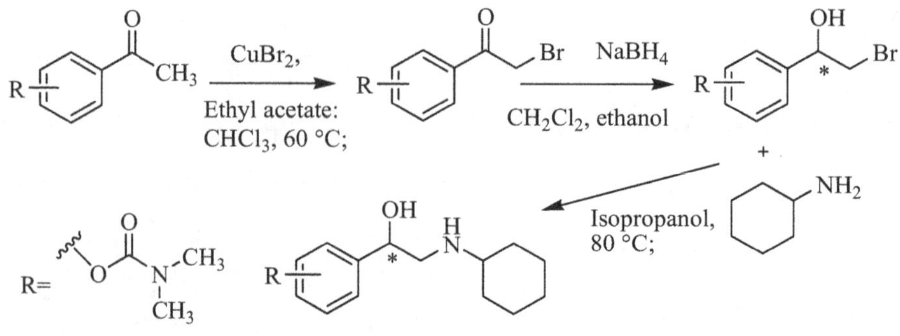

Scheme 7.7 Synthesis of compounds **14** and **15**

the reported series, compound **14** (IC_{50} values of BuChE 2.2 nM and AChE 792 nM) and compound **15** (IC_{50} values of BuChE = 10.6 nM and AChE = 266 nM) were found to be the most potent compounds. It was interesting to note that both the enantiomers of these compounds **14** and **15** were found to be active against the enzyme BuChE.

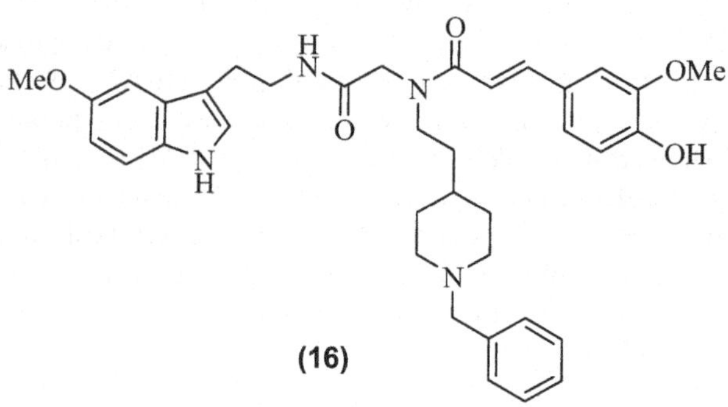

(16)

The synthesis of compounds **14** and **15** was carried out (Scheme 7.7) by simple bromination of the substituted acetophenones to convert them into α-haloketones, followed by reduction of the carbonyl group to hydroxyl, ultimately substituting the bromo group with cyclohexylamine.

7.6 Interaction of Donepezil with Cholinesterase Enzymes

Donepezil (**1**) was approved in the year 1997 by the US FDA. It is an AChE inhibitor containing 5,6-dimethoxyindanone and *N*-benzylpiperidine moieties, which are connected through a methylene bridge. The molecular docking of donepezil revealed that benzylpiperidine part of the structure exhibited interactions at the CAS site of the enzyme, binding to the catalytic site composed of amino acids Tyr70, Asp72, Trp86, Tyr121, and Tyr334, while the indanone part of the molecule showed interactions with the PAS site of the receptor eliciting interaction with Trp286 (Saxena et al. 2003).

Donepezil has a very low selectivity for BuChE due to the presence of a larger gorge of BuChE, and therefore it is unable to bind with PAS and CAS simultaneously. The carbonyl oxygen of the indanone moiety shows hydrogen bond interactions in the gorge, thereby increasing the inhibitory activity, while there is not much discussion about the effects of benzylpiperidine and its interaction with the BuChE enzyme (Agatonovic-Kustrin et al. 2018).

On the basis of in silico studies as discussed above, a number of AChE and BuChE inhibitors have been designed and synthesized having structure similarity to donepezil. Efforts have been made to retain one of the two active components of the drug while substituting the other part with some other moiety. The results of the biological evaluation of such compounds have been found to be highly satisfying. In the following section, the most active molecules with the highest potency along with their syntheses are discussed in order to give an idea about the current scenario of research on anti-AD activities of the resulting compounds.

7.6.1 Design and Development of Donepezil-Based Compounds

In order to develop a series of multi-targeted drugs for the treatment of Alzheimer's disease, Benchekroun et al. (2015) reported hybrids of donepezil-ferulic acid. The compounds were synthesized by one-pot U-4CR technique at room temperature for 24 h. Out of the 12 derivatives reported, compound **16** has shown the most potent activity against AChE and BuChE. Compound **16** showed activity similar to donepezil having IC_{50} value of 29.3 nM. The derivatives were also tested for their oxygen radical scavenging ability, and it has been observed that compound **16** showed free radical scavenging activity at 8.71 μM (Benchekroun et al. 2015).

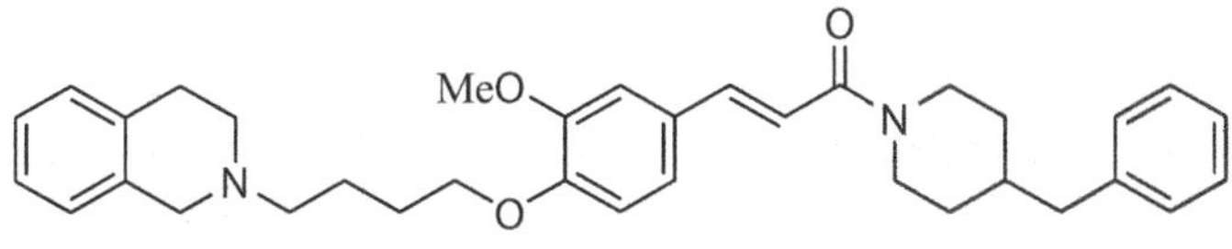

(17)

The compound **16** was synthesized using one-pot reaction by stirring a mixture of 3-(2-isocyanoethyl)-5-methoxy-1*H*-indole, ferulic acid, and N-benzylated piperidine derivative in DCM and methanol at room temperature overnight (Scheme 7.8).

Similarly, in 2016, Sang et al. (2017) reported ferulic acid-*O*-alkylamine derivatives as anti-Alzheimer agents. All the derivatives were screened for the AChE and BuChE inhibitory activity along with Aβ self-induced aggregation and antioxidant profiles. Among the synthesized derivatives, compound **17** showed promising results as AChE ($IC_{50} = 0.39$ nM) and BuChE ($IC_{50} = 76$ nM) inhibiting

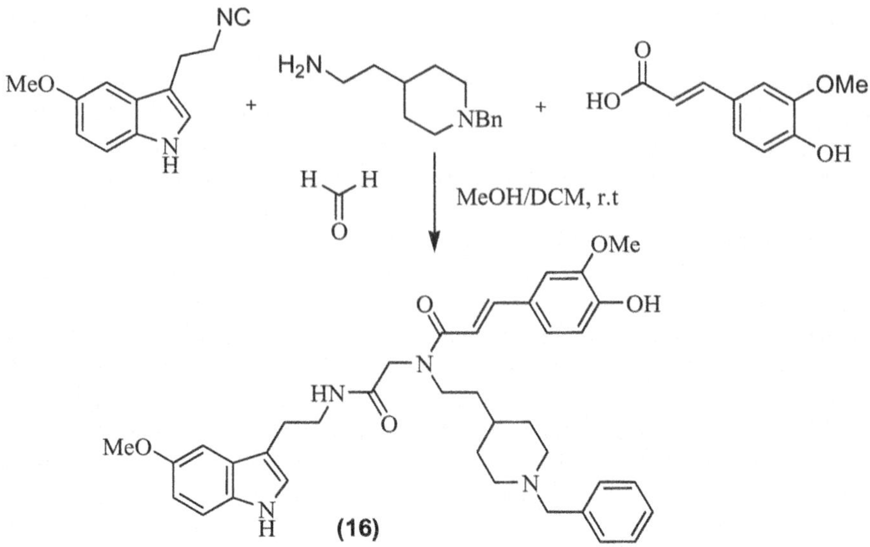

Scheme 7.8 Synthesis of compound **16**

agent. This compound has also shown low toxicity along with the antioxidant activity equivalent to 0.55 eq. of Trolox (Sang et al. 2017).

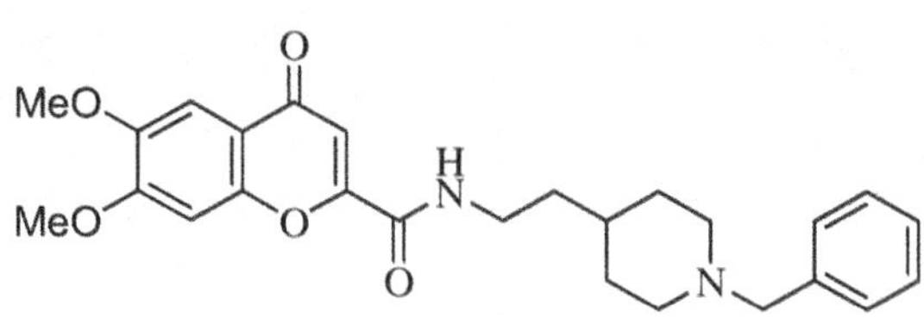

(18)

The most potent compound **17** was synthesized (Scheme 7.9) by reacting ferulic acid with 4-benzylpiperidine in the presence of HOBt and EDC using THF as the solvent. The reaction mixture was kept at room temperature under stirring. The solvent was removed under vacuum after completion of the reaction and diluted with water and the mixture extracted with dichloromethane. The compound so obtained was reacted with 1,4-dibromobutane under an inert atmosphere by heating the reaction mixture for 8–10 h at 60–65 °C in the presence of potassium carbonate as a base and acetonitrile as the solvent. The reaction mixture was extracted with DCM after completion of the reaction mixture and the solvent removed to obtain a compound, which was then reacted with tetrahydroisoquinoline (THIQ) to yield the final product, in the presence of K_2CO_3 and CH_3CN at 60–65 °C and continuous stirring under argon atmosphere.

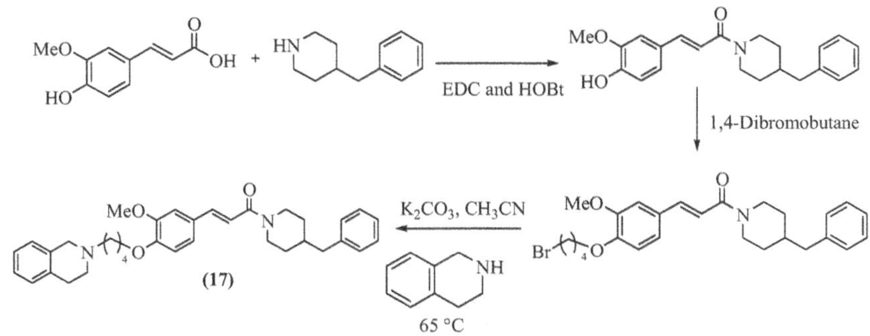

Scheme 7.9 Synthesis of compound **17**

In 2018, donepezil-flavonoid hybrids were explored by Estrada Valencia et al. (2018) and his group with an aim to synthesize multi-targeted derivatives. The compounds were synthesized with an aim to have affinity towards sigma-1 receptor along with inhibition of AChE, 5-lipoxygenase, and monoamine oxidase. Among the synthesized compounds, the benzylpiperidine derivative (**18**) has shown activity in nanomolar range against all the specified targets. This compound also showed good affinity for *h*AChE with IC_{50} value of 46 nM (Estrada Valencia et al. 2018).

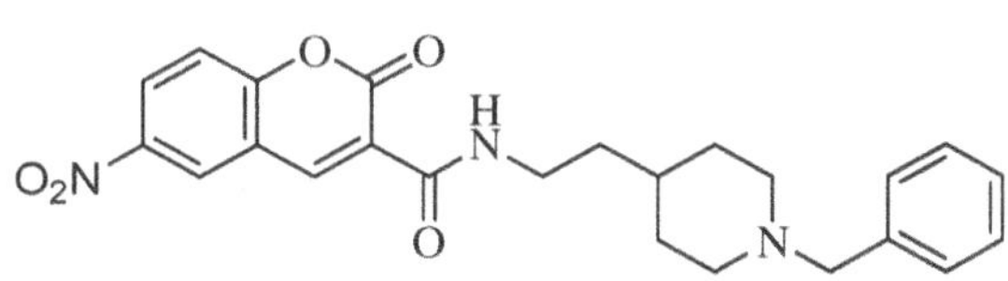

(19)

The most active compound **18** was synthesized (Scheme 7.10) using the chromone derivative 6,7-dimethoxy-4-oxo-4*H*-chromene-2-carboxylic acid as the starting material. The solvent used for the reaction was DMF under nitrogen environment. To the reaction mixture, BOP and triethylamine were added with continuous stirring. After 5 min, 2-(1-benzylpiperidin-4-yl)ethan-1-amine was added to the reaction mixture at room temperature and the reaction mixture stirred overnight. On completion of the reaction, the reaction mixture was sequentially washed with dilute hydrochloric acid and aqueous solution of $NaHCO_3$ followed by brine and purified using column chromatography.

In 2013, potent acetylcholinesterase inhibitors were reported by Asadipour et al. (2013) having coumarin-3-carboxamide and benzylpiperidine moieties. A number of derivatives were synthesized using different linkers. Compound **19** having N-ethylcarboxamide linker and a nitro substituent at sixth position has shown potent activity with high selectivity and 46-fold higher potency against AChE with IC_{50} value of 0.3 nM as compared to the standard drug donepezil (Asadipour et al. 2013).

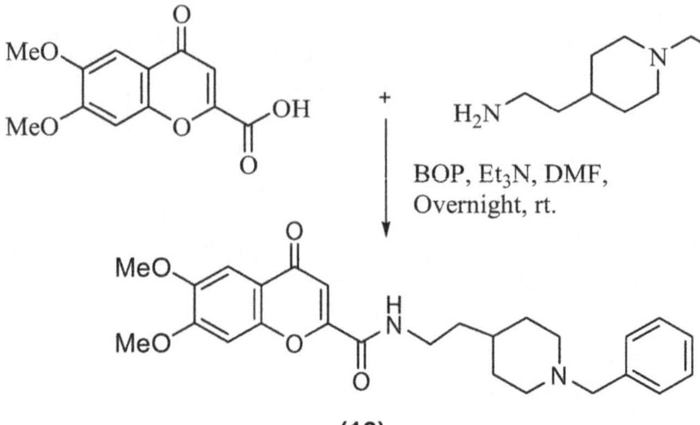

(18)

Scheme 7.10 Synthesis of compound **18**

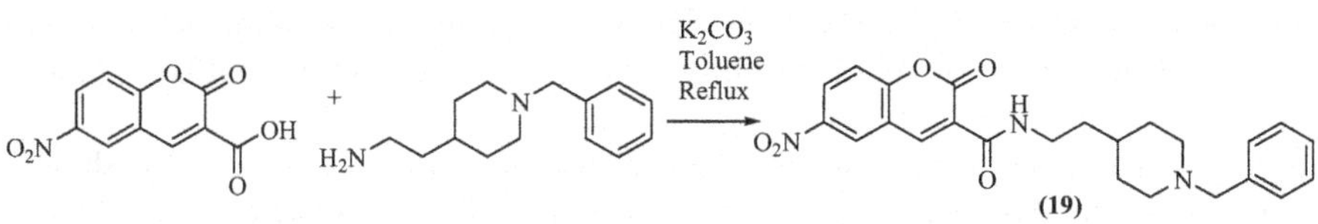

Scheme 7.11 Synthesis of compound **19**

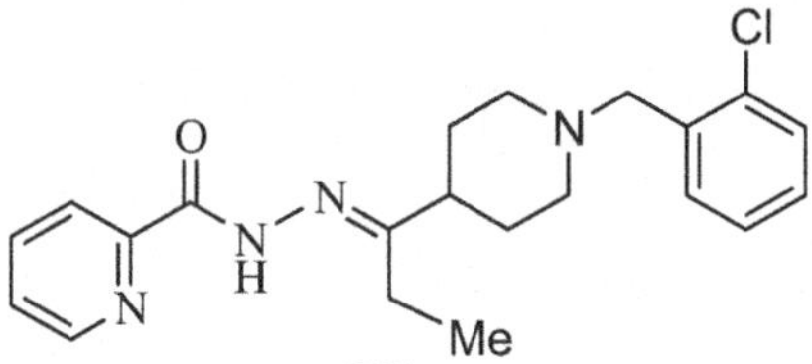

(20)

The compound **19** was synthesized (Scheme 7.11) using substituted coumarin-3-carboxylic acid and amine. Firstly, the coumarin-3-carboxylic acid was refluxed with thionyl chloride for 3–5 h. After completion of the reaction, excess thionyl chloride was removed and 2-(1-benzylpiperidin-4-yl)ethanamine was added to the reaction mixture along with anhydrous potassium carbonate using dry toluene as a solvent. The reaction was refluxed for 6–12 h and monitored by TLC for the completion of the reaction. The reaction mixture was processed, the solvent was removed under vacuum, and the crude product so obtained was purified by column chromatography.

In 2019, Zhou et al. (2019) reported that the main cause of dementia and disability in AD patients is cholinergic depletion, and AChE enzyme is mainly responsible for the cholinergic deficit. Thus, a novel series of compounds were designed having different chemical fragments like pyridine, acylhydrazone, and N-benzylpiperidine. Among the synthesized compounds, compound **20** has shown selectivity and

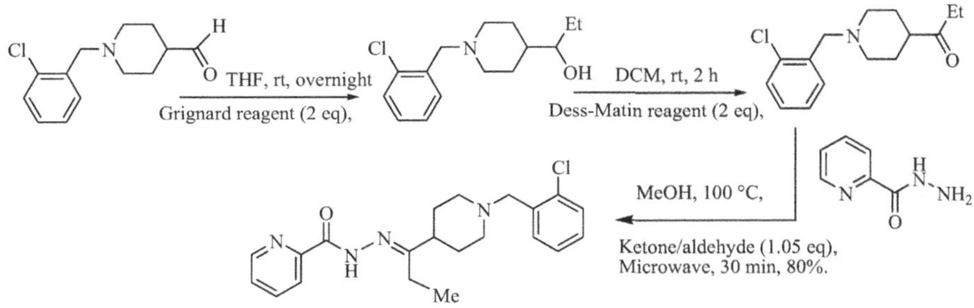

(20)

Scheme 7.12 Synthesis of compound **20**

activity against AChE with IC_{50} value of 6.0 nM, while no activity was observed against BuChE. Drug-like property through ADMET predication and PAMPA permeability evaluation has also been reported for the compound (Zhou et al. 2019).

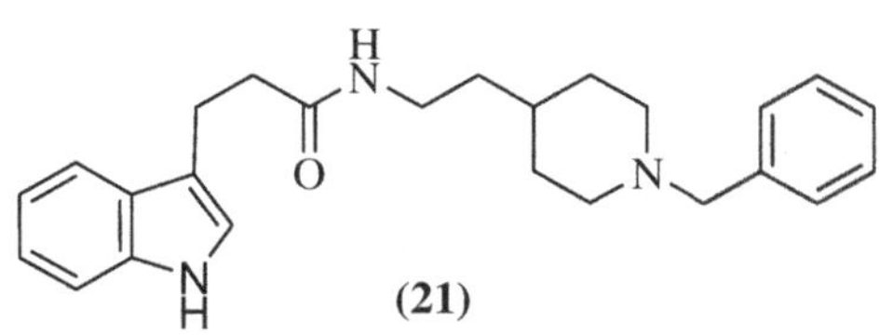

(21)

The synthesis of compound **20** involved green synthesis (Scheme 7.12) and was conducted under microwave conditions using methanol as a solvent. All the required chemicals along with the solvent were irradiated in a microwave reactor for 30 min at 100 °C. The excess solvent was removed from the reaction mixture after the completion of the reaction. The residue so obtained was purified by column chromatography to obtain the final product (**20**).

In 2016, Wang et al. (2016) reported a novel series of multi-targeted compounds. Carboxamide-linked N-benzylpiperidine–indole-based hybrids were reported. Among the synthesized compounds, compound **21** has shown effective antioxidant activity at 20 uM and 56.3% inhibition of Aβ aggregation. Compound **21** successfully crossed blood-brain barrier and decreased the death of PC12 cells by chelating the metal ions, which led to decrease in oxidative stress in the cells. It has also shown effective binding affinity with CAS and PAS sites of AChE.

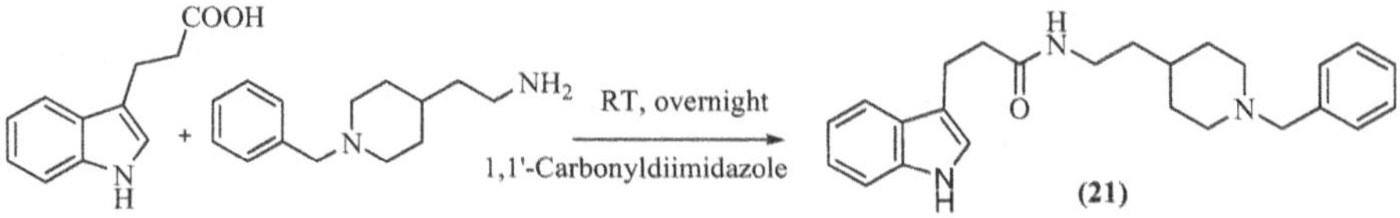

Scheme 7.13 Synthesis of compound **21**

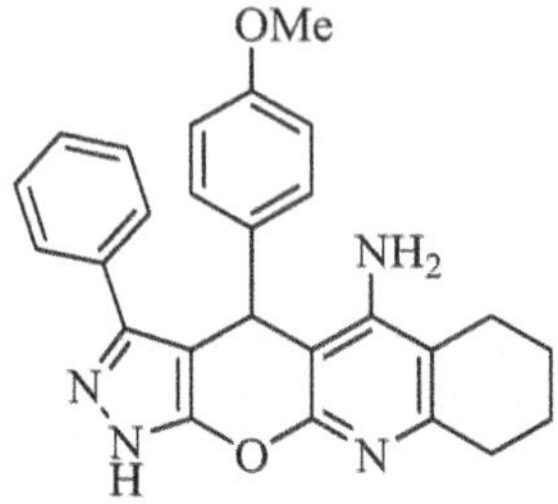

(22)

The most potent compound **21** was synthesized using 1,1-carbonyldiimidazole as the dehydrating agent and the acid, 3-(1*H*-indol-3-yl)propanoic acid, as the starting material in tetrahydrofuran as the solvent at room temperature. 1-Benzylpiperidin-4-ethylamine was added to the reaction mixture after 1-h stirring, and the reaction mixture was continued to be stirred overnight. After completion of the reaction, the reaction mixture was diluted with water, extracted with ethyl acetate, and processed further to obtain the desired compound **21** (Scheme 7.13).

7.7 Interaction of Tacrine with Cholinesterases

Tetrahydroaminoacridine or tacrine was the first marketed drug for AD patients, introduced in 1993 under the name of Cognex, which acted as reversible AChE and BuChE inhibitor, but unfortunately it was withdrawn from the market in the year 2013 due to its persistent hepatotoxicity (Heilbronn 1961). From the in silico studies, it has been seen that tacrine is binding to AChE catalytic triad via double π stacking interaction with the aromatic amino acid residues Trp84 and Phe330, which is responsible for binding of Ach to the active site. A π–π stacking interaction was observed between the indole moiety of Trp84 and the phenyl ring of benzylpiperidine, while the tertiary amine of the piperidine moiety interacted with the phenyl ring of Phe330 to yield a cation–π interaction (Sussman et al. 1991). Tacrine was one of the most widely prescribed drugs for AD, but because of its major side effect of hepatotoxicity, it was stopped from clinical use. Tacrine analogs possess a great scope to be used as the potential treatment of choice if the issue of hepatotoxicity is taken care and the potency is maintained. In the following section, efforts have been made to discuss some recently reported potent tacrine derivatives,

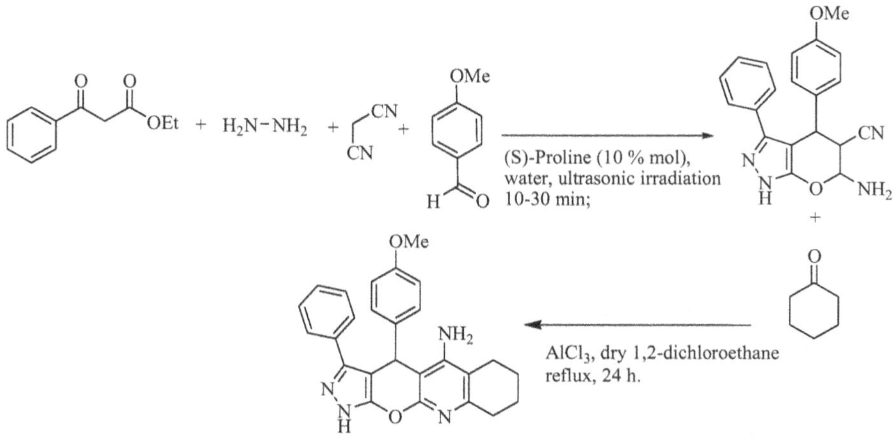

(22)

Scheme 7.14 Synthesis of compound **22**

which have shown potency higher than tacrine without the side effect of hepatotoxicity.

Pourabdi et al. (2016) in 2016 designed and synthesized a series of novel pyrano [2,3-*c*]pyrazoles by taking tacrine as the lead molecule. The designed compounds were prepared considering AChE/BuChE as the targets. Among the synthesized compounds, one compound (**22**) produced inhibition in nanomolar concentration against both BuChE and AChE offering IC_{50} values of 0.007 and 0.081 μM, respectively, against the enzymes. Additionally, compound **22** was also found to show potent activity against 15-LOX with IC_{50} value of 31 μM. Most importantly, the compound was found to be nontoxic in comparison to tacrine in cytotoxicity assay on HepG2 cells even at concentrations as high as 50 μM.

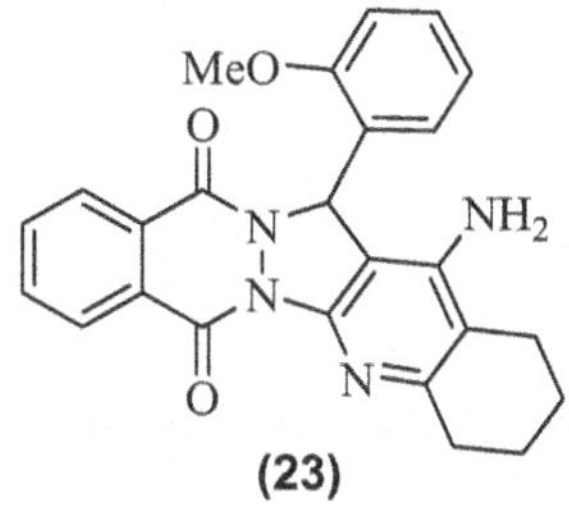

(23)

For the synthesis of compound **22**, the pyrazole-carbonitrile intermediate was synthesized by reacting ethyl benzoylacetate with hydrazine hydrate, *anisaldehyde*, and malononitrile in water under microwave irradiation for 10–30 min. The intermediate so formed is then reacted with cyclohexanone in the presence of anhydrous aluminum chloride under reflux to produce compound **22** (Scheme 7.14).

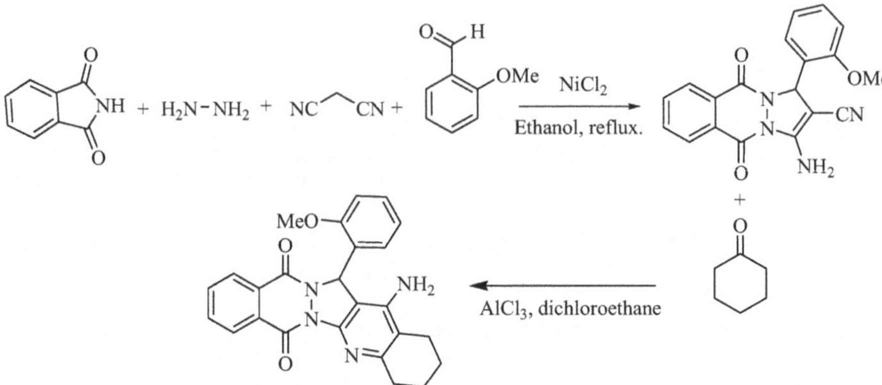

(23)

Scheme 7.15 Synthesis of compound **23**

Jalili-Baleh et al. (2017) in 2017 reported fused pyrazolo[1,2-*b*]phthalazines as anti-AD agents. Compound **23** exhibited higher selectivity for AChE over BuChE, and it was found to be sevenfold more potent ($IC_{50} = 49$ nM) than tacrine. Using neuronal cell line (PC12) and hepatocytes (HepG2) for the cell-based assay, compound **23** was found to show significantly low hepatotoxic effect and provided additional neuroprotective activity against hydrogen peroxide-induced cell damage in PC12 cells. Overall, it was seen that the compound was highly potent, selective, lowly toxic, and neuroprotective in comparison to the standard drug tacrine.

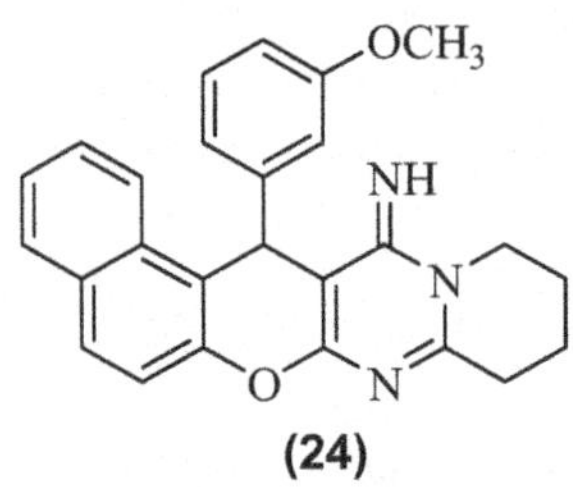

(24)

For the synthesis of compound **23**, phthalimide, hydrazine hydrate, malononitrile, and 2-methoxybenzaldehyde were refluxed in ethanol in the presence of nickel chloride to offer an intermediate (Scheme 7.15). The intermediate on refluxing with cyclohexanone in dichloromethane in the presence of anhydrous $AlCl_3$ offered the desired compound **23**.

Dgachi et al. (2016) in 2016 prepared analogs of tacrine with an intent to inhibit the cholinesterase enzymes. Compound **24** showed the highest activity with AChE inhibition having IC_{50} value of 5.3 nM. Compound **24** was also found to be possessing good antioxidant and amyloid aggregation inhibitory properties. Six-membered ring attached to the pyrimidine ring when replaced with five- or

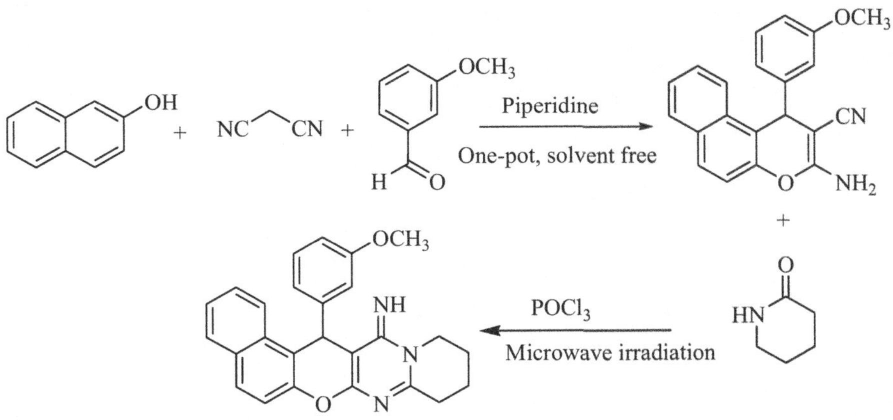

(24)

Scheme 7.16 Synthesis of compound **24**

seven-membered rings, and the activity was found to decrease; therefore, it could be inferred that a six-membered ring is essential for good AChE inhibitory activity.

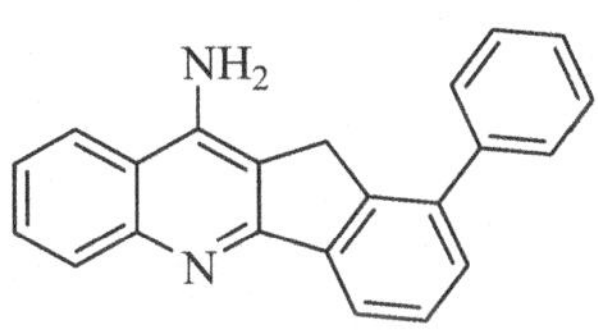

(25)

A one-pot multicomponent reaction starting from malononitrile, 3-methoxybenzaldehyde, and β-naphthol in the presence of piperidine yielded the carbonitrile derivative. Condensation of the intermediate carbonitrile with the lactam for 15 min under microwave conditions in the presence of phosphoryl chloride as the Lewis acid and bromobenzene as the solvent offered the desired compound **24** (Scheme 7.16).

Ekiz et al. (2018) in 2018 synthesized bromoindenoquinolines using Friedlander reaction along with Suzuki coupling to produce compounds with cholinergic inhibitory activity. Out of all the synthesized compounds, compound **25** bearing a phenyl substituent offered the most desirable biological activity yielding BuChE inhibition (IC_{50} value of 93 nM) and AChE inhibition (IC_{50} value of 37 nM), thereby making it a strong cholinesterase enzyme inhibitor. It was also found that the compound was very less hepatotoxic compared to tacrine showing higher potency. Therefore, compound **25** is considered to be a good lead for further drug designing and optimization to offer newer active molecules.

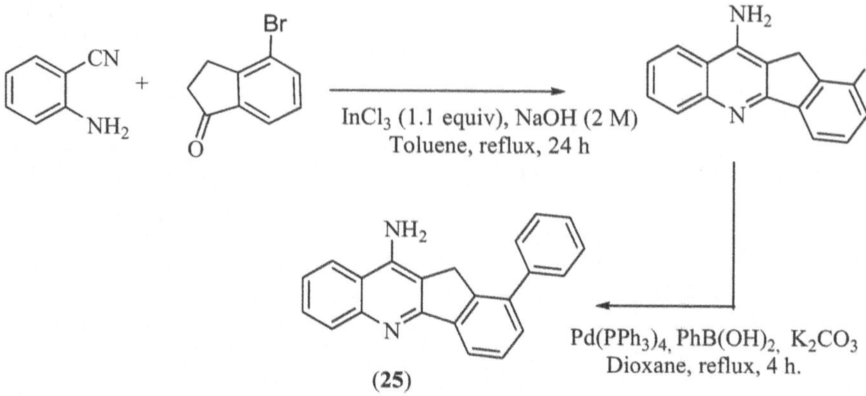

Scheme 7.17 Synthesis of compound **25**

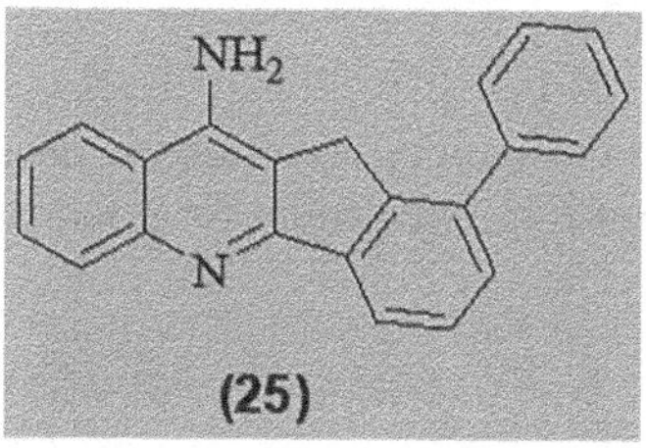

The synthesis of compound **25** was done (Scheme 7.17) using Friedlander reaction in which the carbonitrile compound was taken in toluene and added to bromoindanone to produce the intermediate. The bromo group was replaced with phenyl ring using Suzuki coupling reaction to yield the desired compound **25**.

7.8 Conclusion

Alzheimer's disease, a neurodegenerative disease, has offered multiple targets in its pathophysiology, which could be exploited in order to either slow down or inhibit the progression of the disease. Inhibition of the most likely and popular targets, i.e., AChE and BuChE, has remained a choice for researchers and medicinal chemists, which involved targeting the active sites PAS and CAS of the enzymes. In order to design and synthesize anti-Alzheimer's drugs which act by cholinesterase inhibition, these sites are to be targeted to produce compounds with high potency. Therefore, already existing US FDA-approved marketed drugs donepezil and rivastigmine, which have the potential to bind to both CAS and PAS sites of the cholinesterase enzymes, have been considered to be the primary choice for novel drug designing and for which tremendous amount of work has been done in the past, and research work is still ongoing for the development of potent anticholinesterase inhibitors to treat AD. Furthermore, when we consider rivastigmine as a standard, the carbamate part along with the tertiary amine group becomes important as these groups bind to the respective CAS and PAS sites of the enzyme, whereas benzylpiperidine (binding to CAS) and 5,6-dimethoxyindanone (binding to PAS) of donepezil are considered to be the other two important moieties for drug designing. On the basis of the information acquired from in silico studies conducted for donepezil and rivastigmine on AChE and BuChE enzymes, a number of reports in literature showed similar interactions which mimic the standard drugs wherein different moieties like ferulic acid, flavonoids, coumarins, quinolines, tetrahydroisoquinoline, and many heterocycles were attached to the active parts of either donepezil or rivastigmine, offering highly potent compounds when compared to the standard drugs (Yadav et al. 2023b).

Among the compounds discussed above, compound **6** which was designed on the basis of rivastigmine moiety containing the active carbamate component was synthesized by Jiang et al. (2021). Compound **6** had shown AChE inhibition with IC_{50} value of 5.3 nM and BuChE with IC_{50} value of 13 nM, along with good neuroprotective and antioxidant activities, with the capacity to permeate through the BBB. Compound **6** possesses drug-like properties, and therefore it could be considered as an important lead molecule to further optimize into novel anticholinergic drug molecules based on carbamate component of rivastigmine. The compounds discussed under donepezil-based designing, coumarin-3-carboxamide derivative attached to benzylpiperidine part of donepezil, compound **19** (Asadipour et al. 2013), showed highly potent activity with high selectivity and 46-fold more potency than its lead molecule donepezil as a standard drug against AChE with IC_{50} value of 0.3 nM. This proves that there is still ample scope for research to produce potent compounds bearing the benzylpiperidine fragment of donepezil. For the tacrine-based derivatives, compound **24** (Dgachi et al. 2016) possessed AChE inhibition activity offering IC_{50} value of 5.3 nM, which was more potent than the lead compound tacrine and was comparatively very less hepatotoxic; therefore, such series of compounds can be considered for further development as anti-AD drugs, where modifications on the tricyclic ring system result in reduced hepatotoxicity and increased potency.

In a nutshell, it can be concluded that rivastigmine and donepezil as lead molecules offer tremendous opportunities for the development of newer molecules to treat AD by targeting the cholinesterase enzymes, wherein the active components of these two marketed drugs are attached to various heterocyclic and other chemical moieties and the resulting compounds are evaluated for anticholinesterase activity. It can also be concluded that MTDL approach is one of the most preferred approaches of structure-based drug designing (Murumkar et al. 2023) when it comes to designing of newer anti-AD drugs.

References

Agatonovic-Kustrin S, Kettle C, Morton DW (2018) A molecular approach in drug development for Alzheimer's disease. Biomed Pharmacother 106:553–565

Asadipour A, Alipour M, Jafari M, Khoobi M, Emami S, Nadri H, Sakhteman A, Moradi A, Sheibani V, Homayouni Moghadam F, Shafiee A, Foroumadi A (2013) Novel coumarin-3-carboxamides bearing N-benzylpiperidine moiety as potent acetylcholinesterase inhibitors. Eur J Med Chem 70:623–630. https://doi.org/10.1016/j.ejmech.2013.10.024

Ashraf GM, Tarasov VV, Makhmutova A, Chubarev VN, Avila-Rodriguez M, Bachurin SO, Aliev G (2019) The possibility of an infectious etiology of Alzheimer disease. Mol Neurobiol 56(6): 4479–4491. https://doi.org/10.1007/s12035-018-1388-y

Barmade M, Shidore M, Rajyaguru S, Machhi J, Murumkar P, Yadav MR (2017) Design and development of pramipexole-donepezil hybrids as potential therapeutics for Alzheimer's disease. Amer Chemical Soc 254

Bar-On P, Millard CB, Harel M, Dvir H, Enz A, Sussman JL, Silman I (2002) Kinetic and structural studies on the interaction of Cholinesterases with the anti-Alzheimer drug Rivastigmine. Biochemistry 41(11):3555–3564

Benchekroun M, Ismaili L, Pudlo M, Luzet V, Gharbi T, Refouvelet B, Marco-Contelles J (2015) Donepezil-Ferulic acid hybrids as anti-Alzheimer drugs. Future Med Chem 7(1):15–21. https://doi.org/10.4155/fmc.14.148

Darreh-Shori T, Kumar R, Kumar A, Murumkar P, Nag S, Jia Z, Arakawa R, Leuzy A, Lemoine L, Nordberg A, Yadav MR, Halldin C, Långström B (2019) O5-06-06: ligands of the core acetylcholine biosynthesizing enzyme, choline acetyltransferase, as novel and potential therapeutic agents and in vivo pet tracers for early diagnosis of Alzheimer's disease. Alzheimers Dement 15(7S_Part_31):P1630–P1630. https://doi.org/10.1016/j.jalz.2019.06.4869

Dgachi Y, Ismaili L, Knez D, Benchekroun M (2016) Synthesis and biological assessment of racemic benzochromenopyrimidinimines as antioxidant, cholinesterase, and a b 1 À 42 aggregation inhibitors for Alzheimer's disease. Therapy 11:1318–1327. https://doi.org/10.1002/cmdc.201500539

Ekiz M, Tutar A, Ökten S, Koçyi ÜM (2018) Synthesis, characterization, and SAR of arylated Indenoquinoline-based cholinesterase and carbonic anhydrase inhibitors. Arch Pharm (Weinheim) 351(9):e1800167. https://doi.org/10.1002/ardp.201800167

Estrada Valencia M, Herrera-Arozamena C, de Andrés L, Pérez C, Morales-García JA, Pérez-Castillo A, Ramos E, Romero A, Viña D, Yáñez M, Laurini E, Pricl S, Rodríguez-Franco MI (2018) Neurogenic and neuroprotective donepezil-flavonoid hybrids with sigma-1 affinity and inhibition of key enzymes in Alzheimer's disease. Eur J Med Chem 156:534–553. https://doi.org/10.1016/j.ejmech.2018.07.026

Fu J, Bao F, Gu M, Liu J, Zhang Z, Ding J, Xie SS, Ding J (2020) Design, synthesis and evaluation of quinolinone derivatives containing dithiocarbamate moiety as multifunctional AChE inhibitors for the treatment of Alzheimer's disease. J Enzyme Inhib Med Chem. 35(1): 118–128. https://doi.org/10.1080/14756366.2019.1687460

Galdeano C, Viayna E, Arroyo P, Bidon-Chanal A, Ramon Blas J, Munoz-Torrero D, Javier Luque F (2012) Structural determinants of the multifunctional profile of dual binding site acetylcholinesterase inhibitors as anti-Alzheimer agents. Curr Pharm Des 16(25):2818–2836. https://doi.org/10.2174/138161210793176536

Ge S, Cai M, Pei G (2022) Frequency distribution of the hereditary Alzheimer's disease-related genes seems to fit Poisson distribution, why? Cell Discov 8(1):73. https://doi.org/10.1038/s41421-022-00444-9

Giacobini E (2004) Cholinesterase inhibitors: new roles and therapeutic alternatives. Pharmacol Res 50(4):433–440

Gupta RC (2014) Tacrine. In: Wexler PBT (ed) Encyclopedia of toxicology (third edition). Academic Press, Oxford, pp 466–467. https://doi.org/10.1016/B978-0-12-386454-3.00198-6

Harel M, Schalk I, Ehret-Sabatier L, Bouet F, Goeldner M, Hirth C, Axelsen PH, Silman I, Sussman JL (1993) Quaternary ligand binding to aromatic residues in the active-site gorge of acetylcholinesterase. Proc Natl Acad Sci 90(19):9031–9035. https://doi.org/10.1073/pnas.90.19.9031

Heilbronn E (1961) Inhibition of cholinesterases by tetrahydroaminacrin. Acta Chem Scand 15(6): 1386–1390

Hoffmann M, Stiller C, Endres E, Scheiner M, Gunesch S, Sotriffer C, Maurice T, Decker M (2019) Highly selective butyrylcholinesterase inhibitors with tunable duration of action by chemical modification of transferable carbamate units exhibit pronounced neuroprotective effect in an Alzheimer's disease mouse model. J Med Chem 62(20):9116–9140. https://doi.org/10.1021/acs.jmedchem.9b01012

Jalili-baleh L, Nadri H, Moradi A, Nasir S, Burkhart A, Shakibaie M, Jafari M, Golshani M, Moghadam FH, Firoozpour L, Asadipour A, Emami S, Khoobi M, Foroumadi A (2017) New racemic annulated pyrazolo[1,2-b]phthalazines as tacrine-like AChE inhibitors with potential use in Alzheimer's disease. Eur J Med Chem 139:280. https://doi.org/10.1016/j.ejmech.2017.07.072

Jiang X, Zhang Z, Zuo J, Wu C, Zha L, Xu Y, Wang S, Shi J, Liu XH, Zhang J, Tang W (2021) Novel cannabidiol−carbamate hybrids as selective BuChE inhibitors: docking-based fragment reassembly for the development of potential therapeutic agents against Alzheimer's disease. Eur J Med Chem 223:113735. https://doi.org/10.1016/j.ejmech.2021.113735

Johnson JL, Cusack B, Davies MP, Fauq A, Rosenberry TL (2003) Unmasking tandem site interaction in human acetylcholinesterase. Substrate activation with a cationic acetanilide substrate. Biochemistry 42(18):5438–5452

Kumar A, Murumkar P, Kumar R, Yadav MR, Darreh-Shori T (2019) P3-108: chat potentiating ligands (Cpls): a novel therapeutic strategy for the treatment of Alzheimer's disease. Alzheimers Dement 15(7S_Part_18):P971–P971. https://doi.org/10.1016/j.jalz.2019.06.3136

Lane RM, He Y (2009) Emerging hypotheses regarding the influences of Butyrylcholinesterase-K variant, APOE E4, and Hyperhomocysteinemia in neurodegenerative dementias. Med Hypotheses 73(2):230–250

Leng F, Edison P (2021) Neuroinflammation and microglial activation in Alzheimer disease: where do we go from here? Nat Rev Neurol 17:157–172. https://doi.org/10.1038/s41582-020-00435-y

Lin G, Chen G, Lu C, Yeh S (2005) QSARs for peripheral anionic site of butyrylcholinesterase with inhibitions by 4-Acyloxy-biphenyl-4′-N-butylcarbamates. QSAR Comb Sci 24(8):943–952

Martins FCOL, Batista AD, Melchert WR (2021) Current overview and perspectives in environmentally friendly microextractions of carbamates and dithiocarbamates. Compr Rev Food Sci Food Saf 20(6):6116–6145

Martorana A, Esposito Z, Koch G (2010) Beyond the cholinergic hypothesis: do current drugs work in Alzheimer's disease? CNS Neurosci Ther 16(4):235–245

Mishra S, Pang S, Zhang W, Lin Z, Bhatt P, Chen S (2021) Insights into the microbial degradation and biochemical mechanisms of carbamates. Chemosphere 279:130500

Murumkar P, Sharma M, Gupta P, Patel N, Yadav MR (2023) Selection of suitable protein structure from protein data bank: an important step in structure-based drug design studies. Mini-Rev Med Chem 23(3):246–264. https://doi.org/10.2174/1389557522666220512151454

Nachon F, Masson P, Nicolet Y, Lockridge O, Fontecilla-Camps JC (2003) Comparison of the structures of butyrylcholinesterase and acetylcholinesterase. butyrylcholinesterase its. Funct Inhib:39–54

Pourabdi L, Khoobi M, Nadri H, Moradi A, Homayouni F (2016) SC. Eur J Med Chem 123:298. https://doi.org/10.1016/j.ejmech.2016.07.043

Rampa A, Bartolini M, Bisi A, Belluti F, Gobbi S, Andrisano V, Ligresti A, Di Marzo V (2012) The first dual ChE/FAAH inhibitors: new perspectives for Alzheimer's disease? ACS Med Chem Lett 3(3):182–186. https://doi.org/10.1021/ml200313p

Sang Z, Pan W, Wang K, Ma Q, Yu L, Yang Y, Bai P, Leng C, Xu Q, Li X, Tan Z, Liu W (2017) Design, synthesis and evaluation of novel Ferulic acid-O-Alkylamine derivatives as potential multifunctional agents for the treatment of Alzheimer's disease. Eur J Med Chem 130:379–392. https://doi.org/10.1016/j.ejmech.2017.02.039

Saxena A, Redman AMG, Jiang X, Lockridge O, Doctor BP (1997) Differences in active site gorge dimensions of Cholinesterases revealed by binding of inhibitors to human Butyrylcholinesterase. Biochemistry 36(48):14642–14651

Saxena A, Redman AMG, Jiang X, Lockridge O, Doctor BP (1999) Differences in active-site gorge dimensions of Cholinesterases revealed by binding of inhibitors to human butyrylcholinesterase. Chem Biol Interact 119:61–69

Saxena A, Fedorko JM, Vinayaka CR, Medhekar R, Radić Z, Taylor P, Lockridge O, Doctor BP (2003) Aromatic amino-acid residues at the active and peripheral anionic sites control the binding of E2020 (Aricept®) to cholinesterases. Eur J Biochem 270(22):4447–4458

Scheiner M, Hoffmann M, He F, Poeta E, Chatonnet A, Monti B, Maurice T, Decker M (2021) Selective pseudo-irreversible butyrylcholinesterase inhibitors transferring antioxidant moieties to the enzyme show pronounced neuroprotective efficacy in vitro and in vivo in an Alzheimer's disease mouse model. J Med Chem 64(13):9302–9320. https://doi.org/10.1021/acs.jmedchem.1c00534

Shidore M, Machhi J, Shingala K, Murumkar P, Sharma MK, Agrawal N, Tripathi A, Parikh Z, Pillai P, Yadav MR (2016) Benzylpiperidine-linked diarylthiazoles as potential anti-Alzheimer's agents: synthesis and biological evaluation. J Med Chem 59(12):5823–5846

Sussman JL, Harel M, Frolow F, Oefner C, Goldman A, Toker L, Silman I (1991) Atomic structure of acetylcholinesterase from torpedo Californica: a prototypic acetylcholine-binding protein. Science 253(5022):872–879. https://doi.org/10.1126/science.1678899

Szegletes T, Mallender WD, Thomas PJ, Rosenberry TL (1999) Substrate binding to the peripheral site of acetylcholinesterase initiates enzymatic catalysis. Substrate inhibition arises as a secondary effect. Biochemistry 38(1):122–133

Toublet FX, Lalut J, Hatat B, Lecoutey C, Davis A, Since M, Corvaisier S, Freret T, Sopková-de Oliveira Santos J, Claeysen S, Boulouard M, Dallemagne P, Rochais C (2021) Pleiotropic prodrugs: design of a dual butyrylcholinesterase inhibitor and 5-HT6 receptor antagonist with therapeutic interest in Alzheimer's disease. Eur J Med Chem:210. https://doi.org/10.1016/j.ejmech.2020.113059

Venneri A, McGeown WJ, Shanks MF (2005) Empirical evidence of neuroprotection by dual cholinesterase inhibition in Alzheimer's disease. Neuroreport 16(2):107–110

Wang J, Wang Z-M, Li X-M, Li F, Wu J-J, Kong L-Y, Wang X-B (2016) Synthesis and evaluation of multi-target-directed ligands for the treatment of Alzheimer's disease based on the fusion of donepezil and melatonin. Bioorg Med Chem 24(18):4324–4338. https://doi.org/10.1016/j.bmc.2016.07.025

Wu J, Pistolozzi M, Liu S, Tan W (2020) Design, synthesis and biological evaluation of novel carbamates as potential inhibitors of acetylcholinesterase and Butyrylcholinesterase. Bioorganic Med Chem 28(5):115324. https://doi.org/10.1016/j.bmc.2020.115324

Yadav MR, Barmade MA, Chikhale RV, Murumkar PR (2018) Computational modelling of kinase inhibitors as anti-Alzheimer agents. In: Roy K (ed) Computational modeling of drugs against Alzheimer's disease, Neuromethods, vol 132. Humana Press, New York, NY. https://doi.org/10.1007/978-1-4939-7404-7_14

Yadav MR, Murumkar PR, Barot R, Yadav R, Joshi K, Chauhan M (2023a) Role of computational modeling in drug discovery for Alzheimer's disease. In: Kar S, Leszczynski J (eds) Current trends in computational modeling for drug discovery. Challenges and advances in computational chemistry and physics, vol 35. Springer, Cham. https://doi.org/10.1007/978-3-031-33871-7_3

Yadav MR, Murumkar PR, Yadav R, Joshi K (2023b) Chapter 3 - Structure-based virtual screening in drug discovery. In: Roy K (ed) Cheminformatics, QSAR and machine learning applications for novel drug development. Academic Press, pp 69–88. https://doi.org/10.1016/B978-0-443-18638-7.00006-2. isbn:9780443186387

Zhang H, Wang Y, Wang Y, Li X, Wang S, Wang Z (2022) Recent advance on carbamate-based cholinesterase inhibitors as potential multifunctional agents against Alzheimer's disease. Eur J Med Chem 240:114606. https://doi.org/10.1016/j.ejmech.2022.114606

Zhou Y, Sun W, Peng J, Yan H, Zhang L, Liu X, Zuo Z (2019) Design, synthesis and biological evaluation of novel copper-chelating acetylcholinesterase inhibitors with pyridine and N-benzylpiperidine fragments. Bioorg Chem 93:103322. https://doi.org/10.1016/j.bioorg.2019.103322

Memantine-Based Derivatives: Synthesis and Their Biological Evaluation

8

Vinod Kumar, Vijay Kumar, Naveen Kumar, Vinay Kumar, and Kailash Jangid

Abstract

Alzheimer's disease (AD) is a chronic neurological disorder characterized by memory loss and cognitive decline. It is a multifactorial disease, and a number of hypotheses including glutamate hypothesis are associated with the initiation and progression of AD. The ionic glutamate receptors, mainly the N-methyl-D-aspartate receptors (NMDAR), are largely dependent on the single excitatory agonist glutamate. Glutamate is released for milliseconds at the synaptic ending to communicate with other neurons, allowing the nervous system to transmit complex motor commands and sensory information and to form thoughts and memories. Memantine is an open-channel, low-affinity, uncompetitive antagonist of NMDA receptors. It was found that the physiological receptor functions of NMDAR were unhindered by memantine and selectively block ion channels under pathological conditions, and thus maintain an optimal level of NMDA. Memantine has emerged as a promising therapeutic approach for AD. This chapter explores various methodologies for the synthesis of memantine and its pharmacological properties. The chapter also describes selective action of memantine on NMDAR and recent advancements in the field of novel drug discovery based on memantine to target NMDAR and develop an effective candidate, which may help ameliorate AD symptoms and potentially slow down the progression of disease. Understanding memantine's mechanisms of action could pave the way for novel strategies in the fight against Alzheimer's disease.

V. Kumar (✉) · V. Kumar · N. Kumar · V. Kumar · Kailash Jangid
Laboratory of Organic and Medicinal Chemistry, Department of Chemistry, Central University of Punjab, Bathinda, Punjab, India
e-mail: vinod.kumar@cup.edu.in

A. Sharma, G. P. Modi (eds.), *Natural Product-Based Synthetic Drug Molecules in Alzheimer's Disease*, https://doi.org/10.1007/978-981-99-6038-5_8

Keywords

Alzheimer's disease · N-methyl-D-aspartate receptor · NMDAR · Memantine · Drug discovery · Multifunctional agents

8.1 Introduction

Alzheimer's disease (AD) is a chronic neurological disorder associated with an irreversible loss of memory and cognitive abilities. In India, only AD and dementia-related diseases are the 13th leading cause of death. Alzheimer's Disease International stated that 75% of people suffering from dementia-related diseases go undiagnosed. According to the World Alzheimer Report 2021, 55 million people around the world suffer from dementia-related diseases, a number that is expected to increase to 78 million by the end of the third decade of the twenty-first century. Synaptic dysfunction arises from excitatory neuronal damage and death, associated with AD histopathology. Ca^{2+} influx through the ion channel at the synaptic end causes excitatory neuronal damage and triggers neuronal death by excessive concentration of excitatory neurotransmitter, viz. glutamate.

The ionic glutamate receptors, mainly the N-methyl-D-aspartate receptor (NMDAR), are largely dependent on the single excitatory agonist glutamate. Glutamate is present in high concentration at approximately 10 mM in most neurons and glial cells. Glutamate is released for milliseconds at the synaptic ending to communicate with other neurons, allowing the nervous system to transmit complex motor commands and sensory information and to form thoughts and memories. However, after sequestration, excessive glutamate concentration or prolonged glutamate release excites neurons and leads to neuronal cell death. Not only elevated extracellular glutamate is responsible for the excitotoxic mechanism, but hyperactivity of NMDAR can also invoke the excitotoxic mechanism even at physiological levels of glutamate. The NMDAR channel is blocked with Mg^{2+} ions present in the channel and is active only for a brief period during normal synaptic transmission. However, overactivation of the NMDAR leads to excessive Ca^{2+} ions influx into nerves under pathological conditions. Ca^{2+} influx causes mitochondrial Ca^{2+} overload and Ca^{2+}-dependent activation of neuronal nitric oxide synthase (nNOS), resulting in neuronal injury and apoptosis (Mota et al. 2014).

Memantine is an open-channel, low-affinity, uncompetitive antagonist of NMDA receptors (Bormann 1989; Reisberg et al. 2003; Xia et al. 2010). Initially, adamantane, a precursor of memantine, was developed as an anti-influenza drug, as fever severity and influenza symptoms were significantly decreased on the first day of treatment. Adamantane inhibits the viral matrix protein 2 action by blocking the migration of H^{+} ions into the virions. However, nowadays, it is not recommended due to the development of high drug resistance (Cady et al. 2010; Chang et al. 2023). Serendipitously, amantadine, an analogous of memantine, was found to be effective for Parkinson's disease (PD) as a PD patient on amantadine treatment for influenza showed improvement in PD symptoms. In the early days, memantine was considered

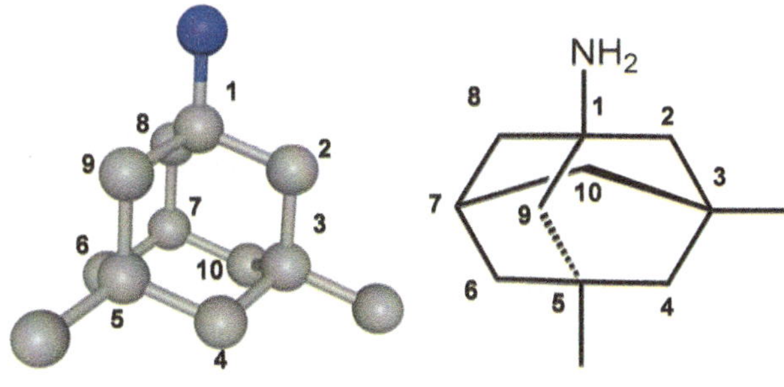

Fig. 8.1 (**a**) 3D structure of memantine; (**b**) 2D structure of memantine

to have the properties of dopaminergic or cholinergic drugs because dopaminergic and cholinergic dysfunctions were the main characteristics of PD. Later in 1989, Joachim Bormann showed that memantine is indeed an NMDAR antagonist as the effect of memantine was like that of dizocilpine maleate (MK-801), another potent non-competitive NMDA antagonist (Bormann 1989) (Fig. 8.1).

Memantine (CID 405) is a member of adamantane family having a three-ring system structure with bridgehead group and two methyl substituents (Fig. 8.1). Under pathophysiological conditions, bridgehead amine becomes positively charged and binds adjacent to the Mg^{2+} site of NMDAR. The side methyl groups play a crucial role in extending the dwell time of the drug and also increasing the affinity towards the channel (Johnson and Kotermanski 2006). It was found that the physiological receptor functions of NMDAR were unhindered by memantine and selectively block ion channels under pathological conditions, and thus maintain an optimal level of NMDA (Chen et al. 1992; Lipton 2004).

8.2 Synthesis of Memantine

Memantine is a derivative of adamantine, which was first isolated from petroleum. IUPAC name of memantine is 3,5-dimethyladamantan-1-amine. In memantine, NH_2 group acts as a hydrogen bond donor. Presence of two methyl groups is very important for binding with NMDA receptor. It is a crystalline solid and show very high water solubility. Melting point and cLogP values of memantine are 258 °C and 3.0, respectively (Alam et al. 2017). Till date, many synthetic routes have been developed for the synthesis of memantine, which are described below in various schemes

In 1963, Gerzon and co-workers reported the synthesis of memantine for the first time. Under reflux conditions, bromination of 1,3-dimethyladamantane (1a) produces 1-bromo-3,5-dimethyladamantane (1b). On acetylation with acetonitrile in sulphuric acid compound, 1b gives *N*-acetamido-3,5-dimethyladamantane (1c), which provides memantine (1d) when reacted with NaOH, and compound 1d is converted to the hydrochloride salt of memantine (1e). Gerzon and co-workers also

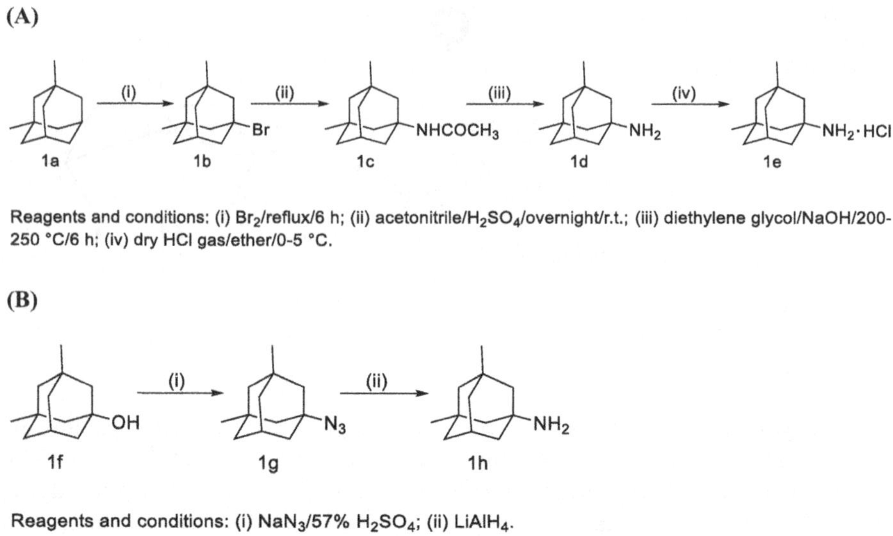

Scheme 8.1 (**a**) Reagents and conditions: (i) Br_2/reflux/6 h; (ii) acetonitrile/H_2SO_4/overnight/r.t.; (iii) diethylene glycol/NaOH/200–250 °C/6 h; (iv) dry HCI gas/ether/0–5 C. (**b**) Reagents and conditions: (i) NaN_3/57% H_2SO_4; (ii) $LiAIH_4$

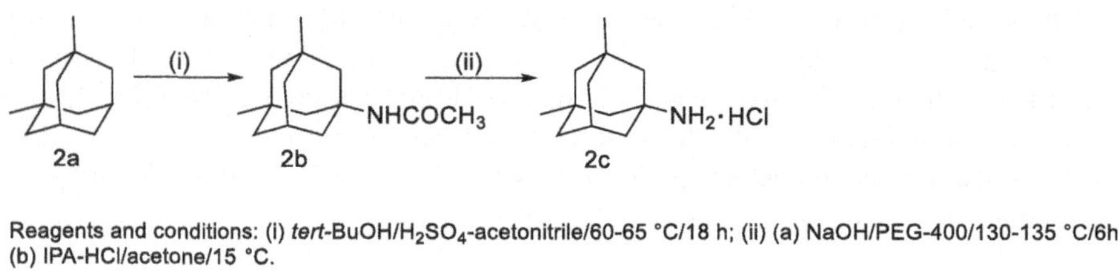

Scheme 8.2 Reagents and conditions: (i) tert-BuOH/H_2SO_4-acetonitrile/60–65 °C/18 h; (ii) (a) NaOH/PEG-400; 130–135 °C/6 h (b) IPA-HCI/acetone/15 °C

reported that 1-hydroxy-3,5-dimethyladamantane (1f) forms corresponding azide (1 g), which on reduction with $LiAlH_4$ forms memantine (1 h) (Gerzon et al. 1963) (Scheme 8.1).

Another three-step process was developed in 2007, and this process also avoided the bromine vapours (by-product) as bromine was not used in this process. When 1,3-dimethyladamantane (2a) reacts with *tert*-BuOH in the presence of H_2SO_4, it gives *N*-acetamido-3,5-dimethyladamantane (2b), which further produces the hydrochloride salt of memantine (2c) on reaction with NaOH and IPA-HCl (Madhra et al. 2007; Reddy et al. 2007) (Scheme 8.2).

Decarboxylative azide (3b) formation of 3,5-dimethyladamantane-1-carboxylic acid (3a) takes place when AgF as catalyst, $PhSO_2N_3$ as azide source and $K_2S_2O_8$ as oxidizing agent are used. Compound 3b is further reduced in the presence of Pd/C and H_2 to get memantine (3c) (Zhu et al. 2015) (Scheme 8.3).

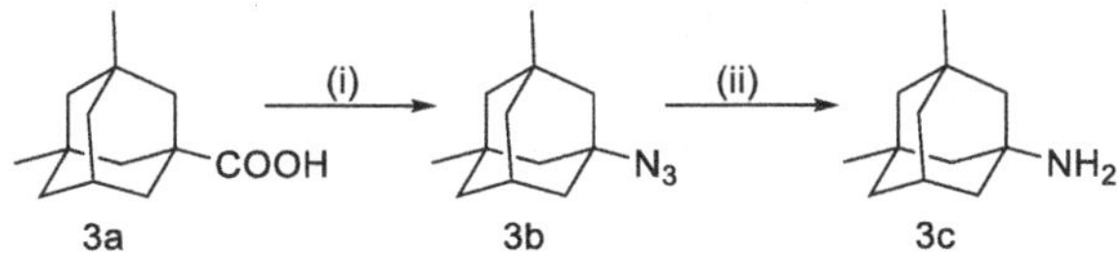

Scheme 8.3 Reagents and conditions: (i) $PhSO_2N_3$/AgF (20 mol%)/$K_2S_2O_8$/CH_3CN/H_2O/55 °C/24 h; (ii) Pd/C (10%)/H_2 (1 atm)/MeOH/r.t./24 h

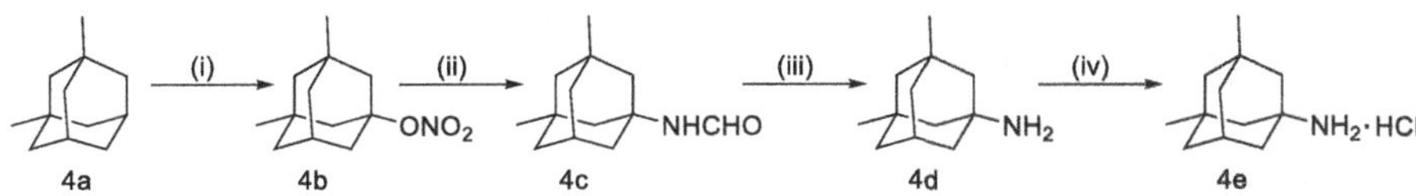

Scheme 8.4 Reagents and conditions: (i) HNO_3/r.t.; (ii) $(CH_3)_2C(OH)(CN)$/r.t.; (iii) (a) H_2O/110 °C (b) aq. NH_3 (25%); (iv) aq. HCI (36%)

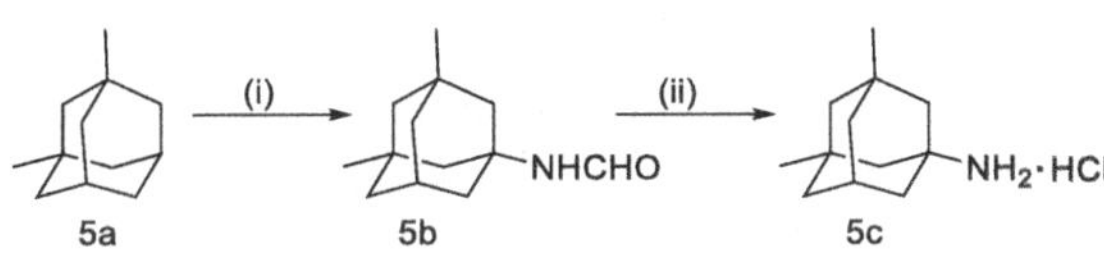

Scheme 8.5 Reagents and conditions: (i) HNO_3, NH_2CHO/85 °C/2 h; (ii) (a) 21% aq. HCI/100 °C/1 h

Ivleva and Klimochkin reported a one-pot strategy for the synthesis of memantine hydrochloride (4e). 1,3-Dimethyladamantane (4a) reacts with HNO_3 and produces 3,5-dimethyladamantane-1-yl nitrate (4b). Compound 4b on reaction with HCN forms 3,5-dimethyladamantane-1-yl formamide (4c), which further on hydrolysis results in the formation of memantine (4d), and reaction of memantine with aq. HCl gives the memantine hydrochloride salt (4e) (Ivleva and Klimochkin 2017) (Scheme 8.4).

Recently in 2020, a two-step synthesis of memantine hydrochloride salt (5c) was introduced by Binh Duong Vu and his group. 3,5-Dimethyladamantane-1-yl formamide (5b) is formed when 1,3-dimethyladamantane (5a) reacts with formamide in the presence of nitric acid. Hydrolysis of compound 5b with 21% HCl produces the memantine hydrochloride salt (5c) (Vu et al. 2020) (Scheme 8.5).

8.3 Structure and Domains of NMDA Receptor

The ability of the nervous system to process and store information is fundamentally dependent on synaptic transmission. Synapses are specific connections between neurons, where neurotransmitter receptors on the membrane of the postsynaptic neuron are activated by the release of neurotransmitter by the presynaptic neuron at synapses. The amino acid glutamate plays major role in mediating excitatory synaptic transmission in mammals by activating two distinct glutamate receptor subtypes: ionotropic and metabotropic. Ionotropic glutamate receptors are ligand-gated ion channels that are classified according to their pharmacological characteristics into the subtypes: GluA (AMPA, 2-amino-3-3-hydroxy-5-methyl-isoxazol-4-yl propanoic acid), GluK (kainate), **GluN (NMDA, N-methyl-D-aspartic acid)** and GluD (δ) receptors (Furukawa et al. 2005).

N-methyl-D-aspartate receptors or NMDARs are the receptors that are principally accountable for learning and memory. Therefore, any interference with the NMDARs' typical signalling pathway could result in CNS damage and emergence of Alzheimer's disease (AD). Human genes GRIN1, GRIN2A, GRIN2B, GRIN2C and GRIN2D code for the NMDARs. NMDARs are selectively mediated by N-methyl-D-aspartate (NMDA). The NMDARs can be divided into two groups: synaptic and extra-synaptic NMDARs. Synaptic NMDA receptors in specific have drawn a lot of interest in recent years due to their crucial role in many types of neural plasticity on one side and their involvement in neuronal excitotoxicity on another side. Finding clinically applicable NMDA receptor antagonists that can preferentially block excitotoxic NMDA receptor activation without affecting NMDA receptor function necessary for normal synaptic transmission and plasticity is a key objective of a lot of recent research (Vyklicky et al. 2014).

Functional NMDA receptors consist of hetero-tetramers containing two necessary GluN1 (NR1) subunits in combination with two optional GluN2 (NR2) and/or GluN3 (NR3) subunits. NMDARs require occupation by two types of agonists for channel activation: a glutamate site agonist at NR2 subunits and a glycine site agonist at NR1 subunits. With a conserved domain organization, all subunits exhibit a high degree of structural similarity and homology. The ion channel is made up of a transmembrane domain (TMD), an extracellular ligand-binding domain (LBD) and an extracellular amino-terminal domain (ATD). The intracellular carboxy-terminal domain is in communication with the transmembrane helices (CTD) (Lee et al. 2014) (Fig. 8.2).

The current channel gating model suggests the following three phases in order: (a) binding of the ligand into its pocket, (b) closing of clamshell-like LBD due to conformational changes and (c) spread of these changes resulting in channel opening. These gating principles have been validated for NMDA receptors, where activation of a typical GluN1/GluN2 receptor requires the simultaneous presence of two distinct ligands. There are two domains in the LBD; in a dimer of two LBDs, the upper domain S1 creates a tight interface with the second upper domain. Both S1 domains are kept relatively stiff by this arrangement. The bottom domain S2, in contrast, is relatively mobile and lacks an obvious dimerization contact (Vyklicky et al. 2014).

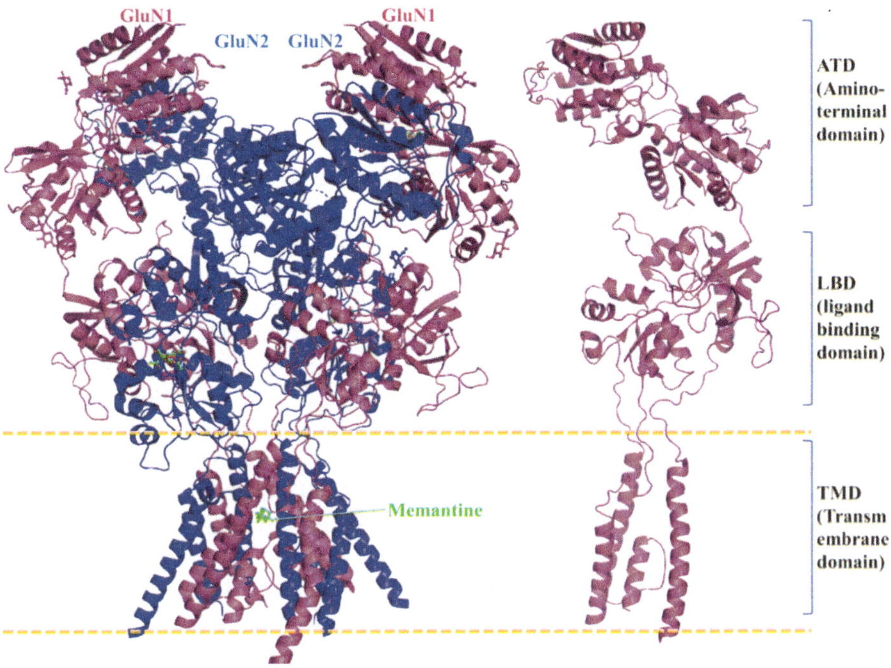

Fig. 8.2 Structure and different domains od NMDAR

8.4 Physiology and Pathogenesis

Memantine, a single targeting drug approved by the FDA in 2003, is used as a therapeutic agent for Alzheimer's disease. It is used as an N-methyl-D-aspartate receptor (NMDAR) antagonist. The common side effects of memantine include nausea, constipation, dizziness and headache, blood clots, etc. (Johnson and Kotermanski 2006). A single, potent excitatory neurotransmitter called glutamate is largely responsible for the nervous system's ability to quickly transmit sensory information and motor impulses from one region of the body to another, as well as for the creation of ideas and memories. Majority of the neurons and glia have glutamate concentrations of at least ~10 mM (Upton and Rosenberg 1995). Glutamate is briefly (milliseconds) released from synaptic vesicles after sequestration in order to connect with neighbouring neurons through synaptic endings.

NMDAR channels have two distinctive features: high Ca^{2+} ions permeability and voltage-dependent channel inhibition by endogenous Mg^{2+} ions (Johnson and Kotermanski 2006). During normal neuronal resting potentials, more than 90% of the current produced by NMDAR is restricted by channel blockage due to endogenous Mg^{2+}, which occurs after cation influx into the neuron via AMPA-sensitive glutamate receptor channels. The voltage field across the channel is altered by postsynaptic depolarization, which drives Mg^{2+} out and amplifies NMDAR-mediated currents. Presynaptic glutamate release and postsynaptic depolarization are therefore crucial for successful current conduction via NMDARs, enabling NMDARs to function as 'coincidence detectors'. Long-term potentiation (LTP), which is believed to be a cellular-electrophysiological correlate of learning and memory formation, is induced in specific brain regions by normal NMDAR activity. In addition, the brainstem's reticular activating system which controls wakefulness and attention, has several NMDARs. If these receptors do not work regularly, drowsiness and even coma can occur. Therefore, it is crucial to maintain physiological NMDAR activity to maintain these normal functions and prevent undesirable clinical side effects. Second characteristic property of NMDAR channels includes high Ca^{2+} ions permeability which gives NMDAR activity considerable physiological and clinical relevance. However, under pathological conditions excessive NMDAR activation results in an excessive Ca^{2+} ions influx into the nerve cell, which causes cellular injury and death. That is mainly because coincident pre- and postsynaptic activity provokes Ca^{2+} influx through NMDARs, activating a variety of intracellular signalling pathways, which have a variety of potential effect, including stabilizing synaptic connections, long-term depression or long-term potentiation of synaptic strength, and either necrotic or apoptotic neuronal death. These injurious processes include Ca^{2+} overload of the mitochondria, which causes oxygen free radical formation, activation of caspases and release of an apoptosis-inducing factor; Ca^{2+}-dependent activation of the nNOS, which increases the production of nitric oxide (NO) and causes the formation of toxic peroxynitrite (ONOO-) and S-nitrosylated GAPDH; and stimulation of p38 mitogen-activated protein kinase (p38 MAPK), which activates transcription factors that can go into the nucleus to influence neuronal injury and apoptosis (Berliocchi et al. 2005; Bonfoco et al. 1995; Dawson et al. 1991; Lipton et al. 1993). As a result, inhibiting NMDAR activity can have a wide range of effects that might be either negative or positive (Fig. 8.3).

Disorders such as multiple sclerosis, amyotrophic lateral sclerosis (ALS), neuropathic pain, glaucoma, Parkinson's disease, HIV-associated dementia, Huntington's disease, multiple sclerosis and glaucoma are having dissimilar mechanisms for cause; nonetheless, these disorders are connected with mutual pathway of neuronal injury that is caused by the overstimulation of glutamate, specially NMDA subtype of glutamate receptors.

Due to its potential role in the pathophysiology of numerous acute and chronic neurodegenerative disorders, excitotoxicity may be a particularly appealing target for neuroprotective efforts. The obstacle in developing anti-excitotoxicity techniques is that the same processes when overactivated result in excitotoxic cell death, however, when at controlled levels, vitally necessary for the normal neuronal function. Until recently, all treatments that appeared promising as inhibitors of excitotoxicity also affected normal excitatory synaptic activity, resulting in undesirable and severe side effects.

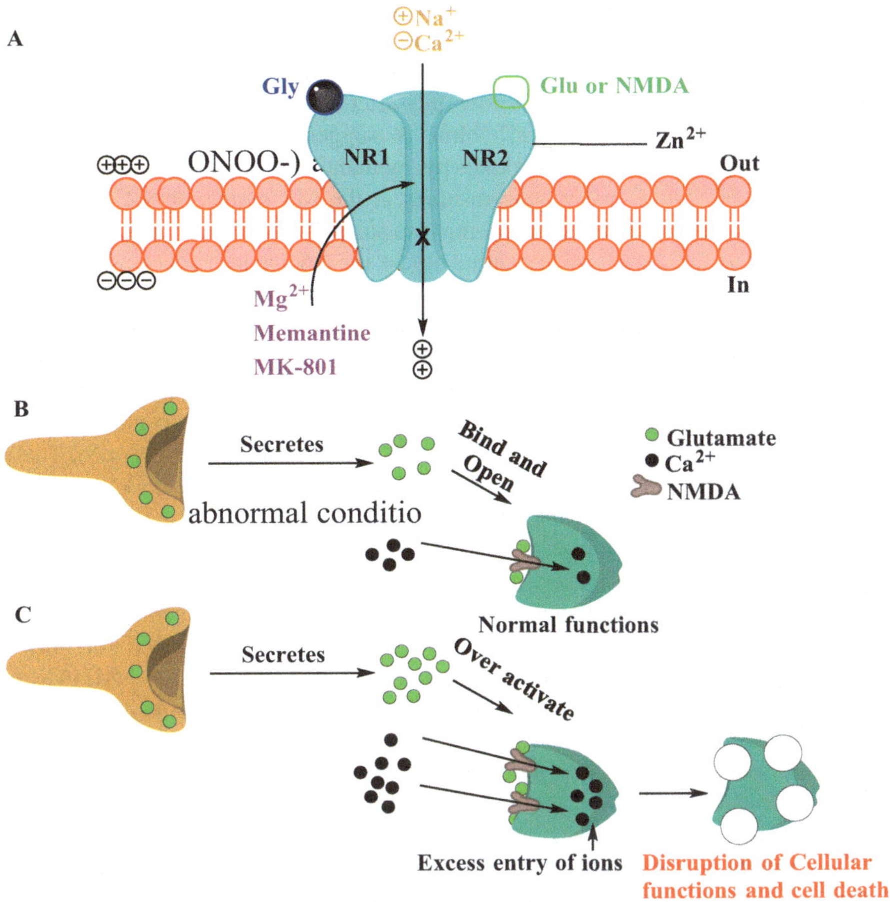

Fig. 8.3 (**a**) NMDAR model illustrating important binding and modulatory sites, (**b**) normal functioning of NMDAR and (**c**) NMDAR under abnormal conditions

8.5 N-Methyl-D-Aspartate Receptor (NMDAR) and Memantine

Memantine was synthesized for the first time in 1960s, it was discovered that it had CNS effects (Parsons et al. 1999). The IC_{50} value of memantine was found to be approximately 1 mM (Parsons et al. 1995), which was in line with the therapeutic concentration range to inhibit NMDARs (Bormann 1989). Memantine blocks the NMDAR channel to exert the pharmacological action. In 2003, the US FDA and the European Union approved memantine for the treatment of moderate-to-severe Alzheimer's disease.

Memantine is considered to be the first neuroprotective medication to achieve clinical approval. Previously, cholinergic medicines had been approved for the treatment of Alzheimer's disease, but they mainly provided symptomatic relief (slowing of cognitive deterioration).

Memantine is categorized as an 'open-channel blocker' because first it can enter in the channel and block the current flow only after channel opening. Overlap of the sites where memantine and Mg^{2+} bind is suggested by the following two observations: Mg^{2+} decreases blockade by memantine (Sobolevsky et al. 1998), and mutation of 'N-site' asparagine residues in the M2 region of NR1 and NR2 subunits that are critical for Mg^{2+} blocking also strongly affects memantine blocking (Kashiwagi et al. 2002). After memantine blocks the NMDAR channel, and closes it which leads to unbinding of agonists, and memantine gets 'trapped' inside the channel (Blanpied et al. 1997). Commonly, memantine is known to be an uncompetitive antagonist. Several other 'trapping channel blockers' of NMDARs have also been designated, which includes ketamine, amantadine, MK-801, and phencyclidine (Palmer 2001). Various other channel blockers are believed to stop channel from closing, and thus they are bound in the NMDAR channel only when it is open. These types of blockers are named as 'sequential' or 'foot-in-the-door' blockers (Bolshakov et al. 2003). Memantine turns into a pretty effective blocker when NMDARs are activated for an extended period of time, as is the case under excitotoxic conditions. However, memantine is unable to build up in the NMDA channels during normal synaptic activity, which results in synaptic activity continuing practically uninterrupted since NMDA channels are open for only a few milliseconds on average during typical synaptic activity. An NMDAR or ionotropic glutamate receptor called GluN2D (PDB ID: 3OEM) was employed to dock against memantine (PubChem CID: 4054) in one investigation. Numerous significant amino acid residues, including ALA644, VAL644, ALA645, MET641, etc., are present at the memantine-binding site as shown by the co-crystallized structure of memantine and NMDAR.

The capacity of memantine to reduce NMDAR responses is consistent with the notion that memantine can decrease the long-term course of AD through neuroprotective activities. After just 2 weeks of treatment, memantine also showed therapeutic outcomes such as enhancements in cognitive function. Memantine's cognitive benefits in AD have been attributed to a number of mechanisms such as reduction in the synaptic 'noise' brought on by excessive NMDAR activation (Parsons et al. 1999), suppression of β-amyloid formation or toxicity (Rogawski and Wenk 2003) and resetting of the balance between inhibition and excitation (Schmitt 2005) (Fig. 8.4).

8.6 Other Inhibitors

With several binding sites for small-molecule ligands that function as subunit-specific allosteric modulators of ion channel activity, the N-terminal domains (NTDs) give NMDARs a special ability for allosteric modulation (Zhu and Paoletti 2015). With different recognition sites for both endogenous and exogenous allosteric inhibitors and potentiators, the GluN2B NTD in particular offers a diverse

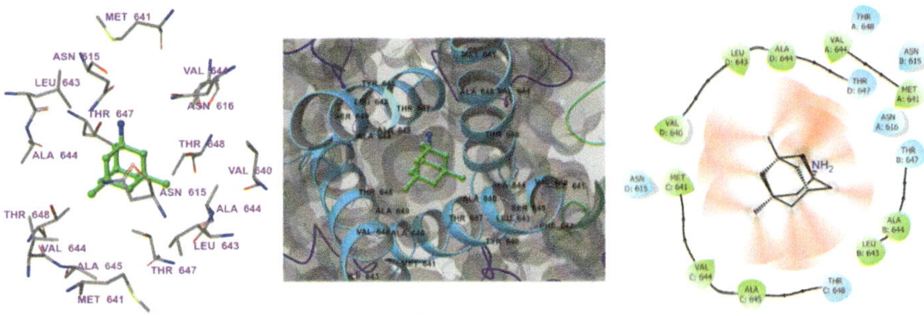

Fig. 8.4 Representation of binding site of memantine in NMDARs

pharmacology (Karakas et al. 2011; Mony et al. 2009). GluN2B-containing NMDARs have a substantial class of synthetic drugs, including ifenprodil and its derivatives, which act as highly selective non-competitive antagonists (Layton et al. 2006). Intriguingly, GluN2B-selective antagonists have demonstrated promising outcomes in several clinical trials and have a lower risk of side effects than pan-NMDAR antagonists like ketamine (Ibrahim et al. 2012). Since it was discovered that ifenprodil inhibits GluN2B receptors only when specifically targeted, a wide variety of GluN2B-selective antagonists have been developed. Many have the same phenylethanolamine scaffold as ifenprodil, including CP101,606 (Traxoprodil (Chenard et al. 1995)), Ro25-6981 (Fischer et al. 1997) and besonprodil (Chizh et al. 2001). Others, such as several extremely effective and orally active GluN2B antagonists like EVT-101 (Kemp and Tasker 2011) and to a lesser extent MK-22 (Layton et al. 2011), have markedly distinct structural motifs, raising doubts about whether these drugs adopt a comparable binding method to ifenprodil. Crystal structures recently established the existence of a 'phenylethanolamine-binding site' at a dimer interface between GluN1 and GluN2B NTDs. David et al. solved the structures of EVT-101 and MK-22 in complex with the GluN1/GluN2B NTD heterodimer using a back-soaking approach to better understand the protein-ligand interactions required for the potency and selectivity of additional nonphenylethanolamine scaffolds (Stroebel et al. 2016). The binding of MK-22 and ifenprodil is virtually superimposable, but EVT-101 occupies a different cavity with relatively slight overlap and forms novel interactions with the pyridazine and imidazole groups (Fig. 8.5).

8.7 Developments on Memantine

Wang and his group synthesized some nitrate derivatives (**2**) (Fig. 8.6) of aminoadamantane. It was planned to combine the properties of memantine and nitroglycerin to increase the efficacy of compounds for neurodegenerative diseases. In vitro tests of the synthesized compounds were performed, but none of the nitrate derivative was found as potent as memantine; however, some derivatives were found

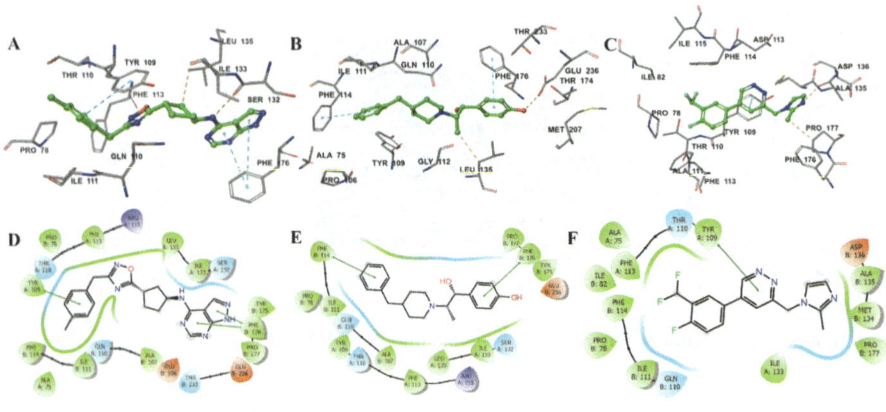

Fig. 8.5 3D and 2D protein-ligand interaction of NMDAR inhibitors at NTD domain: (**a**–**c**) 3D representation of ifenprodil, MK-22 and EVT-101, respectively; (**d**–**f**) 2D representation of ifenprodil, MK-22 and EVT-101, respectively

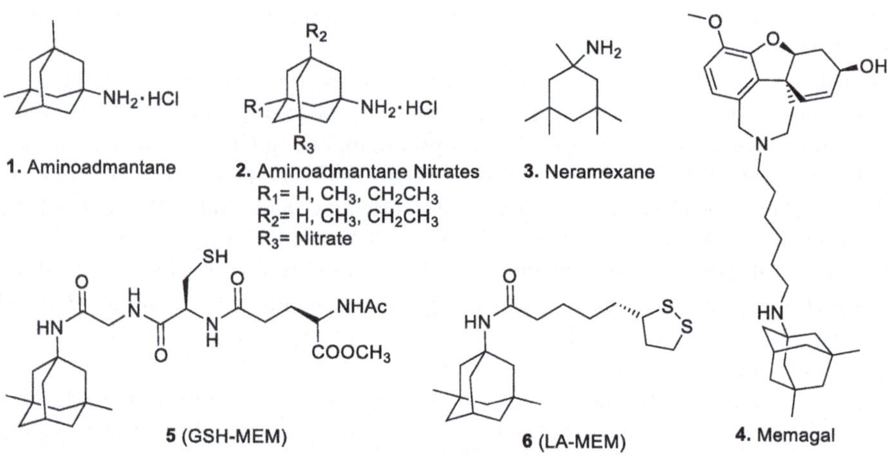

Fig. 8.6 Memantine derivatives as NMDA receptors

as stable as memantine and showed IC_{50} values close to memantine. SAR studies of these compounds can provide some good NMDA receptor antagonists (Wang et al. 2006). Fernanda G. De Felice et al. discovered the relationship between neuronal binding of Aβ oligomers (ADDLs) and oxidative stress. It was found that the ADDLs are responsible for the production of excessive reactive oxygen species by a mechanism of activation of NMDA receptors. NMDA receptors play a significant role in the complete blockage of ROS production and reduce the ADDLs' binding to neurons (De Felice et al. 2007). During the development of memantine derivatives, it was found that the two methyl groups were very important for the activity towards NMDA receptors. A pentamethyl aminocyclohexane (neramexane), **3** (Fig. 8.6), was very close to mimic the memantine. It binds to NMDA receptors at the same site where memantine binds. Compound **3** advanced to clinical trials, but it failed in phase III trails to achieve significance for the treatment of Alzheimer's disease (Alam et al. 2017; Rammes 2009). In 2012, Simoni and group reported a series of novel multitargeted compounds by linking memantine and galantamine together. These compounds were designed on the basis of the crystal structures of acetylcholinesterase (AChE) in complex with galantamine derivatives. Sixteen derivatives were synthesized using different spacers, and these compounds were screened as AChE inhibitors and as N-methyl-D-aspartate receptor (NMDAR) binders. Amongst the synthesized compounds, some displayed inhibitory activity in nanomolar range against AChE and showed affinity in micromolar range towards NMDAR. Selectivity studies of all the synthesized compounds towards NMDAR with 2B subunit (NR2B) were also conducted, and some of the compounds displayed affinity in micromolar range. In the final step, a cell-based assay for the selected compounds was performed to observe their neuroprotective activity. It was found that compound **4** (memagal) (Fig. 8.6) inhibited the neurotoxicity induced due to NMDA with an IC_{50} value of 0.28 nM (Simoni et al. 2012).

A derivative or medication that is metabolized into a pharmacologically active state by metabolic or physico-chemical transformations is often termed as prodrug. It is often inactive in its native pharmacological actions and is often used to study how the drug is absorbed, distributed, metabolized and excreted. Sozio and group developed memantine-based sulphur-containing antioxidant conjugates in 2013 as promising prodrugs to successfully treat AD. They synthesized two memantine derivatives by the covalent linkage via an amide bond with GSH (**5**) (Fig. 8.6) and (R)-α-lipoic acid (**6**) (Fig. 8.6) to give potent NMDA receptor antagonist. Compound **6** was found to be a promising agent which showed free radical scavenging effects to both H_2O_2 and super oxide anion radical (O_2) and significantly inhibited Aβ (1–42) aggregation to 41 ± 2.05%. In PAMPA-BBB assay, compound **6** crossed the blood-brain barrier (BBB) easily, and hence, it may be developed as a promising ligand for both free radical scavenging and inflammatory activities (Sozio et al. 2013). Valverde and group reported some benzopolycyclic amine compounds **7**, **8** and **9** (Fig. 8.7), which showed antagonist activity against NMDA receptor. In comparison to amantadine, all the synthesized compounds were found to be more potent. Antagonist activity of the two compounds was almost similar to memantine, and one compound was found to be more potent than memantine ($IC_{50} = 1.5 \pm 0.1$ μM). It was observed that the introduction of a polar moiety (hydroxyl group, methoxy group, etc.) decreases the potency of the compounds and replacement of primary amines with tertiary amines also reduces activity significantly (Valverde et al. 2014). For the allosteric and direct uncompetitive blockade of NMDA receptors, Jin and group designed and synthesized a memantine nitrate (compound 10) (Fig. 8.7) also known as YQW-035. In the cultured neuronal cells and rat transient stroke model, compound **10** was found to be a better antagonist than memantine and can be used as a promising lead to develop potent NMDA receptor antagonists (Jin et al. 2016). Based on one molecule-multitarget theory, a series of 14 compounds of 7-methoxytacrine-memantine heterodimers have been synthesized by Gazova and group. β-Secretase (BACE-1) activity, Aβ fibrillization inhibition, AChE inhibition

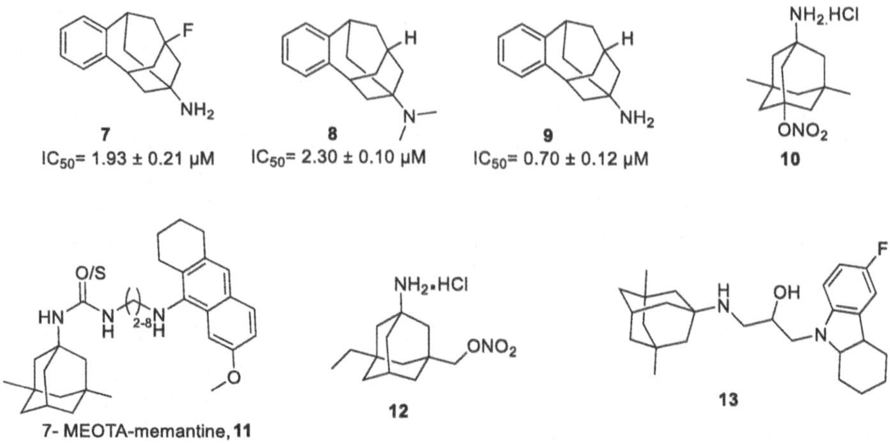

Fig. 8.7 Memantine derivatives as NMDA receptor antagonists

and NMDA receptor antagonistic activities of the synthesized compounds (compound 11) (Fig. 8.7) were evaluated. The hybrid compounds showed better AChE inhibition in micromolar range as compared to parent 7-methoxytacrine and 1-adamantylamine. The hybrid compounds successfully decreased the fibrillization of the Aβ peptide while inhibiting BACE1. On investigating the interactions with GluN1-1a/GluN2B NMDA, the compounds showed almost similar interactions as done by memantine (Gazova et al. 2017).

A series of 16 memantine nitrate derivatives were designed and synthesized by Lui and group in 2016 for the study of neuroprotection and vasodilatory activity for neurodegenerative diseases. In the in vitro studies, these nitrates were found to be effective neuron protection agents against glutamate-induced injuries. Lengthening of the linker between memantine moiety and side chain decreases the potency, while lengthening of the side chain increases the potency in terms of cytoprotective effects. In addition to this, no enhancement of neuroprotective effects was observed on increasing the number of nitrate groups. Compound **12** (Fig. 8.7) was found to be the most significant compound which efficiently reduced the free radical production and maintained the mitochondrial membrane potential by inhibiting Ca^{2+} influx (Liu et al. 2017). Aminoadamantane and carbazole-based new class of conjugates were synthesized and evaluated by Bachurin and group. The compounds were investigated as anti-neurodegenerative agents. These drugs specifically inhibit BuChE while keeping the other characteristics of MK-801-binding site blockers. These also block NMDA receptors that include NR2B subunits. Compound **13** (Fig. 8.7) was found to be the most promising compound exhibiting microtubule-stabilizing properties along with the ability to protect the nerve cells at overloaded calcium conditions (Bachurin et al. 2017).

Kumamoto and group developed a new class of NMDA channel blockers in 2017 with a polyamine-based memantine hybrid structure. A total of 25 derivatives were synthesized varying in alkyl chain length and functional groups. Out of these

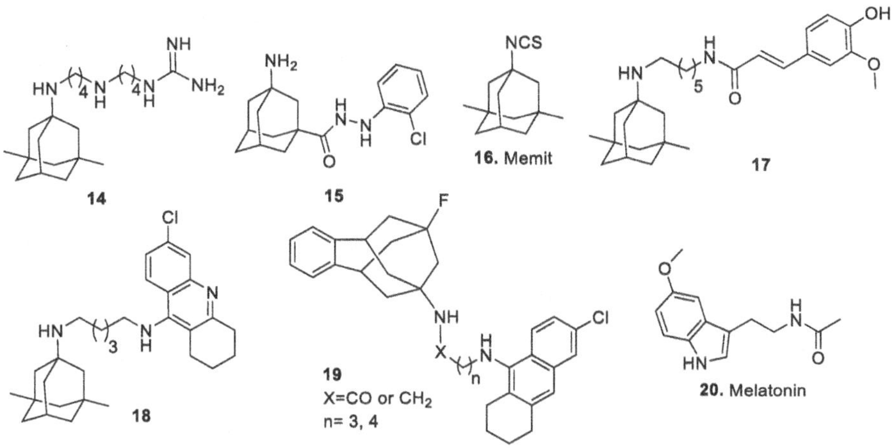

Fig. 8.8 Memantine derivatives as multifunctional agents for AD treatment

synthesized derivatives, diaminoguanidine M44G, compound **14** (Fig. 8.8), showed the most potent antagonistic activity against GluN1/GluN2A and GluN1/GluN2B NMDA subtype receptors. It was concluded that the presence of guanidine group was crucial for the biological activity of the compound (Kumamoto et al. 2018). In 2018, Koola and group studied the comparison of combination of donepezil-memantine over combination of galantamine-memantine drug for the effective emphasis on kynurenic acid and mismatch negativity. Several biomarkers associated with AD like kynurenine pathways (KP), mismatch negativity (MMN), brain-derived neurotrophic factors (BDNF) and oxidative stress were considered for this comparison. Cholinergic and glutaminergic dysfunction and alterations in the KP are responsible for the onset of AD symptoms. So, along with targeting of cholinergic and glutaminergic pathways, KP modulations may be a novel strategy for the treatment of AD. From the studies, it was concluded that the efficacy of the combination of galantamine-memantine was due to the synergistic effects of α7nACh and NMDA receptors (Koola et al. 2018). A novel class of multitarget-directed ligands possessing dual P2X7-NMDA receptor antagonist activity was synthesized and evaluated by Karoutzou and group in 2018. The aim was to delay the neurodegeneration by targeting the glutaminergic NMDA receptors and to reduce the neuroinflammation by targeting P2X7 receptors. NMDA receptor is well recognized, but P2X7 receptor is not much explored as a therapeutic target for the treatment of AD. Compound **15** (Fig. 8.8) displayed the most potent P2X7 antagonistic activity with an IC_{50} value of 6.0 ± 0.6 μM and retaining the NMDA receptor antagonistic activities, and hence **15** was reported as the first dual antagonist of both the receptors (Karoutzou et al. 2018).

Another memantine prodrug (memit, 16) (Fig. 8.8) was synthesized in 2019 by Sestito and group as a new agent for the treatment of Alzheimer's disease. The amine group of memantine was replaced with isothiocyanate functionality for retaining the memantine properties and as a putative H_2S donor moiety. H_2S is reported to play significant anti-inflammatory and neuroprotective roles in the brain, and it is found to exert the neuroprotective effects against inflammation and decrease the ROS production. It was also found to increase autophagy, which is necessary for cellular homeostasis and cell survival, and to lessen the Aβ (1–42) self-induced aggregation. Further, in vivo studies need to be performed to establish the synergistic and cumulative effects of H_2S as neuroprotective and the native moiety memantine as NMDA receptor antagonist (Sestito et al. 2019). Rosini and group merged ferulic acids with memantine to build a connection between NMDA receptors, oxidative stress and Aβ peptide for the treatment of AD. An aromatic conjugated amide was linked with the memantine moiety with a linker aliphatic chain. Compound **17** (Fig. 8.8) was found to be the most promising agent exhibiting voltage-dependant antagonist activity against NMDA receptor with the IC_{50} value of 6.9 μM. In SHSY-5Y cell lines, it showed antioxidant properties at 10 μM concentration. At the same concentration, it was found to follow the APP processing towards α-secretase, the non-amyloidogenic pathway which results in the reduction of Aβ production. So, it may act as a promising NMDA receptor antagonist along with the inhibitor of NMDA-mediated neurotoxic events like ROS formation and Aβ damage (Rosini et al. 2019).

Kaniakova and group designed and synthesized a novel 6-chlorotacrine-memantine hybrid (compound 18) (Fig. 8.8) inspired by memagal (galantamine-memantine hybrid) as a promising multipotent lead for the treatment of AD. It showed better AChE inhibitory activity ($IC_{50} = 9.41 \pm 0.65$ nM) compared to the parent compound, 6-chlorotacrine ($IC_{50} = 20 \pm 1$ nM). It was also predicted to cross the BBB easily and efficiently block the NMDA receptor as done by memantine. Compared to memantine, it showed quantitively better neuroprotective effects in NMDA-induced hippocampal lesions (Kaniakova et al. 2019). An NMDA antagonist fluorobenzohomoadamantanamine was hybridized with potent AChE inhibitor 6-chlorotacrine by using four- or five-membered carbon chain to retain the activities of the parent derivatives along with the development of more potent new class of multitargeting agents. The position of linkage of carbon chain did not show any direct effect on AChE inhibition and NMDA antagonism, but it is apparent in BuChE inhibition. The synthesized hybrids (compound 19) (Fig. 8.8) were found to be almost 44-fold more potent AChE inhibitors than parent 6-chlorotacrine and around two-fold more potent than memantine and fluorobenzohomoadamantanamine. These novel hybrids showed potent AChE inhibitory potential in the nanomolar range (0.33–1.96 nM) and NMDA receptor inhibition in micromolar range (0.89–8.29 μM) (Pérez-Areales et al. 2019). Drug combination is one of the important therapeutic strategies for eradicating the complex diseases like Alzheimer's disease. In a mouse model of AD with an amyloid-dominant brain, combination of memantine and melatonin showed improvement in the memory function and brain neuronal deficits as reported by Jurgenson and colleagues. It was found that drug combination showed improvement in the episodic memory along with the reduction in the number of Aβ aggregates and reactive microgliosis in the brain of 5xFAD mice. It was observed that the combined use of 10 mg/kg memantine and 6 mg/kg melatonin was more beneficial as compared to

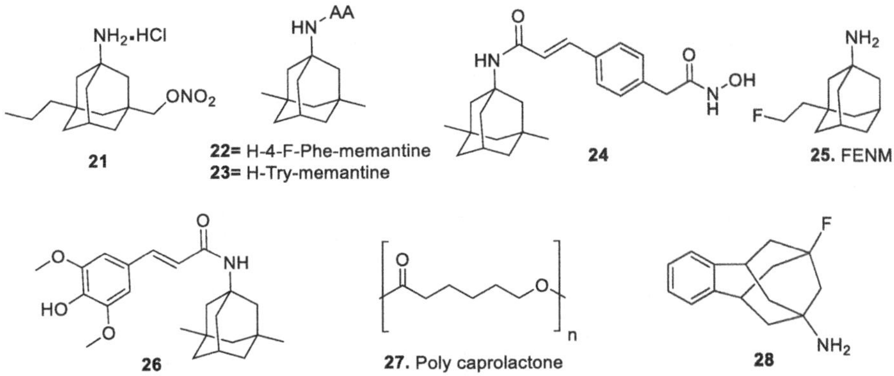

Fig. 8.9 Memantine derivatives as multifunctional agents

administration of single dose of memantine or melatonin, **20** (Fig. 8.8) (Jürgenson et al. 2019). Galantamine is a better positive allosteric modulator of alpha-7 nicotinic acetylcholine receptor (α7nAChR) than donepezil. Several preclinical studies and randomized controlled trial showed that the combination of galantamine-memantine significantly improved the cognition compared to galantamine alone (Koola 2020). Lui and group synthesized and investigated a series of memantine nitrates as glutamate-induced excitotoxicity blockers by inhibiting the calcium influx and impairing the PI3K/Akt/GSK3β pathway. It was found that compound **21** (Fig. 8.9), a new memantine-derived ligand, significantly protected the rat primary cerebellar granule neurons of rat against glutamate-induced excitotoxicity (CGNs). Calcium imaging demonstrates that the NMDA receptor antagonist, compound **21**, prevents calcium influx by inhibiting it. It was found to suppress the glutamate-induced phosphorylated Akt and reverse the activation of GSK3β. It was concluded that compound **21** can be developed as a potent drug candidate for the treatment of AD caused by cerebral vessel dilation and neuroprotection (Liu et al. 2020).

Tencheva and group synthesized amino acid analogues of memantine as neuroprotective agents and to investigate their impact on oxidative stress, neuroinflammation, hypoxia, excitotoxicity and resistance to Aβ toxicity. Most of the compounds provided multifaceted protection against injuries brought on by hypoxia, oxidative stress, glutamate-induced excitotoxicity and copper-triggered Aβ toxicity. H-4-F-Phe-memantine (compound **22**) (Fig. 8.9) and H-Tyr-memantine (compound **23**) (Fig. 8.9) were found to be the most potent compounds and displayed optimum lipophilicity and functional bioactivities similar to memantine hydrochloride. As a result, significant correlations between the biology, structural descriptors and physiochemical characteristics have been revealed. By using merely the structural formula, these correlations assist the different features of novel memantine derivatives (Tencheva et al. 2020). Memantine is a non-competitive NMDA receptor antagonist, while histone deacetylase (HDAC) is also reported to play an important role in neurodegenerative disorders like AD. He and group designed, synthesized and biologically evaluated a series of dual-acting NMDAR and HDAC inhibitors. Amongst the synthesized compounds, compound **24** (Fig. 8.9) exhibited balanced inhibitory activity against both HDAC and NMDAR. It was found to be selective towards HDAC6 with an IC_{50} value of 0.18 μM while retaining the NMDA receptor antagonistic activities (k_i = 0.59 μM). The hydroxamic acid-based compounds with a 7-carbon spacer were found to be most potent HDAC inhibitors. In addition, compound 24 also exhibited 30-fold potent neuroprotection compared to trolox and was found to be safe against H_2O_2-induced injury (He et al. 2020). In 2020, memantine nanoemulsion was proposed for intranasal delivery of memantine to bypass the BBB for the treatment of AD. The nanoemulsion was examined in vitro for antioxidant potential. The in vivo studies were caried out by radiolabelling the memantine with technetium pertechnetate. The nanoemulsion showed 98% cell viability, and the antioxidant potential of the memantine was sustained. The in vitro studies showed that 80% of the drug was released in stimulated nasal fluid. The biodistribution results showed higher uptake of drug in the brain regions via internasal route of administration as compared to oral and intravenous administration in rats. The targeting efficiency through intranasal administration was found to be 207.23%. So, nanoemulsion loaded with memantine could be an effective strategy for the enhanced efficacy of the drug in the treatment of AD (Kaur et al. 2020).

A new derivative of memantine, fluoroethylnormemantine (**FENM**) (compound 25) (Fig. 8.9), was characterized and investigated by Chen and group. An extra fluorine group was attached at an optimized position for the in vivo biomarker labelling. The compound was evaluated for its effectiveness against fear and behavioural despair. It was found that compound **25** decreases immobility in the forced swim test and did not alter any sensorimotor gating or locomotion without producing any non-specific side effects. Additionally, it lessened acquired fear and aided in extinction learning in rats. Compound **25**, as an NMDA receptor antagonist, can be developed as a potent candidate for further preclinical stress-related behavioural studies (Chen et al. 2021). New multitarget chemical entities were reported by Chochkova and group. These compounds were obtained by the framework combination of substituted cinnamic acids and memantine. The neuroprotective and free radical scavenging properties of these synthesized memantine hybrids were investigated. The free radical scavenging activities were evaluated against 1,1-diphenyl-2-picrylhydrazyl (DPPH), hydroxyl (OH) and superoxide (O_2-.) radicals with respect to standard antioxidants and sinapic and ferulic acids. In these studies, compound **26** (Fig. 8.9) was found to be the most potent antioxidant amongst the synthesized compounds. It decreased the formazan production to around 60% at 50 μM concentration and around 75% at 100 μM concentration. In addition, in the in vitro studies, these compounds also displayed moderate neuroprotection (Chochkova et al. 2021). Polycaprolactone (compound 27) (Fig. 8.9) nanocapsule (MEM@PCL) based memantine-releasing system was synthesized, and its physiochemical characteristics were examined by Mahmoudi and his research group. After 1 month of examination of the nanocapsule solution, there was no discernible change in the original free drug concentration. There was no aggregation or decomposition observed in these small-sized nanocapsules (size

<200 nm), and they were found to be stable at 25 °C up to a period of over 40 days. The PCL nanocapsules showed more than 80% association efficiency of the drug. So, MEM@PCL nanocapsules may be the promising systems that could be used as a strategic tool for the treatment of AD (Mahmoudi et al. 2021).

To comprehend the overall impact of the medicine and the electrochemical behaviour of memantine hydrochloride, Natekar and Dey synthesized four metal complexes of the compounds with iron, copper, nickel and cobalt. They also conducted voltametric and spectroscopic studies on the compounds. Different concentrations of the drug have been studied with various instrumental variables like scan rate. Presence of functional group makes the drug more susceptible for binding to the transition metals to facilitate the drug delivery. Different techniques like IR, atomic absorption spectroscopy and cyclic voltammetry have been used for understanding the electrochemical behaviour of the drug (Natekar and Dey 2022). Turcu and group synthesized a series of memantine analogues possessing a benzohomoadamantane scaffold and carried out in vitro and in vivo characterization for the treatment of AD. Majority of the synthetic derivatives effectively inhibited NMDA receptors at micromolar concentrations. The most potent derivative i.e. compound **28** (Fig. 8.9) exhibited electrophysical properties that were identical to those of memantine, including potencies, voltage dependence and blocking sites. Studies using the 5XFAD mouse model of AD showed that compound **28** exhibited similar effects as done by memantine (Turcu et al. 2022).

Memantine, NMDA receptor antagonist, is an FDA-approved medication for the treatment of AD. The amount of drug reaching the target is a major challenge in terms of efficacy and effectiveness including its oral administration. In a new strategy, the drug is administered with the help of some carrier to the target. To achieve this, Radwan and group designed and studied the gamma-irradiated chitosan nanoparticles for the controlled delivery of memantine and reported it to be a promising approach in Alzheimer's therapeutics (Radwan et al. 2020). Pertaining to the memantine's regulated release using the non-solvent-induced phase separation (N-TIPS) technique, Rani and Chawla designed and synthesized memantine-loaded polylactic-co-glycolic acid (PLGA) nano-scaffolds. A total of 15 formulations were proposed to quantify the medication loading and percentage porosity. These memantine-loaded scaffolds diminish the accumulation of amyloid β plaques and result in the improved efficacy than memantine. These nano-scaffolds are reported to reduce the complications involved in AD including improved drug release, retention of the drug within the brain and maintaining the neuronal signalling (Rani and Chawla 2022). Four polymorphic forms of insoluble salts of memantine with a unique fluorescence phenomenon were synthesized by reacting memantine with pamoic acid (Fig. 8.10) through polymorphic screening. Pamoic acid helped in achieving the sustained release of drug by reducing the release rate. The synergistic effect of both pamoic acid and memantine on fluorescence helped in predicting the real-time monitoring in the controlled and continuous release of the drug (Kuang et al. 2022). Two types of carbon dots (CDs) (Fig. 8.10), namely carbon nitride dots (CNDs) and black carbon dots (B-CDs), were synthesized and biochemically characterized by Zhang and group who conjugated them with memantine

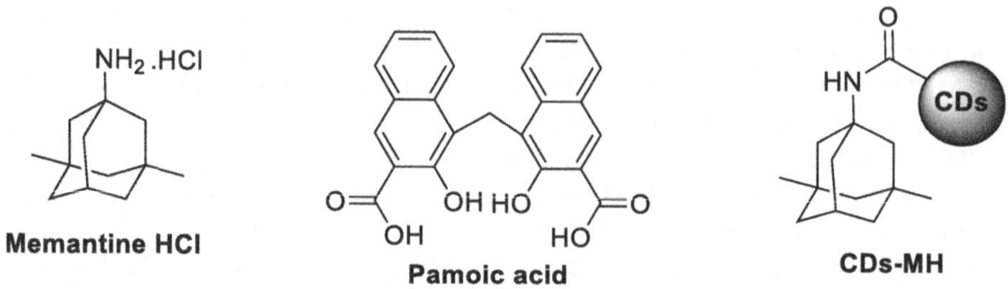

Fig. 8.10 Memantine and derivatives as drug delivery nanocarriers

hydrochloride (MH) as a drug delivery system through penetration of BBB for the treatment of AD. The tau aggregation inhibition was measured in the presence of CDs. B-CDs-MH and B-CDs alone were found to be the most effective tau aggregation inhibitors with the IC_{50} values of 1.5 ± 0.3 and 1.6 ± 1.5 μg/mL, respectively. In addition, it was reported that CNDs were successful in delivering the MH through BBB penetration because of the smaller size of these nano-dots (Zhang et al. 2022).

8.8 Conclusion and Future Perspectives

The aetiology of Alzheimer's disease is complex in nature. Several targets/hypotheses are involved in the progression of the disease. A therapeutic strategy is desired that can simultaneously target multiple cellular pathways responsible for the initiation and progression of AD. Hence, combination therapy or multitarget directed ligand approaches are proposed as effective strategies for the complete eradication of the disease. FDA has already approved a combination of donepezil and memantine in 2014 (Deardorff and Grossberg 2016). Some other combination drugs of memantine like galantamine-memantine, memantine-melatonin and rivastigmine-memantine are also being investigated by many research groups (Dantoine et al. 2006; Jürgenson et al. 2019; Koola 2020; Koola et al. 2018). Combination therapy also has its own drawbacks/side effects for the treatment of complex diseases like AD (Schmitt et al. 2004; Shenfield 1982). Ligands, that can simultaneously target AChE and NMDA receptor, are under various stages of development. NMDA is a specific amino acid derivative that acts as an agonist at glutaminergic receptors. It is understood to critically regulate the memory functions of the brain. Memantine is used as an NMDA receptor antagonist as alone and in combination with donepezil for the treatment of AD. We have described the synthetic, mechanistic and physiological approaches in the drug development process of memantine starting with adamantane. In the first three sections, we described the synthetic strategies starting from adamantane and its derivatives to the structural domains of memantine and showed how memantine is linked to NMDA receptors through glutamate ion channels. The next three sections explained the mechanistic approaches underling the pathology and pathogenesis, along with other similar inhibitors as memantine that functions as subunit-specific allosteric modulators of glutamate ion channel activity. In the last section, we have discussed various recent developments made on memantine throughout the drug development approaches required for better inhibition of the ion channels and effective treatment of the disease. Various inhibitors possessing different scaffolds and varied approaches including the developments on combination and multitargeting modulations have been discussed with their mechanistic targets and inhibitory potentials.

Memantine had shown promising results in preclinical studies, but there were no clear positive effects in clinical applications. This could be due to late administration of the drug after the complete disruption of neurons or maybe there is a need to administer the memantine with other antioxidants, anti-inflammatory, anti-aggregate or other potent drugs. The amount of memantine reaching the targets may be other criteria for the ineffectiveness of the drug. So, designing of nanocarriers or drug delivery systems which carry the effective concentration of the drug to the target might help in the development of potent drug molecule. It is presumed that memantine is an NMDAR-directed structure, so it can be conjugated with other pharmacophoric moieties of known functionalities like antioxidants, AChEIs, and Aβ aggregation inhibitors to get the desired multitargeting effect for the effective cure of the AD. So, the well-designed derivatives of memantine could help in improving the drug development process for the treatment of Alzheimer's disease.

References

Alam S, Lingenfelter KS, Bender AM, Lindsley CW (2017) Classics in chemical neuroscience: memantine. ACS Chem Neurosci 8(9):1823–1829

Bachurin SO, Shevtsova EF, Makhaeva GF, Grigoriev VV, Boltneva NP, Kovaleva NV, Redkozubova OM (2017) Novel conjugates of aminoadamantanes with carbazole derivatives as potential multitarget agents for AD treatment. Sci Rep 7(1):1–15

Berliocchi L, Fava E, Leist M, Horvat V, Dinsdale D, Read D, Nicotera P (2005) Botulinum neurotoxin C initiates two different programs for neurite degeneration and neuronal apoptosis. J Cell Biol 168(4):607–618

Blanpied TA, Boeckman FA, Aizenman E, Johnson JW (1997) Trapping channel block of NMDA-activated responses by amantadine and memantine. J Neurophysiol 77(1):309–323

Bolshakov K, Gmiro V, Tikhonov D, Magazanik L (2003) Determinants of trapping block of N-methyl-d-aspartate receptor channels. J Neurochem 87(1):56–65

Bonfoco E, Krainc D, Ankarcrona M, Nicotera P, Lipton SA (1995) Apoptosis and necrosis: two distinct events induced, respectively, by mild and intense insults with N-methyl-D-aspartate or nitric oxide/superoxide in cortical cell cultures. Proc Natl Acad Sci 92(16):7162–7166

Bormann J (1989) Memantine is a potent blocker of N-methyl-D-aspartate (NMDA) receptor channels. Eur J Pharmacol 166(3):591–592

Cady SD, Schmidt-Rohr K, Wang J, Soto CS, Degrado WF, Hong M (2010) Structure of the amantadine binding site of influenza M2 proton channels in lipid bilayers. Nature 463(7281): 689–692

Chang C, Ramphul K. Amantadine. [Updated 2023 Apr 24]. In: StatPearls [Internet]. Treasure Island (FL): StatPearls Publishing; 2023 Jan-. Available from: https://www.ncbi.nlm.nih.gov/books/NBK499953/

Chen H, Pellegrini J, Aggarwal S, Lei S, Warach S, Jensen F, Lipton S (1992) Open-channel block of N-methyl-D-aspartate (NMDA) responses by memantine: therapeutic advantage against

NMDA receptor-mediated neurotoxicity. J Neurosci 12(11):4427–4436. https://doi.org/10.1523/JNEUROSCI.12-11-04427.1992

Chen BK, Le Pen G, Eckmier A, Rubinstenn G, Jay TM, Denny CA (2021) Fluoroethylnormemantine, a novel derivative of memantine, facilitates extinction learning without sensorimotor deficits. Int J Neuropsychopharmacol 24(6):519–531

Chenard B, Bordner J, Butler T, Chambers L, Collins M, De Costa D, Fox C (1995) (1S, 2S)-1-(4-hydroxyphenyl)-2-(4-hydroxy-4-phenylpiperidino)-1-propanol: a potent new neuroprotectant which blocks N-methyl-d-aspartate responses. J Med Chem 38(16):3138–3145

Chizh BA, Headley PM, Tzschentke TM (2001) NMDA receptor antagonists as analgesics: focus on the NR2B subtype. Trends Pharmacol Sci 22(12):636–642

Chochkova M, Jiang H, Kyoseva R, Stoykova B, Tsvetanova E, Alexandrova A, Dimitrova-Sbirkova H (2021) Cinnamoyl-memantine hybrids: synthesis, X-ray crystallography and biological activities. J Mol Struct 1234:130147

Dantoine T, Auriacombe S, Sarazin M, Becker H, Pere JJ, Bourdeix I (2006) Rivastigmine monotherapy and combination therapy with memantine in patients with moderately severe Alzheimer's disease who failed to benefit from previous cholinesterase inhibitor treatment. Int J Clin Pract 60(1):110–118

Dawson VL, Dawson TM, London ED, Bredt DS, Snyder SH (1991) Nitric oxide mediates glutamate neurotoxicity in primary cortical cultures. Proc Natl Acad Sci 88(14):6368–6371

De Felice FG, Velasco PT, Lambert MP, Viola K, Fernandez SJ, Ferreira ST, Klein WL (2007) Aβ oligomers induce neuronal oxidative stress through an N-methyl-D-aspartate receptor-dependent mechanism that is blocked by the Alzheimer drug memantine. J Biol Chem 282(15):11590–11601

Deardorff WJ, Grossberg GT (2016) A fixed-dose combination of memantine extended-release and donepezil in the treatment of moderate-to-severe Alzheimer's disease. Drug Des Devel Ther 10:3267

Fischer G, Mutel V, Trube G, Malherbe P, Kew J, Mohacsi E, Kemp J (1997) Ro 25–6981, a highly potent and selective blocker of N-methyl-D-aspartate receptors containing the NR2B subunit. Characterization in vitro. J Pharmacol Exp Ther 283(3):1285–1292

Furukawa H, Singh SK, Mancusso R, Gouaux EJN (2005) Subunit arrangement and function in NMDA receptors. Nature 438(7065):185–192

Gazova Z, Soukup O, Sepsova V, Siposova K, Drtinova L, Jost P, Fedunova D (2017) Multi-target-directed therapeutic potential of 7-methoxytacrine-adamantylamine heterodimers in the Alzheimer's disease treatment. Biochim Biophys Acta 1863(2):607–619

Gerzon K, Krumkalns EV, Brindle RL, Marshall FJ, Root MA (1963) THE ADAMANTYL GROUP IN MEDICINAL AGENTS. I. HYPOGLYCEMIC N-ARYLSULFONYL-N'-ADAMANTYLUREAS. J Med Chem 6:760–763

He F, Ran Y, Li X, Wang D, Zhang Q, Lv J, Xu A (2020) Design, synthesis and biological evaluation of dual-function inhibitors targeting NMDAR and HDAC for Alzheimer's disease. Bioorg Chem 103:104109

Ibrahim L, DiazGranados N, Jolkovsky L, Brutsche N, Luckenbaugh DA, Herring WJ, Zarate CA Jr (2012) A randomized, placebo-controlled, crossover pilot trial of the oral selective NR2B antagonist MK-0657 in patients with treatment-resistant major depressive disorder. J Clin Psychopharmacol 32(4):551

Ivleva EA, Klimochkin YN (2017) Convenient synthesis of memantine hydrochloride. Org Prep Proced Int 49(2):155–162

Jin X, Liu Z, Guo B, Zhang G, Zhang Z, Xu L, Sun Y (2016) A promising dual-functional neuroprotective derivative of memantine. J Pharm Biomed Sci 6(6)

Johnson JW, Kotermanski SE (2006) Mechanism of action of memantine. Curr Opin Pharmacol 6(1):61–67

Jürgenson M, Zharkovskaja T, Noortoots A, Morozova M, Beniashvili A, Zapolski M, Zharkovsky A (2019) Effects of the drug combination memantine and melatonin on impaired memory and brain neuronal deficits in an amyloid-predominant mouse model of Alzheimer's disease. J Pharm Pharmacol 71(11):1695–1705

Kaniakova M, Nepovimova E, Kleteckova L, Skrenkova K, Holubova K, Chrienova Z, Vales K (2019) Combination of memantine and 6-chlorotacrine as novel multi-target compound against Alzheimer's disease. Curr Alzheimer Res 16(9):821–833

Karakas E, Simorowski N, Furukawa H (2011) Subunit arrangement and phenylethanolamine binding in GluN1/GluN2B NMDA receptors. Nature 475(7355):249–253

Karoutzou O, Kwak S-H, Lee S-D, Martínez-Falguera D, Sureda FX, Vázquez S, Barniol-Xicota M (2018) Towards a novel class of multitarget-directed ligands: dual P2X7–NMDA receptor antagonists. Molecules 23(1):230

Kashiwagi K, Masuko T, Nguyen CD, Kuno T, Tanaka I, Igarashi K, Williams K (2002) Channel blockers acting atN-methyl-D-aspartate receptors: differential effects of mutations in the vestibule and Ion Channel pore. Mol Pharmacol 61(3):533–545

Kaur A, Nigam K, Srivastava S, Tyagi A, Dang S (2020) Memantine nanoemulsion: a new approach to treat Alzheimer's disease. J Microencapsul 37(5):355–365

Kemp JA, Tasker T (2011) Methods for treating disorders using nmda nr2b-subtype selective antagonist: Google Patents

Koola MM (2020) Galantamine-Memantine combination in the treatment of Alzheimer's disease and beyond. Psychiatry Res 293:113409

Koola MM, Nikiforuk A, Pillai A, Parsaik AK (2018) Galantamine-memantine combination superior to donepezil-memantine combination in Alzheimer's disease: critical dissection with an emphasis on kynurenic acid and mismatch negativity. J Geriatr Care Res 5(2):57

Kuang W, Liu J, Lin X, Wu S, Gong J, Yin Q, Wang J (2022) Insoluble salt of Memantine with a unique fluorescence phenomenon. Mol Pharm 19(5):1389–1399

Kumamoto T, Nakajima M, Uga R, Ihayazaka N, Kashihara H, Katakawa K, Igarashi K (2018) Design, synthesis, and evaluation of polyamine-memantine hybrids as NMDA channel blockers. Bioorg Med Chem 26(3):603–608

Layton ME, Kelly MJ III, Rodzinak KJ (2006) Recent advances in the development of NR2B subtype-selective NMDA receptor antagonists. Curr Top Med Chem 6(7):697–709

Layton ME, Kelly MJ III, Rodzinak KJ, Sanderson PE, Young SD, Bednar RA, Mosser SD (2011) Discovery of 3-substituted aminocyclopentanes as potent and orally bioavailable NR2B subtype-selective NMDA antagonists. ACS Chem Neurosci 2(7):352–362

Lee C-H, Lü W, Michel JC, Goehring A, Du J, Song X, Gouaux E (2014) NMDA receptor structures reveal subunit arrangement and pore architecture. Nature 511(7508):191–197. https://doi.org/10.1038/nature13548

Lipton SA (2004) Failures and successes of NMDA receptor antagonists: molecular basis for the use of open-channel blockers like memantine in the treatment of acute and chronic neurologic insults. NeuroRx 1(1):101–110

Lipton SA, Choi Y-B, Pan Z-H, Lei SZ, Chen H-SV, Sucher NJ, Stamler JS (1993) A redox-based mechanism for the neuroprotective and neurodestructive effects of nitric oxide and related nitroso-compounds. Nature 364(6438):626–632

Liu Z, Yang S, Jin X, Zhang G, Guo B, Chen H, Wang Y (2017) Synthesis and biological evaluation of memantine nitrates as a potential treatment for neurodegenerative diseases. Medchemcomm 8(1):135–147

Liu Z, Qiu X, Mak S, Guo B, Hu S, Wang J, Zhang G (2020) Multifunctional memantine nitrate significantly protects against glutamate-induced excitotoxicity via inhibiting calcium influx and attenuating PI3K/Akt/GSK3beta pathway. Chem Biol Interact 325:109020

Madhra MK, Sharma M, Khanduri C (2007) New synthetic approach to Memantine hydrochloride starting from 1,3-dimethyl-adamantane. Org Process Res Dev 11(5):922–923

Mahmoudi M, Saeidian H, Mirjafary Z, Mokhtari J (2021) Preparation and characterization of memantine loaded polycaprolactone nanocapsules for Alzheimer's disease. J Porous Mater 28(1):205–212

Mony L, Kew JN, Gunthorpe MJ, Paoletti P (2009) Allosteric modulators of NR2B-containing NMDA receptors: molecular mechanisms and therapeutic potential. Br J Pharmacol 157(8): 1301–1317. https://doi.org/10.1111/j.1476-5381.2009.00304.x
Mota SI, Ferreira IL, Rego AC (2014) Dysfunctional synapse in Alzheimer's disease—a focus on NMDA receptors. Neuropharmacology 76:16–26
Natekar A, Dey R (2022) Voltammetric and spectroscopic studies of memantine hydrochloride and its transition metal complexes. Phys Chem Res 10(1):13–21
Palmer GC (2001) Neuroprotection by NMDA receptor antagonists in a variety of neuropathologies. Curr Drug Targets 2(3):241–271
Parsons C, Quack G, Bresink I, Baran L, Przegalinski E, Kostowski W, Danysz W (1995) Comparison of the potency, kinetics and voltage-dependency of a series of uncompetitive NMDA receptor antagonists in vitro with anticonvulsive and motor impairment activity in vivo. Neuropharmacology 34(10):1239–1258
Parsons C, Danysz W, Quack G (1999) Memantine is a clinically well tolerated N-methyl-D-aspartate (NMDA) receptor antagonist-a review of preclinical data. Neuropharmacology 38(6): 735–767
Pérez-Areales FJ, Turcu AL, Barniol-Xicota M, Pont C, Pivetta D, Espargaro A, Pérez B (2019) A novel class of multitarget anti-Alzheimer benzohomoadamantane-chlorotacrine hybrids modulating cholinesterases and glutamate NMDA receptors. Eur J Med Chem 180:613–626
Radwan RR, Abdel Ghaffar AM, Ali HE (2020) Gamma radiation preparation of chitosan nanoparticles for controlled delivery of memantine. J Biomater Appl 34(8):1150–1162
Rammes G (2009) Neramexane: a moderate-affinity NMDA receptor channel blocker: new prospects and indications. Expert Rev Clin Pharmacol 2(3):231–238
Rani V, Chawla R (2022) Design, fabrication, optimization and characterization of Memantine loaded biodegradable PLGA nanoscaffolds for treatment of Alzheimer's disease. Biomed Mater 17:065024
Reddy JM, Prasad G, Raju V, Ravikumar M, Himabindu V, Reddy GM (2007) An improved synthesis of memantine hydrochloride: anti-Alzheimer's drug. Org Process Res Dev 11(2): 268–269
Reisberg B, Doody R, Stöffler A, Schmitt F, Ferris S, Möbius HJ (2003) Memantine in moderate-to-severe Alzheimer's disease. N Engl J Med 348(14):1333–1341. https://doi.org/10.1056/NEJMoa013128
Rogawski MA, Wenk GL (2003) The neuropharmacological basis for the use of memantine in the treatment of Alzheimer's disease. CNS Drug Rev 9(3):275–308
Rosini M, Simoni E, Caporaso R, Basagni F, Catanzaro M, Abu IF, Albani D (2019) Merging memantine and ferulic acid to probe connections between NMDA receptors, oxidative stress and amyloid-β peptide in Alzheimer's disease. Eur J Med Chem 180:111–120
Schmitt HP (2005) On the paradox of ion channel blockade and its benefits in the treatment of Alzheimer disease. Med Hypotheses 65(2):259–265
Schmitt B, Bernhardt T, Moeller H-J, Heuser I, Frölich L (2004) Combination therapy in Alzheimer's disease. CNS Drugs 18(13):827–844
Sestito S, Daniele S, Pietrobono D, Citi V, Bellusci L, Chiellini G, Rapposelli S (2019) Memantine prodrug as a new agent for Alzheimer's disease. Sci Rep 9(1):1–11
Shenfield GM (1982) Fixed combination drug therapy. Drugs 23(6):462–480
Simoni E, Daniele S, Bottegoni G, Pizzirani D, Trincavelli ML, Goldoni L, Piomelli D (2012) Combining galantamine and memantine in multitargeted, new chemical entities potentially useful in Alzheimer's disease. J Med Chem 55(22):9708–9721
Sobolevsky AI, Koshelev SG, Khodorov BI (1998) Interaction of memantine and amantadine with agonist-unbound NMDA-receptor channels in acutely isolated rat hippocampal neurons. J Physiol 512(1):47–60
Sozio P, Cerasa LS, Laserra S, Cacciatore I, Cornacchia C, Di Filippo ES, Grilli M (2013) Memantine-sulfur containing antioxidant conjugates as potential prodrugs to improve the treatment of Alzheimer's disease. Eur J Pharm Sci 49(2):187–198

Stroebel D, Buhl DL, Knafels JD, Chanda PK, Green M, Sciabola S, Pandit J (2016) A novel binding mode reveals two distinct classes of NMDA receptor GluN2B-selective antagonists. Mol Pharmacol 89(5):541–551

Tencheva A, Liu R, Volkova TV, Chayrov R, Mitrev Y, Štícha M, Stankova I (2020) Synthetic analogues of memantine as neuroprotective and influenza viral inhibitors: in vitro and physico-chemical studies. Amino Acids 52(11):1559–1580

Turcu AL, Companys-Alemany J, Phillips MB, Patel DS, Griñán-Ferré C, Loza MI, Sureda FX (2022) Design, synthesis, and in vitro and in vivo characterization of new memantine analogs for Alzheimer's disease. Eur J Med Chem 236:114354

Upton S, Rosenberg P (1995) Mechanisms of disease. Excitatory amino acids as a final common pathway for neurological disorders. New England J Med 330:613–622

Valverde E, Sureda FX, Vázquez S (2014) Novel benzopolycyclic amines with NMDA receptor antagonist activity. Bioorg Med Chem 22(9):2678–2683

Vu BD, Ho Ba NM, Pham VH, Phan DC (2020) Simple two-step procedure for the synthesis of memantine hydrochloride from 1, 3-dimethyl-adamantane. ACS Omega 5(26):16085–16088

Vyklicky V, Korinek M, Smejkalova T, Balik A, Krausova B, Kaniakova M et al (2014) Structure, function, and pharmacology of NMDA receptor channels. Physiol Res 63:S191

Wang Y, Eu J, Washburn M, Gong T, Vincent Chen H-S, Larrick James W, Porter S (2006) The pharmacology of aminoadamantane nitrates. Curr Alzheimer Res 3(3):201–204

Xia P, Chen HS, Zhang D, Lipton SA (2010) Memantine preferentially blocks extrasynaptic over synaptic NMDA receptor currents in hippocampal autapses. J Neurosci 30(33):11246–11250. https://doi.org/10.1523/jneurosci.2488-10.2010

Zhang W, Kandel N, Zhou Y, Smith N, Ferreira BC, Perez M, Leblanc RM (2022) Drug delivery of memantine with carbon dots for Alzheimer's disease: blood–brain barrier penetration and inhibition of tau aggregation. J Colloid Interface Sci 617:20–31

Zhu S, Paoletti P (2015) Allosteric modulators of NMDA receptors: multiple sites and mechanisms. Curr Opin Pharmacol 20:14–23. https://doi.org/10.1016/j.coph.2014.10.009

Zhu Y, Li X, Wang X, Huang X, Shen T, Zhang Y, Jiao N (2015) Silver-catalyzed decarboxylative azidation of aliphatic carboxylic acids. Org Lett 17(19):4702–4705

Part IV

Natural Molecules Inspired Novel Derivatives for Alzheimer's Disease

Huperzine-Based Derivatives: Design, Synthesis, and Anti-Alzheimer Activity

9

Vedika G. Dadlani, Harshal Ashok Pawar, and Pushpendra K. Tripathi

Abstract

Alzheimer's disease (AD) is a progressive illness of the nervous system typified by degenerative cognitive disability with challenging behavioural changes and dwindling physical activities for day-to-day living and remains one of the major reasons of dementia. As per the worldwide status report released by World Health Organization, the numbers of people with dementia will double every 20 years, reaching about 140 million in 2050. The currently used drugs for AD include galantamine, rivastigmine, and donepezil. These slow the progression of the disease or help to control cognitive and behavioural changes. Recently, aducanumab has been approved as a disease modifying drug targeting beta-amyloid and helps to reduce brain lesions. Another drug, memantine, is effective in regulating levels of glutamate, leading to brain cell death. Thus, the future involves multifunctional and targeted approaches, and multipotent naturally occurring agents have recently drawn attention in this regard. Huperzine A, derived from *Huperzia serrata or Lycopodium serrata* is one amongst these. Chemically, it is an alkaloid containing quinolizidine with good potency and selectivity and reversible inhibition of acetylcholinesterase. Many analogs of huperzine A have been reported but most of them are not able to cross the blood–brain barrier. Structural modifications of huperzine A are in process to achieve analogs with better anticholinesterase activity and improve their method of synthesis. The present chapter is aimed to compile and review the numerous

V. G. Dadlani · H. A. Pawar
Dr. L. H. Hiranandani College of Pharmacy, University of Mumbai, Ulhasnagar, Maharashtra, India
e-mail: vedika.dadlani@dlhhcop.org; harshal.pawar@dlhhcop.org

P. K. Tripathi (✉)
Institute of Pharmaceutical Sciences, University of Lucknow, Lucknow, Uttar Pradesh, India

A. Sharma, G. P. Modi (eds.), *Natural Product-Based Synthetic Drug Molecules in Alzheimer's Disease*, https://doi.org/10.1007/978-981-99-6038-5_9

evidence-based scientific information related to huperzine derivatives to make it easily available to the scientific community and researchers.

Keywords

Alzheimer's disease · Huperzine A · Huperzine analogs · Quinolizidine alkaloid · Dementia

9.1 Introduction

Alzheimer's disease (AD) is a progressive illness of the nervous system typified by degenerative cognitive disability with challenging behavioural changes and dwindling physical activities for day-to-day living and remains one of the major reasons of dementia (Hanns and Gabriele 2003; Reitz et al. 2011). As per the worldwide status report released by World Health Organization, the numbers of people with dementia will double every 20 years, reaching about 140 million in 2050 (Prince et al. 2009).

None of the current drugs on the market can completely reverse AD, but they might be able to lessen or control some of its behavioural and cognitive symptoms, offering some solace, respect, and encouragement. Donepezil, galantamine, rivastigmine are the common prescription medications approved by U.S. FDA for the treatment of symptoms of AD (Batsch et al. 2012). These are acetylcholinesterase (AChE) inhibitors, involved in degradation of acetylcholine, a neurotransmitter important for memory, decision-making, and other cognitive processes. Besides these, memantine is also prescribed and seems to function through controlling a separate chemical messenger's activity in the brain. Tacrine is now rarely recommended due to potential liver damage, and these medications have several common adverse effects (Alzheimer's Association n.d.). Hence, newer mechanisms of drug action are in need for better treatment and management of AD. Recently, aducanumab has been approved as a disease modifying drug targeting beta-amyloid and helps to reduce brain lesions. Another drug, memantine, is effective in regulating levels of glutamate, leading to brain cell death. Thus the future involves multifunctional and targeted approaches and multipotent naturally occurring agents have recently drawn attention in this regard.

Scientists in China first discovered huperzine A (Hup A), produced from the Chinese herb, *Huperzia serrata*. It works similarly to currently available AChE inhibitors. Many preclinical and clinical trial studies have demonstrated the prospects of HupA in the management of AD (Wang et al. 1986). It has been shown that Hup A exhibits low toxicity while reversibly inhibiting acetylcholinesterase. Also, it lessens soluble βA and significantly enhances aged people's memory by delaying the formation of amyloid plaques. It benefits from having a long-lasting effect, a high safety index, strong stability, and few adverse side effects (Yang et al. 2020).

In the present chapter, an effort has been made to discuss the most recent advancements in the synthesis of huperzine A and its potential in the management of Alzheimer's disease.

9.2 Huperzine Alkaloids

9.2.1 Source

Huperzine A, present in *Huperzia serrata*, a firmoss, is a sesquiterpene alkaloid. It also occurs in other species like *H. carinata*, *H. elmeri*, *H. aqualupian*, etc. in varying quantities (Lim et al. 2010).

Often referred to as toothed clubmoss, Huperzia serrata is a widespread species belonging to family Lycopodiaceae and grows well in temperate areas of Asia including India, China, Japan, Korea, and Russia, as well as in Oceania and Central America (Fig. 9.1) (Jaswinder et al. 2016). It is an essential herb, particularly in traditional Chinese medicine. The plant taxonomical details are summarized in Table 9.1.

As the use of natural products is increasing in the pharmaceutical sector, several promising compounds have been extracted from this plant while huperzine A being the most interesting amongst these.

The isolation of HupA took place in 1986 from *Huperzia Serrata* while it was only in 1989 that it was chemically synthesized (Yang et al. 2020). Diverse alkaloids belonging to lycopodium genus have been isolated from *H. serrata*. Majority of these alkaloids are lipid soluble and are classified into four main structural classes—fawcettimine, lycodine, and lycopodine-type and group of compounds of

Fig. 9.1 *Huperzia Serrata*

Table 9.1 Taxonomy classification

Biological source	Huperzia serrata
Family	Lycopodiaceae
Kingdom	Plantae
Division	Lycopodiophyta
Class	Lycopodiopsida
Order	Lycopodiales
Genus	Huperzia
Species	Serrata

miscellaneous-type as shown in Table 9.2. Representative compounds for each class have been presented in (Fig. 9.2). The AChE enzyme inhibition activity has been mostly found in alkaloid compounds of lycodine-type class. These include HupA, HupB, 6β-hydroxyHupA, N-methylHupB, and huperzine. Amongst these, HupA is revealed as the most potent compound (Ferreira et al. 2016) Fig. 9.2 modified.

9.2.2 Chemistry

Naturally, HupA is an alkaloid with quinolizidine nucleus and belongs to lycopodium series with its two stereoisomers, (−)-HupA and (+)-HupA (Fig. 9.3). The characteristic structural features of HupA include aromatic pyridine ring conjugated to bicyclic system with an ethylidene group and a primary amino group. It is white and crystalline in nature with solubility in chloroform and aqueous acids. The L isomer has more active than the D isomer (Yang et al. 2020).

9.3 Huperzine and Alzheimer's Disease

In neuropathology, AD is characterized by synapse loss, cerebral amyloid angiopathy, intraneuronal neurofibrillary tangle deposits, and neuritic plaques in the brain parenchyma. ACh is a neurotransmitter in the brain that is necessary for proper brain activity and plays a crucial role in the formation of memories and AChE is an enzyme that breaks down ACh. Given that acetylcholine (ACh) levels in AD patients' brains are abnormally low, the current treatment for AD involves administering AChEIs to raise the concentration of ACh at the neuronal synaptic cleft by inhibiting AChE.

To date, four AChEIs, Aricept (donepezil or E2020), Exelon (rivastigmine), Cognex (tacrine), and Reminyl (galanthamine hydrobromide) currently are approved as prescription drugs by the United States to treat the symptoms of mild-to-moderate AD (Fig. 9.4). However, the clinical usefulness of AChEIs has been limited due to their short half-lives, excessive side effects caused by activation of peripheral cholinergic systems as well as hepatotoxicity, which is the most frequent and main side effect of tacrine therapy (Zhu et al. 2005).

Table 9.2 Major classes of Lycopodium alkaloids obtained from *H. serrata*

Table 2: Major classes of Lycopodium alkaloids obtained from *H. serrata*

Lycodine	Lycopodine	Fawcettimine	Miscellaneous
HuperzineA, Huperzine B, HuperzineC HuperzineD, HuperzineU, 6α-HydroxyhuperzineA, 6β-HydroxyhuperzineA N,N-Dimethyl huperzineA, N-MethylhuperzineB, De-N-methyl-β-obscurine PhlegmariurineM, Huperzinine, Isofordine[17,18]	Serratidine, 6α-hydroxyserratidine, 4α-hydroxyserratidine, 4α,6α-dihydroxyserratidine Serratezomine C, Clavolonine, HuperzineE, Huperzine F, Huperzine G, Huperzine O, n-oxide huperzine E, 12-hydroxyhuperzine, 2-chlorohuperzine E, N-oxide huperzine F, 12-deoxyhuperzine O, Lycodoline, Lucidioline, 12-epilycodoline N-oxide, 6α-hydroxylycopodine, 4α,6α-dihydroxylycopodine, 7-hydroxylycopodine, lycoposerramine series [19,20]	Huperserratinine, lycothunine, 11α-hydroxyfawcettidine, 8α,11α-dihydroxyfawcettidine, 2α,11α-dihydroxyfawcettidine,, Lycoflexine, macleanine, 7-hyroperoxyphlegmariurine B, 11α-hydroxyphlegmariurine B, Serratine, Serratanidine, Neohuperzinine, Serratinine Serratezomine B, 8-deoxy serratinine, 8-deoxy-13-dehydro serratinine, [21,22]	Huperzine V, Huperzine L, Huperzine K, Huperzine J, Huperzinine B, Serratanine B Phlegmariurine N, Serratanine A[23-25]
			Triterpenoids: Serratene-type 21α-hydroxyserrat14-en-3β-yl dihydrocaffeate, 3α,21α-dihydroxyserrat-14-en-24-oicacid, 21α-hydroxyserrat-14-en-3β-yl p-dihydrocoumarate, 21α-hydroxyserrat-14-en-3β-yl propanedioic acid monoester, 16-oxo-3α,21β-dihydroxyserrat1-en-24-oic acid, 16-oxo-3α,21β-dihydroxyserrat-14-en-24-al, 16-oxo-21β-hydroxyserrat-14-en-3α-yl Acetate[26]
			Phenolic Quercetin, Catechin, ferulic acid, chlorogenic acid[27]
			Flavonoids 5,7,4'-trihydroxy-3',5'-dimethoxyflavone, 5,7,4'-trihydroxy-3'methoxyflavone, Apigenin, 5,7,2',4'-tetrahydroxy-5'-methoxy flavone[28,29]

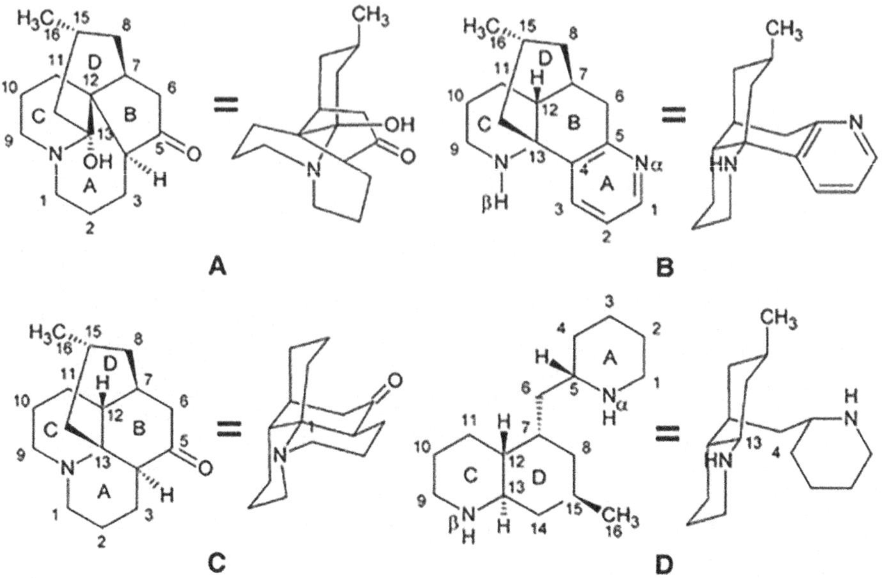

Fig. 9.2 Compounds representative of the fawcettimine (**a**), lycodine (**b**), lycopodine (**c**), and phlegmarine (**d**) class of lycopodium alkaloids obtained from *H. serrata* (Ferreira et al. 2016)

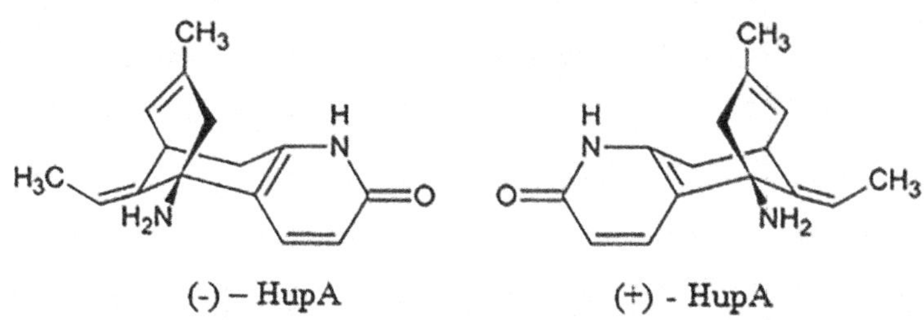

Fig. 9.3 (±) Stereoisomers of huperzine A. HupA (C15H18N2O, M 242) is chemically (1R.9S, 13E)-1amino-13-ethylidene-11-methyl6-azatricyclo[7.3.1.02,7]trideca-2(7),3,10-trien-5-one

AD is a commonly prevalent type of dementia typified by subtle memory loss, related functional impairment in thinking ability, and behavioural abnormalities that affects social and vocational functioning. As the condition worsens, other parts of the brain also experience loss of function and ultimately cell death. A person will eventually pass away from the loss of brain function alone if they do not already have another major condition. The underlying cause is not exactly known. However, it seems to be related to the levels of neurotransmitters in the brain cells (Ma et al. 2007).

Acetylcholine levels in the brain are decreased in AD patients, and this is accompanied with degenerative alterations to the brain tissue. Changes in oxidative

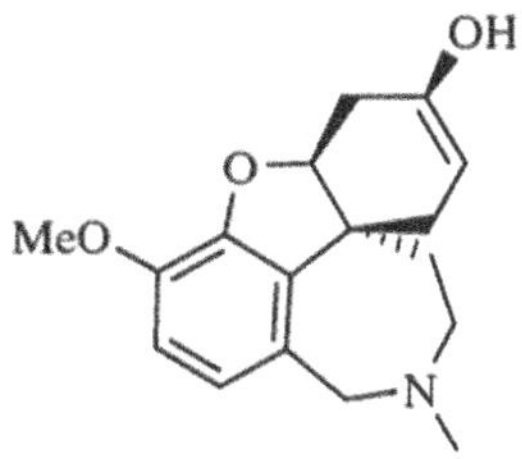

Galathamine

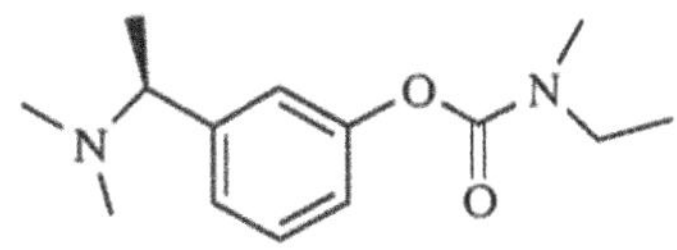

Rivastigmine

Fig. 9.4 Currently approved AChE inhibitors for Alzheimer's disease

metabolism are suspected to be linked to AD. Events leading to AD may involve increased oxidative stress brought on by free radical damage to cellular function. The creation of free radicals, organic and toxic brain damage, as well as other alterations, all contribute to the development of AD. As a result, medications for AD frequently include neuroprotectives, nootropics, and cognitive enhancers.

It has been observed that HupA is better than currently approved drugs as it can effectively cross BBB, has high oral bioavailability and longer duration of action. It is known to improvise the learning ability and memory in humans as well as to restore or reduce loss of cognition in a variety of behavioural models and animal species, including non-human primates (Ma et al. 2007).

HupA is available in U.S. market as dietary supplement to help in improving memory loss and mental disability, but the FDA has not yet approved it. It has quickly developed into a best-selling product and cult favourite in the market for smart-drugs market. It can be purchased online and at 'natural product' stores without prescription. The Journal of American Medical Association once stated in their journal that HupA with its high specificity for AChE and consequently lesser side effects can be beneficial for management of dementia and memory loss (Zangara 2003). Although AChEIs are still the most often prescribed drugs for AD, we are moving into a time where therapies which help in modifying diseases, mainly those which impart neuroprotection, are thought to contribute significantly in prevention of progression of AD. Research suggests that huperzine A may have the ability to alter the course of disease.

In the paragraphs that follow, huperzine A's traditional cholinergic impact and its unique possible non-cholinergic activities are described (Fig. 9.5) (Zhang 2012).

Huperzine A is a reversible inhibitor and displays mixed competitive inhibition against AChE. The strong AChE inhibitory impact may significantly increase synaptic ACh release and, in turn, cholinergic neurotransmission. In addition to the aforementioned classic effects, new pharmacological targets independent of huperzine A's AChE inhibitory effect and expanded cholinergic system effects in neuroprotection are two areas where huperzine A's beneficial properties are being explored with the help of various models (Zhang 2012).

Although ACh is typically thought of as a neurotransmitter, it also has cytokine properties and may be involved in various neuroprotective pathways: In the rat

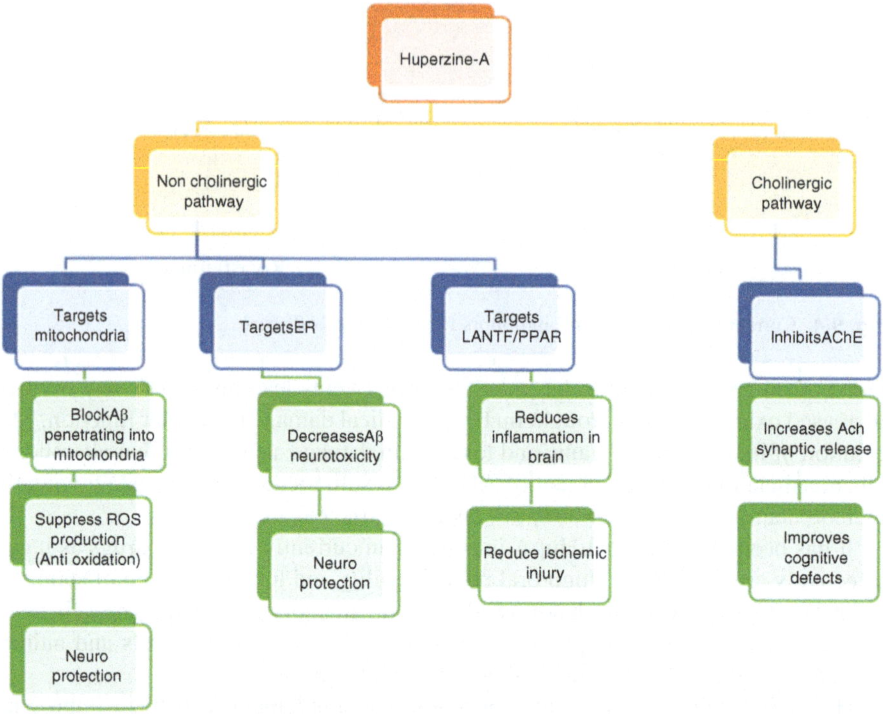

Fig. 9.5 Summary of classical cholinergic and potential non-cholinergic pharmacological targets of huperzine A

hippocampus, there was evidence of a close relationship between ACh and the neurotrophins and BDGF. Non-amyloidogenic APP pathway can be activated by activation of M1 muscarinic ACh receptor. Also α7nAChR seems to be correlated to inflammatory responses and neuronal degeneration in AD. The expression of NGF and its secretion raises the sAPPα levels which activate the non-amyloidogenic pathway in relation to pathway mediated by M1 muscarinic AChR and α7nAChR mediated hypoxia leading to inflammation. The positive effects related to cholinergic system could be due to increase in levels of ACh and not directly related to nAChR activation. As almost similar results are observed with various AChEIs, these neuroprotective benefits are likely potential common outcomes of cholinergic stimulation (Zhang 2012).

According to Fig. 9.5, huperzine A's effects on the cholinergic system may not only aid in treating symptoms but also slowing the progression of AD. Meanwhile, huperzine A has recently been discovered to have additional advantages that set it apart from other AChEIs and seem to be unrelated to AChE inhibition. Mitochondrial dysfunction is one of the major intracellular diseases linked to the development of AD. It was recently found that huperzine A might effectively reduce brain mitochondrial dysfunction caused by Aβ or ischemia insult.

Huperzine A has the ability to treat A-induced mitochondrial dysfunction by reducing the formation of reactive oxygen species (ROS) and increasing the activity of key enzymes of tricarboxylic (TCA) cycle as well as other key respiratory chain components. The fact that huperzine A prevented entry of Aβ into mitochondria and improved Aβ-induced TCA cycle failure in isolated brain cortical mitochondria supports the findings that mitochondrion appears to be an essential target of huperzine A.

As there is no proof that isolated brain mitochondria contain a cholinergic system, HupA effects with mitochondria as the target seem to be unrelated to the cholinergic system. A multitude of physiological processes, such as signal transduction calcium homeostasis, oxidative stress, and apoptosis, are facilitated by mitochondria. Thus, various models can be used for better understanding the effects of huperzine A on oxidative stress and apoptosis with mitochondrion as the target.

It has been studied that there is direct interaction of Aβ with many proteins on the mitochondrial membrane and inside the matrix. It could enter mitochondria through a pore formed by the outer membrane (TOM40) and the inner membrane (TIM22) and interact with mitochondrial proteins from the membrane permeability transition pore (MPTP) and this in turn results in changes in the mitochondrial membrane potential; Aβ also binds with β-amyloid binding alcohol dehydrogenase (ABAD) which results in mitochondrial dysfunction. The molecular target of huperzine A on Aβ induced mitochondrial dysfunction is yet unclear and hence the ability to determine whether and how huperzine A affects the aforementioned A-mitochondrion interactions could be an interesting area of research.

A potent dual PPARα/γ agonist exerted anti-inflammatory and neuroprotective effects, suggesting that ligand-activated PPAR may be a potential drug target against inflammation and brain damage after ischemic injury. Moreover, the endoplasmic reticulum (ER), like the mitochondrion, is a versatile organelle that is crucial to numerous malignant processes in AD. Several studies have demonstrated that huperzine A can improve ischemic damage in many in vivo and in vitro models. Initial studies suggested that Hup A might have an impact on the aberrant ER function linked to Aβ. It is also worthwhile investigating whether huperzine A could target on PPAR or ER to better explain the aforementioned positive effects (Zhang 2012).

9.4 Huperzine: Derivatives and Analogues

With continued interest in AD, design and synthesis of HupA analogs have begun. But changes in the rigid structural framework of HupA mostly resulted in decrease or loss of activity. Few analogues have only shown apparent AChE inhibitory activity (Ferreira et al. 2016). Table 9.3 surmises some of the derivatives of HupA in relation to the changes in structure and its consequent effects on activity (Zhou and Zhu 2000; Ros et al. 2001; Alcalá et al. 2003; Ma and Gang 2004; Jia et al. 2013). Also hybrids of tacrine with HupA have been prepared as AChE inhibitors (Camps et al. 2000a; Gemma et al. 2006). Further newer molecules were designed to create tight

Table 9.3 Summary of huperzine A analogs

Analogs of HupA	Structural changes in relation to HupA	Consequences of the structural changes on compounds activity
(±)-10,10-dimethyl-HupA 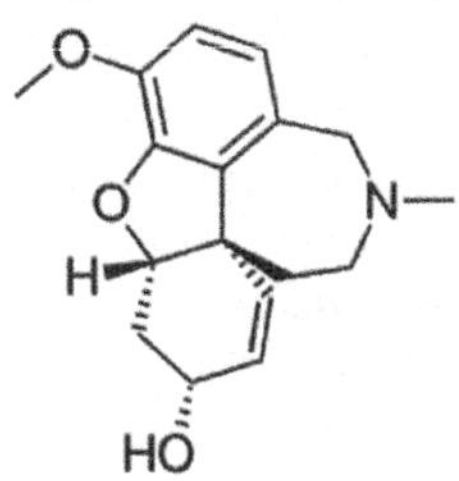 **Galantamine**	Two methyl groups introduced at position C-10	The AChE inhibitory activity increased almost eightfold due to introduction of axial methyl group at C-10 position while it decreased 1.5-fold in the corresponding isomer with equatorial methyl group
H-3 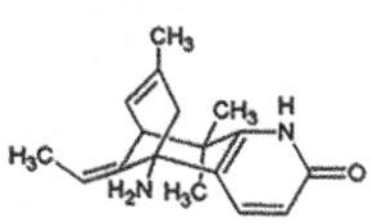	A brings of HupA which are likely to be essential for activity combined with tacrine by an alkane tether	AChE activity was good but less selective than HupA
Huprines 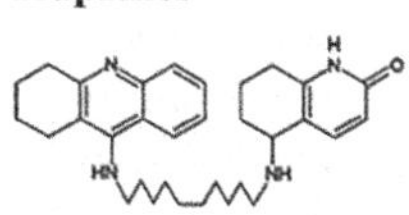	The bridgehead component of HupA combined with tacrine	Better AChE activity for hybrids. AChE levels increase in the synaptic cleft well in comparison to tacrine. Good selectivity ratio of AChE to BuChE. Fluorine in (±)-huprine Z when replaced by chlorine in (±)-huprine Y probably improves binding to AChE and hence potency
Isovan iHupA 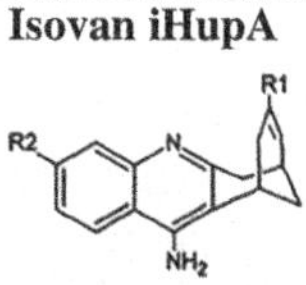	Schiff base with amino group of HupA	AChE activity similar to HupA but problems with stability
(−)-10-Spiro-cyclopropyl-HupA 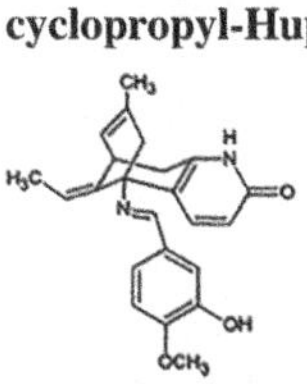	Cyclopropane attached at C-l 0 position of HupA	(−)-HupA and the cyclopropyl derivative similar in in vitro activity
5-substituted analogs 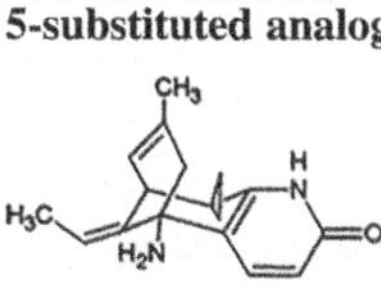	5-substituted HupA derivatives	5-substituted analogues (A) and (**B**) exhibit 50% inhibition of AChE at concentrations of 35 mM and 47 mM, respectively

(continued)

Table 9.3 (continued)

Analogs of HupA	Structural changes in relation to HupA	Consequences of the structural changes on compounds activity
ZT-1 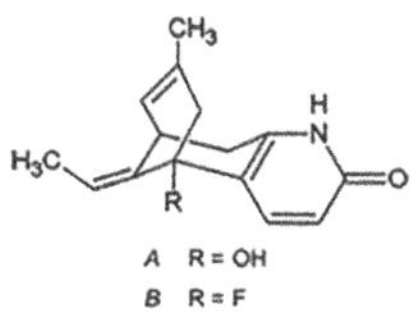	Schiff base of HupA and 5-cl-*O*-vanillin resulting in prodrug with rapid absorption and conversion to HupA	More selective AChE inhibitory activity than HupA and less toxic in mice. Oral bioavailability, ability to cross BBB and duration of action similar to HupA. Good tolerance in humans in phase I studies
Others 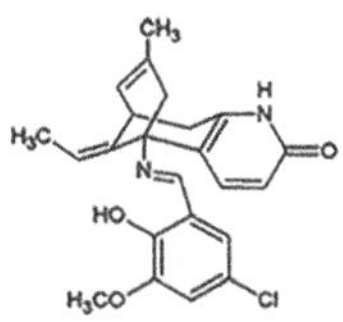	Tacrine-HupA hybrids with characteristic scaffold of 3-methylbicyclo-[3.3.1]non-3-ene	Markedly improved biological profile for hybrids. Multisite inhibition of human AChE and comparable inhibition of AChE and BuChE

interactions with the catalytic or peripheral sites and also midgorge recognition sites of human BuChE and AChE through various binding modes. This resulted in better biological profile in comparison to HupA and tacrine.

9.4.1 Analogs of Huperzine A

The U.S. Food and Drug Administration (FDA) has approved a number of novel AChE inhibitors (AChEIs) since 1993–2001 including tacrine, rivastigmine, and galantamine. However, some side effects were observed with tacrine in about 30% of people causing liver damage. Compared to tacrine, donepezil seems to be better in terms of greater effect and lesser toxicity. Rivastigmine doesn't seem to damage liver unlike tacrine. It can, however, result in side effects like nausea and vomiting. Galantamine is a natural alkaloid derived from Galanthus nivalis L. and allied plants of Amaryllidaceae family and has received FDA approval. Besides AD, it has been shown to be effective in treating vascular dementia also. However, HupA seems to better than the above-mentioned AChEIs in terms of blood–brain barrier penetration, oral bioavailability, and duration of AChE inhibitory effect (Ma et al. 2007).

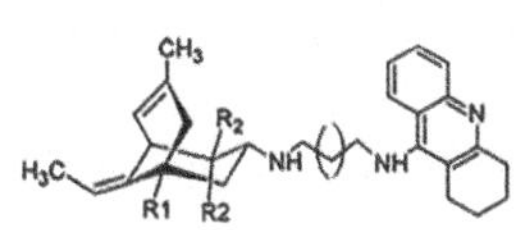

Zhu and colleagues synthesized a semi-synthetic derivative of HupA, ZT-1 (Kaneko et al. 1997). Experimental results showed that ZT-1 has more selectivity for AChE than BChE and was found to be less toxic in mice in contrast to HupA.

Regarding its oral bioavailability, capacity to cross BBB, and duration of action, ZT-1 is similar to HupA.

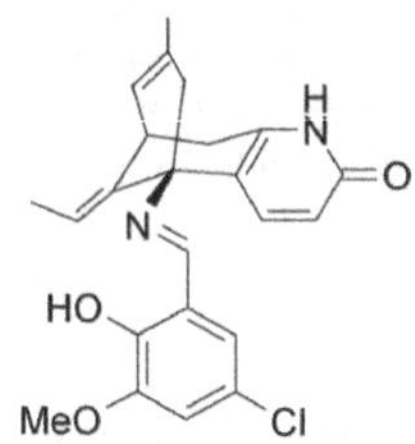

ZT-1

The development and assessment of potential AChE inhibitors with the strategy of hybridization of HupA and tacrine and derivatives of the same have been the focus of a multidisciplinary research consortium. Huperzine X (HupX) is one such hybrid which exhibits selective and potent AChE inhibitory activity. However, the multitarget biological profile is the most salient feature of HX. In fact, HX inhibits AChE induced amyloid aggregation, acts as an agonist on nicotinic and muscarinic M1 receptors, and shows tight-binding inhibition of AChE in reversibility assays. Chronic administration of HX to 3 Tg-AD transgenic mice had positive effects on the disease's hallmarks at the various levels including β amyloid, synaptophysin, and α7 nicotinic ACh receptors resulting in neuropathological effects. This validates that AChE inhibitors have disease-modifying effects.

9.4.2 Huperine X, Hup X (HupA-Tacrine Hybrid)

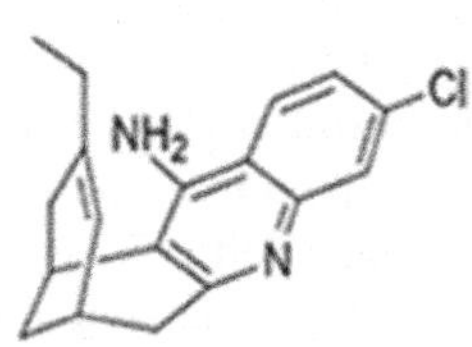

HX

AVCRI104P3 is another example of hybridization. It is a hybrid of HupA and donepezil with multiple targets. It not only inhibits Aβ aggregation induced by AChE but strongly inhibits BuChE and BACE-1 also.

Both HX and AVCRI104P3 have demonstrated potential effects on neuropsychiatric symptoms in a study on group of aged people with AD. Investigations are being done to determine the mechanism causing these beneficial outcomes (Giménez-Llort et al. 2017).

9.4.3 AVCRI104PO3 Hup Y (HupA-Donepezil Hybrid)

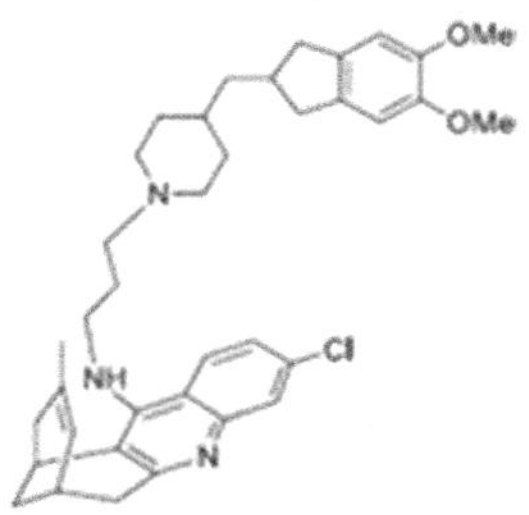

Y-Hybrid

It has been discovered that the donepezil hybrid is less efficient. HupA-tacrine hybrids, Huprines Y and X, have been found to amplify AChE more effectively than tacrine. The inhibitory activity of Huprine Y and Huprine Z was found to be superior than both parent compounds.

9.4.4 Design of HupA Derivatives

Ashani et al. showed that the type and number of aromatic amino acid residues in the catalytic pocket of AChE and BuChE determine the stability of HupA-cholinesterase complex thermodynamically. Computer-aided docking studies coupled with X-ray crystallography studies have shown that the potency and selectivity of AChE inhibition by HupA. It was observed that HupA binds to the in the catalytic domain above Typ 84 at the bottom of the gorge and near the opening of the gorge to the peripheral site with its ammonium group interacting partially with the indole ring of Trp 279 HupA (Fig. 9.6a). The active site also contains Ser 200 and His440 residues (Yamada et al. 1991).

In fact, HupA shows high affinity for AChE enzyme and this can be demonstrated from the orientation of HupA in the active site and its strong binding interactions with the active site residues. Also the hydrophobic interactions of HupA with the aromatic residues in the active-site gorge of AChE play an important role for better binding. The pyridine oxygen in (−)-HupA forms a strong H-bonding with a residue and has three potential H-bond donor and acceptor sites.

The principal interactions between AChE and (−)-HupA are as under:

1. Catalytic triad as seen by strong H-binding interactions between OH group of Tyr 130 and HupA carbonyl group and His 440 oxygen and HupA ethylidene methyl group.
2. One or two water molecules within the active-site gorge mediate indirect hydrogen bonds and similarly form H-bonding with other water molecules or to backbone and side chain atoms of the protein.

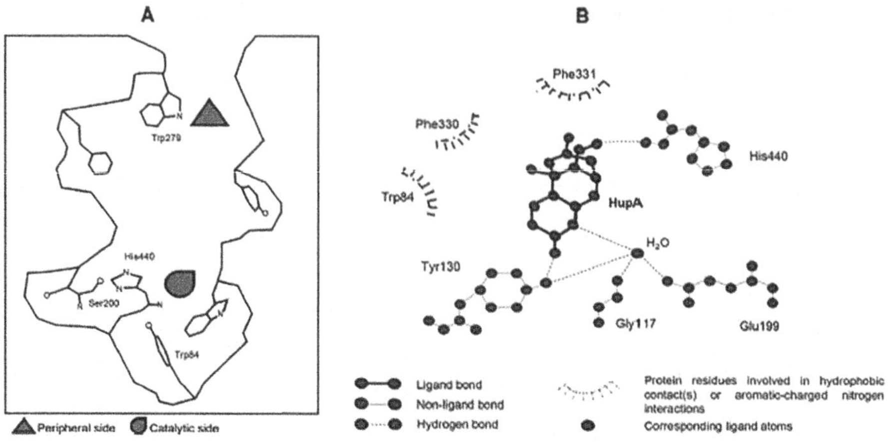

Fig. 9.6 (**a**) Gorge representation with catalytic and peripheral sites and binding pocket with main enzyme–ligand interactions. (**b**) Binding pocket of AchE showing the main interactions between the ligand HupA and the amino acid residues in the enzyme

3. The HupA contains a primary amine which is observed to interact with Trp 84 and Phe 330 aromatic rings by cation-pi interactions and with Asp 72 and Glu 199 carboxyl groups by ionic interactions.
4. HupA binds directly to the active-site opening and prevents the endogenous substrate from entry in AchE. The main chain and side chain atoms of Trp84, Phe331, and His440 also form key hydrophobic interactions (Fig. 9.6b) (Ha et al. 2011).

The NMDA receptor consists of three important sites—recognition site, cation-selective ion channel and binding site for Gly, Zn, and phencyclidine-like (PCP) compounds. (+)-5-methyl-10,11-dihydro-5H-dibenzo[a,d]cyclo hepten-5,10-imine maleate (MK-801) also bids to PCP site and Mg ions are able to block this channel (Fig. 9.7).

It was shown that HupA interacts with NMDA ion channels, causing a dose-dependent suppression of the binding of two additional NMDA antagonists, [3H] thienyl cyclohexyl piperidine, and [3H](+)-5-methyl-10,11-dihydro-5H-dibenzo [a,d]cyclohepten-5,10-imine maleate. As a result, HupA was found to interact with the NMDA receptor ion channel complex and to bind in the plasma membranes of brain synaptosomes, but not to the binding site for glycine, poly-amine, or NMDA ligands. The noncompetitive binding results indicate that HupA binds and blocks the NMDA receptor ion channel with subsequent calcium mobilization at or near the phencyclidine (the parent compound of thienylcyclohexylpiperidine) and (+)-5-methyl-10,11-dihydro-5H-dibenzo [a,d] cyclohepten-5,10-imine maleate ligand sites without producing psychotomimetic side effects (Gordon et al. 2001).

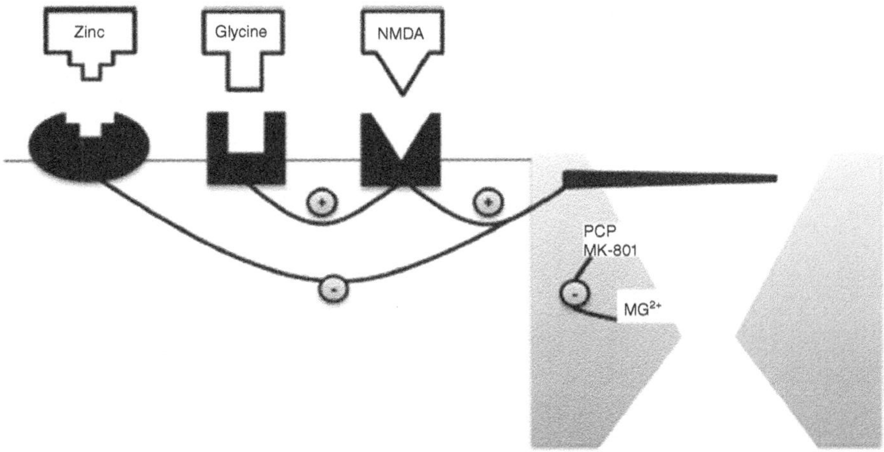

Fig. 9.7 The regulatory sites of NMDA receptor, representing binding site for phencyclidine (PCP)-like compounds where MK-801also binds and can be blocked by Mg^{+2}. The cation-selective ion channel, glycine, and zinc binding sites in NMDA receptor have also been shown

Derivatives of Hup A containing aromatic rings have recently been shown to offer potential therapeutic effects for AD symptoms. However, further studies are needed to determine the possible advantages of Hup A. Hybrids of tacrine and HupA (HX) exhibiting diverse functional moieties with basic nuclei of compounds have been synthesized as cholinesterase agonist for treatment of AD and showed good results for various positions. In comparison to tacrine, the halogen moiety demonstrated better activity and greater therapeutic efficacy. It also shows inhibition of BChE to a limited extent and can cross BBB (Hassan et al. 2018).

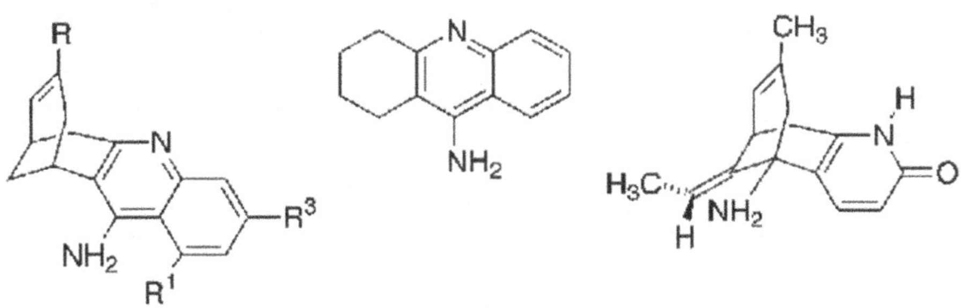

9.4.5 Structure–Activity Relationship

Superimposition of HupA and acetylcholine has been carried out with *in-silico* studies with (Fig. 9.8) The carbonyl groups, nitrogen, and oxygen of ACh share structural similarities with Hup A's homologous carbonyl, nitrogen, and amino-nitrogen (Bai et al. 2000). Since the amino-nitrogen atom and carbonyl group of

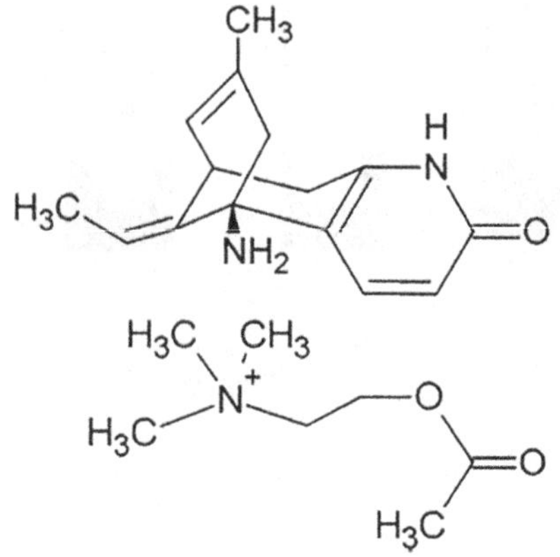

Fig. 9.8 Similarities in the structural features of huperzine A (HupA) and acetylcholine (ACh)

the pyridine ring in HupA and the equivalent quaternary nitrogen atom and ester carbonyl group in Ach are spaced similarly, the 5-aminomethyl-2(1H)-pyridone forms the pharmacophore of HupA.

The amine group (1), −pyridone ring (2), exocyclic ethylene residue (3), three-carbon bridge and its double bond (4), and methyl group (5) are structural features necessary for the inhibitory effect of HupA and their spatial configuration determines its activity (Fig. 9.9). Hence, elimination or substitution of at least one of these groups will significantly decrease AChE inhibitory activity (Ashani et al. 1992). In fact, it has been found that (−)-HupA is more potent AChE inhibitor than (+)-HupA.

9.5 Synthesis of Huperzine-Based Derivatives

Chemical synthesis technology is being developed to boost HupA production. The crucial part in the synthesis of HupA is the three-carbon bridged ring and has been the focus of recent study efforts. Moreover, the process' industrialization has undoubtedly boosted the amount of HupA produced chemically. Although many methods have been developed to manufacture HupA, the three primary techniques are as follows:

1. Extraction and isolation from natural sources
2. Synthesis of HupA chemically
3. Biotechnological production of HupA

HupA concentration in natural sources is also influenced by the growing area, the time of year, and the type of plant. Higher concentrations of HupA are found in plants with *Huperziaceae* family. Production of HupA by chemical synthesis can assist in achieving higher levels for commercial supply.

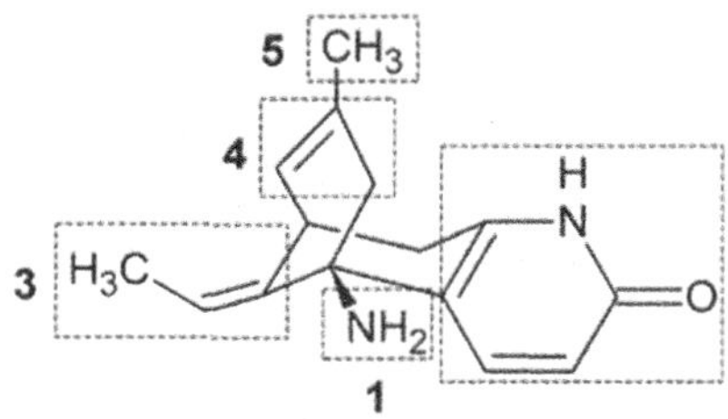

Fig. 9.9 Pharmacophoric features of huperzine: amine group (1), α-pyridone ring (2), exocyclic ethylene residue (3), three-carbon bridge and its double bond (4), and methyl group (5)

In recent years, the manufacturing of HupA has also incorporated standardized biosynthesis by endophytic screening and tissue culture via biotransformation and tissue culture. Despite the various successes, there are still many issues with the synthesis of HupA that needs to be resolved in order to benefit from its therapeutic characteristics (Yang et al. 2020).

An alkaloid with three-carbon bridges called huperzine A is usually present as a racemic combination. Further research has indicated that (+)-HupA activity is inferior to (−)-HupA activity. Hence, building a three-carbon bridge with stereospecificity represents a primary hurdle in synthesis of HupA. Nonetheless, major advancement has been made in the synthesis of (−)-HupA and the methods include:

1. Palladium-catalysed bisallyl substitution reaction
2. Successive Michael addition and aldol reaction
3. Intramolecular Heck reaction
4. Intramolecular α-alkylation

Synthesis

Using commercially plentiful (R)-pulegone, the entire synthesis of the Lycopodium alkaloid (−)-huperzine A was completed in 10 stages with a 17% overall yield. The synthetic pathway includes a highly stereoselective Buchwald-Hartwig coupling reaction-mediated efficient synthesis of **4** followed by diastereoselective alkylation mediated by dianion with enamine product and its intramolecular Heck reaction. The highlights include the Wagner–Meerwein arrangement and stereoselective β-elimination (Ding et al. 2012).

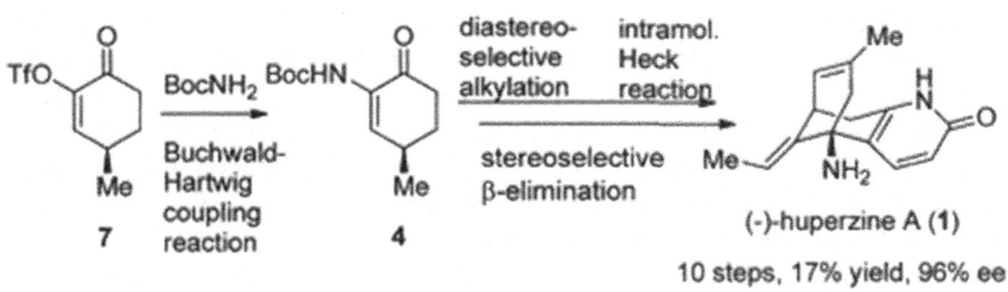

The first total synthesis of racemic HA was separately completed by two groups in 1989 (Qian and Ji 1989; Xia and Kozikowski 1989) The -keto ester **2** or **2a** was used as the starting point for almost the same synthetic approach. Using conventional techniques, Qian and Ji created **2a** with a 17.1% yield from ethyl acetoacetate. Methacrolein and the tandem Michael aldol reaction were used to create the three-carbon bridge ring. Afterwards, a Wittig reaction was performed to create an exocyclic double bond, followed by a MsOH elimination reaction to create an endocyclic double bond, and finally a Curtius rearrangement and deprotection reaction. The significant flaw of the product from Wittig reaction is that the yield of the E isomer is low. (Scheme 9.1) (Qian and Ji 1989).

Xia and Kozikowski used a different approach to create **2a**. It took longer and more expensive reagents, including PhSeCl and Pd(OH), than Qian and Ji's. With the exception of employing PhSH/AIBN to increase the E/Z product ratio from 10: 90 to 90:10, the other processes to produce rac-HA are comparable to those of Qian and Ji (Scheme 9.2) (Xia and Kozikowski 1989).

The preparation of crucial intermediate **4** was enhanced in 1990 by Xia and Kozikowski from 3 (Scheme 9.3) (Kozikwoski et al. 1990). They produced 2-pyridone 6 from a carbonyl molecule, ammonia, and methyl propiolate using an effective one-pot, three-component method that improved the yield of **2a** and did not require those pricey chemicals. A palladium-catalysed bicycloannulation pathway (Scheme 9.4) to racemic HA from **2a** in a 40% overall yield was created by Xia and Kozikowski in 1993 (Campiani et al. 1993). Pd-catalysed alkylation of **2a** with 2-methylenepropane-1,3-diyl diacetate on both sides of the ketone carbonyl was more successfully used to introduce the three-carbon bridge. The racemate showed an IC50 of 0.073 mM in comparison to (−) HA (0.047 mM for AChE inhibition), which is, within error, as expected if the unnatural enantiomer is inactive.

Camps et al. created a pathway to (−)-HA from a keto capamate **7**, which was produced from 1,4-cyclohexanedione monoethylene ketal **3** in an overall yield of 22% (Scheme 9.5) (Kozikowski et al. 1994; Camps et al. 1996, 2000b). The pyridone moiety of racemic HA is elaborated in a late step in this novel method of producing HA. In this manner, the pyridone moiety in HA can be accessed in place of the various heterocyclic analogues. This process did not result in improved yield, and separation of isomers was also tedious.

The ability to produce natural (−)-HA has received a lot of attention since it has a 38-fold greater AChE inhibitory impact than its enantiomer (+)-HA. The first to describe the path to optically pure (−)-HA was Yamada et al. Yamada et al. decided to incorporate absolute stereochemistry at the stage of the Michael-aldol reaction, which generates the bridging ring of HA, based on the pathway established to (−)-HA. According to (Scheme 9.6), over the course of 2 days at room temperature (r.t.), **2a** was trans esterified with (−)-8-phenylmenthol and **8** interacted with methacrolein in the presence of tetramethylguanidine. The isolation of combination **9** has a 90% yield. Further **9** was added in olefin **10** (Yamada et al. 1991). The optical intermediate (5S, 9R) **4** can be easily changed into optical (−)-HA by following stages in the method by Chen and Yang that are comparable to those used in the methods by Qian and Ji and Xia and Kozikowski (Chen and Yang 1995). Over 10 days at room temperature, **2a** and methacrolein under the presence of 0.1 equivalents of quinine produced the (5S,9R) isomer **4**.

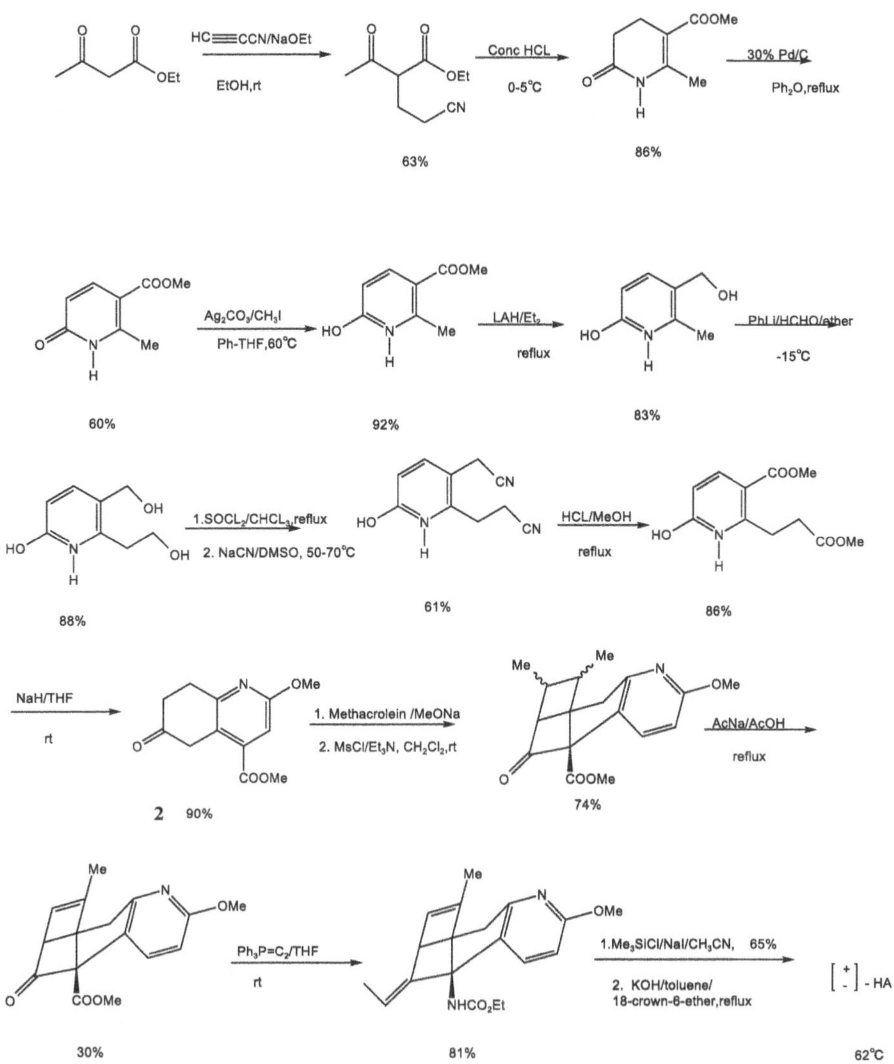

Scheme 9.1 Synthetic route to rac-HA by Qian and Ji

According to Kaneko et al., the crucial intermediate (+)-**12** of (−)-HA was made with β-keto ester **2** on bicycloannulation with 2-methylene-1,3-propanediol diacetate **11** under the catalysis of asymmetric Pd (Schemes 9.6, 9.7, and 9.8) (Kaneko et al. 1997). Enantioselectivity for the chiral ferrocenylphosphine ligand **13** was 64% ee.

These encouraging outcomes led to the creation of numerous novel chiral ferrocenylphosphine ligands (He et al. 1998, 2001). Using ligand **14**, the bicycloannulation's enantioselectivity was obviously enhanced, resulting in **12** in 81% ee. It was clear that adjusting the N-substituent of the ligand with the proper

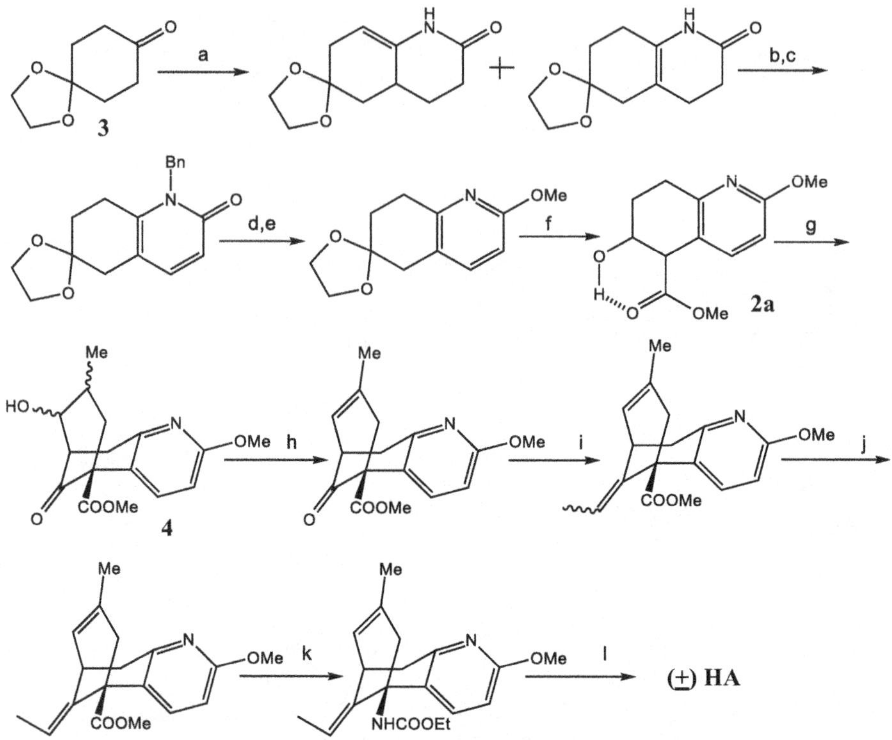

Scheme 9.2 Synthetic approaches to (−)-HA by Xia and Kozikow ski

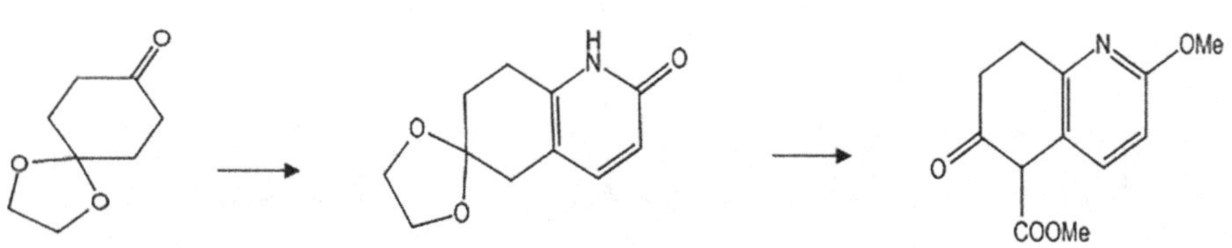

3 6 2a

Scheme 9.3 One-pot process to prepare 6

chain length had a significant impact on the reaction's enantioselectivity. With the help of the (R,S)-ferrocenylphosphine ligand **15**—which has a cyclopentyl group at the nitrogen-enantioselectivity of 90.3% ee for **12** was attained. The chiral nonracemic product **12** was produced in the required configuration for the synthesis of natural (−)-HA using the most effective chiral ligand **15** (He et al. 2001).

By using a Mn(III)-mediated oxidative radical cyclization to create the skeleton of HA from the allylic molecule **16** produced from **2**, Lee et al. created a novel

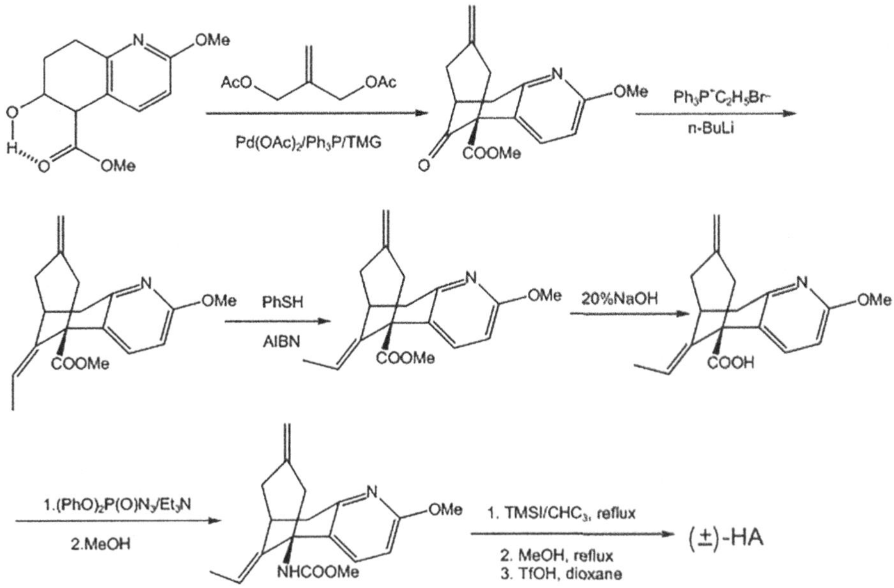

Scheme 9.4 Pd-catalysed route to (−)HA

technique. According to reports in the literature, the triflic acid treatment might quickly isomerize the thermodynamically unstable exo double bond product **17** into the endo olefin **18** (Scheme 9.9) (Lee et al. 2002).

9.6 Analysis of Huperzine-Based Derivatives

9.6.1 HPLC Method

Researchers have created a quick, easy, and trustworthy HPLC method for analysis of alkaloids in TCM herb *Huperzia Serrata*. The chromatographic conditions for the RP-HPLC method are C18 column and methanol: ammonium acetate (pH 6) in 30: 70 v/v as mobile phase with PDA detector combined with MS and in conjunction with UV-Vis detector. This resulted in detection of nine alkaloids in the extracts of *H. serrata*. Their content was measured using HPLC in conjunction with UV-Visible detector. The technique was verified. RSD was 2.44%, and the recovery rates ranged from 96.8 to 97.7%. The intraday and interday precision results in RSD values ranging from 0.53 to 1.51% and linearity in the range of 5–100 μg/mL and $r^2 = 0.9994$ (Wu and Gu 2006).

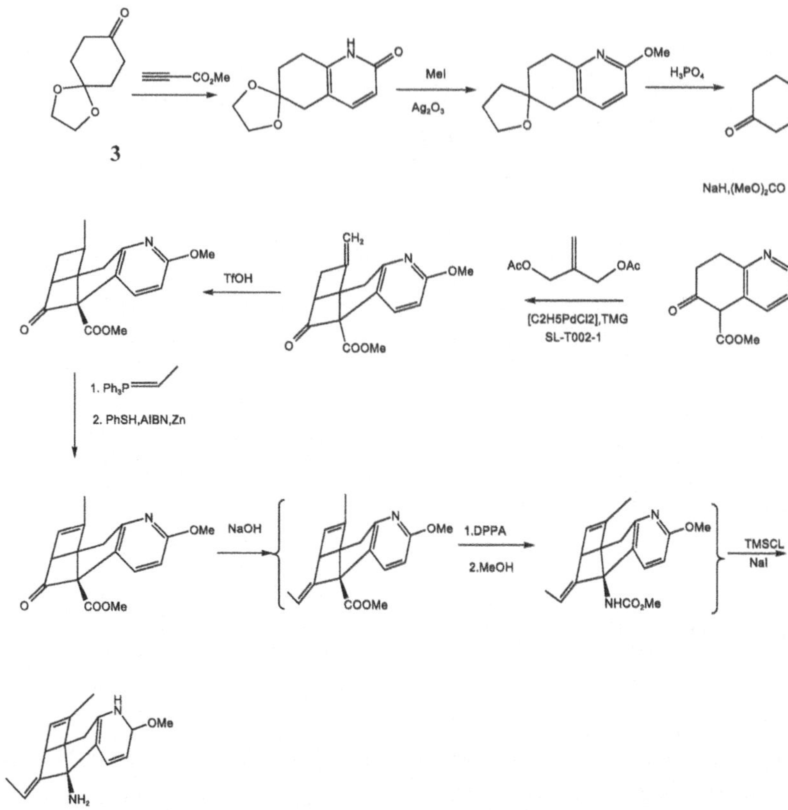

Scheme 9.5 Camps et al. approach to preparing racemic HA

9.6.2 Capillary Electrophoretic Method

Nguyen NC et al. created a dependable capillary zone electrophoresis technique to assess the detection of Hup A in *H. serrata*. The analytical conditions for the technique include hydrodynamic injection for 5 sec at 50 mbar, acetate buffer pH 6, 20 kV voltage, 25 °C temperature, uncoated fused-silica capillary with inner diameter of 70 m and 50 cm actual length and wavelength 310 nm. The recovery rates had a R.S.D. of 2% and ranged from 98.05% to 100.64% and linearity in the range of 1–500 g/mL with $r^2 = 0.9994$. The thresholds for quantification and detection were 0.33 and 1.0 g/mL, respectively (Nguyen et al. 2021).

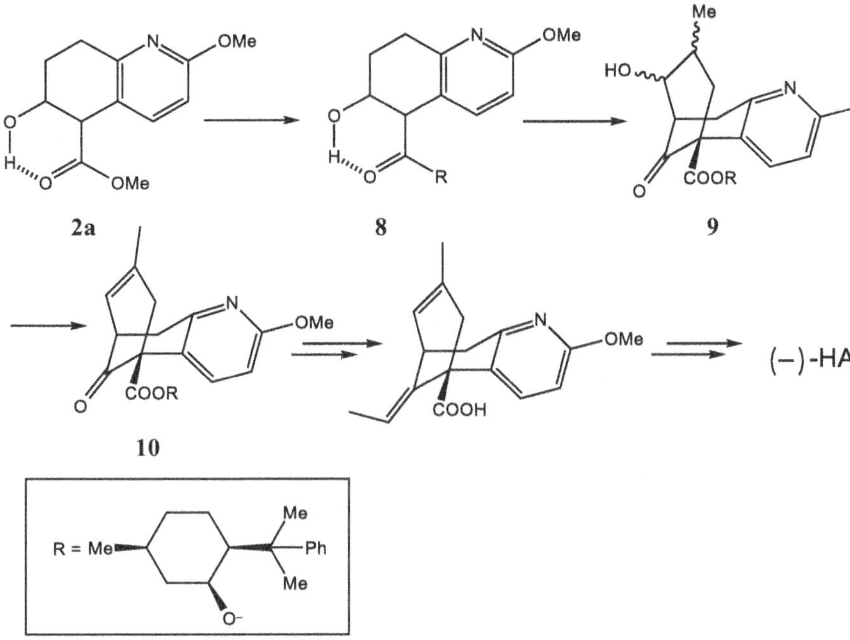

Scheme 9.6 Route to optical pure (_)-HA by Yamada et al.

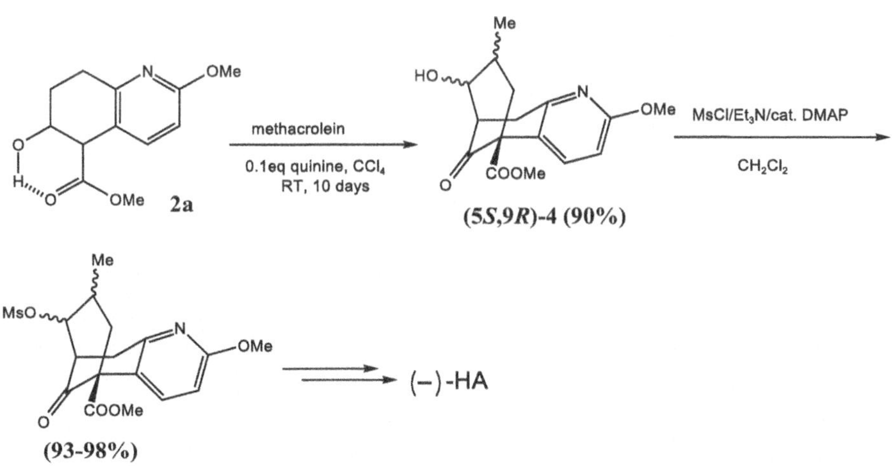

Scheme 9.7 The route to (5S,9R)-4 by Chen and Yang

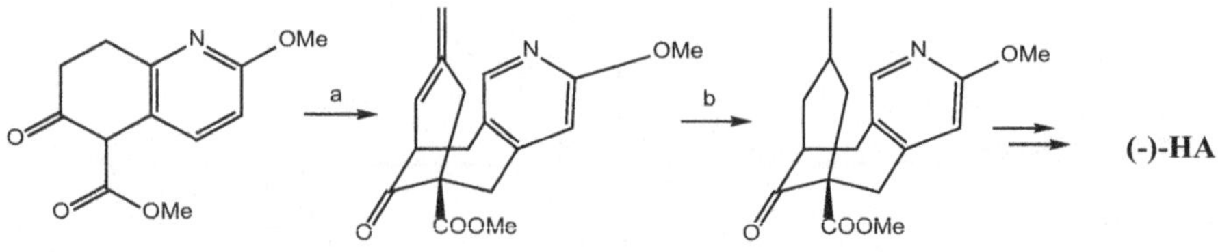

2 12

Reagents: a) **11**, (η3-allyl)Pdcl, TMG, chiral ligand, toluene; b) triflic acid, dioxane.

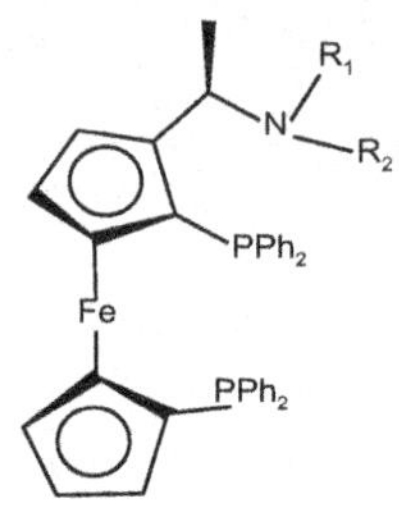

13: R_1=Me, R_2=$(CH_2)_4OH$
14: R_1=Et, R_2=$(CH_2)_4OH$
15: R_1=cyclopentyl, R_2=$(CH_2)_5OH$

Scheme 9.8 The preparation of (+)-12 by Kaneko et al.

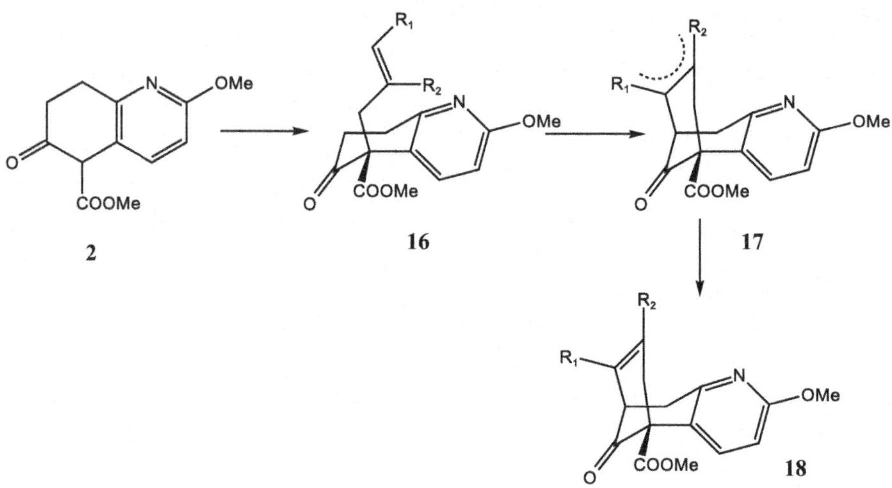

Scheme 9.9 The new method constructing the skeleton of HA by Lee et al.

9.6.3 NMR Assignments of Huperzine A

Huperzine A displayed indications for two methyls, two methylenes, five methines, and six quaternary carbons in its 1H, 13C, and APT spectra. The HETCOR spectrum simplifies to identify protons with carbon bonds (Lin et al. 1993) (Table 9.4).

Table 9.4 1H NMR of Huperzine A

^{1}H	Huperzine A
2a	6.38 (d, 9.4)
3a	7.86 (d, 9.4)
6a	2.85 (dd, 14.5, 5.0)
6b	2.71 (dd, 15.4, 0.9)
7	3.57 (hs)
8a	5.37 (d, 5.0)
10a	1.64 (d, 6.8)
11a	5.44 (q, 6.8)
14a	2.09 (s)
14b	2.09 (s)
16	1.50 (s)

Compound Huperzine A was recorded in CDCl3, chemical shift values are reported as δ values (ppm) from TMS at 299.9 MHz or 500.12 MHz. Signal multiplicity and coupling constants (Hz) are shown in parentheses

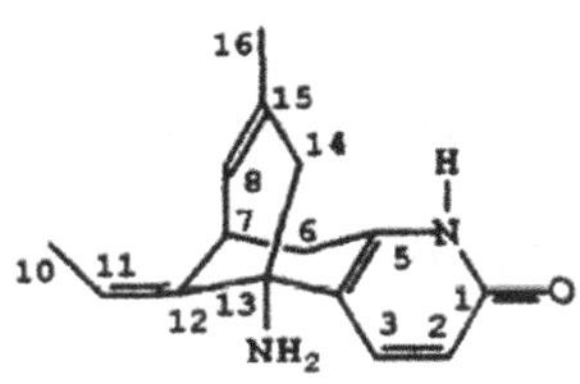

1

Selective INEPT spectra of Huperzine A allowed quaternary carbon atoms to be assigned by demonstrating long-range correlations between 1H and 13C through three bonds and two bonds. Selective irradiation of the H-2 signal enhanced the C4 signals, irradiation of H-3 enhanced the C-1, C-5, and C-13 signals, irradiation of H-6a at δ2.85 enhanced the C-4, C-5, C-7, and C-8 signals, irradiation of H-6b at δ2.71 enhanced the C-4, C-5, C-7, C-8, and C-12 signals, irradiation of H-7 enhanced the C-5, C-6, C-8, C-11, C-12, C-13, and C-14 signals, irradiation of H-8 enhanced the C-7, C-12, C-13, C-14, and C-16 signals, irradiation of the H-10 methyl signal enhanced the C-11 and C-12 signals, irradiation of the H-11 enhanced the C-7, C-8, C-10, and C-13 signals, irradiation of the H-14 methylene signal at δ 2.09 enhanced the C-4, C-8, C-12, C-13, and C-15, signals, and irradiation of the H-16 methyl signal at δ 1.50 enhanced the C-8, C-14, and C-15 signals. Hence, a clear assignment of all 13C NMR signal parameters was made (Table 9.5). The designations of C-5 and C-12 should be reversed, it was discovered when these data were compared to the literature. Irradiations of H-3 and H-10, respectively, were crucial in achieving these reassignments (Lin et al. 1993).

Table 9.5 13C NMR data of 1

^{13}C	Huperzine A
1	165.4 (s)
2	116.9 (d)
3	140.2 (d)
4	122.8 (s)
5	143.1 (s)
6	35.2 (t)
7	32.8 (d)
8	124.2 (d)
9	49.9 (t)
10	12.3 (t)
11	111.2 (d)
12	142.4 (s)
13	54.3 (s)
14	49.1 (t)
15	134.0 (s)
16	22.6 (4)

9.7 Pharmacokinetic and Associated Toxicities

9.7.1 Pharmacokinetic Studies

Healthy humans and rodents have both been used to study the pharmacokinetics of HupA. HupA was quickly absorbed, widely dispersed throughout the body, and moderately excreted (Table 9.6). In rats, the dispersion phase and elimination phase of [3H] HupA led to a two-phase fall in blood levels of HupA. There was 96.9% oral bioavailability. On injection of [3H]HupA, the kindly and liver of mice showed the highest radioactivity. However, most of it was eliminated in urine after 24 h. Only 2.4% of the faces were covered again. Urine analysis by chromatography showed that [3H]HupA and its metabolite were eliminated in part. Autoradiographic study indicated presence of HupA in all areas of the mouse brain, but after intravenously administration, it was particularly concentrated in the frontoparietal cortex, striatal cortex, hippocampus, and nucleus accumbens (Wang et al. 2006).

HupA plasma levels were measured in young, healthy participants using reverse phase HPLC with a spectrophotometric detector. Following oral administration of 0.99 mg HupA, the absorption showed first order kinetics with plasma concentrations taken at regular time intervals in one compartment open model system. In vivo, HupA was quickly absorbed and broadly disseminated (Table 9.7). Tacrine or physostigmine has half-lives that are at least 4–17 times shorter than those of HupA (Tang and Han 1999).

Table 9.6 Pharmacokinetic parameters of [^{3}H]HupA 13.9 mEq/kg in each group of three rats

Parameters	i.v. (mean ± S.D.)	i.g. (mean ± S.D.)
α (min^{-1})	0.107 = 0.016	0.08 ± 0.04
β (min^{-1})	0.006 ± 0.003	0.004 ± 0.001
K_a (min^{-1})	–	0.6 ± 0.07
$T_{1/2\alpha}$ (min)	6.6 = 1.1	10 = 6
$T_{1/2\beta}$ (min)	149 ± 96	203 ± 53
$T_{1/2Ka}$ (min)	–	5.1 ± 3
K_{12} (min^{-1})	0.047 ± 0.02	0.05 ± 0.03
K_{22} (min^{-1})	0.05 ± 0.04	0.024 ± 0.004
K_{10} (min^{-1})	0.014 ± 0.006	0.013 ± 0.006
V_c (L/kg)	1.6 ± 0.9	2.4 ± 0.7
V_d (L/kg)	3.6 ± 1.0	7.8 ± 2.3
Cl_p (L/(kg min))	0.020 ± 0.006	0.028 ± 0.014
AUC (10^{-7} × (dpm × min)/mL)	2.6 ± 0.9	1.8 ± 0.8
T_{max} (min)	–	21 ± 12
C_{max} (dpm/mL)	–	98,569 ±12,153

$F = Cl_{p.ig} .AUC_{ig}/Cl_{p.iv} .AUC_{iv} = 96.9\%$. Data from Ref. Wang et al. (1988)

Table9.7 Pharmacokinetic parameters of HupA after 0.99 mg p.o. (tablet) in 6 healthy volunteers

Parameter	Mean ± S.D.
K_a (min^{-1})	0.061 ± 0.017
K_e (min^{-1})	0.0025 ± 0.0006
$T_{1/2Ka}$ (min)	13 ± 5
$T_{1/2Ke}$ (min)	288 ± 63
T_{max} (min)	80 ± 9
C_{max} (μg/L)	8.4 ± 0.9
T_{lag} (min)	25.4 ± 1.8
V_d/F (L/kg)	0.108 ± 0.008
AUC (mg/(L/min))	4.1 ± 1.2

HupA concentrations in plasma were determined by reverse phase HPLC (Tang and Han 1999)

9.7.2 Clinical Trials

There have been several clinical investigations with HupA described up to this point. In China, the treatment of 447 patients with dementia or age-related cognitive loss revealed the positive efficacy of HupA. In a preliminary investigation, all rating scores assessed by the Buschke Selective Reminding task significantly improved in 100 individuals with probable AD who were given oral HupA (0.15–0.25 mg, t.i.d.) (Tang and Han 1999).

Table 9.8 Antidotal effect of atropine on HupA intoxication

Dose	No. of mice	
(mg/kg, i.p.)	Test	Death
	HupA	
7.0	10	10
	Atropine + HupA	
0.5 + 7.0	10	2
2.5 + 7.0	10	0

9.7.3 Toxicology

According to toxicological tests carried out on several animal species, cholinergic activation of HupA is associated with less severe negative side effects in comparison to other cholinesterase inhibitors. 4.6 mg (po), 3.0 mg (sc), 1.8 mg (ip), and 0.63 mg were the LD50 values for mice (iv). On administration of HupA to dogs intramuscularly (0.6 mg/kg) or to rats orally (1.5 mg/kg) for 180 days, histopathological analysis revealed no major abnormalities. Rats showed no mutagenicity and neither mice nor rabbits showed any teratogenic effects.

Toxicity studies have been carried out on animals. Salivary dose response curves showed that HupA was less effective than other ChE inhibitors. In comparison to donepezil and tacrine, HupA produced less severe cholinergic hyperactivity-related symptoms in rats. Oral HupA, 0.48 mg, did not result in fasciculation or any other cholinergic symptoms. The HupA dosages for mice for determination of LD50 were 4.5 mg, 3 mg, 1.7 mg, and 0.63 mg for different routes, i.e. oral, subcutaneous, intravenous, and intramuscular, respectively. Prior administration of atropine significantly counteracted the toxicity due to HupA (Table 9.8). Acute toxicity studies in animals revealed that after 180 days of administration of HupA, there were no histopathological abnormalities and no teratogenic effect was seen (Tang and Han 1999).

9.7.4 Drug Interactions

Huperzine A may interact with following drugs:

- **Anticholinergic drugs** including antihistamines, some antidepressants, Parkinson medications like Cogentin (benztropine), Akineton (biperiden), Artane (trihexyphenidyl), and Kemadrin (procyclidine), and anti-nausea medication scopolamine.
- **Acetylcholinesterase (AChE)** inhibitors are drugs that treat Alzheimer's disease and include Phospholine Iodide (echothiophate), Prostigmin (neostigmine), and Antilirium (physostigmine).

- **Cholinergic drugs** are used to treat glaucoma, Alzheimer's disease, and other conditions and include Pilocar (pilocarpine), Aricept (donepezil), and Cognex (tacrine) (Zhang 2012).

9.8 Patents Published

Publication No.	Date of publication	Title
CN1781907A	2006-06-07	Method for producing high purity huperzine A
CN101130520B	2010-05-12	Novel method for splitting and producing natural (−)-huperzine and non-natural (+)-huperzine by racemate of diastereoisomer salt
CN101485640B	2011-03-23	Huperzine A mono-layer osmotic pump-controlled release tablets
CN101333190A	2008-12-31	Asymmetric synthesis for chiral huperzine A
CN104016918B	2016-04-13	Huperzine A polymorph, its preparation method, comprise the medical composition and its use of described polymorphs body
US5177082A	1993-01-05	Huperzines and analogs
CN102675204B	2015-04-01	Intermediate for synthesizing (I)-huperzine A, synthesis method and usage thereof
CN103570621B	2015-04-29	Preparation method of (−)-huperzine A
CN103224467A	2013-07-31	Preparation method of (−)-huperzine A
CN1101381C	2003-02-12	Method for extracting huperzine A as acetylcholinesterase depressant
CN1383824A	2002-12-11	Nasal cavity administrated huperzine preparation

9.9 Conclusion

There is proof that huperzine A may outperform currently available cholinesterase inhibitors in terms of symptomatic effectiveness. Also, the antioxidant and neuroprotective qualities of huperzine A raise the possibility that it could be effective as a disease-modifying treatment for Alzheimer's disease (AD). (verify this line. . . .) Although HupA offers therapeutic advantages in the treatment of AD, our knowledge of the molecular processes underlying the restoration of neuronal function following HupA administration is currently limited. The burden associated with extremely prevalent illnesses that are expensive and damaging for not only those are affected but also their carers may be lessened with further study and development of HupA as a therapy for AD.

Conflict of Interest The authors declare no conflict of interest.

References

Alcalá M, Vivas NM, Hospital S et al (2003) Characterisation of the anticholinesterase activity of two new tacrine-huperzine A hybrids. Neuropharmacology 44:749–755

Alzheimer's Association (n.d.) FDA-Approved Treatment for Alzheimer's. http://www.alz.org

Ashani Y, Peggins JO, Doctor BP (1992) Mechanism of inhibition of cholinesterases by Huperzine A. (Reannouncement with new availability information). No. AD-A-254440/1/XAB. Walter Reed Army Inst. of Research, Washington, DC

Bai DL, Tang XC, He XC (2000) Huperzine A, a potential therapeutic agent for treatment of Alzheimer's disease. Curr Med Chem 7(3):355–374

Batsch NL, Mittelman MS, Alzheimer's Disease International (2012) World Alzheimer Report 2012: overcoming the stigma of dementia. Alzheimer's Disease International, London, UK

Campiani G, Sun LQ, Kozikowski AP, Aagaard P, McKinney M (1993) A palladium catalyzed route to huperzine A and its analogues and their anticholinesterase activity. J Org Chem 58: 7660–7669

Camps P, Contreras J, Fontbardia M, Solans X (1996) Improved synthesis of methyl 7,7-ethylenedioxy-3-methyl-9-oxobicyclo[3.3.1]non-3-ene-1-carboxylate intermediate for the synthesis of huperzine A analogues. Synthetic Commun 26(1):9–18

Camps P, El Achab R, Morral J et al (2000a) New tacrine-huperzine A hybrids (huprines): highly potent tight-binding acetylcholinesterase inhibitors of interest for the treatment of Alzheimer's disease. J Med Chem 43:4657–4666

Camps P, Contreras J, El Achab R, Morral J, Munoz-Torrero D, Font-Bardia M, Solans X, Badia A, Vivas NM (2000b) New syntheses of rac-huperzine A and its rac-7-ethyl-derivative. Evaluation of several huperzine A analogues as acetylcholinesterase inhibitors. Tetrahedron 56(26): 4541–4553

Chen WP, Yang FQ (1995) Asymmetric total synthesis of optically active huperzine A. Chinese J Med Chem 5(1):10–17

Ding R, Sun B-F, Lin G-Q (2012) An efficient total synthesis of (−)-huperzine A. Org Lett 14(17): 4446–4449

Ferreira A, Rodrigues M, Fortuna A, Falcão A, Alves G (2016) Huperzine A from Huperzia serrata: a review of its sources, chemistry, pharmacology and toxicology. Phytochem Rev 15(1):51–85

Gemma S, Gabellieri E, Huleatt P et al (2006) Discovery of huperzine A-tacrine hybrids as potent inhibitors of human cholinesterases targeting their midgorge recognition sites. J Med Chem 49: 3421–3425

Giménez-Llort L, Ratia M, Pérez B, Camps P, Muñoz-Torrero D, Badia A, Clos M (2017) Behavioural effects of novel multitarget anticholinesterasic derivatives in Alzheimer's disease. Behav Pharmacol 28(2):124–131

Gordon RK, Nigam SV, Weitz JA et al (2001) The NMDA receptor ion channel: a site for binding of Huperzine A. J Anal Toxicol 21(Suppl 1):S47–S51

Ha GT, Wong RK, Zhang Y (2011) Huperzine A as potential treatment of Alzheimer's disease: an assessment on chemistry, pharmacology, and clinical studies. Chem Biodivers 8:1189–1204

Hanns H, Gabriele N (2003) The discovery of Alzheimer's disease. Dialogues Clin Neurosci 51: 101–108

Hassan M, Abbas Q, Seo S-Y, Shahzadi S, Al Ashwal H, Zaki N, Iqbal Z, Moustafa AA (2018) Computational modeling and biomarker studies of pharmacological treatment of Alzheimer's disease. Mol Med Rep 18(1):639–655

He XC, Wang B, Bai DL (1998) Studies on asymmetric synthesis of huperzine A-1. Palladium-catalyzed asymmetric bicycloannulation of 5,6,7,8-tetrahydro-2-methoxy-6-oxo-5-quinolinecarboxylic esters. Tetrahedron Lett 39(5–6):411–414

He XC, Wang B, Yu GL, Bai DL (2001) Studies on the asymmetric synthesis of huperzine A. Part 2: highly enantioselective palladium-catalyzed bicycloannulation of the beta-keto-ester using new chiral ferrocenylphosphine ligands. Tetrahedron-Asymmetry 12(23):3213–3216

Jaswinder K, Rajmeet S, Gurinder S, Harpreet K, Jasvir K, Manpreet K, Parminder S, Jaspreet K (2016) A systematic review on Huperzia serrata. Int J Pharmac Phytochem Res 8(8):1250–1255
Jia J-Y, Zhao Q-H, Liu Y et al (2013) Phase I study on the pharmacokinetics and tolerance of ZT-1, a prodrug of huperzine A, for the treatment of Alzheimer's disease. Acta Pharmacol Sin 34:976–982
Kaneko S, Yoshino T, Katoh T, Terashima S (1997) A novel enantioselective synthesis of the key intermediate of (_)-huperzine A employing asymmetric palladium-catalyzed bicycloannulation. Tetrahedron-Asymmetry 8(6):829–832
Kozikowski AP, Campiani G, Tuckmantel W (1994) An approach to open-chain and modified heterocyclic-analogs of the acetylcholinesterase inhibitor huperzine-A through a bicyclo[3.3.1] nonane intermediate. Heterocycles 39(1):101–116
Kozikwoski AP, Reddy ER, Miller CP (1990) A simplified route to a key intermediate in the synthesis of the Chinese noortropic agent huperzine A. J Chem Soc Perkin Trans 1:195–197
Lee IYC, Jung MH, Lee HW, Yang JY (2002) Synthesis of huperzine intermediates via Mn(III)-mediated radical cyclization. Tetrahedron Lett 43(13):2407–2409
Lim WH, Goodger JQ, Field AR, Holtum JA, Woodrow IE (2010) Huperzine alkaloids from Australasian and southeast Asian Huperzia. Pharm Biol 48(9):1073–1078
Lin L-J, Lin L-Z, Cordell GA, Zhou B-N, Zhu D-Y, Huang M-F, Han X-Y (1993) NMR assignments of huperzine A, serratinine and lucidioline. Phytochemistry 34(5):1425–1428
Ma X, Gang DR (2004) The lycopodium alkaloids. Nat Prod Rep 21:752–772
Ma X, Tan C, Zhu D, Gang DR, Xiao P (2007) Huperzine A from Huperzia species—an ethnopharmacological review. J Ethnopharmacol 113(1):15–34
Nguyen NC, Vinh D, Nguyen DT, Nguyen HVT, Tran CL, Tran MH (2021) Development of a capillary electrophoretic method for the determination of huperzine A concentration in vietnamese huperzia serrata. Nat Prod Commun 16(9). https://doi.org/10.1177/1934578X211033225
Prince M, Jackson J, Alzheimer's Disease International (2009) World Alzheimer Report 2009. Alzheimer's Disease International
Qian LG, Ji RY (1989) A total synthesis of (−)Huperzine A. Tetrahedron Lett 30:2089–2090
Reitz C, Brayne C, Mayeux R (2011) Epidemiology of Alzheimer disease. Nat Rev Neurol 7(3): 137–152
Ros E, Aleu J, Go'mez de Aranda I et al (2001) The pharmacology of novel acetylcholinesterase inhibitors, (±)-huprines Y and X, on the torpedo electric organ. Eur J Pharmacol 421:77–84
Tang XC, Han YF (1999) Pharmacological profile of huperzine A, a novel acetylcholinesterase inhibitor from Chinese herb. CNS Drug Rev 5(3):281–300
Wang YE, Yue DX, Tang XC (1986) Anti-cholinesterase activity of Huperzine-A. Acta Pharmacology Sinica 7:110–113
Wang YE, Feng J, Lu WH, Tang XC (1988) Pharmacokinetics of huperzine A in rats and mice. Acta Pharmacol Sin 9:193–196
Wang R, Yan H, Xi-can TANG (2006) Progress in studies of huperzine A, a natural cholinesterase inhibitor from Chinese herbal medicine 1. Acta Pharmacol Sin 27(1):1–26
Wu Q, Gu Y (2006) Quantification of huperzine A in Huperzia serrata by HPLC-UV and identification of the major constituents in its alkaloid extracts by HPLC-DAD-MS-MS. J Pharm Biomed Anal 40(4):993–998. https://doi.org/10.1016/j.jpba.2005.07.047. Epub 2005 Dec 7
Xia Y, Kozikowski AP (1989) A practical synthesis of the Chinese "nootropic" agent huperzine A: a possible lead in the treatment of Alzheimer's disease. J Am Chem Soc 111:4116–4117
Yamada F, Kozikowski AP, Reddy ER, Pang YP, Miller JH, Mckinney M (1991) A route to optically pure (−)-huperzine-A-molecular modeling and in vitro pharmacology. J Am Chem Soc 113(12):4695–4696
Yang HL, Ma YS, Wang XL, Zhu D (2020) Huperzine A: a mini-review of biological characteristics, natural sources, synthetic origins, and future prospects. Russ J Org Chem 56(1):148–157

Zangara A (2003) The psychopharmacology of huperzine A: an alkaloid with cognitive enhancing and neuroprotective properties of interest in the treatment of Alzheimer's disease. Pharmacol Biochem Behav 75(3):675–686

Zhang H-y (2012) New insights into huperzine A for the treatment of Alzheimer's disease. Acta Pharmacol Sin 33(9):1170–1175

Zhou GC, Zhu DY (2000) Synthesis of 5-substituted analogues of huperzine A. Bioorg Med Chem Lett 10:2055–2057

Zhu D-Y, Tan C-H, Li Y-M (2005) The overview of studies on huperzine A: a natural drug for the treatment of Alzheimer's disease. Med Chem Bioactive Nat Prod:143–182

Polyphenol: Development of Polyphenol-Inspired Derivatives Targeting Pathological Factors of AD

10

Bhanuranjan Das, Anurag T. K. Baidya, and Rajnish Kumar

Abstract

Alzheimer's disease (AD) is the most common type of dementia having a complex pathophysiology with no disease modifying treatments available in clinics. Polyphenolic compounds, a class of natural plant compounds, have been extensively studied for their potential therapeutic use in AD. The neuroprotective, antioxidant, as well as anti-inflammatory properties exhibited by these compounds are important in the treatment and management of AD. Phenolic compounds can inhibit the formation and aggregation of toxic proteins that are implicated in the disease, as well as promote their clearance from the brain. The anti-inflammatory effects of phenolic compounds are significant because chronic inflammation in the brain is thought to contribute to the development and advancement of AD. Additionally, phenolic compounds have been shown to improve cerebral blood flow and enhance neuronal signaling, which can help to improve cognitive function in AD patients. Due to their potential therapeutic efficacy, polyphenols are a key focus in medicinal chemistry for designing multi-targeted as well as single-targeted drug ligands against different therapeutic targets of AD. In this chapter, we have aimed to provide an overview of the synthesis and pre-clinical evaluation of polyphenol derivatives for their potential applications in the management of AD. A short summary of clinical trials performed on polyphenols is also given, followed by design and synthesis of novel derivatives using multiple chemical modification strategies, including naturally inspired derivatives, to further enhance the therapeutic potential of these compounds. Overall, the potential therapeutic effects of phenolic

B. Das · A. T. K. Baidya · R. Kumar (✉)
Department of Pharmaceutical Engineering & Technology, Indian Institute of Technology (B.H.U.), Varanasi, Uttar Pradesh, India
e-mail: rajnish.phe@iitbhu.ac.in

A. Sharma, G. P. Modi (eds.), *Natural Product-Based Synthetic Drug Molecules in Alzheimer's Disease*, https://doi.org/10.1007/978-981-99-6038-5_10

compounds in ameliorating the AD pathophysiology make them an attractive class of natural products holding great promise for the development of therapeutics for this deadly disease.

Keywords

Alzheimer's disease · Natural polyphenols · Naturally inspired poly-phenolic compounds · Amyloid beta · Clinical trials · Multi-targeted drug ligands · Neuroprotective · Chemical modification

Abbreviations

AChE	Acetylcholinesterase
AD	Alzheimer's disease
Aβ	Amyloid Beta
BChE	Butyrylcholinesterase
CAS	Catalytic active site
DHA	Docosahexaenoic acid
eeAChE	Electric eel acetylcholinesterase
EGCG	Epigallocatechin-3-gallate
MAO	Mono amine oxidase
MTDL	Multi-targeted drug ligands
ORAC-FL	Oxygen radical absorbance capacity fluorescein assay
PAS	Peripheral anionic site

10.1 Introduction

Alzheimer's disease (AD) is a major form of dementia that affects the brain, resulting in a gradual decline in memory and cognitive abilities (Mao and Zhang 2022). World Alzheimer Report of 2022 revealed that dementia is a syndrome caused by neurodegeneration and is a leading cause of dependence, disability, as well as mortality (Fekadu et al. 2022). One of the primary causes of AD is considered to be the agglomeration of beta-amyloid (Aβ) plaques and neurofibrillary tangles in the brain. Aβ is a protein that can aggregate and form sticky plaques outside neurons, while neurofibrillary tangles are composed of twisted fibers of the hyperphosphorylated tau protein which accumulate inside neurons (Kidd 2008; Wang et al. 2021). These abnormal protein accumulations disrupt normal brain function, ultimately leading to neurodegeneration and cognitive decline in patients with AD. Other than the abnormal protein accumulation, inflammation, oxidative stress, and impaired synaptic function are also present in the diseased brain and complicate the pathology of AD. Chronic inflammation and oxidative stress can damage neurons and impair their ability to communicate with one another, leading to cognitive decline (Kumar and Singh 2015).

Unfortunately, existing treatment options for AD are limited to managing symptoms and no disease-modifying therapeutics is available in the market despite the extensive research efforts made by the researchers in last couple of decades (Mueller et al. 2005; Sperling et al. 2011). A cure for AD is still elusive, and researchers are exploring various approaches, such as rational synthesis, designing of multi-targeted drug ligands (MTDL), structural modification or semisynthetic modification of different natural products, and computational approaches like AI- and ML-based algorithms for screening large libraries to develop potential therapeutics (Das et al. 2023a).

Polyphenols, a class of natural compounds found in plant-based foods and beverages, are known for their antioxidant and anti-inflammatory properties, which make them promising candidates for preventing or alleviating symptoms of AD (Pasinetti et al. 2015; Syarifah-Noratiqah et al. 2018). Several studies have explored the biological activity of polyphenols in AD in both animal and humans (Rigacci and Stefani 2016). One of the most popularly studied polyphenols is resveratrol, which is found in grapes and red wine. Resveratrol has been reported to enhance cognitive function and lessen the accumulation of Aβ plaques, a hallmark of AD, in animal models. Other polyphenols that have been studied for their potential benefits in AD include curcumin, found in turmeric, and green tea containing epigallocatechin-3-gallate (EGCG) (Dragicevic et al. 2011). Curcumin has been shown to lessen inflammation and enhance cognitive function in animal models of AD, while EGCG decreases the accumulation of Aβ plaques (Betts and Wareham 2014; Tang and Taghibiglou 2017). While promising, more research is needed to better understand the potential benefits of polyphenols for AD and how they may be incorporated into prevention or treatment strategies (Maczurek et al. 2008). Additionally, it is important to remember that while dietary polyphenols may have some benefits, they are not a replacement for medical treatments for AD and should not be used as such.

Polyphenols have been extensively studied for their potential therapeutic effects, particularly their antioxidant and anti-inflammatory properties. These properties are attributed to the chemical structure of polyphenols, which enables them to scavenge free radicals, thereby preventing oxidative stress as well as inflammation in the body. One of the key chemical features of polyphenols is their ability to donate hydrogen atoms to reactive oxygen species (ROS) and other free radicals (Losada-Barreiro and Bravo-Diaz 2017). This is due to the occupancy of hydroxyl (-OH) groups which can act as electron donors (Durazzo et al. 2019). This mechanism of action allows polyphenols to shut out the oxidative damage to cells and reduce neuroinflammation, which are key factors in the pathogenesis of AD. In addition to their antioxidant activity, polyphenols have also been shown to inhibit the formation of Aβ plaques and tau neurofibrillary tangles, which are hallmark pathological features of AD. For example, the flavonoid EGCG present in green tea has been found to inhibit the aggregation of Aβ and reduce neuroinflammation in experiments (Youn et al. 2022).

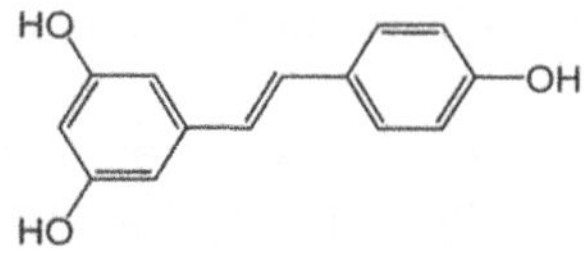

(1) Trans-Resveratrol

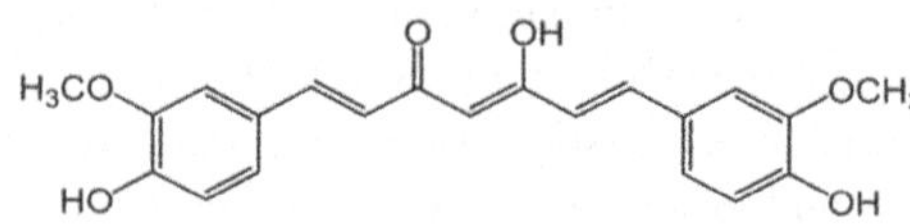

(2) Curcumin

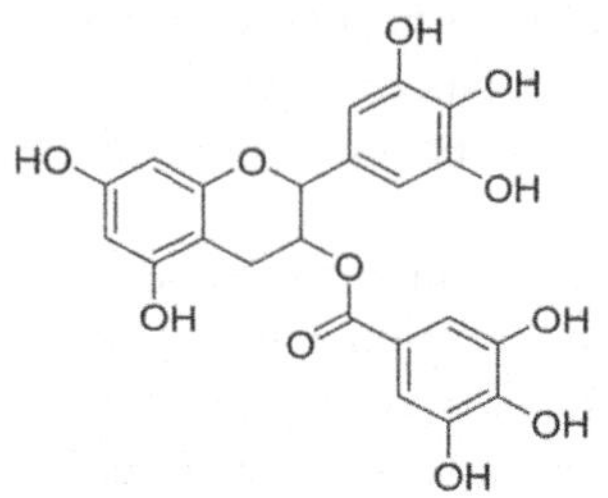

(3) Epigallocatechin-3-Gallate

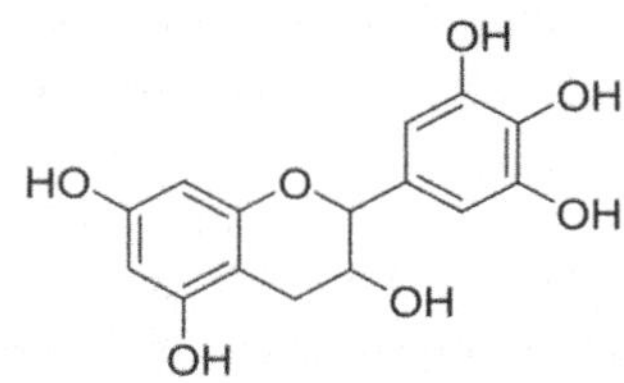

(4) Epigallocatechin

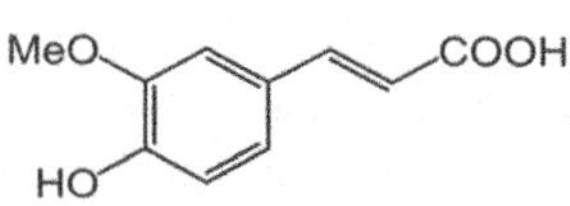

(5) Ferulic Acid

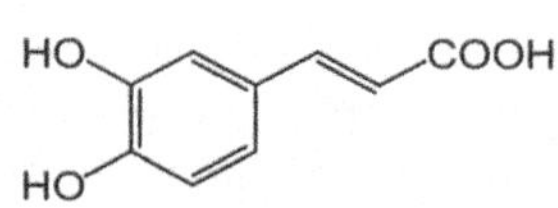

(6) Caffeic Acid

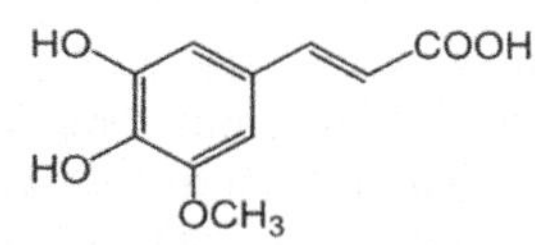

(7) Sinapinic Acid

Fig. 10.1 Chemical structures of some of polyphenols studies for their potential for the treatment of AD

Chemical structures of some of polyphenols studies for their potential for the treatment of AD are shown in Fig. 10.1.

In this chapter, we have provided an overview of the synthesis and pre-clinical evaluation of polyphenol derivatives for their potential application in the management of AD. A short summary of clinical trials performed on polyphenols is also given, followed by design and synthesis of novel derivatives using multiple chemical modification strategies, including naturally inspired derivatives, to further enhance the therapeutic potential of these compounds. Overall, the potential therapeutic effects of phenolic compounds in ameliorating the AD pathophysiology make them an attractive class of natural products holding great promise for the development of therapeutics for this deadly disease.

10.2 Natural Polyphenols Reported for the Treatment of AD

Natural polyphenols have been reported to possess variety of valuable biological effects for the treatment of AD (Syarifah-Noratiqah et al. 2018; Choi et al. 2012). One of the imported biological effect of polyphenols is neuroprotection, by which it

Table 10.1 Examples of important natural polyphenols, their source and role in management of AD

Natural polyphenols	Source	Use in AD
Resveratrol	Grapes, berries, and red wine	Neuroprotective effects, improve cognitive function, and reduce Aβ levels in animal models of AD (Komorowska et al. 2020)
Curcumin	Turmeric	Anti-inflammatory, antioxidant properties, reduce Aβ levels in AD. Clinical trials are underway to investigate its potential as a treatment for AD (Mishra and Palanivelu 2008)
Epigallocatechin gallate (EGCG)	Green tea	Antioxidant, anti-inflammatory, and metal chelating properties (Choi et al. 2001; Reznichenko et al. 2006)
Quercetin	Fruits, vegetables, and grains	Antioxidant and anti-inflammatory, Aβ properties
Apigenin	Parsley, chamomile, and other plants	Antioxidant neuroprotective, anti-inflammatory, and metal chelating
Anthocyanins	Blueberries, blackberries, and raspberries	Antioxidant, Improve cognitive function and reduce the accumulation of Aβ plaques IN AD
Tannic acid	Fruits, nuts, teas and coffee	Antioxidant, anti-inflammatory, and metal-chelating properties (Ono et al. 2004)
Stilbenes	Grapes, peanut	Aβ inhibitors (Rivière et al. 2007)

can lessen and moderate the progression of AD (Spagnuolo et al. 2016). Another property that polyphenols possess is antioxidant which help to protect neurons from oxidative damage and preserve cognitive function (Choi et al. 2012). Among other properties, anti-inflammatory effect of polyphenols can help to reduce inflammation in the brain and protect it against neuronal damage (Henríquez et al. 2020).

The accumulation of extracellular Aβ plaques in the brain is a prominent hallmark of AD, which ultimately results in cognitive decline and neurodegeneration (Das et al. 2023b). Some natural polyphenols have exhibited to inhibit the formation of Aβ plaques and reduce their accumulation in the brain (Freyssin et al. 2018). In animal models of AD, several reports have reported that natural polyphenols can enhance cognitive function and memory. This may be due to their neuroprotective and anti-inflammatory effects, as well as their ability to increase blood flow to the brain and enhance the activity of neurotransmitters involved in cognitive function.

Overall, polyphenols are a group of compounds with a broad range of potential health benefits. While polyphenols have been studied for decades, they have not been able to reach clinics for the treatment of AD. However, their multi-targeting property can be explored to develop polyphenol derived/inspired therapeutics for the treatment of AD. A summary of some of the polyphenols explored for their utility in the treatment of AD is given in Table 10.1.

10.3 Polyphenols in Clinical Trials for AD

The effects of polyphenolic compounds on AD patients have been investigated through several clinical trials, either as single agents or in combination with other drugs, with the aim of exploring their potential usage in the treatment of AD (Molino et al. 2016). Curcumin and resveratrol have been studied in several stages of clinical trials (Mazzanti and Di Giacomo 2016). Quercetin and Fisetin have undergone several stages of clinical trials for AD, according to the previous reports and while some preliminary studies have suggested that these flavonoids have shown promising activity in the treatment of AD (Maher 2021). Table 10.2 summarizes the polyphenolic compounds studied in clinical trials. The results of these clinical trials are promising but not conclusive. Extensive studies are needed to determine the optimal dose, duration, bioavailability, safety, and efficacy of polyphenolic compounds for the treatment of AD. Additionally, the mechanisms of action and interactions of these compounds with other drugs and biological factors need to be further elucidated. Here we have reported some of the polyphenols that have underwent clinical trials and it is reviewed in detail elsewhere and not in the scope of this chapter (Molino et al. 2016).

10.4 Polyphenol-Inspired Derivatives Targeting Pathological Features of AD

Polyphenolic compounds have shown great promise in alleviating the AD pathology in experimental studies. However, none of them reached to the clinics after passing the clinical trials. Keeping in the view, the multiple properties such as potential

Table 10.2 Summary of some of the important polyphenols studied in clinical trials for the management of AD

Natural polyphenol	Clinical trial status	Title of the study	Clinical trial NO.	Disease
Curcumin	Completed	Efficacy and safety of curcumin formulation in AD	NCT01001637	AD (Farkhondeh et al. 2019)
Cycloastragenol	Completed	Telomerase activator and retinal amyloid	NCT02530255	AD (Yu et al. 2018)
Resveratrol	Completed	Resveratrol for AD	NCT01504854	AD (Berman et al. 2017)
Epigallocatechin-gallate	Completed	Effects of epigallocatechin gallate (EGCG) in healthy, young adults	NCT01357681	AD (Lange et al. 2022)
Quercetin + Dasatinib	Completed	Senolytic therapy to modulate progression of AD (SToMP-AD)	NCT04063124	AD (Gonzales et al. 2022)

therapeutic efficacy and good pharmacokinetic properties, polyphenols have become the key frame in designing various molecules against different neurodegenerative disorders. Therefore, researchers are now focusing on exploring the polyphenolic-derived/inspired molecules for the therapeutic management of AD.

Several reports on solubility study suggest that alcohol group is a key functional group in maintaining solubility and other pharmacokinetic properties by forming hydrogen bonding with water and proteins inside the biological system (Das et al. 2022). Researchers are now designing multi-target directed ligands (**MTDLs**) by using different multi-targeted strategies such as merging the natural scaffold with another, connecting the natural scaffold with another through a linker, and incorporating natural products as the primary scaffold (Ramalakshmi et al. 2021). In order to develop potential ligands against AD, medicinal chemists synthesize different derivatives from natural ligands by using various synthetic methods and evaluate their activity against different targets of AD. Various synthesized ligands have been found to possess therapeutic potential against AD. In the following section, several examples of natural product-inspired molecules have been discussed with their biological potency in AD.

10.4.1 Caffeic Acid Inspired Molecules

Caffeic acid is a hydroxycinnamic class of natural compounds and is commonly found in various plant-based foods such as coffee, fruits, vegetables, and whole grains (Godos et al. 2021; Sova and Saso 2020). It has antioxidant, anti-inflammatory, and neuroprotective properties and has been studied for its potential role in AD (Tolba et al. 2013). Several research reports have suggested that caffeic acid plays an important biological significance in engaging various targets of AD (Habtemariam 2017).

The biological activity of caffeic acid is expected due to the presence of both hydroxyl group and α,β-unsaturated carbonyl group. Medicinal chemists have focused on designing derivatives of caffeic acid in order to improve the potency against the target as well as to improve their pharmacokinetic profile. Several hybrids of caffeic acids and their biological activity against AD targets have been extensively reviewed in previous literature (Zhao et al. 2022). In this section, we have reported some of the examples of caffeic acid inspired compounds studied for their potential against AD drug targets.

In 2012, Chao and his co-workers synthesized different MTDLs composed of tacrine-caffeic acid hybrids. After careful biological evaluation, compound **8** (Fig. 10.2) showed selective inhibition towards the acetylcholinesterase (AChE) in comparison to butyrylcholinesterase (BChE). The compound **8** showed binding to both catalytic and peripheral anionic sites (PAS) of AChE as deduced from the enzyme kinetic study. Moreover, compound **8** also showed inhibition of self or AChE induced $A\beta_{1-40}$ aggregation and neuroprotection against glutamate and H_2O_2 induced cell death (Chao et al. 2012). Following that, in 2019 Wan and co-workers synthesized caffeic acid phenethyl ester 4-*O*-glucoside (**compound 9**) and after

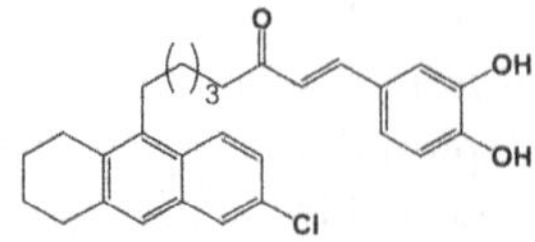

(**8**) AChE: IC_{50} 0.3 ± 0.11 μM
BChE: IC_{50} 29.5 ± 0.5 μM

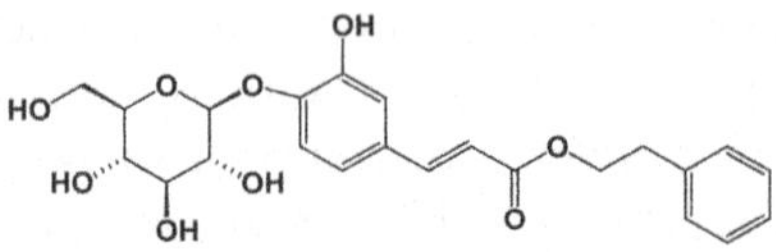

(**9**) Reducing ROS production
Neuroprotective & Antioxidant activities

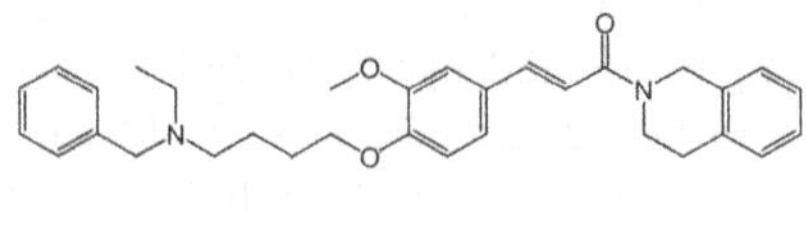

(**10**) BChE: IC_{50} 8.9 μM
MAO A: IC_{50} 6.3 μM
MAP B: IC_{50} 8.6 μM

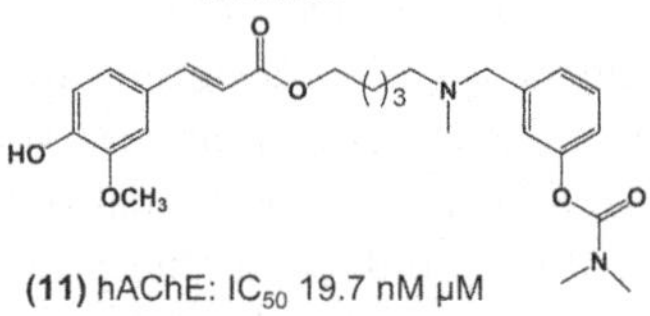

(**11**) hAChE: IC_{50} 19.7 nM μM
hBuChE : IC_{50} 0.66 μM
Aβ aggregation inhibition (49.2% at 20 μM)

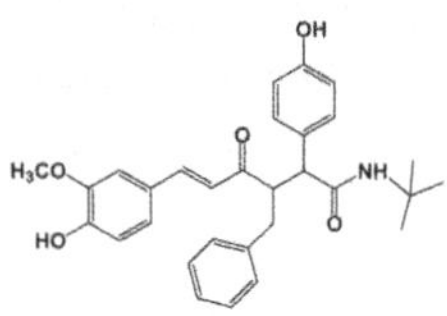

(**12**) Inhibits detrimental effects Aβ in In vivo

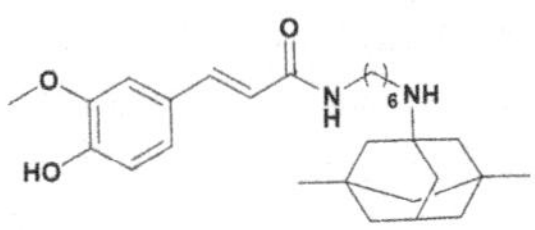

(**13**) NMDAR: IC_{50} 6.9 μM
Good Aβ inhibitor

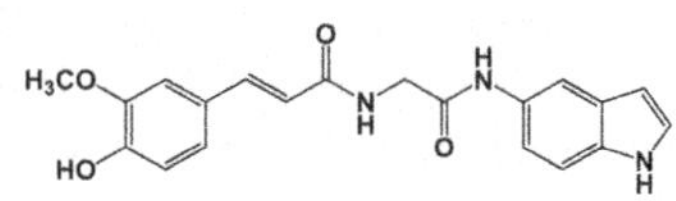

(**14**) AChE: IC_{50} 5.74 ± 0.3 μM
BChE: IC_{50} 14.05 ± 0.10 μM
Antioxidant

Fig. 10.2 Caffeic and ferulic acid inspired lead molecules possessing activity against AD drug targets

biological evaluation it was revealed that the compound 9 showed protection against oxidative stress-mediated neuronal cell apoptosis in an in vitro assay using SH-SY5 and PC12 cell lines (Wan et al. 2019).

10.4.2 Ferulic Acid Inspired Molecules

Ferulic acid, the cinnamic acid derivative found in plants, has become the key frame in medicinal chemistry in designing various lead compounds against different drug targets of AD (Chen et al. 2017; Singh et al. 2021). It has been reported to possess antioxidant, metal chelating and anti-inflammatory properties along with inhibition of Aβ aggregation and cholinesterases inhibition. Similar to caffeic acid, the biological effects of ferulic acid are expected due to the adjacent methoxy and hydroxyl group as well as α,β-unsaturated carbonyl group. Researchers are exploring the potential of ferulic acid derivatives against the AD drug targets and the design, synthesis and biological potential of ferulic acid derivatives/hybrids has been nicely reviewed in previous literature (Singh et al. 2021; Zhang et al. 2018). In this section, we discussed some of the examples of ferulic acid-inspired molecules and their potential as AD therapeutics.

In 2018, Zhipei Sang and co-workers synthesized several ferulic acid derivatives, and on further biological evaluation it was found that the compound 10, also known as TM-10, demonstrated potent inhibitory activity against BChE (IC_{50} = 8.9 nM) and monoamine oxidase A and B (IC_{50} = 6.3 μM and 8.6 μM, respectively). Additionally, compound **10** showed potent antioxidant activity as well as neuroprotective effect against $A\beta_{1-42}$ induced neurotoxicity in SH-SY5Y cells (Sang et al. 2018). In 2019, Lan and co-workers synthesized a series of ferulic acid derivatives and evaluated them against different therapeutic targets of AD. Among all the compounds, compound **11** showed significant ability to inhibit hAChE (IC_{50} = 19.7 nM) and hBChE (IC_{50} = 0.66 μM) and good inhibitory activity towards Aβ aggregation. Moreover, compound **11** showed antioxidant activity comparable to Trolox. From the enzyme kinetic and molecular modelling studies, it was noted that compound **11** acts as a mixed inhibition towards AChE by binding to both the catalytic active site (CAS) and the peripheral anionic site (PAS) (Lan et al. 2020). In the same year, Simona and co-workers synthesized six different derivatives of polyphenolic compounds by using a smart synthetic approach and performed in vivo experiments by using $A\beta_{42}$ oligomer-induced acute mouse model in order to evaluate their therapeutic effect in AD. It was found that compound **12** (PP04) showed potent inhibition of the detrimental effects of $A\beta_{42}$ oligomers (Tomaselli et al. 2019). In 2019, Michaela Rossini and co-workers synthesized molecules by conjugation of ferulic acid with memantine using a multifunctional approach, and the compound **13** obtained was found to have NMDA receptor antagonistic activity (IC_{50} 6.9 μM). Additionally, compound **13** also showed potent antioxidant activity at 10 mM concentration (Rosini et al. 2019).

In 2020, Yash and his co-workers synthesized diamide derivatives of ferulic acid in order to find potential lead candidates against AD drug targets. Among all of them, compound **14** was identified as an AChE (IC_{50} 5.7 ± 0.3 μM) and BChE (IC_{50} 14.05 ± 0.10 μM) inhibitor with significant antioxidant activity as indicated by DPPH assay (Singh et al. 2020).

10.4.3 Sinapic Acid Naturally Inspired Molecules

Sinapic acid is a natural compound that is primarily found in orange, grapefruit, and cranberry, as well as in certain herbs such as canola, mustard seed as well as rapeseed. It is a derivative of cinnamic acid and is characterized by its 3,5-dimethoxyl and 4-hydroxy substitutions on the phenyl group (**7**, Fig. 10.1). This compound has demonstrated significant biological activity and exhibited potent antioxidant, anti-inflammatory, anti-cancer, and hepatoprotective properties (Nićiforović and Abramovič 2014; Lee et al. 2012; Verma et al. 2020). Inspired by the biological activity of sinapic acid, researchers have developed several molecules containing sinapic acid scaffolds and explored them for potential therapeutic applications in AD (Pandi and Kalappan 2021). In this section, we have summarized a couple of examples of naturally inspired molecules containing sinapic acid scaffold.

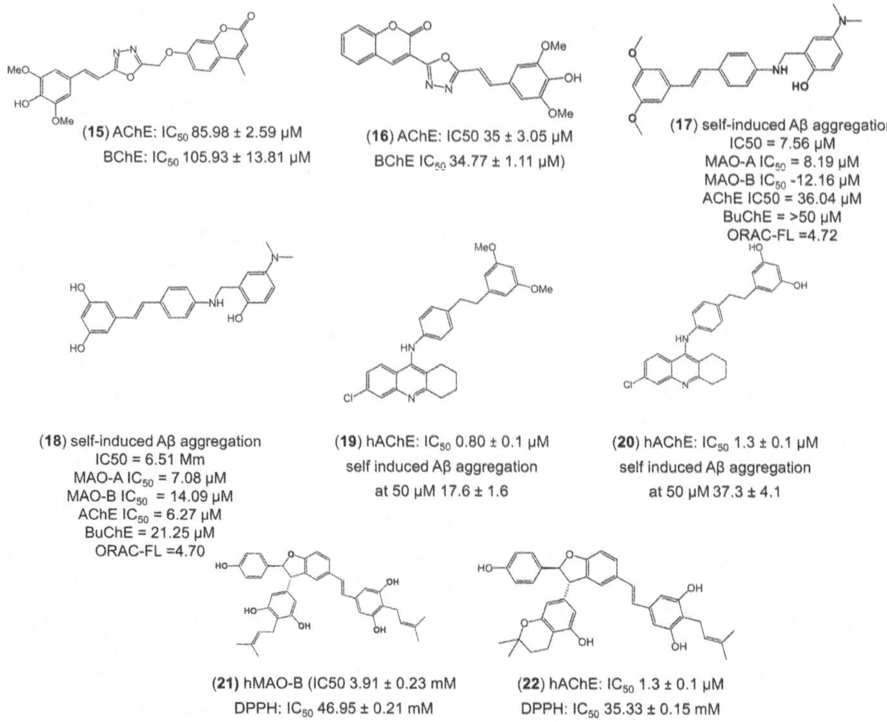

Fig. 10.3 Sinapic and resveratrol inspired lead molecules possessing activity against AD drug targets

In 2022, George and his coworker synthesized several coumarone and sinapic acid derivatives and evaluated them against in vitro inhibition of AChE and BChE enzymes. It was observed that the compound **15** (Fig. 10.3) showed good inhibitory activity towards the enzyme with AChE (IC_{50} 85.98 ± 2.59 μM) and BChE (105.93 ± 13.81 μM). Similarly compound **16** also showed inhibition of AChE (IC_{50} 35 ± 3.05 μM) and BChE (34.77 ± 1.11 μM) (George et al. 2022).

10.4.4 Resveratrol-Based Natural Inspired Molecules

Resveratrol is present in different plants including grapes and berries (Shrikanta et al. 2015). Some studies have reported that resveratrol may have potential benefits in the treatment of AD (Sawda et al. 2017). Resveratrol has been found to have antioxidant and anti-inflammatory properties, which may help to reduce oxidative stress as well as other inflammation in the brain, both of which are believed to play a role in the advancement of AD (Ahmed et al. 2017). Several studies have investigated the effects of resveratrol on AD in animal models and in vitro studies. Some studies have suggested that resveratrol may help to reduce Aβ accumulation in the brain and improve cognitive function in mice with AD (Pasinetti et al. 2015). Other studies have suggested that resveratrol may help to protect neurons from oxidative damage and reduce inflammation in the brain (Labban et al. 2021; Jia et al. 2017). However, the results of human studies investigating the effects of resveratrol on AD have been mixed. Some small studies have suggested that resveratrol supplementation may help to improve cognitive function and reduce markers of inflammation in older adults with mild cognitive decline (Choi et al. 2012). However, further clinical trials of greater magnitude are required to substantiate these findings and ascertain the ideal dosage and duration of resveratrol supplementation for the treatment of AD (Albani et al. 2010). Overall, while the potential benefits of resveratrol for the treatment of AD are promising, further research is needed to understand its effects and determine its clinical utility fully. In the following section, resveratrol derived compounds have been summarized.

In 2013, Chuanjun Lu and co-workers reported several resveratrol derivatives as MTDLs for AD. Among all the synthesized and tested compounds, compound **17** and **18** (Fig. 10.3) displayed the moderate inhibitory activity against self-induced Aβ aggregation having IC_{50} values of 7.56 μM and 6.51 μM, respectively. Compounds **17** and **18** also displayed strong antioxidant activity in vitro having ORAC-FL values of 4.72 and 4.70, respectively. Furthermore, both the compounds displayed well balanced MAO inhibitory activity, for compound **17** IC_{50} value for MAO-A was 8.19 μM and for MAO-B was 12.16 μM, whereas for compound **18**, IC_{50} value for MAO-A was 7.08 μM and for MAO-B was 14.09 μM. The compounds also exhibited moderate cholinesterase inhibitory activity, for compound **17** IC_{50} value for electric eel acetylcholinesterase (eeAChE) was 36.04 μM and eqBuChE was >50 μM, likewise for compound **18**, IC_{50} value for eeAChE was 6.27 μM and eqBuChE was 21.25 μM. Furthermore, the compound **17** displayed good blood–brain barrier permeability in vitro thus making compound **17** the most promising lead compound of the series for its development as lead candidate for AD therapeutics (Lu et al. 2013). In 2016, Jakub and co-workers synthesized several fused hybrids of tacrine-resveratrol as MTDLs against AD drug targets. Among all the derivatives, compounds **19** (IC_{50} 0.80 ± 0.1 μM) and **20** (IC_{50} 1.3 ± 0.1 μM) showed good potency against hAChE and self-induced Aβ aggregation (Jeřábek et al. 2017). In 2018, Tang and co-workers synthesized a series of resveratrol dimers and evaluated their MAO-B inhibition potential. The compounds were also evaluated for their antioxidant activity by using ABTS and FRAP methods. Among all the synthesized derivatives, compounds **21** and **22** showed moderate inhibition of hMAO-B showing an IC_{50} value of 3.91 ± 0.23 mM and 0.90 ± 0.01 mM, respectively. In the antioxidant activity assays, compound **21** demonstrated an IC_{50} value of 46.95 ± 0.21 mM for DPPH, and 1.43 and 1.74 trolox equivalent by ABTS and FRAP methods, respectively. Similarly, compound **22** showed an IC_{50} value of 35.33 ± 0.15 mM for DPPH, and 1.70 and 1.97 which is equivalent to trolox by ABTS and FRAP methods, respectively. These results indicate that both compound **21** and **22** exhibit significant antioxidant effects (Tang et al. 2019). In 2018, Cheng and co-workers synthesized different hybrids of resveratrol and maltol as novel MTDLs against AD. All compounds were evaluated against $A\beta_{1-42}$ aggregation,

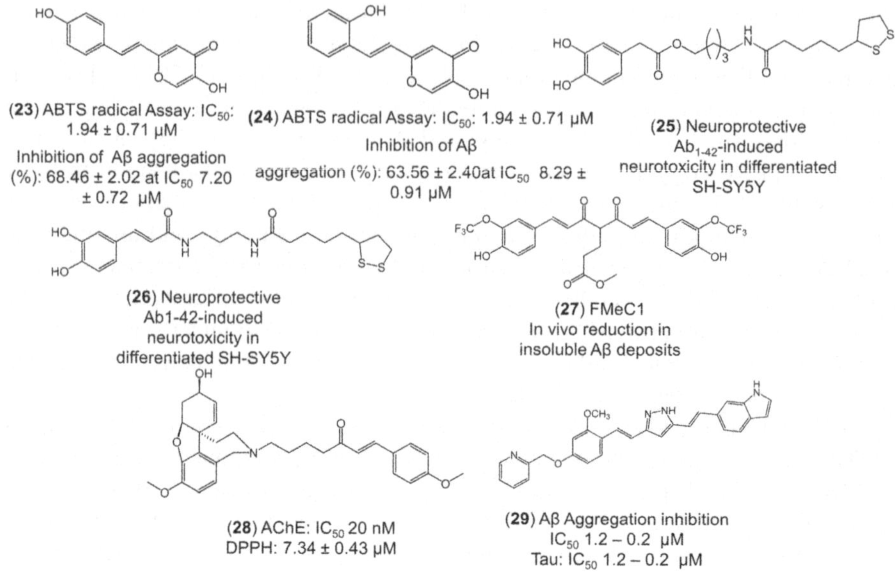

Fig. 10.4 Naturally inspired molecules containing resveratrol and curcumin scaffolds

and compound **23** (inhibition % 68.46 ± 2.02 at IC_{50} 7.20 ± 0.72 μM) and **24** (inhibition % 63.56 ± 2.40 at IC_{50} 8.29 ± 0.91 μM) (Fig. 10.4) showed a good percentage of inhibition towards Aβ aggregation (Cheng et al. 2018). Additionally, both compounds also showed good antioxidant and metal chelating property. In 2019, Aikaterini and co-workers reported compounds **25** and **26** which showed neuroprotective activity against $A\beta_{1-42}$ by reducing neurotoxicity in differentiated SH-SY5Y cells (Pagoni et al. 2020).

10.4.5 Curcumin-Based Natural Inspired Molecules

Curcumin, a polyphenolic compound that is extracted from the rhizomes of *Curcuma longa* L., has been found to have therapeutic potential in the treatment of neurodegenerative disorders, including AD (Askarizadeh et al. 2020). Curcumin possesses a remarkable safety profile and exhibits pleiotropic effects with the potential to provide neuroprotective benefits, such as its anti-inflammatory, antioxidant, and anti-protein-aggregation properties (Cole et al. 2007). The lower therapeutic efficacy of curcumin is due to the low BBB permeability (Subedi and Gaire 2021). So in order to improve the BBB permeability and to increase the potency, several curcumin derivatives have been synthesized and evaluated for potential use in the management of AD. One curcumin derivative that has been studied extensively is called "CNB-001" developed to improve its bioavailability and enhance its neuroprotective effects (Jayaraj et al. 2014). Preclinical studies have suggested that CNB-001 may have a potential advantage in the management of AD by reducing inflammation and oxidative stress in the brain, inhibiting the formation of Aβ plaques, and improving cognitive function (Panzhinskiy et al. 2014; Jayaraj et al. 2013). A second curcumin derivative that has been studied is called "Curcumin-DHA" which is a combination of curcumin and docosahexaenoic acid (DHA) (Guerzoni et al. 2017). Preclinical studies have suggested that Curcumin-DHA showed good therapeutic activity against AD by reducing inflammation and Aβ accumulation in the brain, and improving cognitive function. While the preclinical data on these curcumin derivatives are promising, clinical studies in humans have yielded mixed results, and larger, well-designed clinical trials are needed to determine their clinical use in the treatment of AD. The potential biological significance of curcumin is expected due to the presence of o-methoxy phenolic and α,-β-unsaturated β-diketone groups (Nelson et al. 2017). It also believes that α,-β-unsaturated carbonyl known as Michael acceptor plays a major role in therapeutic activity of curcumin in AD (Anand et al. 2011; Thapa et al. 2021). Different curcumin-based naturally inspired molecules have been well discussed in the previous (Noureddin et al. 2019). In this section, we have summarized some of the curcumin derivatives with their biological activity in AD.

In 2015, Daijiro Yanagisawa and co-workers synthesized curcumin derivatives having substitutions at C-4 position, of which compound **27** (FMeC1) in vivo studies displayed by reduced cognitive deficits, thereby modulating the formation of Aβ aggregates which were comparable to the standard curcumin (Yanagisawa et al. 2015). In 2021, Simeonova and coworkers synthesized MTDLs from galantamine and curcumin and investigated them for their AChE inhibitory activity and antioxidant activity using in vitro DPPH assay. The compound **28** emerged as a potent low nanomolar AChE inhibitor as well as exhibited moderate antioxidant activity at low micromolar concentration (Simeonova et al. 2021). In order to develop potential lead molecules against AD drug targets, in 2016, different derivatives of curcumin were synthesized by Okuda and colleagues, who evaluated their inhibitory activity against tau and Aβ. The compound **29** was identified as a potent inhibitor of tau as well as Aβ aggregation. The compound **29** also showed better pharmacokinetic as well as good therapeutic efficacy in vivo as compared to curcumin (Okuda et al. 2016).

10.4.6 Other Natural Product-Inspired Molecules

Polyphenolic compounds such as galantamine, genistein, and quercetin have demonstrated significant activity against different AD drug targets (de Moura et al. 2021). In light of these findings, researchers are exploring the therapeutic potential of these compounds by synthesizing novel polyphenolic compounds, not only through semi-synthetic modifications of natural polyphenols but also through synthetic means. This approach has yielded a range of polyphenolic compounds inspired by natural compounds but exhibit unique properties. In this section, we have covered some of the natural compound-inspired polyphenolic compounds and their potential as AD therapeutic.

In 2001, Manabu Node and co-workers reported an improved phenolic oxidative coupling synthesis of (±)-narwedine (**30**) and (±)-galanthamine (**31**). Galanthamine is an alkaloid obtained from the Amaryllidaceae family and is a known selective inhibitor of acetylcholinesterase, it has been recognized by Austria and United Kingdom for the treatment of AD (Node et al. 2001). In 2012, Kenji Sugimoto and co-workers reported some novel dihydrofuran-fused perhydrophenanthrenes having a phenolic hydroxyl group inspired from mono-protected catechol of perhydrophenanthrenes. Among the synthesized compounds, compound **32** (DF-10) displayed the best potency with regard to the dendritic and axonal regeneration activities comparable to the previously reported, potent anti-AD drug candidates sominone and denosomin (Sugimoto et al. 2012).

In 2016, Angela Rampa and co-workers synthesized different compounds by using a pharmacophore combination strategy connecting BACE-1 inhibitory scaffold with the metal chelating pharmacophore having chalcone. Among all the compounds, they found conjugate **33** showing good neuroprotective effects by inhibiting BACE-1 at low-micromolar concentration. Additionally, the compound **33** also showed neuroprotective effect by inhibiting the ROS formation induced by the Cu^{2+} ions (Rampa et al. 2016).

In 2016, Lais Flavia Nunes Lemes and co-workers designed and synthesized some novel cardanol derivatives as acetylcholinesterase inhibitors. Cardanol is a byproduct phenolic lipid compound obtained from the cashew nutshell food processing. The in vitro testing against eeAChE as well as eqBuChE revealed that compound **34** displayed moderate inhibition of cholinesterase, i.e., eeAChE (IC_{50} of 6.6 μM) and eqBuChE (IC_{50} of 5.0 μM) as well as had similar inhibitory results against the human isoform AChE having IC_{50} of 5.7 μM. Moreover, compound **34** showed good BBB permeability determined using PAMPA-BBB assay and good cholinesterase inhibitory activity as well (Motta et al. 2016). In 2017, Riho Taguchi and co-workers reported some rosmarinic acid derivatives having good inhibitory activity against Aβ aggregation and antioxidant properties. Among the synthesized compounds **35** and **36** displayed moderate Aβ (Mao and Zhang 2022; Fekadu et al. 2022; Kidd 2008; Wang et al. 2021; Kumar and Singh 2015; Mueller et al. 2005; Sperling et al. 2011; Das et al. 2023a, b, 2022; Pasinetti et al. 2015; Syarifah-Noratiqah et al. 2018; Rigacci and Stefani 2016; Dragicevic et al. 2011; Betts and Wareham 2014; Tang and Taghibiglou 2017; Maczurek et al. 2008; Losada-Barreiro and Bravo-Diaz 2017; Durazzo et al. 2019; Youn et al. 2022; Choi et al. 2001, 2012; Spagnuolo et al. 2016; Henríquez et al. 2020; Freyssin et al. 2018; Komorowska et al. 2020; Mishra and Palanivelu 2008; Reznichenko et al. 2006; Ono et al. 2004; Rivière et al. 2007; Molino et al. 2016; Mazzanti and Di Giacomo 2016; Maher 2021; Farkhondeh et al. 2019; Yu et al. 2018; Berman et al. 2017; Lange et al. 2022; Gonzales et al. 2022; Ramalakshmi et al. 2021; Godos et al. 2021; Sova and Saso 2020; Tolba et al. 2013) aggregation inhibitory activity by MSHTS assay having EC50 value of 2.58 μM and 2.82 μM, respectively. Furthermore, the DPPH radical scavenging activity of compound **35** and **36** at 100 μM concentration were 46% and 48%, respectively. Thus, making compound **35** and **36** the most promising lead compound for development (Taguchi et al. 2017). In 2017, Chiara Lambruschini synthesized some polyphenol-containing peptidomimetics with an objective to inhibit the Aβ oligomerization. A short fragment-based approach and Ugi multicomponent reaction mechanism for synthesis of phenol containing simple substrates were utilized. In thioflavin T (ThT) fluorescence assay, it was found that the compound **37** displayed best inhibition of $A\beta_{1-42}$ fibrillation process, whereas for AβpE3–42, it was found that the compound **38** displayed the best inhibitory effect during the aggregation process (Lambruschini et al. 2017).

In 2017, Zhipei Sang and co-workers reported some novel 2-acetyl-5-*O*-(aminoalkyl)phenol analogues as MTDLs for the treatment of AD. The synthesized compound **39** displayed selective inhibition of AChE having IC_{50} value of 0.96 μM. Enzyme inhibition kinetics and molecular modelling analysis suggested that it has mixed-type inhibition and ability to bind with both CAS and PAS site of AChE. Compound **39** also displayed selective MAO-B inhibitory activity having IC_{50} value of 6.8 μM, along with moderate antioxidant, neuroprotective, and selective metal chelating activity. The compound also exhibited good BBB permeability as well as other pharmacokinetic properties (Sang et al. 2017).

In 2018, Alberto Martinez and co-workers synthesized some ionophoric polyphenols inspired novel MTDLs in order to target multiple drug targets involved in the development of AD. The synthesized compounds on evaluation against BACE 1 enzyme inhibition and Aβ aggregation inhibition revealed that compounds **40** (IC_{50} 4.4 ± 0.3 μM) and **41** (IC_{50} 1.7 ± 0.3 μM) displayed moderate to good inhibition of BACE1 enzyme activity performed using FRET assay which was in agreement with the molecular docking analysis having the best binding affinities for compound **40** and **41** (Fig. 10.5). The compounds also had selective Cu^{2+} metal ion chelation forming 2:1 complexes resulting in the complete inhibition of $A\beta\text{-}Cu^{2+}$ catalyzed hydroxyl radical production (Martínez et al. 2018).

10.5 Future Prospects

Polyphenols exhibit potent activity against a plethora of drugs targets of AD making it a key scaffold in design and development of potential therapeutics against AD (Henríquez et al. 2020; Cassidy et al. 2020). While the current evidence is promising, more research is needed to determine the potential therapeutic use of polyphenols in AD. Future studies may include clinical trials in humans to investigate the safety as well as therapeutic efficacy of polyphenol-based treatments for AD. Additionally, researchers are also focusing on developing several novel polyphenolic derivatives and also focusing to increase the bioavailability of existed natural polyphenolic compounds.

Polyphenols have gained significant attention for their potential health benefits. However, their low bioavailability due to poor solubility, stability, and absorption in the gastrointestinal tract can significantly limit their therapeutic potential. To conquer these limitations, various techniques have been employed to enhance their bioavailability and potency. Chemical modification is one such technique that involves esterification, glycosylation, and incorporation of other heterocyclic

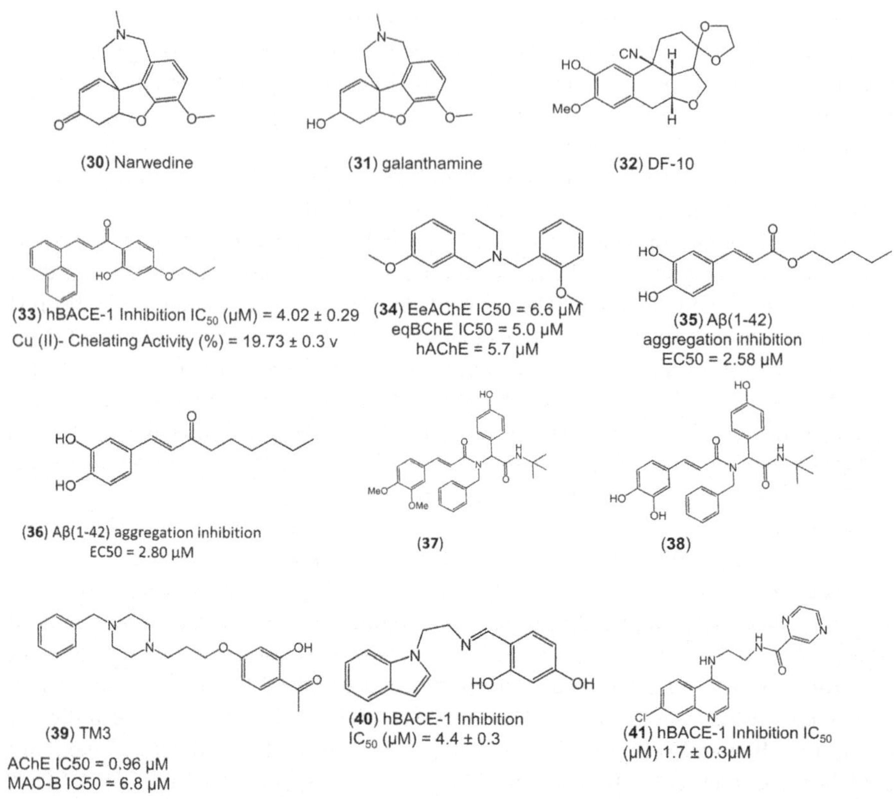

Fig. 10.5 Overview of some other naturally inspired that have demonstrated biological activity

scaffolds to improve the aqueous solubility of polyphenols. Formulation of polyphenols with other compounds, such as lipids, proteins, or polysaccharides, has also been investigated to enhance their solubility and stability, thus preventing their degradation in the gastrointestinal tract and increasing their bioavailability. Nanotechnology is another promising approach that involves encapsulating polyphenols in nanoparticles to enhance their absorption and protect them from degradation. This method offers targeted delivery of polyphenols to specific tissues, further improving their therapeutic potential. Polyphenols can also degrade during food processing and storage, which can lead to a reduction in their bioactivity and health benefits. Hence, gentle processing techniques, such as freeze-drying or air-drying, are employed to minimize polyphenol degradation. Further research is required to optimize these approaches and develop effective polyphenol-based therapies for various health conditions. By overcoming the limitations of polyphenols, researchers can maximize their therapeutic potential and contribute to improving human health.

However, to tackle the limitations associated with polyphenols and to convert them to promising lead molecules, medicinal chemists are using several chemical strategies to synthesize various analogues of polyphenols, which are similar in structure but have different chemical properties. These modifications can improve their bioavailability, stability, and potency. For example, chemical modifications can be made to increase the solubility of polyphenols in water, which can improve their absorption in the body. Moreover, these modifications can increase their ability to cross the BBB, improve their binding affinity for specific protein targets, and their ability to activate specific signaling pathways.

Overall, designing polyphenol-based drugs to treat AD offers a promising avenue for developing effective and affordable therapies for this devastating disease. Despite the promising findings, further research is necessary to gain a comprehensive understanding of the mechanisms underlying the effects of polyphenols in AD and to develop optimized compounds with better bioavailability and efficacy. Researchers are using machine learning and computational methods to identify new polyphenol-based drug candidates. By analyzing large databases of chemical compounds and their properties, researchers can identify polyphenol analogs with optimal drug-like properties such as bioavailability, specificity, and safety. Furthermore, novel synthetic phenolic compounds, semi-synthetic phenolic compounds, and combinational compounds currently under study are showing promise in overcoming the present challenges of phenolic compound therapy in AD. Therefore, the future for phenolic compounds in AD treatment provides hope for a better tomorrow.

10.6 Conclusion

Polyphenolic compounds are having tremendous biological significance against different therapeutic targets of AD. This mechanism of action allows polyphenols to prevent oxidative damage to cells and reduce inflammation, which are key factors in the pathogenesis of AD. Due to the presence of hydroxyl groups, these can form good number of hydrogen bonds with various protein and non-protein molecules inside the biological systems. Among various dietary natural polyphenolic compounds, compounds such as ferulic acid, caffeic acid, sinapic acid, resveratrol, and several others were found to act on multiple drug targets of AD. Phenolic compounds, including curcumin, resveratrol, and EGCG, have been extensively researched, showing promising results in preclinical and clinical trials. Natural product inspired molecules were synthesized to increase the bioavailability and therapeutic efficacy of existing natural polyphenolic compounds.

There have been studies on the protective effects of polyphenols on AD; however, the results were not confirmative and need further large trails. Results from some studies suggest the possibility of developing a "combination" of dietary polyphenolic compounds for AD prevention and/or therapy by modulating multiple drug targets of AD. However, exploration of these scaffolds and their optimization into a lead drug candidate for the treatment of AD is possible and researchers are working on it with full efforts, and polyphenolic derived compounds can be possible (see the clinics in near future for the treatment of AD).

References

Ahmed T, Javed S, Javed S, Tariq A, Šamec D, Tejada S et al (2017) Resveratrol and Alzheimer's disease: mechanistic insights. Mol Neurobiol 54:2622–2635

Albani D, Polito L, Signorini A, Forloni G (2010) Neuroprotective properties of resveratrol in different neurodegenerative disorders. Biofactors 36(5):370–376

Anand P, Sung B, Kunnumakkara AB, Rajasekharan KN, Aggarwal BB (2011) RETRACTED: Suppression of pro-inflammatory and proliferative pathways by diferuloylmethane (curcumin) and its analogues dibenzoylmethane, dibenzoylpropane, and dibenzylideneacetone: role of Michael acceptors and Michael donors. Elsevier

Askarizadeh A, Barreto GE, Henney NC, Majeed M, Sahebkar A (2020) Neuroprotection by curcumin: a review on brain delivery strategies. Int J Pharm 585:119476

Berman AY, Motechin RA, Wiesenfeld MY, Holz MK (2017) The therapeutic potential of resveratrol: a review of clinical trials. NPJ Precis Oncol 1(1):35

Betts JW, Wareham DW (2014) In vitro activity of curcumin in combination with epigallocatechin gallate (EGCG) versus multidrug-resistant Acinetobacter baumannii. BMC Microbiol 14:1–5

Cassidy L, Fernandez F, Johnson JB, Naiker M, Owoola AG, Broszczak DA (2020) Oxidative stress in Alzheimer's disease: a review on emergent natural polyphenolic therapeutics. Complement Ther Med 49:102294

Chao X, He X, Yang Y, Zhou X, Jin M, Liu S et al (2012) Design, synthesis and pharmacological evaluation of novel tacrine–caffeic acid hybrids as multi-targeted compounds against Alzheimer's disease. Bioorg Med Chem Lett 22(20):6498–6502

Chen Y, Lin H, Zhu J, Gu K, Li Q, He S et al (2017) Design, synthesis, in vitro and in vivo evaluation of tacrine–cinnamic acid hybrids as multi-target acetyl-and butyrylcholinesterase inhibitors against Alzheimer's disease. RSC Adv 7(54):33851–33867

Cheng G, Xu P, Zhang M, Chen J, Sheng R, Ma Y (2018) Resveratrol-maltol hybrids as multi-target-directed agents for Alzheimer's disease. Bioorg Med Chem 26(22):5759–5765

Choi Y-T, Jung C-H, Lee S-R, Bae J-H, Baek W-K, Suh M-H et al (2001) The green tea polyphenol (−)-epigallocatechin gallate attenuates β-amyloid-induced neurotoxicity in cultured hippocampal neurons. Life Sci 70(5):603–614

Choi D-Y, Lee Y-J, Hong JT, Lee H-J (2012) Antioxidant properties of natural polyphenols and their therapeutic potentials for Alzheimer's disease. Brain Res Bull 87(2–3):144–153

Cole GM, Teter B, Frautschy SA (2007) Neuroprotective effects of curcumin. Adv Exp Med Biol 595:197–212

Das B, Baidya AT, Mathew AT, Yadav AK, Kumar R (2022) Structural modification aimed for improving solubility of lead compounds in early phase drug discovery. Bioorg Med Chem 56: 116614

Das B, Mathew AT, Baidya AT, Devi B, Salmon RR, Kumar R (2023a) Artificial intelligence assisted identification of potential tau aggregation inhibitors: ligand-and structure-based virtual screening, in silico ADME, and molecular dynamics study. Mol Divers. https://doi.org/10.1007/s11030-023-10645-3

Das B, Baidya AT, Devi B, Rom T, Paul AK, Thakur B et al (2023b) Synthesis, single crystal X-ray, DFT, spectroscopic, molecular docking studies and in vitro biological evaluation of compound N-benzyl-4-(4-chlorophenyl)-2-oxobutanamide. J Mol Struct 1276:134782

de Moura ÉP, Monteiro AFM, Dias N, Fernandes HIRM, dos Santos IJ, Nascimentos MTS et al (2021) Polyphenol compounds as potential therapeutic agents in Alzheimer's disease. Front Clin Drug Res Dementia 2:204

Dragicevic N, Smith A, Lin X, Yuan F, Copes N, Delic V et al (2011) Green tea epigallocatechin-3-gallate (EGCG) and other flavonoids reduce Alzheimer's amyloid-induced mitochondrial dysfunction. J Alzheimers Dis 26(3):507–521

Durazzo A, Lucarini M, Souto EB, Cicala C, Caiazzo E, Izzo AA et al (2019) Polyphenols: a concise overview on the chemistry, occurrence, and human health. Phytother Res 33(9): 2221–2243

Farkhondeh T, Samarghandian S, Pourbagher-Shahri AM, Sedaghat M (2019) The impact of curcumin and its modified formulations on Alzheimer's disease. J Cell Physiol 234(10): 16953–16965

Fekadu B, Tareke M, Tadesse M, Anbesaw T (2022) Neurocognitive impairment and associated factors among elderly in the Bahir Dar City Administration, Northwest Ethiopia. Front Aging Neurosci 14:888704

Freyssin A, Page G, Fauconneau B, Bilan AR (2018) Natural polyphenols effects on protein aggregates in Alzheimer's and Parkinson's prion-like diseases. Neural Regen Res 13(6):955

George N, Al Sabahi B, AbuKhader M, Al Balushi K, Akhtar MJ, Khan SA (2022) Design, synthesis and in vitro biological activities of coumarin linked 1,3,4-oxadiazole hybrids as potential multi-target directed anti-Alzheimer agents. J King Saud Univ Sci 34(4):101977

Godos J, Caraci F, Micek A, Castellano S, D'Amico E, Paladino N et al (2021) Dietary phenolic acids and their major food sources are associated with cognitive status in older Italian adults. Antioxidants 10(5):700

Gonzales MM, Garbarino V, Zilli EM, Petersen R, Kirkland J, Tchkonia T et al (2022) Senolytic therapy to modulate the progression of Alzheimer's disease (SToMP-AD): a pilot clinical trial. J Prev Alzheimers Dis 9(1):22–29

Guerzoni LP, Nicolas V, Angelova A (2017) In vitro modulation of TrkB receptor signaling upon sequential delivery of curcumin-DHA loaded carriers towards promoting neuronal survival. Pharm Res 34:492–505

Habtemariam S (2017) Protective effects of caffeic acid and the Alzheimer's brain: an update. Mini Rev Med Chem 17(8):667–674

Henríquez G, Gomez A, Guerrero E, Narayan M (2020) Potential role of natural polyphenols against protein aggregation toxicity: in vitro, in vivo, and clinical studies. ACS Chem Neurosci 11(19):2915–2934

Jayaraj RL, Tamilselvam K, Manivasagam T, Elangovan N (2013) Neuroprotective effect of CNB-001, a novel pyrazole derivative of curcumin on biochemical and apoptotic markers against rotenone-induced SK-N-SH cellular model of Parkinson's disease. J Mol Neurosci 51: 863–870

Jayaraj RL, Elangovan N, Manigandan K, Singh S, Shukla S (2014) CNB-001 a novel curcumin derivative, guards dopamine neurons in MPTP model of Parkinson's disease. Biomed Res Int 2014:236182

Jeřábek J, Uliassi E, Guidotti L, Korábečný J, Soukup O, Sepsova V et al (2017) Tacrine-resveratrol fused hybrids as multi-target-directed ligands against Alzheimer's disease. Eur J Med Chem 127:250–262

Jia Y, Wang N, Liu X (2017) Resveratrol and amyloid-beta: mechanistic insights. Nutrients 9(10): 1122

Kidd PM (2008) Alzheimer's disease, amnestic mild cognitive impairment, and age-associated memory impairment: current understanding and progress toward integrative prevention. Altern Med Rev 13(2):85–115

Komorowska J, Wątroba M, Szukiewicz D (2020) Review of beneficial effects of resveratrol in neurodegenerative diseases such as Alzheimer's disease. Adv Med Sci 65(2):415–423

Kumar A, Singh A (2015) A review on Alzheimer's disease pathophysiology and its management: an update. Pharmacol Rep 67(2):195–203

Labban S, Alghamdi BS, Alshehri FS, Kurdi M (2021) Effects of melatonin and resveratrol on recognition memory and passive avoidance performance in a mouse model of Alzheimer's disease. Behav Brain Res 402:113100

Lambruschini C, Galante D, Moni L, Ferraro F, Gancia G, Riva R et al (2017) Multicomponent, fragment-based synthesis of polyphenol-containing peptidomimetics and their inhibiting activity on beta-amyloid oligomerization. Org Biomol Chem 15(44):9331–9351

Lan J-S, Zeng R-F, Jiang X-Y, Hou J-W, Liu Y, Hu Z-H et al (2020) Design, synthesis and evaluation of novel ferulic acid derivatives as multi-target-directed ligands for the treatment of Alzheimer's disease. Bioorg Chem 94:103413

Lange KW, Lange KM, Nakamura Y (2022) Green tea, epigallocatechin gallate and the prevention of Alzheimer's disease: clinical evidence. Food Sci Human Wellness 11(4):765–770

Lee HE, Kim DH, Park SJ, Kim JM, Lee YW, Jung JM et al (2012) Neuroprotective effect of sinapic acid in a mouse model of amyloid β1–42 protein-induced Alzheimer's disease. Pharmacol Biochem Behav 103(2):260–266

Losada-Barreiro S, Bravo-Diaz C (2017) Free radicals and polyphenols: the redox chemistry of neurodegenerative diseases. Eur J Med Chem 133:379–402

Lu C, Guo Y, Yan J, Luo Z, Luo H-B, Yan M et al (2013) Design, synthesis, and evaluation of multitarget-directed resveratrol derivatives for the treatment of Alzheimer's disease. J Med Chem 56(14):5843–5859

Maczurek A, Hager K, Kenklies M, Sharman M, Martins R, Engel J et al (2008) Lipoic acid as an anti-inflammatory and neuroprotective treatment for Alzheimer's disease. Adv Drug Deliv Rev 60(13–14):1463–1470

Maher P (2021) Preventing and treating neurological disorders with the flavonol fisetin. Brain Plasticity (Amsterdam, Netherlands) 6(2):155–166

Mao K, Zhang G (2022) The role of PARP1 in neurodegenerative diseases and aging. FEBS J 289(8):2013–2024

Martínez A, Zahran M, Gomez M, Cooper C, Guevara J, Ekengard E et al (2018) Novel multi-target compounds in the quest for new chemotherapies against Alzheimer's disease: an experimental and theoretical study. Bioorg Med Chem 26(17):4823–4840

Mazzanti G, Di Giacomo S (2016) Curcumin and resveratrol in the management of cognitive disorders: what is the clinical evidence? Molecules 21(9):1243

Mishra S, Palanivelu K (2008) The effect of curcumin (turmeric) on Alzheimer's disease: an overview. Ann Indian Acad Neurol 11(1):13

Molino S, Dossena M, Buonocore D, Ferrari F, Venturini L, Ricevuti G et al (2016) Polyphenols in dementia: from molecular basis to clinical trials. Life Sci 161:69–77

Motta F, Silva R, Castro GD, Boni S, Jorge M, Guimar R et al (2016) Chemistry cardanol-derived AChE inhibitors: towards the development of dual binding derivatives for Alzheimer's disease Laís Fl a be. Eur J Med 108:687–700

Mueller SG, Weiner MW, Thal LJ, Petersen RC, Jack CR, Jagust W et al (2005) Ways toward an early diagnosis in Alzheimer's disease: the Alzheimer's Disease Neuroimaging Initiative (ADNI). Alzheimers Dement 1(1):55–66

Nelson KM, Dahlin JL, Bisson J, Graham J, Pauli GF, Walters MA (2017) The essential medicinal chemistry of curcumin: miniperspective. J Med Chem 60(5):1620–1637

Nićiforović N, Abramovič H (2014) Sinapic acid and its derivatives: natural sources and bioactivity. Compr Rev Food Sci Food Saf 13(1):34–51

Node M, Kodama S, Hamashima Y, Baba T, Hamamichi N, Nishide K (2001) An efficient synthesis of (±)-narwedine and (±)-galanthamine by an improved phenolic oxidative coupling. Angew Chem 113(16):3150–3152

Noureddin SA, El-Shishtawy RM, Al-Footy KO (2019) Curcumin analogues and their hybrid molecules as multifunctional drugs. Eur J Med Chem 182:111631

Okuda M, Hijikuro I, Fujita Y, Teruya T, Kawakami H, Takahashi T et al (2016) Design and synthesis of curcumin derivatives as tau and amyloid β dual aggregation inhibitors. Bioorg Med Chem Lett 26(20):5024–5028

Ono K, Hasegawa K, Naiki H, Yamada M (2004) Anti-amyloidogenic activity of tannic acid and its activity to destabilize Alzheimer's β-amyloid fibrils in vitro. Biochim Biophys Acta Mol Basis Dis 1690(3):193–202

Pagoni A, Marinelli L, Di Stefano A, Ciulla M, Turkez H, Mardinoglu A et al (2020) Novel anti-Alzheimer phenol-lipoyl hybrids: synthesis, physico-chemical characterization, and biological evaluation. Eur J Med Chem 186:111880

Pandi A, Kalappan VM (2021) Pharmacological and therapeutic applications of Sinapic acid—an updated review. Mol Biol Rep 48(4):3733–3745

Panzhinskiy E, Hua Y, Lapchak PA, Topchiy E, Lehmann TE, Ren J et al (2014) Novel curcumin derivative CNB-001 mitigates obesity-associated insulin resistance. J Pharmacol Exp Ther 349(2):248–257

Pasinetti GM, Wang J, Ho L, Zhao W, Dubner L (2015) Roles of resveratrol and other grape-derived polyphenols in Alzheimer's disease prevention and treatment. Biochim Biophys Acta Mol Basis Dis 1852(6):1202–1208

Ramalakshmi N, Remya SR, Nalini CN (2021) Multitarget directed ligand approaches for Alzheimer's disease: a comprehensive review. Mini Rev Med Chem 21(16):2361–2388

Rampa A, Tarozzi A, Mancini F, Pruccoli L, Di Martino RMC, Gobbi S et al (2016) Naturally inspired molecules as multifunctional agents for Alzheimer's disease treatment. Molecules 21(5):643

Reznichenko L, Amit T, Zheng H, Avramovich-Tirosh Y, Youdim M, Weinreb O et al (2006) Reduction of iron-regulated amyloid precursor protein and β-amyloid peptide by (−)-epigallocatechin-3-gallate in cell cultures: implications for iron chelation in Alzheimer's disease. J Neurochem 97(2):527–536

Rigacci S, Stefani M (2016) Nutraceutical properties of olive oil polyphenols. An itinerary from cultured cells through animal models to humans. Int J Mol Sci 17(6):843

Rivière C, Richard T, Quentin L, Krisa S, Mérillon J-M, Monti J-P (2007) Inhibitory activity of stilbenes on Alzheimer's β-amyloid fibrils in vitro. Bioorg Med Chem 15(2):1160–1167

Rosini M, Simoni E, Caporaso R, Basagni F, Catanzaro M, Abu IF et al (2019) Merging memantine and ferulic acid to probe connections between NMDA receptors, oxidative stress and amyloid-β peptide in Alzheimer's disease. Eur J Med Chem 180:111–120

Sang Z, Wang K, Wang H, Wang H, Ma Q, Han X et al (2017) Design, synthesis and biological evaluation of 2-acetyl-5-O-(amino-alkyl) phenol derivatives as multifunctional agents for the treatment of Alzheimer's disease. Bioorg Med Chem Lett 27(22):5046–5052

Sang Z, Wang K, Han X, Cao M, Tan Z, Liu W (2018) Design, synthesis, and evaluation of novel ferulic acid derivatives as multi-target-directed ligands for the treatment of Alzheimer's disease. ACS Chem Neurosci 10(2):1008–1024

Sawda C, Moussa C, Turner RS (2017) Resveratrol for Alzheimer's disease. Ann N Y Acad Sci 1403(1):142–149

Shrikanta A, Kumar A, Govindaswamy V (2015) Resveratrol content and antioxidant properties of underutilized fruits. J Food Sci Technol 52:383–390

Simeonova R, Zheleva D, Valkova I, Stavrakov G, Philipova I, Atanasova M et al (2021) A novel galantamine-curcumin hybrid as a potential multi-target agent against neurodegenerative disorders. Molecules 26(7):1865

Singh YP, Tej GNVC, Pandey A, Priya K, Pandey P, Shankar G et al (2020) Design, synthesis and biological evaluation of novel naturally-inspired multifunctional molecules for the management of Alzheimer's disease. Eur J Med Chem 198:112257

Singh YP, Rai H, Singh G, Singh GK, Mishra S, Kumar S et al (2021) A review on ferulic acid and analogs based scaffolds for the management of Alzheimer's disease. Eur J Med Chem 215: 113278

Sova M, Saso L (2020) Natural sources, pharmacokinetics, biological activities and health benefits of hydroxycinnamic acids and their metabolites. Nutrients 12(8):2190

Spagnuolo C, Napolitano M, Tedesco I, Moccia S, Milito A, Luigi Russo G (2016) Neuroprotective role of natural polyphenols. Curr Top Med Chem 16(17):1943–1950

Sperling RA, Aisen PS, Beckett LA, Bennett DA, Craft S, Fagan AM et al (2011) Toward defining the preclinical stages of Alzheimer's disease: recommendations from the National Institute on Aging-Alzheimer's Association workgroups on diagnostic guidelines for Alzheimer's disease. Alzheimers Dement 7(3):280–292

Subedi L, Gaire BP (2021) Neuroprotective effects of curcumin in cerebral ischemia: cellular and molecular mechanisms. ACS Chem Neurosci 12(14):2562–2572

Sugimoto K, Tamura K, Ohta N, Tohda C, Toyooka N, Nemoto H et al (2012) Synthesis of dihydrofuran-fused perhydrophenanthrenes having a phenolic hydroxyl group as a novel anti-Alzheimer's disease agent. Bioorg Med Chem Lett 22(1):449–452

Syarifah-Noratiqah S-B, Naina-Mohamed I, Zulfarina MS, Qodriyah H (2018) Natural polyphenols in the treatment of Alzheimer's disease. Curr Drug Targets 19(8):927–937

Taguchi R, Hatayama K, Takahashi T, Hayashi T, Sato Y, Sato D et al (2017) Structure–activity relations of rosmarinic acid derivatives for the amyloid β aggregation inhibition and antioxidant properties. Eur J Med Chem 138:1066–1075

Tang M, Taghibiglou C (2017) The mechanisms of action of curcumin in Alzheimer's disease. J Alzheimers Dis 58(4):1003–1016

Tang Y-W, Shi C-J, Yang H-L, Cai P, Liu Q-H, Yang X-L et al (2019) Synthesis and evaluation of isoprenylation-resveratrol dimer derivatives against Alzheimer's disease. Eur J Med Chem 163: 307–319

Thapa P, Upadhyay SP, Suo WZ, Singh V, Gurung P, Lee ES et al (2021) Chalcone and its analogs: therapeutic and diagnostic applications in Alzheimer's disease. Bioorg Chem 108:104681

Tolba MF, Azab SS, Khalifa AE, Abdel-Rahman SZ, Abdel-Naim AB (2013) Caffeic acid phenethyl ester, a promising component of propolis with a plethora of biological activities: a review on its anti-inflammatory, neuroprotective, hepatoprotective, and cardioprotective effects. IUBMB Life 65(8):699–709

Tomaselli S, La Vitola P, Pagano K, Brandi E, Santamaria G, Galante D et al (2019) Biophysical and in vivo studies identify a new natural-based polyphenol, counteracting Aβ oligomerization in vitro and Aβ oligomer-mediated memory impairment and neuroinflammation in an acute mouse model of Alzheimer's disease. ACS Chem Neurosci 10(11):4462–4475

Verma V, Singh D, Kh R (2020) Sinapic acid alleviates oxidative stress and neuro-inflammatory changes in sporadic model of Alzheimer's disease in rats. Brain Sci 10(12):923

Wan T, Wang Z, Luo Y, Zhang Y, He W, Mei Y et al (2019) FA-97, a new synthetic caffeic acid phenethyl ester derivative, protects against oxidative stress-mediated neuronal cell apoptosis and scopolamine-induced cognitive impairment by activating Nrf2/HO-1 signaling. Oxidative Med Cell Longev 2019:8239642

Wang L, Kumar R, Pavlov PF, Winblad B (2021) Small molecule therapeutics for tauopathy in Alzheimer's disease: walking on the path of most resistance. Eur J Med Chem 209:112915

Yanagisawa D, Ibrahim NF, Taguchi H, Morikawa S, Hirao K, Shirai N et al (2015) Curcumin derivative with the substitution at C-4 position, but not curcumin, is effective against amyloid pathology in APP/PS1 mice. Neurobiol Aging 36(1):201–210

Youn K, Ho C-T, Jun M (2022) Multifaceted neuroprotective effects of (−)-epigallocatechin-3-gallate (EGCG) in Alzheimer's disease: an overview of pre-clinical studies focused on β-amyloid peptide. Food Sci Human Wellness 11(3):483–493

Yu Y, Zhou L, Yang Y, Liu Y (2018) Cycloastragenol: an exciting novel candidate for age-associated diseases. Exp Ther Med 16(3):2175–2182

Zhang X, He X, Chen Q, Lu J, Rapposelli S, Pi R (2018) A review on the hybrids of hydroxycinnamic acid as multi-target-directed ligands against Alzheimer's disease. Bioorg Med Chem 26(3):543–550

Zhao X, Liu Z, Liu H, Guo J, Long S (2022) Hybrid molecules based on caffeic acid as potential therapeutics: a focused review. Eur J Med Chem 243:114745

11 Flavonoid-Based Derivatives for Modulating Various Targets of Alzheimer's Disease

Jyoti Pandey

Abstract

Alzheimer's disease (AD), an advanced multifactorial neurodegenerative disease mainly accompanied by cognitive impairment, is the most conversant form of dementia. Contemporary studies have shown that inhibition of oxidative stress and target enzymes such as cholinesterase, β-secretase, monoamine oxidase, inhibition of hyperphosphorylation of tau protein and aggregation of amyloid-β plagues, inflammatory responses and unfolded protein responses perform a vital role in AD genesis. Flavonoids are polyphenol structured neuroprotecting agents, naturally found in plants and have imperative protective effects against neural dysfunction and injuries. Exhaustive research work has been performed on flavonoids of natural origin and its derivatives, to obtain effective drugs/therapy for AD control. This review is mainly focusing on the flavonoid derivatives which have been suggested to have anti-AD effects and may be utilized for reducing the risk and used for the effective treatment of AD.

Keywords

Flavonoid derivatives · Alzheimer's disease · Aβ-protein · Tau protein · Neuroinflammation · Angiogenesis

J. Pandey (✉)
Department of Chemistry, School for Physical and Decision Sciences, Babasaheb Bhimrao Ambedkar University, Lucknow, Uttar Pradesh, India

A. Sharma, G. P. Modi (eds.), *Natural Product-Based Synthetic Drug Molecules in Alzheimer's Disease*, https://doi.org/10.1007/978-981-99-6038-5_11

11.1 Introduction

Alzheimer's disease (AD) is an ever-increasing global epidemic. It is an advanced multifactorial neurodegenerative disease and the popular form of dementia. Its prevalence among elderly population increases aggressively with age, and it has been evaluated that it increases from 10% to approximately 50% from the age of 65 to 85 years. It has been estimated that globally more than 47 million people were affected in the year 2015 and number get twice in each 20 years, and anticipated to enhance to 115 million by 2050 (Brookmeyer et al. 2011; Prince et al. 2013; Ballard et al. 2016). In addition to language disturbances, behavioural changes, memory, and orientation loss, AD is mainly associated with cognition impairment in activities of daily living (Yiannopoulou and Papageorgiou 2013).

In recent years, two forms of the AD were recognized, a familial and a sporadic form (Kozlov et al. 2017). In more general way, it is supposed that AD is triggered by an amalgamation of three components, i.e., genetic, lifestyle, and environment, where the genetic component is responsible for about 70% cases of AD (Dorszewska et al. 2016). The apolipoprotein E gene (APOE gene) appears to be mainly accountable for sporadic form (Karran et al. 2011), whereas the familial form of AD was related with the genes PSEN1 (presenilin 1), PSEN2 (presenilin 2), and APP (amyloid precursor protein), along with other co-morbidity factors, including gene TREM2, which played an important role in the disorder of the lipid such as ABCA1 and ABCA7, or for bio-thiol metabolism MTHFD1 and gamma-secretase and metabolites transport (BIN1) (Dorszewska et al. 2016; Jayne et al. 2016). Nevertheless, both the forms (familial and sporadic) of AD show analogous molecular and neuropathological manifestations such as accumulation of hyperphosphorylated tau protein intracellularly, deposition of amyloid beta (Aβ) extracellularly, disfigurement of *N*-methyl-D-aspartate receptor-related signalling pathways, abnormalities in metal ion metabolism and lipid metabolism, oxidative stress, and disturbances in mitochondrial functions with structural distortion (Kozlov et al. 2017; Swerdlow et al. 2014; Karran et al. 2011; Šimić et al. 2016; Agostinho et al. 2010a; Bush 2013). AD is recognized by the occurrence of beta amyloid (Aβ) deposition as Aβ plaques and the presence of neurofibrillary tangles (NFTs) in the aggrieved patient brain region and also by the selective neuronal cell death (Selkoe 2001, 2005; Lahiri et al. 2003; Meleleo et al. 2019; Hardy and Selkoe 2002). Neurochemically, the AD is related with a rapid loss of cholinergic neurons and decrease of presynaptic markers in the cholinergic system, principally in brain regions associated with memory and learning (Bartus et al. 1982; Davies and Maloney 1976; Perry et al. 1978).

Previous years studies have indicated that inhibition of oxidative stress and target enzymes such as cholinesterase, monoamine oxidase, β-secretase, and inhibition of hyperphosphorylation of tau protein and aggregation of amyloid-β plagues, inflammatory responses, and unfolded protein responses have a vital role in AD genesis. So far, no cure nor treatment is available to cure AD or alter its progression. The foremost reason for the ineffective treatment of AD relies on the fact that the disease is of multifactorial in nature, therefore, a single molecule graced with a definite target might not be adequately operative for the AD treatment. Consequently, the recent approaches are generally founded on the concept of molecular hybridization which involve the combination of two or more pharmacophores with specific properties to reach array of characteristics in a single molecule to overcome the multifarious complex nature of AD (Rampa et al. 2011; León et al. 2013; Agis-Torres et al. 2014; Wang et al. 2016a). In this regard, nowadays huge number of naturally occurring or bioactive molecules have been applied for the development of hybrid molecules to combat AD pathogenesis (Rampa et al. 2011).

Flavonoids are polyphenol structured neuroprotectant and have significant protective effects against neural dysfunction and injuries. These naturally occurring compounds are widespread in vegetables, fruits, cocoa, and certain beverages including tea and coffee, to directly inhibit uncontrolled pathological events responsible for neurodegeneration (Gu et al. 2010; Solfrizzi et al. 2011).

The higher intakes of vegetables and fruits result in overall decrease in the risk of cognitive decline which is more likely referable to an increase consumption of specific flavonoids present in vegetables and fruits (Barberger-Gateau et al. 2007). It also has been found that a higher total intake of flavonoids in middle-aged adults results in better language performance and episodic memory (Kesse-Guyot et al. 2012) and greater cognitive performance in the older-aged adults (Letenneur et al. 2007). These findings have inspired further investigation of flavonoids and its derivatives for their possible health benefits and to explore the underneath mechanisms. Therefore, significant research work has been executed to modify naturally occurring flavonoid scaffolds, to develop such drugs which manage AD effectively. This review is mainly focusing on different molecular mechanism, targets, and the flavonoid derivatives which have been suggested to have anti-AD effects and may be utilized for reducing the risk and used for the effective medical care of AD.

11.2 Alzheimer's Disease (AD): Molecular Targets

The various targets and molecular mechanism which are involved in Alzheimer's disease treatment are shown in Fig. 11.1.

11.2.1 Amyloid Beta (Aβ) Protein

Amyloid beta (Aβ) proteins recognised as well-known hallmark of AD, is an important constituent of amyloid plaques, typically 39–42 residues long, and formed by the proteolysis of amyloid precursor protein (APP) (Wilquet and De Strooper 2004). The C-terminal and N-terminal of transmembrane protein APP are connecting to the neuronal cells through its phospholipid bilayer (Beyreuther et al. 1991). Proteolytic cleavage of APP causes its fragmentation and a number of events have been occurred which results in the formation of fibrils, oligomers, and aggregation of the neurotoxic forms of Aβ into plaques, all of which causes neuronal dysfunction in the AD patients (Haass and Selkoe 2007).

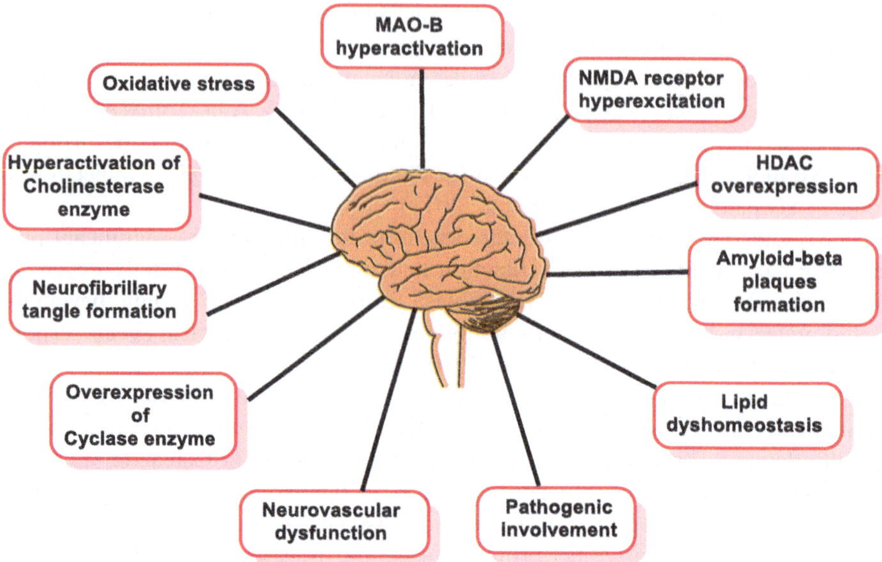

Fig. 11.1 AD molecular mechanism and major targets

The three major enzymes playing a crucial role in the processing of APP are α-secretase, β-secretase, and γ-secretase enzymes (Silva et al. 2014). The former two enzymes are principally responsible for the breakdown of APP into Aβ protein (Lichtenthaler 2011; Vassar et al. 2009). The non-amyloidogenic pathway has been accompanied by α-secretase enzyme to generate subunit of APPα, however, β-secretase followed amyloidogenic pathway to produce subunit of APPβ (Nistor et al. 2007). Furthermore, γ-secretase enzyme is accountable for cleavage of β subunit which produces toxic Aβ42 isoforms. Aβ42 is the most abundant isoforms of Aβ, which is highly prone to aggregate and produce oligomers. These oligomers are highly toxic and recognized as vital neurotoxin for this disease. The mutations within the APP gene and accumulation of excessive Aβ isoforms, followed by the self-assembly results in the generation of insoluble Aβ aggregates. Various factors such as improved ratio of Aβ42/Aβ40, enhanced Aβ aggregation, and less clearance of Aβ protein are responsible for the development of amyloid plaques around the neurons (see Fig. 11.4). The amyloid plaques accumulation and formation of neurofibrils are principally accountable for the AD pathogenesis progression (De Strooper et al. 2010). This process is acknowledged as the amyloid cascade hypothesis. From the last 25 years, this hypothesis dominated in AD research and drug development and has been considered as the most probable cause of Alzheimer disease. The senescent amyloid plaque development may be incarcerated with the utilization of specific secretase inhibitors in clinical therapies. Although the relation between AD development and amyloid plaques is still poorly unspecified, however, Aβ accumulations have been found to persuade a higher neurotoxicity than the insoluble neurofibrils (Tseng 2014).

Additionally, genetic mutation of Presenilin 1 and 2 (genes encode PSEN 1 and PSEN 2) is correspondingly accountable to increase Aβ level. Presenilin, the catalytic portion of the γ-secretase enzyme is accountable for the breakage of APP derived proteins (De Strooper 2007). Therefore, nowadays, it is considered that several Aβ isoforms may be liable for initiating AD pathogenesis and to develop novel therapeutic drugs with high potential necessitating a lot of persistence, knowledge, and research associations from various related research areas.

11.2.2 Cholinesterase

In mid-1970s, it was found in the study of post-mortem brain of Alzheimer diseased individual that the level of choline acetyltransferase (CAT) was lessen. ACh production is dependent on CAT (Wilcock et al. 1982); being neurotransmitter ACh has a vital role in memory function (Lombardo and Maskos 2015), cognitive functions (Wallace and Bertrand 2013), and learning (Hasselmo 2006). The uncontrolled release and transport of ACh, malfunctioning of allied receptor result in behavioural changes and impairment of cognitive functions in AD individuals (Winkler et al. 1995). These results suggested that AD is accompanied with an intense decay of different types of neurons, together with cholinergic neurons (Davies and Maloney 1976). Cholinesterase, extracellular serine hydrolase enzyme, primarily found in the central nervous system (CNS) and at peripheral neuromuscular junctions and accountable for the breakdown of ACh present in the hippocampus region of basal forebrain's (Terry and Buccafusco 2003). Their key function is to hydrolyse neurotransmitter ACh into choline and acetic acid (Čolović et al. 2013). The mechanism involved in the hydrolysis has an important role for normal cell signalling, and for the inhibition of neurons overstimulation. Therefore, a cholinergic replacement may serve as an initial goal to vanquish AD. The acetylcholinesterase (AChE) and butyrylcholinesterase (BuChE) are the two enzymes existing in the cholinergic nervous system (Pejchal et al. 2011). It is well documented that the AChE levels significantly increased around amyloid plaques and neurofibrillary tangles (NFTs) throughout every stage of AD, and it has also been anticipated that direct interaction of AChE with Aβ may promote the formation of plaques (Čolović et al. 2013). BuChE, a pseudo-neuronal cholinesterase enzyme intensifies the catalytic process of choline ester hydrolysis (Darvesh et al. 2003). AChE possesses two distinct binding sites within the enzyme gulley: one is peripheral anionic site (PAS) present at the entry of the gulley, and the catalytic site (CAS) present at the bottommost of gulley (Čolović et al. 2013). It has been found that AChE inhibitors mostly bind to these binding sites at one or more places in order to inhibit the AChE activity and also decrease the disintegration rate of acetylcholine, hence enhance the cholinergic neurotransmission (Čolović et al. 2013).

All the clinical drugs presently accessible in the market are AChE inhibitors (donepezil, galantamine, and rivastigmine), except memantine which is an antagonist of *N*-methyl-D-aspartate (NMDA) (Chen and Pan 2015).

While existing clinical AChE inhibitor drugs only provide symptomatic relief and show improvement in cognitive function, however, none of these treatments are efficient for arresting the AD progression (Rosini et al. 2014), which further illustrates the multifactorial nature of AD pathogenesis. Several other approaches such as stimulation of M1 and M4 muscarinic ACh receptors (Foster et al. 2014) and nicotinic ACh receptor (Parri et al. 2011) also increase the cognitive functions directly or indirectly but they are unsuccessful at certain extent during the process of drug discovery and development. Hence, for the cure of AD pathogenesis effectively, optimization and development of multi-targeted new cholinergic inhibitors with drug remodelling strategies are urgently required.

11.2.3 Tau Protein

Tau is a microtubule-related neuronal phosphoprotein, generally present in the central and peripheral nervous system and imperative part of cytoskeleton in neurons. Six isoforms of this protein with three foremost binding sites involving C-terminal microtubule binding domain, N-terminal projection, and shorter tailing sequence are present (Lee et al. 1989). Tau protein plays a crucial role in microtubule stability by maintaining cell integrity. The microtubules present in neurons are important for the protection of synaptic plasticity and neuronal structure and also for the maintenance of axonal transport of organelles, for example, ER, mitochondria, lysosomes, and vesicles comprising neurotransmitters and proteins. The microtubules properties present in axons and dendrites determine the neuronal polarity. Tau protein directs and controls the uniform orientation of microtubules in axons (Kosik 1993; Shahani and Brandt 2002; Lindwall and Cole 1984).

There is a phosphorylation gradient in different regions of brain, particularly along the axon (Buée et al. 2000; Hernández and Avila 2007). In a normal condition, tau protein present in the axonal membrane in its phosphorylated form. Alteration in the Tau phosphorylation state happened during cytoskeleton transformation process. The mechanisms which are involved in regulation of phosphorylation of Tau have become important for enhancement of synaptic plasticity. Hyperphosphorylated Tau impairs synaptic metabolism and axonal transport, causing microtubule instability which leads to NFT generation and bring about lack of cell viability and eventually causes microtubular cytoskeleton collapse and neuronal death. The phosphorylation and dephosphorylation of tau are vital regulatory events in neuronal homeostasis at serine and threonine phosphoepitopes.

There are various factors responsible for hyperphosphorylation of tau protein in AD such as tumour necrosis factor-alpha (TNF-α) (Xia and Hyman 2002), nuclear factor-kB (NF-kB) (Alawdi et al. 2017), and interleukins (ILs) (Ghosh et al. 2013), which are ultimately responsible for neurotic injury in the brain.

Several protein kinases and proteases such as glycogen synthase kinase-3-beta (GSK-3β), tau protein kinase-1 (TPK-1), AMP-activated protein kinase (AMPK), mitogen-activated protein kinase (MAPK), microtubule-affinity regulating kinase (MARK), and dual-specificity tyrosine phosphorylation-regulated kinase 1A

(Dyrk1A) are found to engaged in phosphorylation of tau and their hyperactivation creates a number of incidents which leads to hyperphosphorylation of tau (Ferrer et al. 2005). Among all, glycogen synthase kinase-3β (GSK3β) is an important tau kinase in neurons (Iqbal et al. 2005). Kinase-induced hyperphosphorylation of tau is upregulated by silencing of phosphatases protein (PP). The expression of PP1, PP2A, and PP5 get diminished in cerebral tissues of AD patients (Buée et al. 2000; Wang et al. 2007). Tau kinases are among the family members of the proline-directed kinases (Wang et al. 2007), such as MAP kinases (MAPK), GSK, and cyclin-dependent kinases (CDK5) (Lovestone and Reynolds 1997) and are efficient for in vitro tau phosphorylation and found to be appeared in the brain of AD patient. Other proline-directed kinases including protein kinase A and C (PKA and PKC), calcium calmodulin-kinase II (CaMPK-II) (Lovestone and Reynolds 1997; Johnson and Stoothoff 2004), casein-kinases I and II (Drechsel et al. 1992), all have been recognized in neurofibrillary tangles (NFTs) and also found to regulate phosphorylation of tau.

During the course of development, hyperphosphorylated neuronal tau is predominant in the embryonic stages, as there is an excessive need for neuroplastic changes in neurons and synapses of the CNS (Lovestone and Reynolds 1997). However, the mature CNS has dephosphorylated state of neuronal tau, this makes cytoskeleton steady to maintain neuronal homeostasis (Johnson and Stoothoff 2004). Neurodegeneration occurs when hyperphosphorylated tau has been found to transformed into insoluble intracellular NFTs. The insolubility of NFTs displays minimized clearance and consequences into cognitive impairment. Hence, the demand to design and development of new approaches either by inhibiting the NFTs formation or by intensifying their clearance is upsurge.

11.2.4 Oxidative Stress

Oxidative stress is accompanied by many pathological conditions, including diabetes, cancer, asthma, and neurological diseases (Jenner 2003; Lyras et al. 1997; Sayre et al. 2001). The disproportion between prooxidants and antioxidants, whereby there is either excessive production of oxidants or deficiency of antioxidants, is the root cause of oxidative stress. This imbalance results in overproduction of reactive oxygen species (ROS), for example, nitric oxide, hydrogen peroxide, superoxide anions, and monoxide and hydroxyl radicals and other free radicals (Federico et al. 2012; Nunomura et al. 2012) which can potentially impair cell components such as DNA, proteins, and lipids and can result in stimulation of various stress-induced transcription factors and generation of proinflammatory and anti-inflammatory cytokines (Phaniendra et al. 2015). As ROS are highly reactive species and have capability to begin a series of oxidation reactions in the presence of essential cellular molecules, which generate toxic side products and result in cell death. ROS initiate the lipid peroxidation which may be accompanied by the majority of neurodegenerative diseases. The lipids present in the bilayer of cellular membranes undergo free radical attacks in this process and result in the deactivation of membrane-bound enzymes and receptors and the generation of highly reactive molecules which have the capability to damage proteins and DNA (Jenner 2003; Lyras et al. 1997; Sayre et al. 2001). Alternately, it can be inherently produced by the electron transport chain in mitochondria, respiratory bursts from inflammatory cells, and some enzyme activities (for example, NADH oxidase) (Jenner 2003; Lyras et al. 1997; Sayre et al. 2001).

An immoderate consumption of oxygen in the brain might result in excessive free radical generation and hence brain is highly vulnerable for oxidative stress because it holds high levels of unsaturated fatty acids which are prone to attack by free radicals and lipid peroxidation (Phaniendra et al. 2015). Immoderate genesis of free radicals and reduced number of enzymes related to antioxidant properties may become the reason for neurodegeneration (Gandhi and Abramov 2012).

The abnormal energy metabolism, presence of trace elements such as zinc (Zn) and iron (Fe), and nucleic acid oxidase are also accountable for oxidative stress. This mitochondrial hypothesis illustrates one of the most significant factors for the AD pathogenesis (Swerdlow 2018). Initial studies on AD patients have shown that amyloid beta protein (Aβ) entered into the mitochondria, interrupted the electron transport chain (ETC) mechanism and resulted in oxidative stress due to excessive free radical generation, and eventually lead to neurodegeneration (Butterfield and Boyd-Kimball 2004). The neuronal death is only happened when Aβ is capable to activate a vital phase of the cell death cascade allied to mitochondrial permeability transition (MPT) pore opening (Du and ShiDu Yan 2010; Shevtzova et al. 2001). Therefore, MPT may aid as a most probable target to develop new and effective AD drugs which can be preferably applicable for the formation of drugs with multi-target and accompanied with various inhibitory activities such as AChE and BuChE (Makhaeva et al. 2019; Shevtsova et al. 2021). Various environmental factors may also be accountable for oxidative stress which include chemicals, UV radiation (Singh et al. 2004), and infectious organisms. Some enzymes such as NADPH oxidase (Kleinschnitz et al. 2010), lipoxygenase (LOX) (Praticò et al. 2004), xanthine oxidase (XO) (Gandhi and Abramov 2012), nitric oxide oxidase (Brown 1997), cytochrome p450 (Halliwell 2006), cyclooxygenase (COX) (Phillis et al. 2006) may also lead to free radical generation. Hence, there is an urgent need to investigate and elaborate the oxidative stress mechanism and its AD correlation.

11.2.5 Monoamine Oxidases (MAO)

Flavin-containing enzymes, monoamine oxidases (MAO), are occurring externally on mitochondrial membrane and catalyse the biogenic monoamines oxidation by substituting the amine group (NH_2) with oxygen. The MAO enzymatic activity comprises of catalytic deamination of various neurotransmitters (noradrenaline, dopamine, and serotonin) and exogenous amines (tyramine) which results in the generation of hydrogen peroxide, which may serve as a principal source of oxidative stress (Melo et al. 2011).

Based on inhibitor specificity and coding at the genetic level, MAO can be categorized into two categories of isozymes: (1) MAO-A and (2) MAO-B (Kennedy et al. 2003). Inhibition of MAO-A is principally accountable for the advancement of antidepressant therapy (Meyer et al. 2009). MAO-A takes part in the deamination of serotonin and sensitive for Clorgyline, MAO-A inhibitor. MAO-B inhibitors are currently applicable for the treatment and alleviation of Parkinson's disease as they stop the cleavage of dopamine in the brain (Ucar et al. 2005). The enzymatic activity of MAO-B has been found to be enhanced in the AD patient brains. However, their role in AD pathogenesis is not clear yet. MAO-B isozyme deaminate β-phenylethylamine, is sensitive for Selegiline, specific MAO-B inhibitors (Carreiras and Marco 2004). Now a days, scientists and researchers are exploring more about MAO physiopathology and its AD corelation, for the development of new MAO inhibitors.

11.2.6 NMDA Receptor

Glutamate, an excitatory neurotransmitter, situated in the cortex and hippocampal region of the brain, after tie up to its specific receptor, assists the maintenance of Na^+ and Ca^{2+} ions by controlling their entry into the neuronal cell (Green and LaFerla 2008). Hyperactivation of glutamate may result in excitotoxicity and finally causes the death of neuronal cell (Zott et al. 2019). In brain, glutamate receptors are of two types: (1) G protein coupled (GPCR) receptor, (2) Ionotropic (ion-channel) receptor (Barnes and Slevin 2003). *N*-Methyl-D-aspartate (NMDA) receptor is considered as an ionotropic receptor which after attaching with glutamate (agonist) neurotransmitters has a crucial role in learning and memory via maintaining Na^+ and Ca^{2+} ions influx in the neuronal cell (Lynch et al. 1983). It has been suggested that in AD patients, either there is hyperactivation of the NMDA receptor or there is an impairment of glutamate uptake mechanisms which consequently increases the glutamate availability to attach to NMDA receptor and results in the abrupt and frequent Na^+ and Ca^{2+} ions influx producing suppression of neuronal signalling, neurodegeneration, and finally neuronal cell death (Liu et al. 2019). Many NMDA receptor antagonists have proven unsuccessful because of severe adverse effects.

Further, it was found that, while inhibiting hyperactivation of NMDA receptor, it was anticipated to maintain the normal function of the receptor (Liu et al. 2019) Till date, memantine is the solitary NMDA receptor antagonist which was clinically approved for the AD treatment, it functions by blocking the NMDA receptor channel; in the meantime, it continues to permit the essential physiological activation important for normal brain function (Liu et al. 2019). Therefore, further investigation of NMDA receptor may provide key information which could help for the better understanding of AD and also help to develop new competitive inhibitors against glutamate receptors for the treatment of AD.

11.2.7 Histone Deacetylases (HDACs)

Histone deacetylases (HDACs), the enzyme removes the acetyl (CH_3CO-) group present on amino acid ε-*N*-acetyl-lysine, therefore, also known as deacetylase lysine (Van Dyke 2014). There are 11 HDACs and founded on their sequence homology to yeast equivalents, it can be classified into four major classes: class I, class II, class III, and class IV (de Ruijter et al. 2003).

Class I comprises of HDAC 1, 2, 3, and 8, which mainly confines in the nucleus. Class II is further divided into two subtypes: class IIa which comprises HDAC 4, 5, 7, and 9 and class IIb which comprises HDAC 6 and 10. In response to cellular signals, class IIa HDACs transport among the nucleus and the cytoplasm, while class IIb of HDACs primarily present in cytoplasm. Class III of HDAC, known as "sirtuins", includes SirT1–7 (Michishita et al. 2005). Class IV HDAC has only one member, HDAC11, which locates in the cellular nucleus (Gao et al. 2002). Among all the classes, zinc-dependent enzymes belong to class I, class II, and class IV, while nicotinamide adenine dinucleotide (NAD^+) dependent enzymes belong to class III (Chuang et al. 2009; Carey and la Thangue 2006). HDAC 6, as a cytosolic enzyme, is appeared in the cytoplasm of hippocampal neuronal cells. HDAC-6 is used to catalyse various nonhistone proteins, such as HSP90 deacetylase and α-acetyl-tubulin (Bali et al. 2005; Zhao et al. 2010) to keep microtubule stability. Consequently, HDAC-6 overexpression results in dysfunction of microtubule. HDAC-6 also affects the phosphorylated tau protein level (Hubbert et al. 2002). In AD patients, HDAC6 protein level is markedly improved in hippocampus and cortex compared with the normal brains (Zhang et al. 2013). The selective inhibition of HDAC-6 increases the mitochondrial movement inside the hippocampal neuronal cells. Glycogen synthase kinase-3β (GSK-3β) actively regulates HDAC-6 by phosphorylation pathway. Therefore, deregulation of HDAC-6 by GSK-3β may result in abnormal mitochondrial transportation (Chen et al. 2010). Furthermore, selective inhibition of HDAC-6 provides protection against oxidative stress-induced neurodegeneration and intensifies neurite overgrowth in neurons of cortical region (Rivieccio et al. 2009).

HDAC-4 of class IIa also plays a key role in nerve function. Abnormal expression of nuclear-localized HDAC-4 assists neuronal apoptosis; however, HDAC-4 inactivation suppresses neuronal cell death (Bolger and Yao 2005). Hence, continuous efforts are required for the development of selective and specific HDACs inhibitors for AD management and cure.

11.2.8 Cyclase

Recently, it has been found that AD patients have increased amounts of pyroglutamates, namely (PE) Aβ3–40/42 and (PE) Aβ11–40/42. Pyroglutamate amino acid residues such as Glu-3 and Glu-11 are linked with various N-truncated Aβ (Schlenzig et al. 2009). Glutamate residues with truncated Aβ undergo cyclization and result in pyroglutamate (PE) Aβ, cyclase enzyme is responsible for this cyclization process. These modified Aβ play a vital role in the development of Aβ plaques in AD patients. Modified Aβ elevates the Aβ accumulation, increases the hydrophobic nature of Aβ plaques and hence creates them indolent towards various proteolytic degrading enzymes. It has also been found that the cyclization of glutamate residues having truncated Aβ is enhanced in the presence of glutaminyl cyclases (QCs, QPCT, EC 2.5.2.3), majorly found at cortex and hippocampal region of the brain. Therefore, QCs overexpression directly increase PE Aβ3–42 and other PE Aβ levels in AD patient's brain. It has been observed that precise inhibition of QCs remarkably decreases the PE Aβ3–42 level (Morawski et al. 2014) and hence enhancing the enzymatic degradation of PE Aβ3–42 by proteolysis along with its clearance may serve as a favourable approach for the development of novel drug candidates for AD treatment.

11.2.9 Exopeptidase

A protease enzyme, Cathepsin B (CTSB), performs a crucial part in the pathophysiology of number of diseases, for example, AD and cancer. It occurs in lysosomes and endosomes where Aβ is produced, therefore, this is related to the production and aggregation of Aβ plaques in AD. It has also been found that CTSB splits APP at β-sites more effectively than BACE-1. The research work on CTSB gene knockout AD mice models revealed that the application of specific CTSB inhibitor results in decreased pGlu A production. N-terminally truncated species of Aβ, namely pGlu A, have an important role in Aβ oligomers aggregation (Hook et al. 2020).

Aminopeptidase A is an important exopeptidase involved in the N-terminal truncation of Aβ, and research findings have shown that it performs a crucial part in the control of structural and behavioural abnormalities in AD patients. Therefore, development of specific inhibitor for aminopeptidase A might affect its proteolytic activity which directly impacts the AD pathology.

11.2.10 Lipid Dys-Homeostasis

The bulk of the brain dry mass is constituted of lipids and therefore, lipid is associated with healthy functioning of the brain and could also be used for predicting the general pathological conditions of the brain. Lipids are at the centre of AD pathology, established on their participation in the blood–brain barrier (BBB) function, receptor signalling, amyloid precursor protein (APP) processing, remodelling, membrane inflammation, myelination, oxidation, and energy balance. Therefore, lipid disturbances may contribute to dysfunction in brain and may serve as the hallmark of AD. Under healthy conditions, a balanced cellular environment is achieved by lipid homeostasis, which further allows the normal functioning of brain cells. The major factors that interrupt lipid metabolism and responsible for lipid disruption in AD are demographic factors, genetic factors, and lifestyles. In addition, the most common genetic risk factor of AD is apolipoprotein E (APOE) ϵ4 genotype, which engaged in lipid metabolism and transport through the whole body including central nervous system (CNS) and in free form, it has a tendency for Aβ aggregation; however, lipidated-ApoE has a tendency to prevent Aβ aggregation (Kim et al. 2009). Periphery high density lipids (HDLs) and cholesterol withdrawn in CNS through blood–brain barrier (BBB) are metabolized with the help of HMG-CoA reductase and hence its inhibition results in lower production of Aβ aggregates (Vance et al. 2005). The enzyme sterol *O*-acetyltransferase 2 (ACAT) is responsible for transformation of free cholesterol into a cholesterol ester, and selective inhibition also impacts the Aβ accumulation by limiting their aggregation (Bhattacharyya and Kovacs 2010). Therefore, development of selective novel inhibitors for ACAT enzyme and HMG-CoA reductase could improve AD therapy (Di Paolo and Kim 2011).

11.2.11 Neurovascular Dysfunction

Blood vessels are accountable for providing oxygen and essential nutrients to nerve cell and remove other toxic metabolic remains from the brain interstitial fluid. AD and other neurodegenerative diseases are related with neurovascular disintegration, malfunctioning of BBB and neurovascular system. All these neurovascular deficiencies worsen the supply of blood into the brain and sequentially inhibit the oxygen flow, nutrients, and energy supplements to the cerebral region. Due to improper blood supply, the excretion of other neurotoxic metabolites is also decreased, which results in accumulation of toxic residues and causes cerebral β-amyloidosis, neurotoxicity, angiopathy, and cerebral amyloid (Kalaria 2010; Zipser et al. 2007; Zlokovic 2005). By the consideration of the specific pathogenesis related with blood–brain barrier impairment in AD, researchers may develop an effective and preventive therapeutic approach against these neurovascular dysfunctions.

11.2.12 Cysteine Protease

Cysteine proteases, a class of protease enzymes (Richard 2005), are generally distributed in all living beings (Grzonka et al. 2001) and play a key role in AD pathology (Lee et al. 2013; Urbanelli et al. 2008; Rohn 2010). It includes caspases, cathepsins, calpains, small ubiquitin-like modifier (SUMO) protease, and deubiquitinating enzymes (Richard 2005). Based on abnormal protein turnover, all the members of cysteine protease family are important in the development and advancement of many diseases (Grzonka et al. 2001). Cystatin C (CysC), a cysteine protease inhibitor, acts as protectors in AD and in related neurodegenerative diseases (Kaur and Levy 2012). AD pathology is characterized by Aβ aggregations (Aβ oligomers) in walls of cerebral vessel, and NFTs made up of mainly hyperphosphorylated tau. From in vitro studies, it has been shown that CysC inhibits Aβ aggregation and fibril formation due to binding with Aβ. Therefore, design and development of new potent and selective inhibitors for cysteine protease can serve as an area of interest for AD treatment.

11.2.13 Glycogen Synthase Kinase-3

Glycogen synthase kinase 3 (GSK3), proline-directed serine/threonine kinase, that regulates several physiological processes extending from glycogen synthesis and metabolism to gene transcription (Welsh and Proud 1993; Troussard et al. 1999). GSK3 has a crucial role in the AD pathogenesis, cancer, and diabetes. Overexpression of GSK3 accounts for increased Aβ production, hyperphosphorylation of tau, memory impairment, and microglial-mediated inflammatory responses associated with local plaque along with the loss of cholinergic neurons, apoptosis, synaptic loss, and inflammation as well; all these characteristics are of hallmark for AD (Aplin et al. 1996). Evidence suggested that GSK-3 is upregulated by the hippocampus and peripheral lymphocytes of AD patient's brains (Welsh and Proud 1993; Troussard et al. 1999). Therefore, the GSK3 inhibitors would be able to open new avenues for therapeutic intervention in AD.

11.2.14 Pathogen *Porphyromonas gingivalis*

Epidemiological studies have recognized an association between periodontitis and AD. Brain occupied by the periodontal pathogen *Porphyromonas gingivalis* may associated with these two inflammatory and degenerative conditions. In chronic periodontitis (CP) condition, *P. gingivalis* gets translocated and collected in extra-neuronal spaces (Kamer et al. 2015). Evidence of *P. gingivalis* permeation has been detected in autopsy specimens obtained from the cerebrospinal fluid of AD patients. *P. gingivalis* colonized themselves and oozes out gingipains, a class of *P. gingivalis* proteases, comprising of arginine-gingipain A (RgpA), arginine-gingipain B (RgpB), and lysine-gingipain (Kgp) in extra-neuronal spaces which results in neurotoxicity. It is found associated with neurons, beta amyloid, and tau tangles in the individual's brain suffering from AD. Collection and colonization of *P. gingivalis* activate several immunological phenomena such as microglial activation, abnormal cytokines and cause neuroinflammation (Gui et al. 2016). Therefore, the gingipain inhibition may provide a good approach to the treatment of both periodontitis and AD (Dominy et al. 2019).

11.3 Flavonoids, Their Cellular and Molecular Aspects Towards Cognitive Effects

Flavonoids are polyphenolic compounds embracing of a benzo-γ pyrone moiety, synthesized in plants via phenylpropanoid pathway to protect themselves from microbial infections and in most cases, their function is structure dependent

(Mahomoodally et al. 2005; Pandey 2007). They display diverse biological activities including antioxidants, neuroprotection, anti-inflammatory, and vasodilatory. Flavonoids are made up of a C_{15} skeleton with two (A and B) benzene rings, connected by a heterocyclic pyran ring (C). Plants remain the solitary source of flavonoids and those occurred in humans or animals are also of plant origin. Generally, the chemical nature of flavonoids relies on their structure, various patterns of substitutions or conjugations, degree of hydroxylation, and polymerization (Heim et al. 2002).

Different classes of flavonoids along with their examples and dietary sources have been depicted in Table 11.1.

Evidence from in vitro studies suggests that among all the depicted mechanisms of action associated with flavonoids, reduced cell damage is the most probable one and it has been interrelated to their intensive inherent capacity to serve as free radical scavenger and also eliminates various other nitrogen species (Halliwell 2006; Pannala et al. 1997; Russo et al. 2000; Visioli et al. 1998). The poor bioavailability and extensive metabolism of flavonoids restrict them to reach the threshold for action (Williams et al. 2004). However, recent studies revealed that small dosages of flavonoids are also sufficient for their pharmacological effects on neuronal cells. Therefore, various other probable mechanisms have also been proposed for the explanation of their action (Mandel et al. 2008; Schroeter et al. 2007a). The other probable mechanisms through which flavonoids provide prevention against neurodegeneration are as follows:

11.3.1 By Reducing Accumulation of Neuropathological Protein

The neuropathological hallmarks of AD include the aggregation of amyloid plaques extracellularly as well as accumulation of hyperphosphorylated tau proteins intracellularly (Savonenko et al. 2012). Most preclinical studies exploring that the flavonoids effects have been concentrated on the models where Aβ production has been increased (Walsh and Selkoe 2007). The one proposed pathway through which flavonoids may break the Aβ accumulation is by stopping the neuronal apoptosis through activation of α-secretase (ADAM10) (Williams and Spencer 2012) and inhibition of β-secretase (BACE-1) (Mori et al. 2012). In animal studies, it has found that administration of epigallocatechin-3-gallate (EGCG) for 6 months substantially reduced cognitive decline and Aβ accumulation in an AD mouse model (Rezai et al. 2008), and administration of green tea catechin helps to improve memory and spatial learning in senescence-prone mice (Li et al. 2009), the improvement in memory is associated with proteins upregulation linked to synaptic plasticity in the brain's hippocampus region and reduced Aβ accumulation. The mechanisms through which flavonoids may stop Aβ accumulation include inhibition of NFTs formation as well as amyloid aggregation (Ono et al. 2008), either by simplifying non-toxic oligomers production (Ehrnhoefer et al. 2008) or by metal chelation activity (Amit et al. 2008) in addition to the α-secretase protein upregulation through variation of metalloproteinase domain-containing protein 10 (ADAM10) and A

Table 11.1 Classification of flavonoids, examples, and their nutritional sources

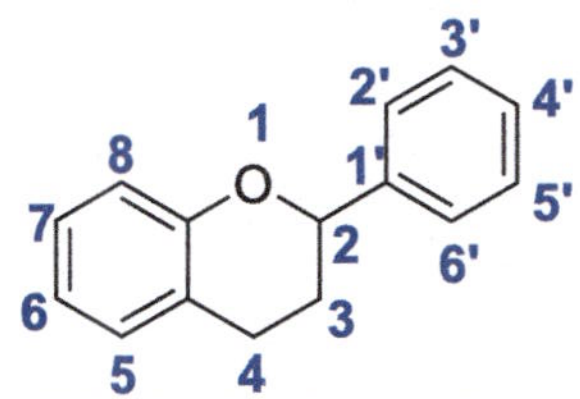

Class of flavonoids	Examples	Structural backbone	Nutritional sources	References
Flavone	Rutin, apigenin, luteolin, nobiletin	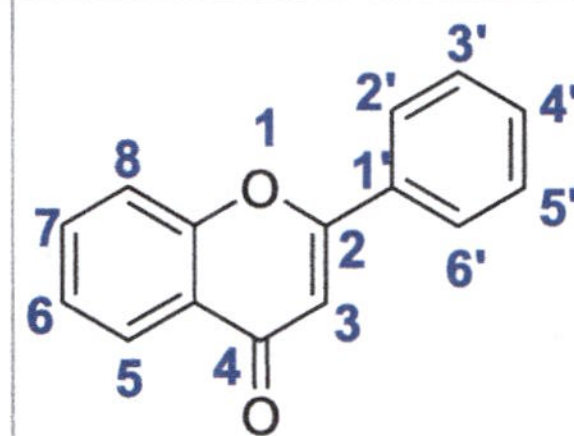	Red wine, fruit skin, parsley, celery, tomato skin	Jang et al. (2008), Feng et al. (2016), Onozuka et al. (2008)
Flavonol	Kaempferol, quercetin, *myricetin*	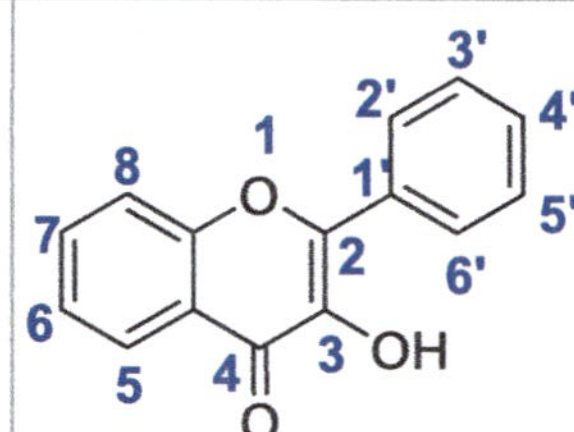	Red wine, broccoli, *onion, grapefruit*	Darbandi et al. (2016), Kouhestani et al. (2018), Lee et al. (2010), Spencer et al. (2003), Wang et al. (2014), Pluta et al. (2021)
Flavanone	Hesperidin, naringin, naringenin, taxifolin		Citrus fruits, tomatoes, lemon, orange, grapefruit	Hajialyani et al. (2019), Lee et al. (2022), Vafeiadou et al. (2009), Tanaka et al. (2019)

(continued)

Table 11.1 (continued)

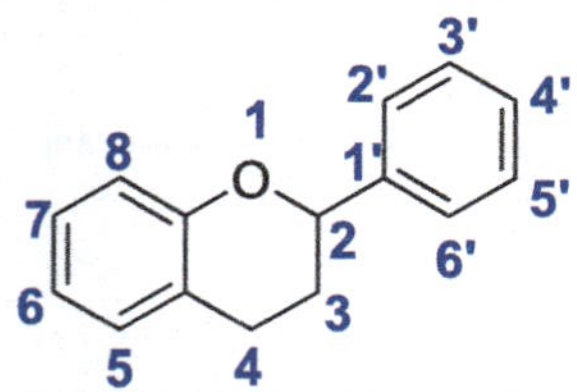

Class of flavonoids	Examples	Structural backbone	Nutritional sources	References
		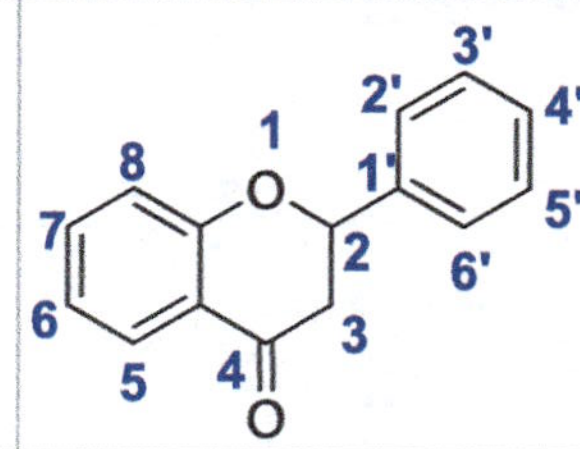		
Flavanol	Epigallocatechin-3-gallate, epigallocatechin, epicatechin, catechin	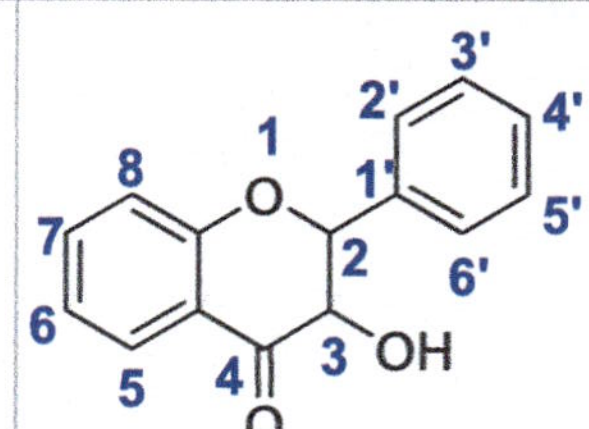	Tea, chocolate, red wine	Lee et al. (2009, 2000), Choi et al. (2001), Weinreb et al. (2009, 2004), Rezai et al. (2008)
Isoflavones	Genistein, daidzein		Soyabean	Wang and Murphy (1994), Mas-Bargues et al. (2022), Duan et al. (2021), Heo et al. (2006)

		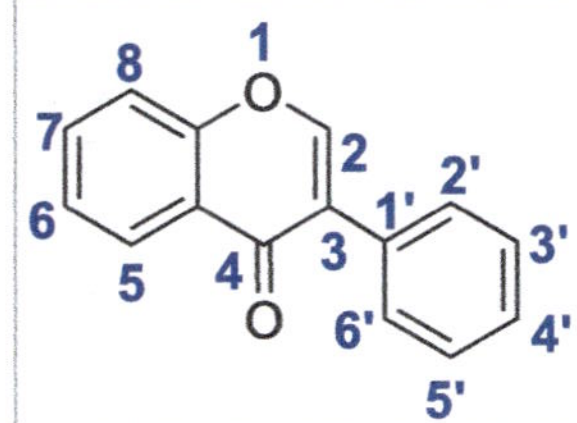		
Anthocyanidins	Cyanidin, malvidin, pelargonidin	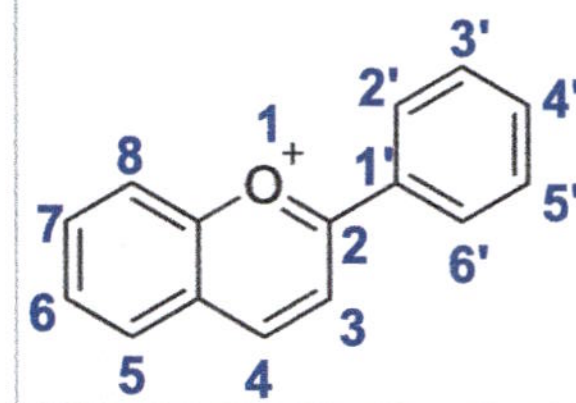	Blueberry, strawberry, cherry, red wine	Vepsäläinen et al. (2013), Qin et al. (2013), Afzal et al. (2019)

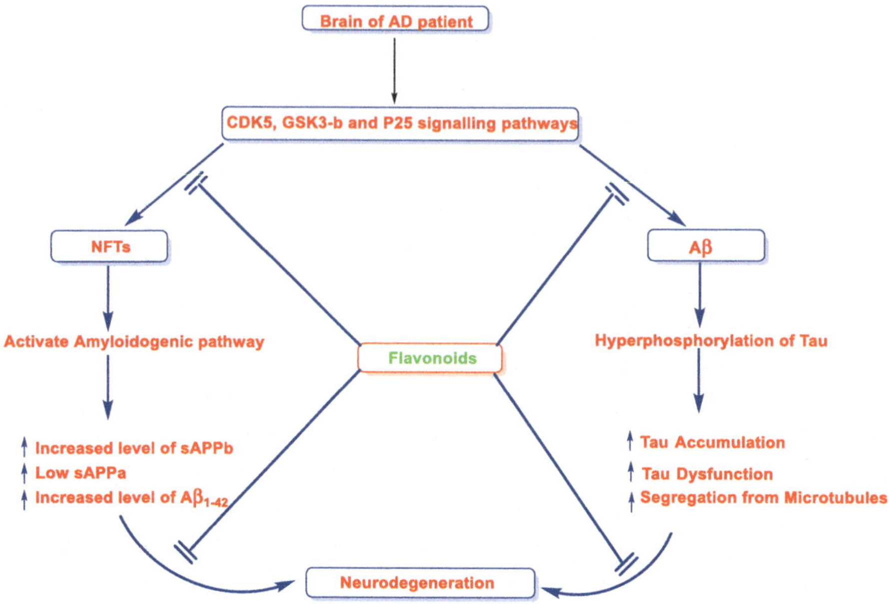

Fig. 11.2 The plausible mechanisms of flavonoids in impeding different signalling pathways

disintegrin (Obregon et al. 2006). The flavonoids, for example, (−)-epicatechin and hesperetin inhibit the development of NFTs through the activation of protein kinase B to reduce the GSK3β-driven hyperphosphorylation of tau (Fig. 11.2) (Schroeter et al. 2007b; Vauzour et al. 2007).

11.3.2 By Stimulating Neuronal Signalling Pathways and Synaptic Plasticity

The specific positions of interconnections among flavonoids and neuronal signalling pathways although ambiguous, however, prevailing evidences suggested that flavonoids may employ their effects through three processes: (1) affecting mitochondrial activity, (2) modulating specific genes expression, and (3) modifying signalling cascades which ultimately regulate neuronal apoptosis (Vauzour 2012). The current studies suggested that flavonoids affect the signal-regulated kinase (ERK) pathway extracellularly (Schroeter et al. 2007b), facilitated by interactions of mitogen-activated protein kinase (MEK) 1 and 2, and most probable membrane receptors (Schroeter et al. 2002). For example, flavanol (−)-epicatechin (Schroeter et al. 2001) and citrus flavanone hesperetin (Vauzour et al. 2007) have been found to involved in ERK1/2 activation. In cortical neurons, the sub-micromolar (μm) and nanomolar (nm) concentrations of EGCC were reported for restoration of activities of ERK1/2 in serum-deprived or 6-hydroxydopamine treated neurons (Levites et al. 2002). The CREB transcription factor, vital for assisting synaptic plasticity (Impey et al.

1998) and managing neuronal existence by controlling a number of significant genes expression, for example, brain-derived neurotrophic factor (BDNF), may get activated by ERK activation (Tully et al. 2003). Flavonoids are also well known to control the Akt/PKB enzyme activity, by regulating phosphoinositide 3-kinase (PI3K) enzyme. A citrus flavanone hesperetin is found to activate Akt/PKB enzyme system which is followed by deactivation of proteins accountable for neuronal apoptosis along with inhibition of apoptosis signal-regulating kinase 1 (ASK1), caspase-3, caspase-9, and Bad (Vauzour et al. 2007). Moreover, flavonoid-elicited stimulation of Akt in hippocampal neurons results in higher activity-regulated cytoskeletal-associated protein (Arc/Arg3.1) for mRNA translation (Ramirez et al. 2005). Enhanced arc expression enables changes in morphology and dendritic spines synaptic strength (Waltereit et al. 2001). In vitro studies claimed that flavonoid supplementation is responsible for morphological changes in neurons and also affects the growth of dendrites (Reznichenko et al. 2005).

11.3.3 Neuroinflammation

Normal inflammatory response plays an important part for supporting health, more particularly, the brain's defence system against impairment. A chronic upregulation of neuroinflammation, shown by enhanced level of circulatory pro-inflammatory cytokines including tumour necrosis factor (TNF-α) and biomarkers, could result in a number of events which results in progressive neuronal injury (Agostinho et al. 2010b; Calder et al. 2009). Chronic neuroinflammation may interrupt proper neuronal functioning, obstruct episodic memory encoding, and activate Aβ plaques aggregation. Individuals with mild cognitive impairment (MCI) were found to have excess amounts of circulating serum TNF-α (Trollor et al. 2010) and are overexpressed in cerebrospinal fluid (CSF) (Tarkowski et al. 2003) and in affected neural regions of AD brain (Akiyama et al. 2000). Moreover, the excess serum C-reactive protein (CRP), also known as acute phase protein, is related to high risk of dementia inception (Kravitz et al. 2009), memory impairment, and co-localized with accumulation of Aβ plaques and NFTs present in AD patient brains (Duong et al. 1997; Wood et al. 1993).

Increased concentrations of CRP in plasma have been constantly occurred in MCI (Roberts et al. 2009) and AD (Zaciragic et al. 2007) patients. Neuroinflammation also contributes to AD pathology via enhanced stimulation of microglial and subsequently increased stimulation of acetyl cholinesterase (AChE) and generation of free radicals (Williams et al. 2011). Usage of non-steroidal anti-inflammatory drugs (NSAIDs) (Szekely et al. 2004) for long-duration indeed lowers the AD risk and helps in development of anti-inflammatory agents to lessen the neuroinflammation effect on brain disease. Usage of flavonoids may stop neuroinflammation through various anti-inflammatory mechanisms, which include (1) inhibition of microglial stimulation of inflammatory cytokines, for example, TNF-α and IL-1β; (2) inhibition of NADPH oxidase activation and consequently ROS production in activated glia; (3) inhibition of iNOS induction and consequently production of nitric oxide with

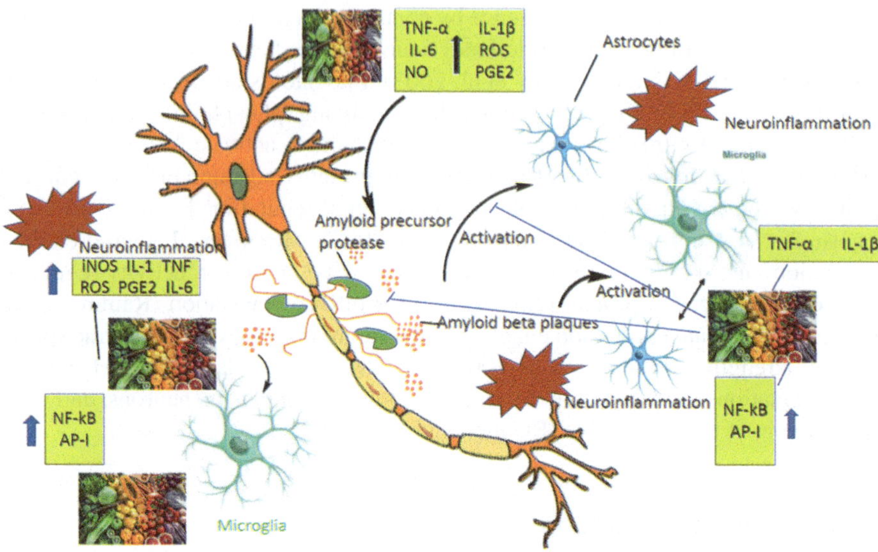

Fig. 11.3 Neuro-inflammation and flavonoids

response to stimulation of microglial; and (4) downregulation of pro-inflammatory transcription factors, for example, NF-κB via modification of glial and neuronal signalling pathways (Fig. 11.3) (Spencer et al. 2012; Bakhtiari et al. 2017).

The flavonols epicatechin and catechin have been found to inhibit TNF-α release, but did not show any effect on expression of iNOS or production of nitric oxide in primary glial cells (Vafeiadou et al. 2009), which provide satisfactory evidence of anti-inflammatory effects of flavanols at physiologically relevant concentrations. Human-based studies have also shown that increase in a flavonoid-rich diet lowers the level of inflammatory biomarkers such as IL-6, CRP, and adhesion factors (Nanri et al. 2008). Moreover, the specific flavonols-rich diet such as anthocyanidins and isoflavones have been associated with lowering of CRP amounts in blood (Chun et al. 2008). Blueberry anthocyanin also helps to reduce NF-kB-related pro-inflammatory cytokines and chemokines (IL-4, IL-8, IL-13, and IFN-α) in plasma (Karlsen et al. 2007). Therefore, the protective anti-inflammatory effects of flavonoid supplement strengthen the requirement of more investigation including larger sample sizes with random trials.

11.3.4 Vascular Function and Angiogenesis

There are several threats for reduced vascular health and function, for example, hypertension, diabetes mellitus, arteriosclerosis, smoking, in conjunction with different kinds of dementia including AD (Breteler 2000). The various findings have shown that flavonoid-rich diets help to lower the blood pressure and hence decreasing cardiovascular risk by improving arterial flow mediated dilation (Balzer et al. 2008) and increasing nitric oxide bioavailability (Grassi et al. 2008). In brain, vascular functions are important to inhibit ischemic events and also responsible to regulate continuous cerebral blood flow to support cognitive functions. Research studies suggest that flavonoids have some effect on the cerebral and peripheral vascular system through which it helps to improve cerebrovascular blood flow (Sorond et al. 2008). Neuroimaging studies have claimed that the cocoa consumption, rich in flavanol, remarkably enhances the cerebral blood flow in regard to cognitive tasks in both healthy older (Sorond et al. 2008, 2013; Lamport et al. 2015; Brickman et al. 2014) and younger adults (Francis et al. 2006). Additionally, studies have also shown that after drinking of a flavonoid-rich citrus drink by healthy adults result in significant increase in vascularization in the middle and inferior right frontal gyrus (Lamport et al. 2016). The increased cerebrovascular function facilitates adult neurogenesis (Gage 2000) and enhances vascularisation (Zhao et al. 2008), both the events are important for the cognitive performance. Therefore, flavonoid-rich diet might have necessary capacities to decrease the development of neurodegeneration and cognitive decline and at the same time, it potentially reverses the cognitive impairment by stimulating neural growth in the hippocampus region of the brain.

11.3.5 By Interactions with the Microbiome

The gastrointestinal (GI) tract has a crucial part in protection and promotion of health, and it also supports immune system by controlling energy metabolism and serving as a barricade against potentially toxic compounds present in the consumed diet. Flavonoids and their metabolites directly come into contact with the gut microbiota which regulate the GI tract functioning. The molecular structure of flavonoids is considered to be the determining factor for its impact on specific strains of bacteria (Xie et al. 2015), for e.g., A-type pro-anthocyanidins found in cranberries prevent the adhesion of *E. coli* in the human urinary tract (Howell et al. 2005). Similarly, quercetin has potential to break the development of *B. galacturonicus*, *R. gauvreauii*, and *Lactobacillus* sp. strains (Duda-Chodak 2012). Flavonoids found in berries also show potential towards inhibitory effects against *C. perfringens*, *B. cereus*, *H. pylori*, *S. aureus*, *C. albicans*, *S. epidermidis*, and *C. jejuni* (Nohynek et al. 2006).

There are bidirectional relations between microbiota and flavonoids, as microbiota present in the GI tract have capacity to control the absorption and bioavailability of flavonoids found in consumed diet, for example, pro-anthocyanidins, which are badly absorbed in the GI tract, were easily metabolized by the microbiota into phenolic acids which get easily absorbed (Zhang et al. 2016).

Flavonoid metabolites have been found to impact health in a different way with respect to their parent flavonoid compounds (Verbeek et al. 2005; Wang et al. 2015; di Gesso et al. 2015). More particularly, flavonoid metabolites in their specific compositions result in reduction of IL-Iβ secretion, which further suggest its potential application as anti-inflammatory agents (di Gesso et al. 2015). Moreover, elaborate research studies are essential to completely perceive the underneath interactive mechanism between flavonoids and the microbiota, which help to explore and recognize the particular nature of their local and systemic health benefits.

11.4 Flavonoids Derivatives in AD Management

In the last few years, various reasons which are responsible for the AD onset have been investigated, however, the etiopathogenesis of AD is remain unclear. Among various factors, accumulation of Aβ plaques, oxidative stress, and low concentration of acetylcholine have a crucial role in the genesis of AD (Citron 2010). To deal with the multifactorial nature of AD pathogenesis, there is an immediate requirement to develop multipotent novel anti-AD drugs to seize and inverse the AD progression, this led to the development of multi-target directed ligand (MTDL) approach, which attract a lot of attention towards the natural product including flavonoids (Cavalli et al. 2008; Bolognesi et al. 2007). Flavonoids are non-toxic, therefore, may serve as important scaffold for AD treatment (Williams and Spencer 2012).

Recently, few research groups have screened flavonoids and modified their chemical structures for the advancement of new anti-AD drugs. Based on their work, Shrivastava et al. have proposed the structure–activity relationship (SAR) for flavonoids as shown in Fig. 11.4 (Verma et al. 2022).

11.4.1 Flavone Analogues

Sang et al. discovered multifunctional agents for AD treatment. They have planned and prepared a series of scutellarin carbamate derivatives and scutellarin-*O*-alkylamine derivatives (Sang et al. 2015a, b, c). They have evaluated these derivatives in vitro for antioxidant activities, metal chelating properties, AChE and BChE inhibitory activities, and neuroprotective effects against hydrogen peroxide induced PC12 cell injury and found that most of the derivatives exhibited good multifunctional activities. Scutellarin carbamate derivative **1** (*N,N-dimethylcarbamic acid-4-(5,6,7-trihydroxy-4-oxo-4H-1-benzopyran-2-yl)phenyl ester*) (Fig. 11.5) exhibited dual inhibitory potency towards AChE ($IC_{50} = 1.2 \pm 0.03$ μM) and BuChE ($IC_{50} = 22.1 \pm 0.15$ μM) and had the strong antioxidant activity (10.3 Trolox equivalents). It has a selective metal chelation and neuroprotective properties. Compound **1** can be used for improvement of memory impairment brought by sodium nitrite, ethanol, and scopolamine (Sang et al. 2015a). Compound **2** (*4-(5-Hydroxy-6,7-dimethoxy-4-oxo-4H-chromen-2-yl)phenyl N, N-diethylcarbamate*) (Fig. 11.5) was obtained, on substituting hydroxyl groups at position 6 and 7 of flavone ring of compound **1** by methoxy groups, has more inhibitory effect towards AChE ($IC_{50} = 0.57$ μM) without remarkable changes in BuChE inhibitory activity ($IC_{50} = 22.6$ μM) and has less antioxidative activity (1.3-fold of Trolox equivalents) (Sang et al. 2015b). Compound **2** also behaved as a

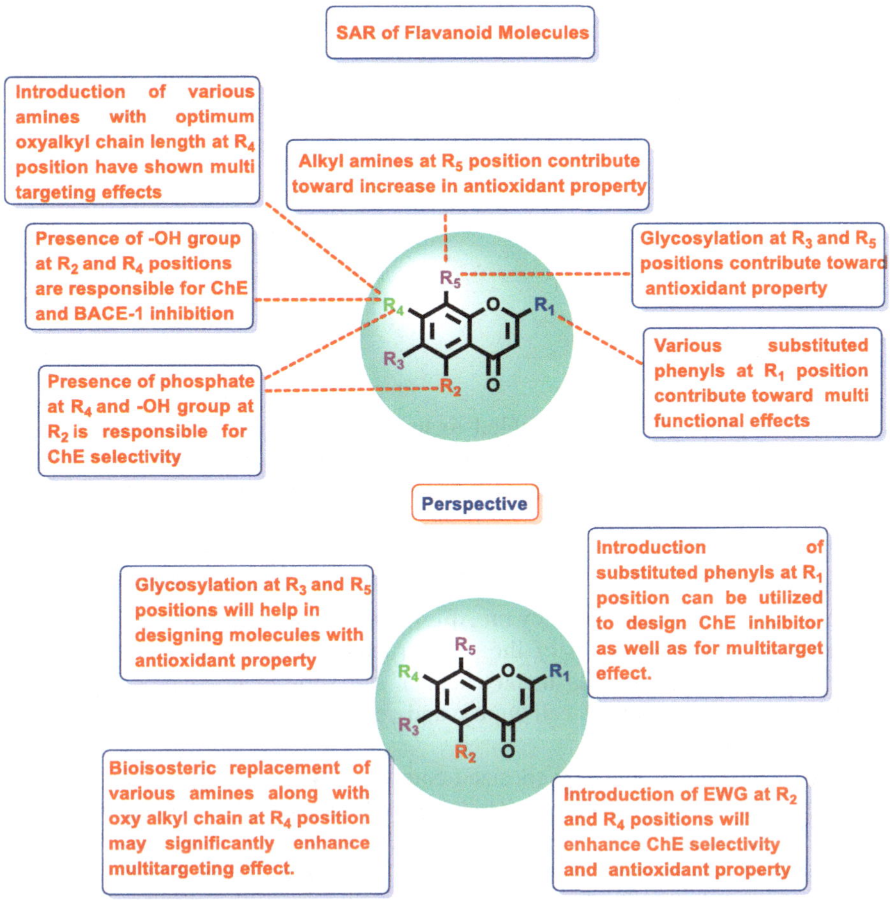

Fig. 11.4 SAR of Flavonoid Molecules

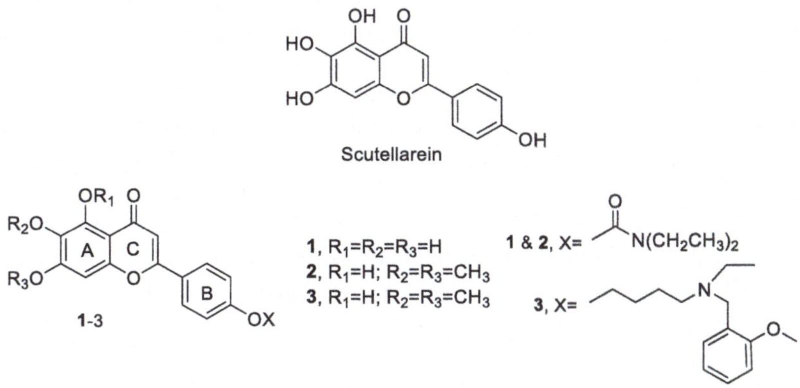

Fig. 11.5

Fig. 11.6

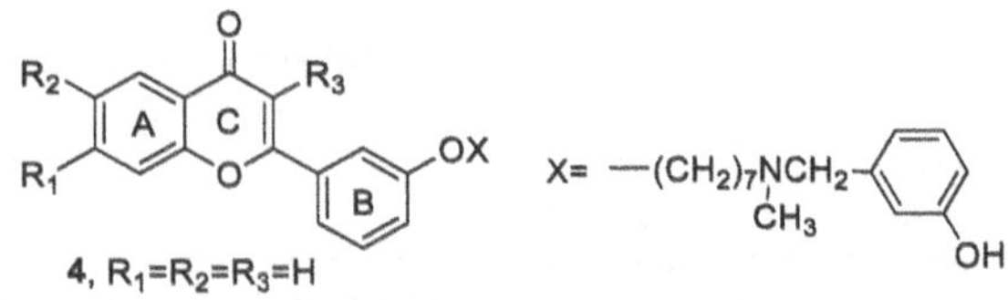

4, $R_1=R_2=R_3=H$

Fig. 11.7

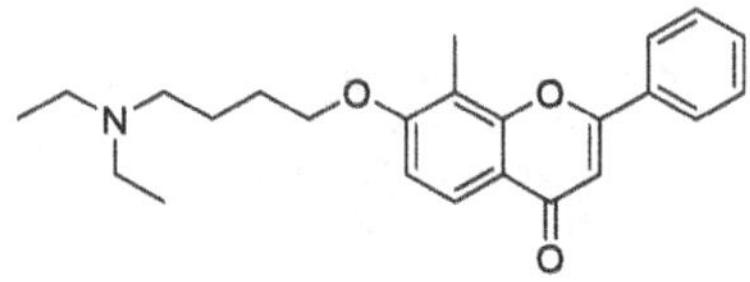

5

selective bio-metal chelator and abled to traverse the blood–brain barrier (BBB) in vitro. It also had imperative neuroprotective effects in scopolamine-induced cognitive impaired mice (Sang et al. 2015b). Similarly, scutellarin-*O*-alkylamine compound **3** (*5-Hydroxy-4′-(3-((ethyl)(2-methoxybenzyl)amino)butoxy)-6,7-dimethoxyflavone*) (Fig. 11.5) having alkyl-*N*-ethyl-benzylamine moiety at position 4′ of ring B, exhibited moderate antioxidative activity and AChE inhibitory potency and an excellent inhibitory potency on $A\beta_{1-42}$ accumulation and disassembled Cu^{2+} induced $A\beta_{1-42}$ accumulation of the well-structured $A\beta_{1-42}$ fibrils. Furthermore, compound **3** also behaved as metal chelator and showed good protective effect against H_2O_2-induced PC12 cell injury and reduced toxicity towards SH-SY5Y cells (Sang et al. 2015c).

Earlier Rampa et al. have synthesized compound **4** (*2-(3-((7-((3-hydroxybenzyl)(methyl)amino)heptyl)oxy)phenyl)-4H-chromen-4-one*) (Fig. 11.6) which is a 3′--modified flavonoid with inhibitory potency towards hAChE and hBuChE (IC_{50} = 0.52 μM and 63.7 μM, respectively) (Rizzo et al. 2009). Thus, the compounds having substitutions at position 4′ have potential as AChE inhibitors.

Li et al. (2013a) have strategically synthesized and biologically screened a novel series of flavonoid derivatives as potential multifunctional AChE inhibitors for the treatment of AD. Most of these derivatives were potent AChE inhibitors and exhibited moderate to good inhibitory activities against self-mediated Aβ42 accumulation. Among all, compound **5** (*7-(4-(diethylamino)butoxy)-8-methyl-2-phenyl-4H-chromen-4-one*) (Fig. 11.7), with a diethyl amino linkage to flavonoid scaffold by a 4-carbon bond spacer, served as most active AChE inhibitor (IC_{50} = 0.13 μM), highly selective for AChE over BuChE, enhanced biometal-chelating properties and found to inhibit Aβ42 aggregation by 38.95% with comparable results to reference compound curcumin which inhibit Aβ42 aggregation by 50.12%. The molecular modelling studies and inhibition kinetic analysis showed that compound **5** attached to AChE through CAS region via cation-π interactions occurred between Phe330 and Trp84 as well as through PAS region via π–π stacking interactions occurred between Tyr334 and Trp279. Overall, compound **5** has strong probability to serve as a multifunctional AChE inhibitor for AD cure and treatment and this type of multifunctional flavonoid compounds with amino-alkyl substituted group might

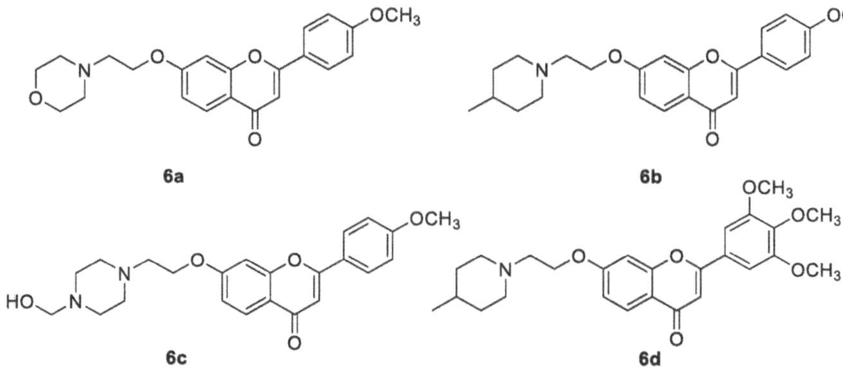

Fig. 11.8

serve as a good lead candidate for the development of new anti-AD agents with multiple activities.

By the use of multi-target directed ligands (MTDLs) approach, Silakari et al. (Singh and Silakari 2016; Singh et al. 2017) have prepared several new flavonoid derivatives and biologically screened all of them as acetylcholinesterases (AChEs) inhibitors, for advanced glycation end product (AGEs) production and free radical scavenging activity. Most of the compounds have acetylcholinesterase (AChE) inhibitory activity with IC_{50} values ranging from low micromolar to nanomolar range as verified from in vitro studies. Among all, derivatives **6a** *(2-(4-Methoxyphenyl)-7-(2-morpholinoethoxy)-4H-chromen-4-one)*, **6b** *(2-(4-Methoxyphenyl)-7-(2-(4-methylpiperidin-1-yl)ethoxy)-4H-chromen-4-one)*, and **6c** *(7-(2-(4-(Hydroxyethyl)piperazin-1-yl)ethoxy)-2-(4-methoxyphenyl)-4H-chromen-4-one)* are strong AChE inhibitors (Fig. 11.4.), having IC_{50} values of 6.33 nM, 7.56 nM, and 11.0 nM, respectively, and found more active in comparison to the reference compound donepezil ($IC_{50} = 12.7$ nM) (Singh and Silakari 2016). Moreover compound **6d** *(7-(2-(4-methylpiperidin-1-yl)ethoxy)-2-(3,4,5-trimethoxyphenyl)-4H-chromen-4-one)* (Fig. 11.8) have 3′,4′,5′-trimethoxy-substituted flavone moiety containing 4-methylpiperidine strongly inhibited AChE ($IC_{50} = 5.87$ nM) and represented most potent inhibitor of the series. Additionally, compound **6d** also displayed the capability to prevent AGEs formation ($IC_{50} = 23.0$ μM) with free radical scavenging property ($IC_{50} = 37.12$ μM) (Singh et al. 2017).

Li et al. (2013b) have planned the synthesis of a new series of tacrine-flavonoid hybrids as potent multifunctional ChE inhibitors for AD treatment. In vitro studies of these hybrids have shown that majority of the molecules have remarkable ChE inhibitory ability and exhibited self-induced amyloid-β ($A\beta_{1-42}$) aggregation. Compound **7a** has shown highest potency with balanced inhibitory profile against ChE such as $IC_{50} = 0.133$ μM for AChE and 0.558 μM for BChE and 79.1% self-induced $A\beta_{1-42}$ aggregation (Fig. 11.9). Furthermore, compound **7a** has also shown strongest

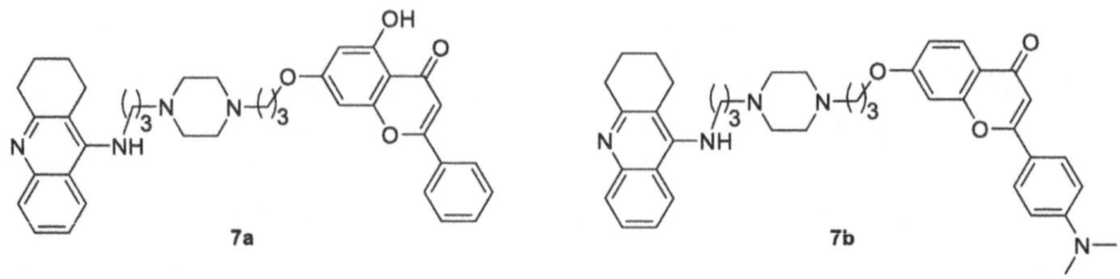

Fig. 11.9

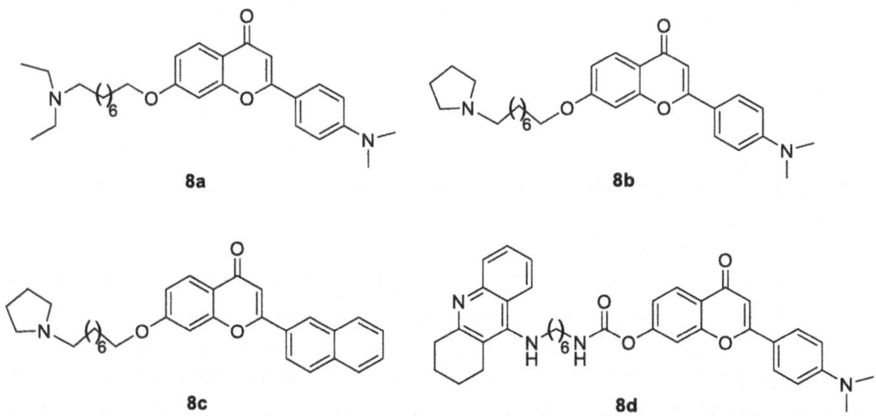

Fig. 11.10

metal chelating property and low cell toxicity. Therefore, compound **7a** might serve as an unrivalled multifunctional agent for AD treatment.

Luo et al. have prepared a number of 4-dimethylamine flavonoid scaffolds and biologically assessed them as potent multifunctional agents for the treatment of Alzheimer patients. They found that majority of the synthesized compounds have AChE and BChE inhibitory potency with IC_{50} values in the micromolar range (IC_{50} = 1.83–33.20 μM for AChE and 0.82–11.45 μM for BChE) (Luo et al. 2013). Among all the derivatives, compound **8a** (*7-(8-(Diethylamino)octyloxy)-2-(4-(dimethylamino)phenyl)-4H-chromen-4-one*) and compound **8b** (*2-(4-(Dimethylamino)phenyl)-7-(8-(pyrrolidin-1-yl)octyloxy)-4H-chromen-4-one*) were most potent compounds with IC50, 1.83, and 1.99 μM for AChE and 5.01 and 2.60 μM for BChE, respectively (Fig. 11.10) (Luo et al. 2013). In continuation of their works (Luo et al. 2013), the substitution of naphthyl group at 2′ position results in increase in AChE and BChE inhibitory activities of compound **8c** (*2-(naphthalen-1-yl)-7-(8-(pyrrolidin-1-yl)octyloxy)-4H-chromen-4-one*) with IC_{50} = 0.72 μM for AChE and 0.42 μM for BChE without any toxic effects on cell viability of PC12 and HepG2 cells, these results were found to be better than their earlier reported

Fig. 11.11

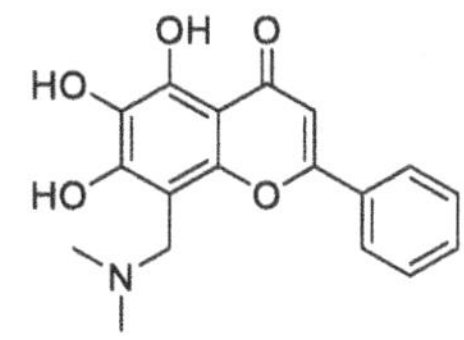

Fig. 11.12

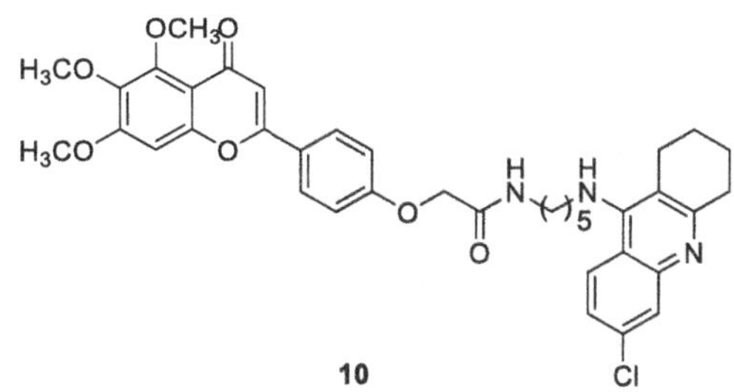

compounds as well as in comparison with the reference cholinergic agent rivastigmine (Fig. 11.10) (Luo et al. 2015). Further, optimization of A-ring by tacrine modification results in most potent compound **8d** (*2-(2-(4-(Dimethylamino)-phenyl)-4-oxo-4H-chromen-7-yloxy)-N-((1,2,3,4-tetrahydroacridin-9-ylamino) methyl)acetamide*) with $IC_{50} = 0.059$ μM for AChE and 0.046 μM for BChE, which have about 5- and 18-fold additional potency in comparison to compound **7b** against AChE and BuChE, respectively and have 79.3% anti-$A\beta_{1-42}$ aggregation activity in comparison to 12.2% for compound **7b** (Figs. 11.9 and 11.10) (Luo et al. 2016).

Cruz et al. reported Mannich base compound **9** (*8-((Dimethylamino)methyl)-5,6,7-trihydroxy-2-phenyl-4H-chromen-4-one*) as dual agents for AChE inhibition (IC_{50} values of 81.99 μM) and antioxidant activity and further, work as an active metal chelating agent (Fig. 11.11) (Cruz et al. 2017).

Deng et al. have synthesized new 5,6,7-trimethoxyflavone-6-chlorotacrine analogues and biologically screened them as potential multifunctional agents for AD cure (Liao et al. 2015). Most of the compounds had remarkable antioxidant activities and displayed good AChE inhibitory activity with high selectivity toward AChE over BuChE and also exhibited notable inhibitory activity toward self-induced Aβ aggregation. Hybrid compound **10** (*N-(((6-chloro-1,2,3,4-tetrahydroacridin-9-yl)amino)methyl)-2-(4-(5,6,7-trimethoxy-4-oxo-4H-chromen-2-yl)phenoxy)acetamide*) has shown the strongest AChE inhibitory activity (IC_{50} = 12.8 nM) and potent inhibition of self-induced $A\beta_{1-42}$ aggregation with inhibition ratio of 33.8% at 25 μM (Fig. 11.12). Additionally, compound **10** represented as an antioxidant, a neuroprotectant and have potency to traverse the blood–brain barrier (BBB) in vitro (Liao et al. 2015).

Shah and co-workers have prepared a number of flavone analogues and biologically screened all for anticholinesterase potential (as AChE and BuChE inhibitors) (Shoaib et al. 2015; Al-Joufi et al. 2022). They found that compound **11a** (*2-(2,5-dichlorophenyl)-4H-chromen-4-one*) was the most active AChE inhibitor

Fig. 11.13

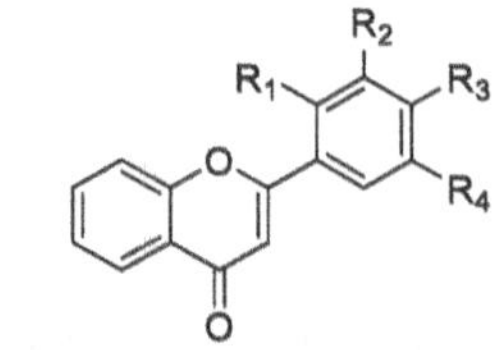

11a, R_1 = Cl, R_2 = H, R_3 = H, R_4 = Cl
11b, R_1 = H, R_2 = Cl, R_3 = Cl, R_4 = H
11c, R_1 = Cl, R_2 = H, R_3 = Cl, R_4 = H

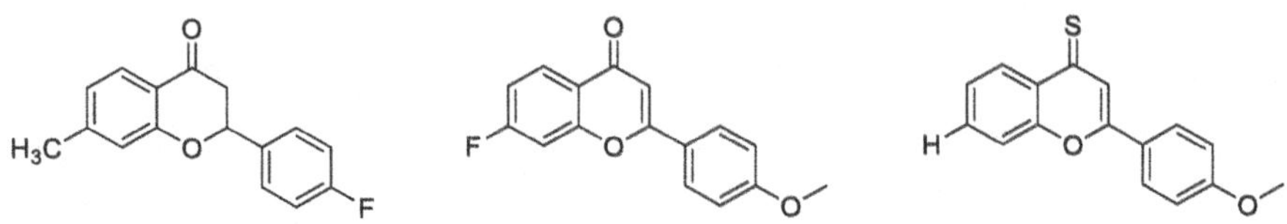

12a (Flavanone) **12b** (Flavone) **12c** (Thioflavone)

Fig. 11.14

(IC_{50} = 98.42 μM) followed by compound **11b** (*2-(3,4-dichlorophenyl)-4H-chromen-4-one*, IC_{50} = 112.33 μM), whereas compound **11c** (*2-(2,4-dichlorophenyl)-4H-chromen-4-one*) was the most promising BuChE inhibitor (IC_{50} = 105.20 μM) (Fig. 11.13). In scopolamine-induced amnesic models, compound **11a** has shown remarkable anti-amnesic effects and helped to upgrade the memory loss in behavioural model studies.

Chimenti et al. have found that flavanone analogues are more potent and selective hMAO-B inhibitors than their flavones, and thio-flavones analogues. Further, they have shown that insertion of double bond in the flavanone framework and replacement of oxygen atom with sulphur atom in the flavones result in the reduction of hMAO-B inhibitory activity such as for **12a** (*2-(4-fluorophenyl)-7-methylchroman-4-one*), IC_{50} = 0.13 μM, for **12b** (*7-fluoro-2-(4-methoxyphenyl)-4H-chromen-4-one*), IC_{50} = 1.34 μM, and for **12c** (*2-(4-methoxyphenyl)-4H-chromene-4-thione*), IC_{50} = 0.48 μM (Fig. 11.14) (Chimenti et al. 2010).

Peng and co-workers synthesized a new number of phosphorylated flavonoids and pharmacologically evaluated these compounds in vitro for pancreatic cholesterol esterase (CEase) and AChE inhibitory activity. They found that majority of the derivatives has shown activity against CEase in nanomolar range and micromolar potency against AChE. The utmost active and selective compound **13** (*5-hydroxy-4-oxo-2-phenyl-4H-chromen-7-yl dimethyl phosphate*) has showed inhibitory activity against CEase (IC50 value of 0.72 nM) and AChE (IC50 = 8.520 μM) (Fig. 11.15). Compound **13** exhibited 11,800-fold selectivity for CEase over AChE. The SAR studies revealed that insertion of hydroxyl (-OH) group at position 5 and phosphate ($-PO_3^{2-}$) group at position 7 of flavonoid moieties are favourable for CEase inhibition (Wei et al. 2014).

Fig. 11.15

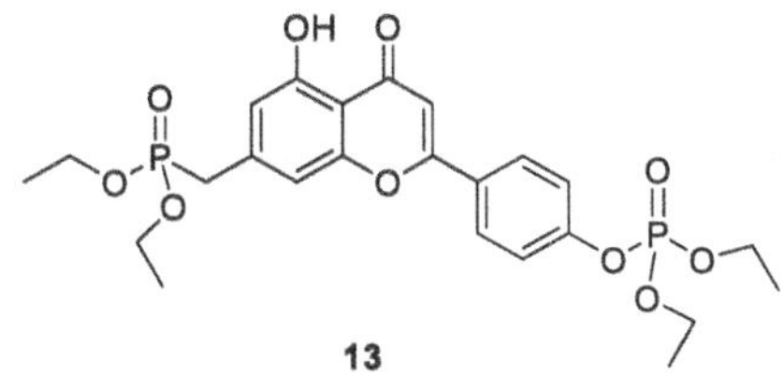

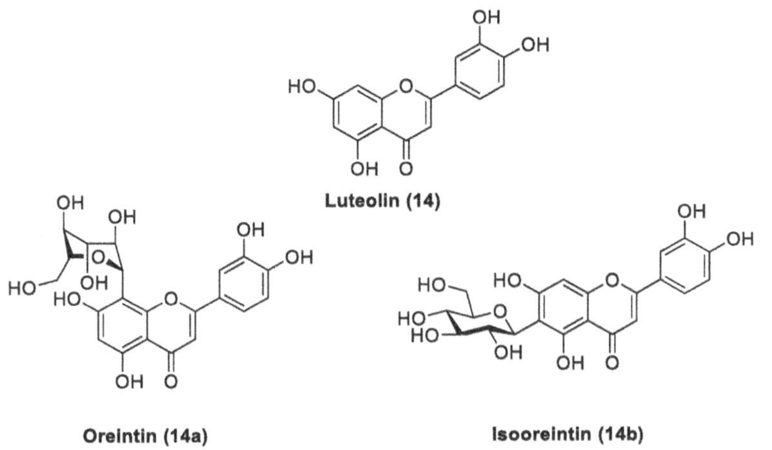

Fig. 11.16

Choi and co-workers have synthesized the C-glycosylated product of luteolin (**14**), namely orientin (**14a**) and isoorientin (**14b**) and evaluated their structure–activity relationships (SAR) (Fig. 11.16) (Choi et al. 2014). They explored the impacts of C-glycosylation of flavonoids, together with antioxidant, anti-inflammatory, and antidiabetic effects of compound **14** and its two C-glycoside analogues **14a** and **14b**, via in vitro assays of AChE, BChE, BACE1, total reactive oxygen species (ROS), 1,1-diphenyl-2-picrylhydraxyl (DPPH), peroxynitrite (ONOO-), nitric oxide (NO), aldose reductase, protein tyrosine phosphatase 1B (PTP1B), cellular assays of NO production, and inducible nitric oxide synthase (iNOS)/cyclooxygenase-2 expression in lipopolysaccharide (LPS)-stimulated RAW 264.7 cells and found that among all the three compounds **14**, **14a,** and **14b**, isoorientin (**14b**) exhibited the maximum free radical scavenging activity against DPPH, ONOO- and NO, whereas luteolin was found to be the most active inhibitor of ROS generation. Additionally, luteolin also exhibited maximum anti-AD activity as it inhibits AChE, BChE, and BACE1 (Choi et al. 2014).

11.4.2 Isoflavone Analogues

In isoflavonoids, the position-3 of the chromone core is occupied by the aromatic ring-B. Genistein (*4′,5,7-trihydroxyisoflavone*) (Fig. 11.17), the omnipresent isoflavone, may serve as a preparatory template for the designing of multifunctional drugs for AD cure. Although the alkyl side chain with amine group has shown significant AChE inhibition; therefore, linking of genistein with suitable secondary amines by varying lengths of carbon spacers results in AChE inhibitors with a dual binding site and also have metal chelating properties, anti-Aβ accumulation, and antioxidant activities (Fig. 11.17).

Deng and co-workers have prepared several genistein derivatives with carbon spacer-linked alkyl benzylamines and screened them as multifunctional agents for AD cure (Qiang et al. 2014). They found that majority of the derivatives showed good AChE inhibitory activity, with moderate to good antioxidative activity. Particularly, compounds **15a** (*7-((6-(Benzyl(ethyl)amino)hexyl)oxy)-5-hydroxy-3-(4-hydroxyphenyl)-4H-chromen-4-one*), **15b** (*3-(4-(4-(Ethyl(2-methoxybenzyl)amino)butoxy)phenyl)-5,7-dihydroxy-4H-chromen-4-one*), and **15c** (*7-(4-(Ethyl(2-methoxybenzyl)amino)butoxy)-3-(4-(4-(ethyl(2-methoxybenzyl)amino) butoxy) phenyl)-5-hydroxy-4H-chromen-4-one*), exhibited remarkable inhibition of Aβ aggregation (at 25 μM) with inhibition ratios of 34.3%, 24.6%, and 35%, respectively (Fig. 11.17), compared with curcumin (43.1% at 25 μM) and exhibited metal chelating properties.

Moreover, both the molecular modelling study and kinetic analysis of AChE inhibition suggested that **15c** binds instantaneously to CAS and PAS of AChE (Qiang et al. 2014).

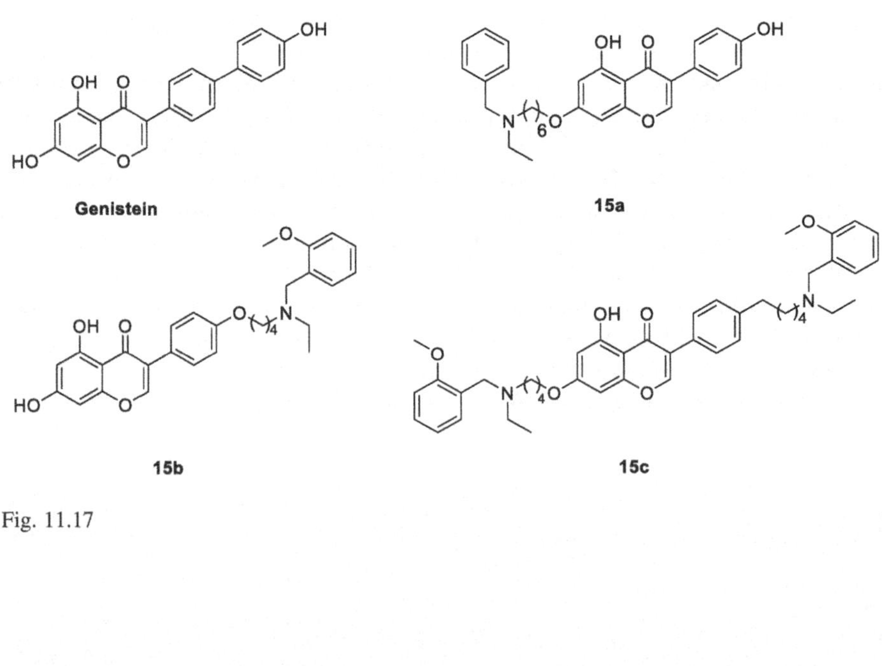

Fig. 11.17

Fig. 11.18

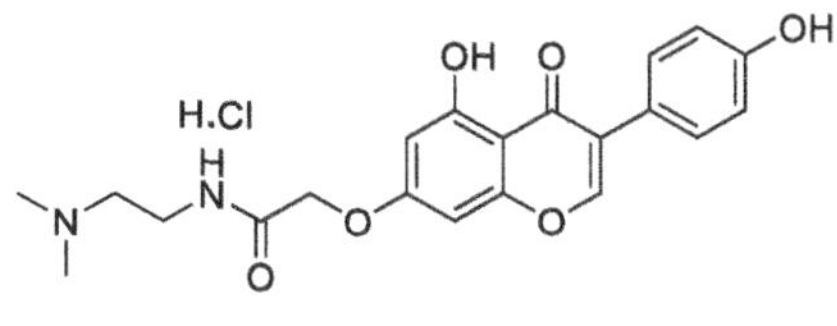

16

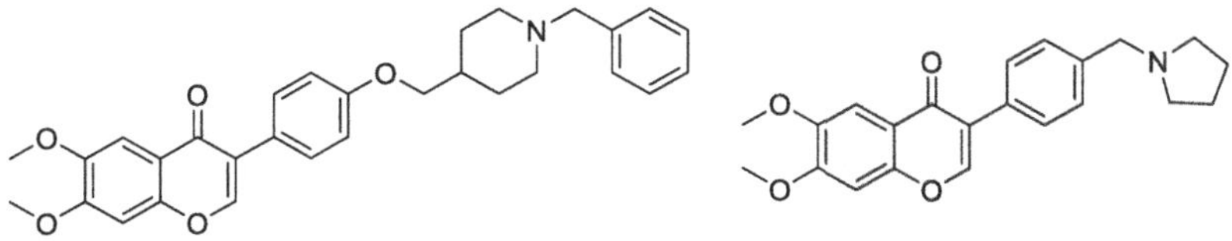

17a **17b**

Fig. 11.19

Wang et al. have planned and prepared genistein-polyamine hybrids and screened all the hybrids as multifunctional agents for the treatment of AD. Compound **16** (*N-(2-(dimethylamino)ethyl)-2-((5-hydroxy-3-(4-hydroxyphenyl)-4-oxo-4H-chromen-7-yl)oxy)acetamide*) displayed the strongest AChE inhibition with $IC_{50} = 2.75$ μM (Fig. 11.18), this result was better in comparison to rivastigmine (5.60 μM) and not affect the HepG-2 cell viability at 10 μM concentration (Zhang et al. 2015).

Hu and co-workers have prepared and evaluated the biological activity of novel flavonoid derivatives as acetylcholinesterase inhibitors (Shen et al. 2009; Sheng et al. 2009). Among them, compound **17a** (*3-(4-(1-Benzylpiperidin-4-yloxy)phenyl)-6,7-dimethoxy-4H-chromen-4-one*) displayed inhibitory activity ($IC_{50} = 0.093$ μM) in the same range as donepezil ($IC_{50} = 0.025$ μM) and compound **17b** (*6,7-Dimethoxy-3-[4-(pyrrolidin-1-ylmethyl)-phenyl]-4H-chromen-4-one*) with pyrrolidine group was highly active AChE inhibitor ($IC_{50} = 0.004$ μM) of the series which exhibited large AChE/BuChE inhibition ratio (4575-fold) (Fig. 11.19). Their study indicated that the isoflavone unit has a significant role in the interaction of these derivatives with AChE as it acts as an anchor in the PAS region of AChE.

Wu et al. have synthesized new genistein derivative by modification at seventh position as a multi-target-directed compound for AD treatment (Shi et al. 2012). They investigated the AChE inhibitory effect, estrogenic activity, and neuroprotective effect of compound **18** (*7-(4-(diethylamino)butoxy)-5-hydroxy-3-(4-hydroxyphenyl)-4H-chromen-4-one*) and found that compound **18** acted as selective AChE inhibitor in vitro with IC_{50} value of 0.17 μM (Fig. 11.20), showed neuroprotective activity against Aβ-induced death in SH-SY5Y cells and at the same time also displayed an estrogenic activity with no undesirable side effects, therefore may be used in place of oestrogen (Shi et al. 2012).

Fig. 11.20

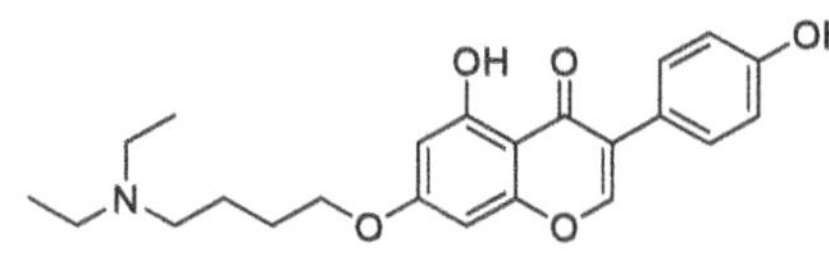

18

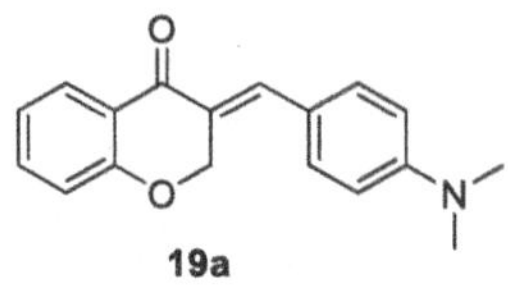

19a

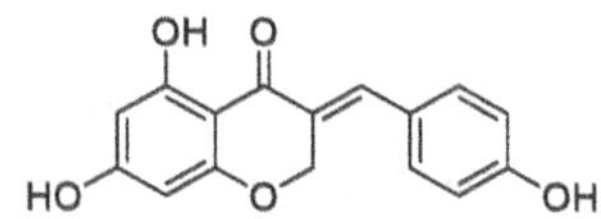

19b

Fig. 11.21

Fig. 11.22

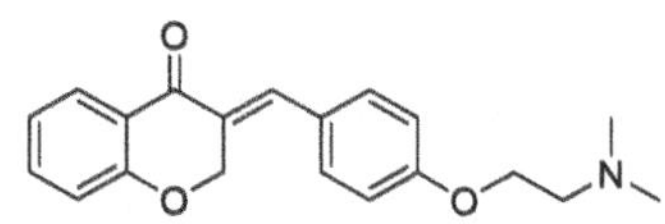

20

11.4.3 Homoisoflavonoids Analogues

Homoisoflavonoids have been reported as an antidiabetic, antioxidant, anti-inflammatory, antibacterial, antifungal (Castelli and López 2017), AChE and MAO-B inhibitors with high activities towards Aβ aggregates (Liu et al. 2017; Desideri et al. 2011; Gan et al. 2014).

Desideri and co-workers have synthesized various homoisoflavonoid derivatives and biologically screened them in vitro as human monoamine oxidase isoforms A and B (hMAO-A and hMAO-B inhibitors). They found that the majority of the compounds were displayed potent activities in the nano- or micromolar range in association with major selectivity against hMAO-B. Among entire series of the compounds, compounds **19a** (*(E)-3-(4-(Dimethylamino)benzylidene)chroman-4-one*) and **19b** (*(E)-5,7-dihydroxy-3-(4-hydroxybenzylidene)chroman-4-one*) have utmost potential, showing hMAO-B affinity with IC_{50} values 8.51 nM and 8.61 nM, respectively. These results were better in comparison to the results obtained for the selective inhibitor selegiline (Fig. 11.21) (Desideri et al. 2011).

Wang and co-workers have developed a new series of homoisoflavonoid analogues and evaluated all these analogues for inhibitory activities towards MAO and AChE. Among them, compound **20** (*(E)-3-(4-(2-(dimethylamino)ethoxy) benzylidene)chroman-4-one*) showed moderate AChE inhibition (eeAChE $IC_{50} = 0.89 \pm 0.02$ μM; hAChE $IC_{50} = 0.657 \pm 0.002$ μM) and significant MAO-B inhibition (hMAO-B IC50 = 0.0372 ± 0.0002 μM) which traverse through the blood–brain barrier in vitro (Fig. 11.22). Above all, when compound **20** was administrated orally, it had not shown any noticeable signs of acute toxicity and it

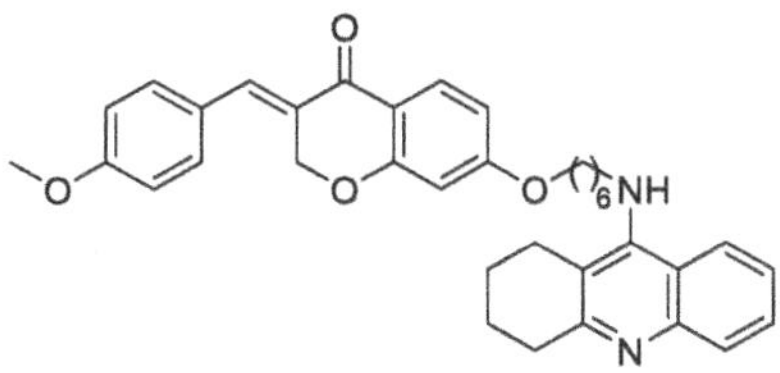

21a

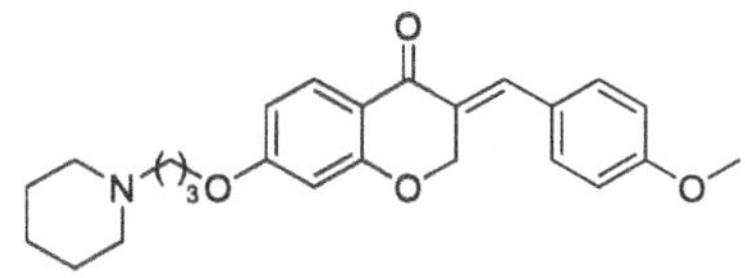

21b

Fig. 11.23

Fig. 11.24

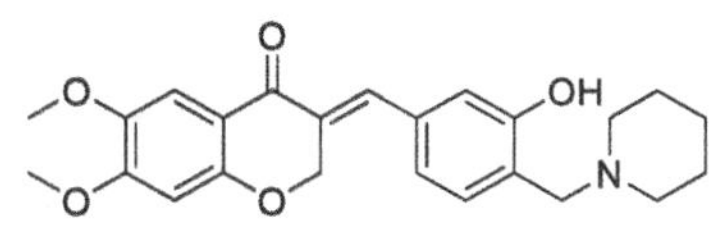

22

had tendency to reverse scopolamine-induced memory impairment in mice (Liu et al. 2017).

Li and co-workers have synthesized a novel tacrine–homoisoflavonoid hybrids and screened them as ChEs and hMAOs inhibitors and found that majority of the compounds have shown potency against both ChEs and MAO-B. Compound **21a** *((E)-3-(4-Methoxybenzylidene)-7-(6-(1,2,3,4-tetrahydroacridin-9-ylamino) hexyloxy)chroman-4-one)*, with six-carbon linkage between tacrine and (*E*)-7-hydroxy-3-(4-methoxybenzylidene)chroman-4-one, has shown highest potency against MAO-B and AChE with IC_{50} values of 0.401 μM and 67.9 nM, respectively (Fig. 11.19) and penetrate BBB in a parallel artificial membrane permeation assay (PAMPA) (Sun et al. 2013). Later on, a required balance between AChE and hMAO-B inhibition activities has been achieved by the synthesis of compound **21b** *((E)-3-(4-methoxybenzylidene)-7-(3-(piperidin-1-yl)propoxy)chroman-4-one)* with IC_{50} = 3.94 μM and 3.44 μM, respectively. The SAR analysis of these compounds revealed that the inhibitory activity against MAO-B increases by increasing carbon spacer length (Fig. 11.23) (Wang et al. 2016b).

Deng and co-workers have developed a new series of homoisoflavonoid Mannich base hybrids and evaluated them as AChE and MAO-B. Among all, compound **22** *((E)-3-(3-hydroxy-4-(piperidin-1-ylmethyl)benzylidene)-6,7-dimethoxychroman-4-one)* displayed an excellent AChE and MAO-B inhibitory activities with IC_{50} values of 2.49 nM and 1.74 μM, respectively, bio-metal chelating ability, antioxidant activity, good self- and Cu^{2+}-induced $A\beta_{1-42}$ aggregation inhibitory potency, and superior BBB permeability (Fig. 11.24). All these multifunctional properties make compound **22** an outstanding template to develop new and effective drugs for AD cure (Li et al. 2017).

11.5 Flavonoid Therapy, Challenges, and Future Studies

Currently, only four drugs for AD patients are approved by FDA which are mostly cholinesterase inhibitors (galantamine, rivastigmine, and donepezil) and *N*-methyl-D-aspartate (NMDA) antagonist (memantine). Among them, cholinesterase inhibitors have been extensively used for mild to moderate symptoms of AD and memantine is mainly used for moderate to severe symptoms of AD. Though FDA has approved the above said drugs, however all of these are used to treat the symptoms only that are too up to certain duration (Schneider et al. 2014; Mangialasche et al. 2010). These medications only improve the quality of life of patients, but they are ineffective to prevent disease progression and could not alter the disease status (Chen and Pan 2015). In this continuation, all the treatments targeting the Aβ accumulation get failed to produce any significant improvement in memory performance of AD patients, even though they are reducing amyloid peptides deposits (Karran and Strooper 2016). Failure of some other clinical trials also led to the scientific community to develop some new hypotheses for AD pathogenesis (Hardy and Strooper 2017; Pugazhenthi 2017; Strooper and Karran 2016; McGeer and McGeer 2013; Sarazin et al. 2013) which may help to explore novel targets that will effectively prevent the progression of AD and alter the underlying causes of memory loss in AD patients.

In recent years, flavonoids have received special attention to be effective against many neurodegenerative pathologies, including halt of symptoms associated with AD (Loera-Valencia et al. 2018) and Parkinson's Disease (PD) (Kemppainen et al. 2018). Flavonoids (Table 11.1) are also well-known for their antioxidant activity, which can be responsible for neuroprotective effect against oxidative stress. However, till date several pharmacological and toxicological aspects of flavonoids have not been well explored that restrict its application as the solution for AD therapy.

Additionally, despite their great abundance in many common food products such as fruits, herbal extracts, and vegetables, they have limited bioavailability and their extract undergoes rapid metabolization that restrict their beneficial effects. This challenge of flavonoid therapy can be overcome by the formation of such derivatives which are more stable and more active than flavonoids. Over the few years, numerous strategies have been applied in the development of multi-target flavonoid derivatives to combat AD, which may serve as a fruitful approach for further explorations and development. Various studies have shown that multi-target flavonoid derivatives could interact with PAS and CAS region of ChEs and act as dual binding inhibitors. Moreover, selection of substitution pattern and type of cross-linker in flavonoid scaffolds may introduce additional pharmacological properties including BACE-1, MAO, BuChE, Aβ aggregation inhibition, metal chelation, radical scavenging, where all play a key role for the AD treatment.

Furthermore, technology improvements from imaging to gene editing help to provide precise evidence behind alternative mechanisms and have opened new lines of research which would be helpful to understand and explain the genesis and progression of AD (Worker et al. 2018; Fraser et al. 2015; Vos et al. 2016; Baptista et al. 2014; Ay et al. 2017).

11.6 Conclusion

Till date, the exact cause of AD is not properly understood, therefore, drug development for AD treatment faces one of the highest failure rates than any other therapeutic area. This may be due to development of drug candidates for the wrong target, or there may be the multiple targets involved in the disease. Therefore, an increased focus on the development of different strategies to treat AD patients is urgently required. A deeper understanding of the underneath biology of innate immune response cells might play a pivotal role towards the development of new drugs. Moreover, the multifunctional flavonoid derivatives might provide a useful prototype for the development of new anti-AD agents with multiple potencies and use to combat pathological neurodegeneration and age-related cognitive decline. The present chapter will provide assistance to the scientific community to address the current difficulties they are facing in AD cure. Our concise information will help to recognize and investigate newly developed flavonoid-based hybrids which might show potential to prevent AD progression and also delivers greater understanding of various targets involved into the pathophysiological mechanism. The present chapter highlights the recent advancements and possible derivatization approaches on every side of the flavonoid molecule that has been exploited to develop anti-AD compounds. Finally, we have tried to include all newly reported flavonoid derivatives that can be serve as either potential candidate or template for the target-specific advancement and for the development of multi-targeting agents for AD management.

References

Afzal M, Redha A, AlHasan R (2019) Anthocyanins potentially contribute to defense against Alzheimer's disease. Molecules 24(23):4255

Agis-Torres A, Sölhuber M, Fernandez M, Sanchez-Montero JM (2014) Multi-target-directed ligands and other therapeutic strategies in the search of a real solution for Alzheimer's disease. Curr Neuropharmacol 12(1):2–36

Agostinho P, Cunha RA, Oliveira C (2010a) Neuro inflammation, oxidative stress and the pathogenesis of Alzheimer's disease. Curr Pharm Des 16(25):2766–2778

Agostinho P, Cunha RA, Oliveira C (2010b) Neuroinflammation, oxidative stress and the pathogenesis of Alzheimer's disease. Curr Pharm Des 16(25):2766–2778

Akiyama H, Barger S, Barnum S, Bradt B, Bauer J, Cole GM, Cooper NR, Eikelenboom P, Emmerling M, Fiebich BL, Finch CE, Frautschy S, Griffin WS, Hampel H, Hull M, Landreth G, Lue L, Mrak R, Mackenzie IR, McGeer PL, O'Banion MK, Pachter J, Pasinetti G, Plata-Salaman C, Rogers J, Rydel R, Shen Y, Streit W, Strohmeyer R, Tooyoma I, Van Muiswinkel FL, Veerhuis R, Walker D, Webster S, Wegrzyniak B, Wenk G, Wyss-Coray T (2000) Inflammation and Alzheimer's disease. Neurobiol Aging 21(3):383–421

Alawdi SH, El-Denshary ES, Safar MM, Eidi H, David M-O, Abdel-Wahhab MA (2017) Neuroprotective effect of nanodiamond in Alzheimer's disease rat model: a pivotal role for modulating NF-κB and STAT3 signaling. Mol Neurobiol 54(3):1906–1918

Al-Joufi FA, Shah SWA, Shoaib M, Ghias M, Shafiullah, Khalil A-AK, Jamal SB, Shah SMH, Zahoor M (2022) Flavonoid derivatives as potential cholinesterase inhibitors in

Scopolamine-induced amnesic mice: an in vitro, in vivo and integrated computational approach. Brain Sci 12(6):731–748
Amit T, Avramovich-Tirosh Y, Youdim MB, Mandel S (2008) Targeting multiple Alzheimer's disease etiologies with multimodal neuroprotective and neurorestorative iron chelators. FASEB J 22(5):1296–1305
Aplin E, Gibb GM, Jacobsen JS, Gallo J-M, Anderton BH (1996) In vitro phosphorylation of the cytoplasmic domain of the amyloid precursor protein by glycogen synthase kinase-3beta. J Neurochem 67(2):699–707
Ay M, Luo J, Langley M, Jin H, Anantharam V, Kanthasamy A, Kanthasamy AG (2017) Molecular mechanisms underlying protective effects of quercetin against mitochondrial dysfunction and progressive dopaminergic neurodegeneration in cell culture and MitoPark transgenic mouse models of Parkinson's disease. J Neurochem 141(5):766–782
Bakhtiari M, Panahi Y, Ameli J, Darvishi B (2017) Protective effects of flavonoids against Alzheimer's disease-related neural dysfunctions. Biomed Pharmacother 93:218–229
Bali P, Pranpat M, Bradner J, Balasis M, Fiskus W, Guo F, Rocha K, Kumaraswamy S, Boyapalle S, Atadja P, Seto E, Bhalla K (2005) Inhibition of histone deacetylase 6 acetylates and disrupts the chaperone function of heat shock protein 90: a novel basis for antileukemia activity of histone deacetylase inhibitors. J Biol Chem 280(29):26729–26734
Ballard C, Brayne C, Brodaty H, Cedazo-Minguez A, Dubois B, Edvardsson D, Feldman H, Fratiglioni L, Frisoni GB, Gauthier S, Georges J, Graff C, Iqbal K, Jessen F, Johansson G, Jönsson L, Kivipelto M, Knapp M, Mangialasche F, Melis R, Nordberg A, Rikkert MO, Qiu C, Sakmar TP, Scheltens P, Schneider LS, Sperling R, Tjernberg LO, Waldemar G, Wimo A, Zetterberg H (2016) Defeating Alzheimer's disease and other dementias: a priority for European Science and Society. Lancet Neurol 15(5):455–532
Balzer J, Rassaf T, Heiss C, Kleinbongard P, Lauer T, Merx M, Heussen N, Gross HB, Keen CL, Schroeter H, Kelm M (2008) Sustained benefits in vascular function through flavanol containing cocoa in medicated diabetic patients a double-masked, randomized, controlled trial. J Am Coll Cardiol 51(22):2141–2149
Baptista FI, Henriques AG, Silva AMS, Wiltfang J, da Cruz e Silva OAB (2014) Flavonoids as therapeutic compounds targeting key proteins involved in Alzheimer's disease. ACS Chem Neurosci 5(2):83–92
Barberger-Gateau P, Raffaitin C, Letenneur L, Berr C, Tzourio C, Dartigues JF, Alpérovitch A (2007) Dietary patterns and risk of dementia: the Three-City cohort study. Neurology 69(20): 1921–1930
Barnes GN, Slevin JT (2003) Ionotropic glutamate receptor biology: effect on synaptic connectivity and function in neurological disease. Curr Med Chem 10(20):2059–2072
Bartus RT, Dean RL III, Beer B, Lippa AS (1982) The cholinergic hypothesis of geriatric memory dysfunction. Science 217(4558):408–414
Beyreuther K, Bush AI, Dyrks T, Hilbich C, König G, Mönning U, Multhaup G, Prior R, Rumble B, Schubert W, Small DH, Weidemann A, Masters CL (1991) Mechanisms of amyloid deposition in Alzheimer's disease. Ann N Y Acad Sci 640:129
Bhattacharyya R, Kovacs DM (2010) ACAT inhibition and amyloid beta reduction. Biochim Biophys Acta 1801(8):960–965
Bolger TA, Yao T-P (2005) Intracellular trafficking of histone deacetylase 4 regulates neuronal cell death. J Neurosci 25(41):9544–9553
Bolognesi ML, Cavalli A, Valgimigli L, Bartolini M, Rosini M, Andrisano V, Recanatini M, Melchiorre C (2007) Multi-target-directed drug design strategy: from a dual binding site acetylcholinesterase inhibitor to a trifunctional compound against Alzheimer's disease. J Med Chem 50(26):6446–6449
Breteler MM (2000) Vascular risk factors for Alzheimer's disease: an epidemiologic perspective. Neurobiol Aging 2(2):153–160

Brickman AM, Khan UA, Provenzano FA, Yeung L-K, Suzuki W, Schroeter H, Wall M, Sloan RP, Small SA (2014) Enhancing dentate gyrus function with dietary flavanols improves cognition in older adults. Nat Neurosci 17(12):1798–1803

Brookmeyer R, Evans D, Hebert AL, Langa KM, Heeringa SG, Plassman BL, Kukull WA (2011) National estimates of the prevalence of Alzheimer's disease in the United States. Alzheimers Dement 7(1):61–73

Brown GC (1997) Nitric oxide inhibition of cytochrome oxidase and mitochondrial respiration: implications for inflammatory, neurodegenerative and ischaemic pathologies. Mol Cell Biochem 174(1–2):189–192

Buée L, Bussière T, Buée-Scherrer V, Delacourte A, Hof PR (2000) Tau protein isoforms, phosphorylation and role in neurodegenerative disorders. Brain Res Brain Res Rev 33(1): 95–130

Bush AI (2013) The metal theory of Alzheimer's disease. J Alzheimers Dis 33(Suppl 1):S277–S281

Butterfield DA, Boyd-Kimball D (2004) Amyloid β-peptide (1–42) contributes to the oxidative stress and neurodegeneration found in Alzheimer disease brain. Brain Pathol 14(4):426–432

Calder PC, Albers R, Antoine J-M, Blum S, Bourdet-Sicard R, Ferns GA, Folkerts G, Friedmann PS, Frost GS, Guarner F, Løvik M, Macfarlane S, Meyer PD, M'Rabet L, Serafini M, van Eden W, van Loo J, Vas Dias W, Vidry S, Winklhofer-Roob BM, Zhao J (2009) Inflammatory disease processes and interactions with nutrition. Br J Nutr 101(Suppl 1):S1–S45

Carey N, la Thangue NB (2006) Histone deacetylase inhibitors: gathering pace. Curr Opin Pharmacol 6(4):369–375

Carreiras MC, Marco JL (2004) Recent approaches to novel anti-Alzheimer therapy. Curr Pharm Des 10(25):3167

Castelli MV, López SN (2017) Homoisoflavonoids: occurrence, biosynthesis, and biological activity. Stud Nat Prod Chem 54:315–354

Cavalli A, Bolognesi ML, Minarini A, Rosini M, Tumiatti V, Recanatini M, Melchiorre C (2008) Multi-target-directed ligands to combat neurodegenerative diseases. J Med Chem 51(3): 347–372

Chen X, Pan W (2015) The treatment strategies for neurodegenerative diseases by integrative medicine. Integr Med Int 1:223–225

Chen S, Owens GC, Makarenkova H, Edelman DB (2010) HDAC6 regulates mitochondrial transport in hippocampal neurons. PLoS One 5(5):e10848

Chimenti F, Fioravanti R, Bolasco A, Chimenti P, Secci D, Rossi F, Yáñez M, Orallo F, Ortuso F, Alcaro S, Cirilli R, Ferretti R, Sanna ML (2010) A new series of flavones, thioflavones, and flavanones as selective monoamine oxidase-B inhibitors. Bioorg Med Chem 18(3):1273–1279

Choi Y-T, Jung C-H, Lee S-R, Bae J-H, Baek W-K, Suh M-H, Park J, Park C-W, Suh S-I (2001) The green tea polyphenol (−)-epigallocatechin gallate attenuates beta-amyloid-induced neurotoxicity in cultured hippocampal neurons. Life Sci 70(5):603–614

Choi JS, Islam MN, Ali MY, Kim YM, Park HJ, Sohn HS, Jung HA (2014) The effects of C-glycosylation of luteolin on its antioxidant, anti-Alzheimer's disease, anti-diabetic, and anti-inflammatory activities. Arch Pharm Res 37(10):1354–1363

Chuang DM, Leng Y, Marinova Z, Kim HJ, Chiu CT (2009) Multiple roles of HDAC inhibition in neurodegenerative conditions. Trends Neurosci 32(11):591–601

Chun OK, Chung S-J, Claycombe KJ, Song WO (2008) Serum C-reactive protein concentrations are inversely associated with dietary flavonoid intake in U.S. adults. J Nutr 138(4):753–760

Citron M (2010) Alzheimer's disease: strategies for disease modification. Nat Rev Drug Discov 9(5):387–398

Čolović MB, Krstić DZ, Lazarević-Pašti TD, Bondžić AM, Vasić VM (2013) Acetylcholinesterase inhibitors: pharmacology and toxicology. Curr Neuropharmacol 11(3):315–335

Cruz I, Puthongking P, Cravo S, Palmeira A, Cidade H, Pinto M, Sousa E (2017) Xanthone and flavone derivatives as dual agents with acetylcholinesterase inhibition and antioxidant activity as potential anti-Alzheimer agents. J Chem 2017:8587260

Darbandi N, Ramezani M, Khodagholi F, Noori M (2016) Kaempferol promotes memory retention and density of hippocampal CA1 neurons in intra-cerebroventricular STZ-induced experimental AD model in Wistar rats. Biologija 62(3):157–168

Darvesh S, Hopkins DA, Geula C (2003) Neurobiology of butyrylcholinesterase. Nat Rev Neurosci 4(2):131–138

Davies P, Maloney AJ (1976) Selective loss of central cholinergic neurons in Alzheimer's disease. Lancet 2(8000):1403

de Ruijter AJM, van Gennip AH, Caron HN, Kemp S, van Kuilenburg ABP (2003) Histone deacetylases (HDACs): characterization of the classical HDAC family. Biochem J 370(Pt 3): 737–749

De Strooper B (2007) Loss-of-function presenilin mutations in Alzheimer disease: talking Point on the role of presenilin mutations in Alzheimer disease. EMBO Rep 8(2):141–146

De Strooper B, Vassar R, Golde T (2010) The secretases: enzymes with therapeutic potential in Alzheimer disease. Nat Rev Neurol 6(2):99–107

Desideri N, Bolasco A, Fioravanti R, Monaco LP, Orallo F, Yáñez M, Ortuso F, Alcaro S (2011) Homoisoflavonoids: natural scaffolds with potent and selective monoamine Oxidase-B inhibition properties. J Med Chem 54(7):2155–2164

di Gesso JL, Kerr JS, Zhang Q, Raheem S, Yalamanchili SK, O'Hagan D, Kay CD, O'Connell MA (2015) Flavonoid metabolites reduce tumor necrosis factor-α secretion to a greater extent than their precursor compounds in human THP-1 monocytes. Mol Nutr Food Res 59(6):1143–1154

Di Paolo G, Kim T-W (2011) Linking lipids to Alzheimer's disease: cholesterol and beyond. Nat Rev Neurosci 12(5):284–296

Dominy SS, Lynch C, Ermini F, Benedyk M, Marczyk A, Konradi A, Nguyen M, Haditsch U, Raha D, Griffin C, Holsinger LJ, Arastu-Kapur S, Kaba S, Lee A, Ryder MI, Potempa B, Mydel P, Hellvard A, Adamowicz K, Hasturk H, Walker GD, Reynolds EC, Faull RLM, Curtis MA, Dragunow M, Potempa J (2019) Porphyromonas gingivalis in Alzheimer's disease brains: evidence for disease causation and treatment with small-molecule inhibitors. Sci Adv 5(1): eaau3333

Dorszewska J, Prendecki M, Oczkowska A, Dezor M, Kozubski W (2016) Molecular basis of familial and sporadic Alzheimer's disease. Curr Alzheimer Res 13(9):952–963

Drechsel DN, Hyman AA, Cobb MH, Kirschner MW (1992) Modulation of the dynamic instability of tubulin assembly by the microtubule-associated protein tau. Mol Biol Cell 3(10):1141–1154

Du H, ShiDu Yan S (2010) Unlocking the door to neuronal woes in Alzheimer's disease: Aβ and mitochondrial permeability transition pore. Pharmaceuticals (Basel) 3(6):1936–1948

Duan X, Li Y, Xu F, Ding H (2021) Study on the neuroprotective effects of Genistein on Alzheimer's disease. Brain Behav 11(5):e02100

Duda-Chodak A (2012) The inhibitory effect of polyphenols on human gut microbiota. J Physiol Pharmacol 63(5):497–503

Duong T, Nikolaeva M, Acton PJ (1997) C-reactive protein-like immunoreactivity in the neurofibrillary tangles of Alzheimer's disease. Brain Res 749(1):152–156

Ehrnhoefer DE, Bieschke J, Boeddrich A, Herbst M, Masino L, Lurz R, Engemann S, Pastore A, Wanker EE (2008) EGCG redirects amyloidogenic polypeptides into unstructured, off-pathway oligomers. Nat Struct Mol Biol 15(6):558–566

Federico A, Cardaioli E, Da Pozzo P, Formichi P, Gallus GN, Radi E (2012) Mitochondria, oxidative stress and neurodegeneration. J Neurol Sci 322(1–2):254–262

Feng X, Weng D, Zhou F, Owen YD, Qin H, Zhao J, WenYu, Huang Y, Chen J, Fu H, Yang N, Chen D, Li J, Tan R, Shen P (2016) Activation of PPARg by a natural flavonoid modulator, apigenin ameliorates obesity-related inflammation via regulation of macrophage polarization. EBioMedicine 9:61–76

Ferrer I, Gomez-Isla T, Puig B, Freixes M, Ribé E, Dalfó E, Avila J (2005) Current advances on different kinases involved in tau phosphorylation, and implications in Alzheimer's disease and tauopathies. Curr Alzheimer Res 2(1):3–18

Foster DJ, Choi DL, Conn PJ, Rook JM (2014) Activation of M1 and M4 muscarinic receptors as potential treatments for Alzheimer's disease and schizophrenia. Neuropsychiatr Dis Treat 10: 183

Francis ST, Head K, Morris PG, Macdonald IA (2006) The effect of flavanol-rich cocoa on the fMRI response to a cognitive task in healthy young people. J Cardiovasc Pharmacol 47(Suppl. 2):S215–S220

Fraser MA, Shaw ME, Cherbuin N (2015) A systematic review and meta-analysis of longitudinal hippocampal atrophy in healthy human ageing. Neuroimage 112:364–374

Gage FH (2000) Mammalian neural stem cells. Science 287(5457):1433–1438

Gan C, Zhao Z, Nan D-D, Yin B, Hu J (2014) Homoisoflavonoids as potential imaging agents for β-amyloid plaques in Alzheimer's disease. Eur J Med Chem 76:125–131

Gandhi S, Abramov AY (2012) Mechanism of oxidative stress in neurodegeneration. Oxidative Med Cell Longev 2012:428010

Gao L, Cueto MA, Asselbergs F, Atadja P (2002) Cloning and functional characterization of HDAC11, a novel member of the human histone deacetylase family. J Biol Chem 277(28): 25748–25755

Ghosh S, Wu MD, Shaftel SS, Kyrkanides S, LaFerla FM, Olschowka JA, O'Banion MK (2013) Sustained interleukin-1β overexpression exacerbates tau pathology despite reduced amyloid burden in an Alzheimer's mouse model. J Neurosci 33(11):5053–5064

Grassi D, Desideri G, Necozione S, Lippi C, Casale R, Properzi G, Blumberg JB, Ferri C (2008) Blood pressure is reduced and insulin sensitivity increased in glucose-intolerant, hypertensive subjects after 15 days of consuming high-polyphenol dark chocolate. J Nutr 138(9):1671–1676

Green KN, LaFerla FM (2008) Linking calcium to Aβ and Alzheimer's disease. Neuron 59(2): 190–194

Grzonka Z, Jankowska E, Kasprzykowski F, Kasprzykowska R, Lankiewicz L, Wiczk W, Wieczerzak E, Ciarkowski J, Drabik P, Janowski R, Kozak M, Jaskólski M, Grubb A (2001) Structural studies of cysteine protease and their inhibitors. Acta Biochim Pol 48(1):1–20

Gu Y, Nieves JW, Stern Y, Luchsinger JA, Scarmeas N (2010) Food combination and Alzheimer disease risk: a protective diet. Arch Neurol 67(6):699–706

Gui MJ, Dashper SG, Slakeski N, Chen Y-Y, Reynolds EC (2016) Spheres of influence: porphyromonas gingivalis outer membrane vesicles. Mol Oral Microbiol 31(5):365–378

Haass C, Selkoe DJ (2007) Soluble protein oligomers in neurodegeneration: lessons from the Alzheimer's amyloid β-peptide. Nat Rev Mol Cell Biol 8(2):101–112

Hajialyani M, Farzaei MH, Echeverría J, Nabavi SM, Uriarte E, Sobarzo-Sánchez E (2019) Hesperidin as a neuroprotective agent: a review of animal and clinical evidence. Molecules 24(3):648

Halliwell B (2006) Oxidative stress and neurodegeneration: where are we now? J Neurochem 97(6): 1634–1658

Hardy J, Selkoe DJ (2002) The amyloid hypothesis of Alzheimer's disease: progress and problems on the road to therapeutics. Science 297(5580):353–356

Hardy J, Strooper BD (2017) Alzheimer's disease: where next for anti-amyloid therapies? Brain 140(4):853–855

Hasselmo ME (2006) The role of acetylcholine in learning and memory. Curr Opin Neurobiol 16(6):710–715

Heim KE, Tagliaferro AR, Bobilya DJ (2002) Flavonoid antioxidants: chemistry, metabolism and structure-activity relationships. J Nutr Biochem 13(10):572–584

Heo HJ, Suh Y-M, Kim M-J, Choi S-J, Mun NS, Kim H-K, Kim E, Kim C-J, Cho H-Y, Kim YJ, Shin D-H (2006) Daidzein activates choline acetyltransferase from MC-IXC cells and improves drug-induced amnesia. Biosci Biotechnol Biochem 70(1):107–111

Hernández F, Avila J (2007) Tauopathies. Cell Mol Life Sci 64(17):2219–2233

Hook V, Yoon M, Mosier C, Ito G, Podvin S, Head BP, Rissman R, O'Donoghue AJ, Hook G (2020) Cathepsin B in neurodegeneration of Alzheimer's disease, traumatic brain injury, and related brain disorders. Biochim Biophys Acta Proteins Proteom 1868(8):140428

Howell AB, Reed JD, Krueger CG, Winterbottom R, Cunningham DG, Leahy M (2005) A-type cranberry proanthocyanidins and uropathogenic bacterial anti-adhesion activity. Phytochemistry 66(18):2281–2291
Hubbert C, Guardiola A, Shao R, Kawaguchi Y, Ito A, Nixon A, Yoshida M, Wang X-F, Yao T-P (2002) HDAC6 is a microtubule-associated deacetylase. Nature 417(6887):455–458
Impey S, Smith DM, Obrietan K, Donahue R, Wade C, Storm DR (1998) Stimulation of cAMP response element (CRE)-mediated transcription during contextual learning. Nat Neurosci 1(7): 595–601
Iqbal K, Adel AC, Chen S, Chohan MO, El-Akkad E, Gong CX, Khatoon S, Li B, Liu F, Rahman A, Tanimuk H, Grundke-Iqbal I (2005) Tau pathology in Alzheimer disease and other tauopathies. Biochim Biophys Acta 1739(2–3):198–210
Jang S, Kelley KW, Johnson RW (2008) Luteolin reduces IL-6 production in microglia by inhibiting JNK phosphorylation and activation of AP-1. Proc Natl Acad Sci U S A 105(21): 7534–7539
Jayne T, Newman M, Verdile G, Sutherland G, Münch G, Musgrave I, Nik SHM, Lardelli M (2016) Evidence for and against a pathogenic role of reduced γ-secretase activity in familial Alzheimer's disease. J Alzheimers Dis 52(3):781–799
Jenner P (2003) Oxidative stress in Parkinson's disease. Ann Neurol 53(Suppl 3):S26–S36
Johnson GVW, Stoothoff WH (2004) Tau phosphorylation in neuronal cell function and dysfunction. J Cell Sci 117(Pt 24):5721–5729
Kalaria RN (2010) Vascular basis for brain degeneration: faltering controls and risk factors for dementia. Nutr Rev 68(Suppl 2):S74–S87
Kamer AR, Pirraglia E, Tsui W, Rusinek H, Vallabhajosula S, Mosconi L, Yi L, McHugh P, Craig RG, Svetcov S, Linker R, Shi C, Glodzik L, Williams S, Corby P, Saxena D, de Leon MJ (2015) Periodontal disease associates with higher brain amyloid load in normal elderly. Neurobiol Aging 36(2):627–633
Karlsen A, Retterstøl L, Laake P, Paur I, Bøhn SK, Sandvik L, Blomhoff R (2007) Anthocyanins inhibit nuclear factor-kappaB activation in monocytes and reduce plasma concentrations of proinflammatory mediators in healthy adults. J Nutr 137(8):1951–1954
Karran E, Strooper BD (2016) The amyloid cascade hypothesis: are we poised for success or failure? J Neurochem 139(Suppl 2):237–252
Karran E, Mercken M, De Strooper B (2011) The amyloid cascade hypothesis for Alzheimer's disease: an appraisal for the development of therapeutics. Nat Rev 10(9):698–712
Kaur G, Levy E (2012) Cystatin C in Alzheimer's disease. Front Mol Neurosci 6(5):79
Kemppainen N, Johansson J, Teuho J, Parkkola R, Joutsa J, Ngandu T, Solomon A, Stephen R, Liu Y, Hänninen T, Paajanen T, Laatikainen T, Soininen H, Jula A, Rokka J, Rissanen E, Vahlberg T, Peltoniemi J, Kivipelto M, Rinne JO (2018) Brain amyloid load and its associations with cognition and vascular risk factors in FINGER Study. Neurology 90(3):e206–e213
Kennedy BP, Ziegler MG, Alford M, Hansen LA, Thal LJ, Masliah E (2003) Early and persistent alterations in prefrontal cortex MAO A and B in Alzheimer's disease. J Neural Transm 110(7): 789–801
Kesse-Guyot E, Fezeu L, Andreeva VA, Touvier M, Scalbert A, Hercberg S, Galan P (2012) Total and specific polyphenol intakes in midlife are associated with cognitive function measured 13 years later. J Nutr 142(1):76–83
Kim J, Basak JM, Holtzman DM (2009) The role of apolipoprotein E in Alzheimer's disease. Neuron 63(3):287–303
Kleinschnitz C, Grund H, Wingler K, Armitage ME, Jones E, Mittal M, Barit D, Schwarz T, Geis C, Kraft P, Barthel K, Schuhmann MK, Herrmann AM, Meuth SG, Stoll G, Meurer S, Schrewe A, Becker L, Gailus-Durner V, Fuchs H, Klopstock T, de Angelis MH, Jandeleit-Dahm K, Shah AM, Weissmann N, Schmidt HHHW (2010) Post-stroke inhibition of induced NADPH oxidase type 4 prevents oxidative stress and neurodegeneration. PLoS Biol 8(9):e1000479
Kosik KS (1993) The molecular and cellular biology of tau. Brain Pathol 3(1):39–43

Kouhestani S, Jafari A, Babaei P (2018) Kaempferol attenuates cognitive deficit via regulating oxidative stress and neuroinflammation in an ovariectomized rat model of sporadic dementia. Neural Regen Res 13(10):1827–1832

Kozlov S, Afonin A, Evsyukov I, Bondarenko A (2017) Alzheimer's disease: as it was in the beginning. Rev Neurosci 28(8):825–843

Kravitz BA, Corrada MM, Kawas CH (2009) Elevated C-reactive protein levels are associated with prevalent dementia in the oldest-old. Alzheimers Dement 5(4):318–323

Lahiri DK, Farlow MR, Sambamurti K, Greig NH, Giacobini E, Schneider LS (2003) A critical analysis of new molecular targets and strategies for drug developments in Alzheimer's disease. Curr Drug Targets 4(2):97–112

Lamport DJ, Pal D, Moutsiana C, Field DT, Williams CM, Spencer JPE, Butler LT (2015) The effect of flavanol-rich cocoa on cerebral perfusion in healthy older adults during conscious resting state: a placebo controlled, crossover, acute trial. Psychopharmacology 232(17): 3227–3234

Lamport DJ, Pal D, Macready AL, Barbosa-Boucas S, Fletcher JM, Williams CM, Spencer JPE, Butler LT (2016) The effects of flavanone-rich citrus juice on cognitive function and cerebral blood flow: an acute, randomised, placebo-controlled cross-over trial in healthy, young adults. Br J Nutr 116(12):2160–2168

Lee G, Neve RL, Kosik KS (1989) The microtubule binding domain of tau protein. Neuron 2(6): 1615–1624

Lee S-R, Suh S-I, Kim S-P (2000) Protective effects of the green tea polyphenol (−)-epigallocatechin gallate against hippocampal neuronal damage after transient global ischemia in gerbils. Neurosci Lett 287(3):191–194

Lee JW, Lee YK, Ban JO, Ha TY, Yun YP, Han SB, Oh KW, Hong JT (2009) Green tea (−)-epigallocatechin-3-gallate inhibits beta-amyloid-induced cognitive dysfunction through modification of secretase activity via inhibition of ERK and NF-kB pathways in mice. J Nutr 139(10): 1987–1993

Lee B-H, Choi S-H, Shin T-J, Pyo MK, Hwang S-H, Kim B-R, Lee S-M, Lee J-H, Kim H-C, Park H-Y, Rhim H, Nah S-Y (2010) Quercetin enhances human α7 nicotinic acetylcholine receptor mediated ion current through interactions with Ca^{2+} binding sites. Mol Cells 30(3):245–253

Lee E, Eom J-E, Kim H-L, Baek KH, Jun K-Y, Kim H-J, Lee M, Mook-Jung I, Kwon Y (2013) Effect of conjugated linoleic acid, μ-calpain inhibitor, on pathogenesis of Alzheimer's disease. Biochim Biophys Acta 1831(4):709–718

Lee D, Kim N, Jeon SH, Gee MS, Ju Y-J, Jung M-J, Cho JS, Lee Y, Lee S, Lee JK (2022) Hesperidin improves memory function by enhancing neurogenesis in a mouse model of Alzheimer's disease. Nutrients 14(15):3125

León R, Garcia AG, Marco-Contelles J (2013) Recent advances in the multitarget directed ligands approach for the treatment of Alzheimer's disease. Med Res Rev 33(1):139–189

Letenneur L, Proust-Lima C, Le GA, Dartigues JF, Barberger-Gateau P (2007) Flavonoid intake and cognitive decline over a 10-year period. Am J Epidemiol 165(12):1364–1371

Levites Y, Youdim MBH, Maor G, Mandel S (2002) Attenuation of 6-hydroxydopamine (6-OHDA)-induced nuclear factor-kappaB (NF-kappaB) activation and cell death by tea extracts in neuronal cultures. Biochem Pharmacol 63(1):21–29

Li Q, Zhao HF, Zhang ZF, Liu ZG, Pei XR, Wang JB, Li Y (2009) Long-term green tea catechin administration prevents spatial learning and memory impairment in senescence-accelerated mouse prone-8 mice by decreasing Abeta1–42 oligomers and upregulating synaptic plasticity-related proteins in the hippocampus. Neuroscience 163(3):741–749

Li R-S, Wang X-B, Hu X-J, Kong L-Y (2013a) Design, synthesis and evaluation of flavonoid derivatives as potential multifunctional acetylcholinesterase inhibitors against Alzheimer's disease. Bioorg Med Chem Lett 23(9):2636–2641

Li S-Y, Wang X-B, Xie S-S, Jiang N, Wang KDG, Yao H-Q, Sun H-B, Kong L-Y (2013b) Multifunctional tacrine-flavonoid hybrids with cholinergic, b-amyloid-reducing, and metal chelating properties for the treatment of Alzheimer's disease. Eur J Med Chem 69:632–646

Li Y, Qiang X, Luo L, Yang X, Xiao G, Zheng Y, Cao Z, Sang Z, Su F, Deng Y (2017) Multitarget drug design strategy against Alzheimer's disease: homoisoflavonoid mannich base derivatives serve acetylcholinesterase and monoamine oxidase B dual inhibitors with multifunctional properties. Bioorg Med Chem 25(2):714–726

Liao S, Deng H, Huang S, Yang J, Wang S, Yin B, Zheng T, Zhang D, Liu J, Gao G, Ma J, Deng Z (2015) Design, synthesis and evaluation of novel 5,6,7-trimethoxyflavone-6-chlorotacrine hybrids as potential multifunctional agents for the treatment of Alzheimer's disease. Bioorg Med Chem Lett 25(7):1541–1545

Lichtenthaler SF (2011) Alpha-secretase in Alzheimer's disease: molecular identity, regulation and therapeutic potential. J Neurochem 116(1):10–21

Lindwall G, Cole RD (1984) Phosphorylation affects the ability of tau protein to promote microtubule assembly. J Biol Chem 259(8):5301–5305

Liu Q-H, Wu J-J, Li F, Cai P, Yang X-L, Kong L-Y, Wang X-B (2017) Synthesis and pharmacological evaluation of multi-functional homoisoflavonoid derivatives as potent inhibitors of monoamine oxidase B and cholinesterase for the treatment of Alzheimer's disease. MedChemComm 8(7):1459–1467

Liu J, Chang L, Song Y, Li H, Wu Y (2019) The role of NMDA receptors in Alzheimer's disease. Front Neurosci 13:43

Loera-Valencia R, Piras A, Ismail MAM, Manchanda S, Eyjolfsdottir H, Saido TC, Johansson J, Eriksdotter M, Winblad B, Nilsson P (2018) Targeting Alzheimer's disease with gene and cell therapies. J Intern Med 284(1):2–36

Lombardo S, Maskos U (2015) Role of the nicotinic acetylcholine receptor in Alzheimer's disease pathology and treatment. Neuropharmacology 96(Pt B):255–262

Lovestone S, Reynolds CH (1997) The phosphorylation of tau: a critical stage in neurodevelopmental and neurodegenerative processes. Neuroscience 78(2):309–324

Luo W, Su Y-B, Hong C, Tian R-G, Su L-P, Wang Y-Q, Li Y, Yue J-J, Wang C-J (2013) Design, synthesis and evaluation of novel 4-dimethylamine flavonoid derivatives as potential multifunctional anti-Alzheimer agents. Bioorg Med Chem 21(23):7275–7282

Luo W, Chen Y, Wang T, Hong C, Chang L-P, Chang C-C, Yang Y-C, Xie S-Q, Wang C-J (2015) Design, synthesis and evaluation of novel 7-aminoalkylsubstituted flavonoid derivatives with improved cholinesterase inhibitory activities. Bioorg Med Chem 24(4):672–680

Luo W, Wang T, Hong C, Yang Y-C, Chen Y, Cen J, Xie S-Q, Wang C-J (2016) Design, synthesis and evaluation of 4-dimethylamine flavonoid derivatives as potential multifunctional anti-Alzheimer agents. Eur J Med Chem 122:17–26

Lynch G, Larson J, Kelso S, Barrionuevo G, Schottler F (1983) Intracellular injections of EGTA block induction of hippocampal long-term potentiation. Nature 305(5936):719–721

Lyras L, Cairns NJ, Jenner A, Jenner P, Halliwell B (1997) An assessment of oxidative damage to proteins, lipids, and DNA in brain from patients with Alzheimer's disease. J Neurochem 68(5): 2061–2069

Mahomoodally MF, Gurib-Fakim A, Subratty AH (2005) Antimicrobial activities and phytochemical profiles of endemic medicinal plants of Mauritius. Pharm Biol 43(3):237–242

Makhaeva GF, Shevtsova EF, Boltneva NP, Lushchekina SV, Kovaleva NV, Rudakova EV, Bachurin SO, Richardson RJ (2019) Overview of novel multifunctional agents based on conjugates of γ-carbolines, carbazoles, tetrahydrocarbazoles, phenothiazines, and aminoadamantanes for treatment of Alzheimer's disease. Chem Biol Interact 308:224–234

Mandel SA, Amit T, Kalfon L, Reznichenko L, Weinreb O, Youdim MBH (2008) Cell signaling pathways and iron chelation in the neurorestorative activity of green tea polyphenols: special reference to epigallocatechin gallate (EGCG). J Alzheimers Dis 15(2):211–222

Mangialasche F, Solomon A, Winblad B, Mecocci P, Kivipelto M (2010) Alzheimer's disease: clinical trials and drug development. Lancet Neurol 9(7):702–716

Mas-Bargues C, Borrás C, Viña J (2022) The multimodal action of genistein in Alzheimer's and other age-related diseases. Free Radic Biol Med 183:127–137

McGeer PL, McGeer EG (2013) The amyloid cascade-inflammatory hypothesis of Alzheimer disease: implications for therapy. Acta Neuropathol 126(4):479–497

Meleleo D, Notarachille G, Mangini V, Arnesano F (2019) Concentration-dependent effects of mercury and lead on Aβ42: possible implications for Alzheimer's disease. Eur Biophys J 48(2): 173–187

Melo A, Monteiro L, Lima RMF, de Oliveira DM, de Cerqueira MD, El-Bachá RS (2011) Oxidative stress in neurodegenerative diseases: mechanisms and therapeutic perspectives. Oxidative Med Cell Longev 2011:467180

Meyer JH, Wilson AA, Sagrati S, Miler L, Rusjan P, Bloomfield PM, Clark M, Sacher J, Voineskos AN, Houle S (2009) Brain monoamine oxidase A binding in major depressive disorder: relationship to selective serotonin reuptake inhibitor treatment, recovery, and recurrence. Arch Gen Psychiatry 66(12):1304–1312

Michishita E, Park JY, Burneskis JM, Barrett JC, Horikawa I (2005) Evolutionarily conserved and nonconserved cellular localizations and functions of human SIRT proteins. Mol Biol Cell 16(10):4623–4635

Morawski M, Schilling S, Kreuzberger M, Waniek A, Jäger C, Koch B, Cynis H, Kehlen A, Arendt T, Hartlage-Rübsamen M, Demuth H-U, Roßner S (2014) Glutaminyl cyclase in human cortex: correlation with (pGlu)-amyloid-β load and cognitive decline in Alzheimer's disease. J Alzheimers Dis 39(2):385–400

Mori T, Rezai-Zadeh K, Koyama N, Arendash GW, Yamaguchi H, Kakuda N, Horikoshi-Sakuraba Y, Tan J, Town T (2012) Tannic acid is a natural beta-secretase inhibitor that prevents cognitive impairment and mitigates Alzheimer-like pathology in transgenic mice. J Biol Chem 287(9):6912–6927

Nanri A, Yoshida D, Yamaji T, Mizoue T, Takayanagi R, Kono S (2008) Dietary patterns and C-reactive protein in Japanese men and women. Am J Clin Nutr 87(5):1488–1496

Nistor M, Don M, Parekh M, Sarsoza F, Goodus M, Lopez GE, Kawas C, Leverenz J, Doran E, Lott IT, Hill M, Head E (2007) Alpha- and beta-secretase activity as a function of age and betaamyloid in Down syndrome and normal brain. Neurobiol Aging 28(10):1493–1506

Nohynek LJ, Alakomi H-L, Kähkönen MP, Heinonen M, Helander IM, Oksman-Caldentey K-M, Puupponen-Pimiä RH (2006) Berry phenolics: antimicrobial properties and mechanisms of action against severe human pathogens. Nutr Cancer 54(1):18–32

Nunomura A, Moreira PI, Castellani RJ, Lee H-G, Zhu X, Smith MA, Perry G (2012) Oxidative damage to RNA in aging and neurodegenerative disorders. Neurotox Res 22(3):231–248

Obregon DF, Rezai-Zadeh K, Bai Y, Sun N, Hou H, Ehrhart J, Zeng J, Mori T, Arendash GW, Shytle D, Town T, Tan J (2006) ADAM10 activation is required for green tea (−)-epigallocatechin-3-gallate-induced alpha-secretase cleavage of amyloid precursor protein. J Biol Chem 281(24):16419–16427

Ono K, Condron MM, Ho L, Wang J, Zhao W, Pasinetti GM, Teplow DB (2008) Effects of grape seed-derived polyphenols on amyloid betaprotein self-assembly and cytotoxicity. J Biol Chem 283(47):32176–32187

Onozuka H, Nakajima A, Matsuzaki K, Shin R-W, Ogino K, Saigusa D, Tetsu N, Yokosuka A, Sashida Y, Mimaki Y, Yamakuni T, Ohizumi Y (2008) Nobiletin, a citrus flavonoid, improves memory impairment and Aβ pathology in a transgenic mouse model of Alzheimer's disease. J Pharmacol Exp Ther 326(3):739–744

Pandey AK (2007) Anti-staphylococcal activity of a pan-tropical aggressive and obnoxious weed Parthenium histerophorus: an in vitro study. Natl Acad Sci Lett 30(11):383–386

Pannala A, Rice-Evans CA, Halliwell B, Singh S (1997) Inhibition of peroxynitrite-mediated tyrosine nitration by catechin polyphenols. Biochem Biophys Res Commun 232(1):164–168

Parri HR, Hernandez CM, Dineley KT (2011) Research update: Alpha7 nicotinic acetylcholine receptor mechanisms in Alzheimer's disease. Biochem Pharmacol 82(8):931–942

Pejchal V, Stepankova S, Padelkova Z, Imramovsky A, Jampilek J (2011) 1,3-Substituted imidazolidine-2,4,5-triones: synthesis and inhibition of cholinergic enzymes. Molecules 16(9): 7565–7582

Perry EK, Perry RH, Blessed G, Tomlinson BE (1978) Changes in brain cholinesterases in senile dementia of Alzheimer type. Neuropathol Appl Neurobiol 4(4):273–277

Phaniendra A, Jestadi DB, Periyasamy L (2015) Free radicals: properties, sources, targets, and their implication in various diseases. Indian J Clin Biochem 30(1):11–26

Phillis JW, Horrocks LA, Farooqui AA (2006) Cyclooxygenases, lipoxygenases, and epoxygenases in CNS: their role and involvement in neurological disorders. Brain Res Rev 52(2):201–243

Pluta R, Januszewski S, Czuczwar SJ (2021) Myricetin as a promising molecule for the treatment of post-ischemic brain neurodegeneration. Nutrients 13(2):342

Praticò D, Zhukareva V, Yao Y, Uryu K, Funk CD, Lawson JA, Trojanowski JQ, Lee VM-Y (2004) 12/15-Lipoxygenase is increased in Alzheimer's disease: possible involvement in brain oxidative stress. Am J Pathol 164(5):1655–1662

Prince M, Bryce R, Albanese E, Wimo A, Ribeiro W, Ferri CP (2013) The global prevalence of dementia: a systematic review and metaanalysis. Alzheimers Dement 9(1):63–75

Pugazhenthi S (2017) Metabolic syndrome and the cellular phase of Alzheimer's disease. Prog Mol Biol Transl Sci 146:243–258

Qiang X, Sang Z, Yuan W, Li Y, Liu Q, Bai P, Shi Y, Ang W, Tan Z, Deng Y (2014) Design, synthesis and evaluation of genistein-O-alkylbenzylamines as potential multifunctional agents for the treatment of Alzheimer's disease. Eur J Med Chem 76:314–331

Qin L, Zhang J, Qin M (2013) Protective effect of cyanidin 3-*O*-glucoside on β-amyloid peptide-induced cognitive impairment in rats. Neurosci Lett 534:285–288

Ramirez MR, Izquierdo I, do Carmo Bassols Raseira M, Zuanazzi JA, Barros D, Henriques AT (2005) Effect of lyophilised Vaccinium berries on memory, anxiety and locomotion in adult rats. Pharmacol Res 52(6):457–462

Rampa A, Belluti F, Gobbi S, Bisi A (2011) Hybrid-based multi-target ligands for the treatment of Alzheimer's disease. Curr Top Med Chem 11(22):2716–2730

Rezai-Zadeh K, Arendash GW, Hou H, Fernandez F, Jensen M, Runfeldt M, Shytle RD, Tan J (2008) Green tea epigallocatechin-3-gallate (EGCG) reduces beta-amyloid mediated cognitive impairment and modulates tau pathology in Alzheimer transgenic mice. Brain Res 1214:177–187

Reznichenko L, Amit T, Youdim MBH, Mandel S (2005) Green tea polyphenol (−)-epigallocatechin-3-gallate induces neurorescue of longterm serum-deprived PC12 cells and promotes neurite outgrowth. J Neurochem 93(5):1157–1167

Richard I (2005) The genetic and molecular bases of monogenic disorders affecting proteolytic systems. J Med Genet 42(7):529–539

Rivieccio MA, Brochier C, Willis DE, Walker BA, D'Annibale MA, McLaughlin K, Siddiq A, Kozikowski AP, Jaffrey SR, Twiss JL, Ratan RR, Langley B (2009) HDAC6 is a target for protection and regeneration following injury in the nervous system. Proc Natl Acad Sci U S A 106(46):19599–19604

Rizzo S, Cavalli A, Ceccarini L, Bartolini M, Belluti F, Bisi A, Andrisano V, Recanatini M, Rampa A (2009) Structure-reactivity relationships and binding mode in the human acetylcholinesterase active site of pseudo irreversible inhibitors related to xanthostigmine. ChemMedChem 4(4): 670–679

Roberts RO, Geda YE, Knopman DS, Boeve BF, Christianson TJH, Pankratz VS, Kullo IJ, Tangalos EG, Ivnik RJ, Petersen RC (2009) Association of C-reactive protein with mild cognitive impairment. Alzheimers Dement 5(5):398–405

Rohn TT (2010) The role of caspases in Alzheimer's disease; potential novel therapeutic opportunities. Apoptosis 15(11):1403–1409

Rosini M, Simoni E, Minarini A, Melchiorre C (2014) Multi-target design strategies in the context of Alzheimer's disease: acetylcholinesterase inhibition and NMDA receptor antagonism as the driving forces. Neurochem Res 39(10):1914–1923

Russo A, Acquaviva R, Campisi A, Sorrenti V, Di Giacomo C, Virgata G, Barcellona ML, Vanella A (2000) Bioflavonoids as antiradicals, antioxidants and DNA cleavage protectors. Cell Biol Toxicol 16(2):91–98

Sang Z-P, Qiang X-M, Li Y, Wu B, Zhang H, Zhao M-G, Deng Y (2015a) Design, synthesis and biological evaluation of scutellarein carbamate derivatives as potential multifunctional agents for the treatment of Alzheimer's disease. Chem Biol Drug Des 86(5):1168–1177
Sang Z-P, Li Y, Qiang X, Xiao G, Liu Q, Tan Z, Dang Y (2015b) Multifunctional scutellarin-rivastigmine hybrids with cholinergic, antioxidant, biometal chelating and neuroprotective properties for the treatment of Alzheimer's disease. Bioorg Med Chem 23(4):668–680
Sang Z-P, Qiang X, Li Y, Yuan W, Liu Q, Shi Y, Ang W, Luo Y, Tan Z, Deng Y (2015c) Design and synthesis and evolution of scutellarein-*O*-alkylamines as multifunctional agent for the treatment of Alzheimer's disease. Eur J Med Chem 94:348–366
Sarazin M, Dorothée G, de Souza LC, Aucouturier P (2013) Immunotherapy in Alzheimer's disease: do we have all the pieces of the puzzle? Biol Psychiatry 74(5):329–332
Savonenko AV, Melnikova T, Hiatt A, Li T, Worley PF, Troncoso JC, Wong PC, Price DL (2012) Alzheimer's therapeutics: translation of preclinical science to clinical drug development. Neuropsychopharmacology 37(1):261–277
Sayre LM, Smith MA, Perry G (2001) Chemistry and biochemistry of oxidative stress in neurodegenerative disease. Curr Med Chem 8(7):721–738
Schlenzig D, Manhart S, Cinar Y, Kleinschmidt M, Hause G, Willbold D, Funke SA, Schilling S, Demuth H-U (2009) Pyroglutamate formation influences solubility and amyloidogenicity of amyloid peptides. Biochemistry 48(29):7072–7078
Schneider LS, Mangialasche F, Andreasen N, Feldman H, Giacobini E, Jones R, Mantua V, Mecocci P, Pani L, Winblad B, Kivipelto M (2014) Clinical trials and late stage drug development for Alzheimer's disease: an appraisal from 1984 to 2014. J Intern Med 275(3):251–283
Schroeter H, Spencer JPE, Rice-Evans C, Williams RJ (2001) Flavonoids protect neurons from oxidized low-density-lipoprotein-induced apoptosis involving c-Jun N-terminal kinase (JNK), c-Jun and caspase-3. Biochem J 358(Pt 3):547–557
Schroeter H, Boyd C, Spencer JPE, Williams RJ, Cadenas E, Rice-Evans C (2002) MAPK signaling in neurodegeneration: influences of flavonoids and of nitric oxide. Neurobiol Aging 23(5): 861–880
Schroeter H, Bahia P, Spencer JPE, Sheppard O, Rattray M, Cadenas E, Rice-Evans C, Williams RJ (2007a) (−) Epicatechin stimulates ERK-dependent cyclic AMP response element activity and up-regulates GluR2 in cortical neurons. J Neurochem 101(6):1596–1606
Schroeter H, Bahia P, Spencer JPE, Sheppard O, Rattray M, Cadenas E, Rice-Evans C, Williams RJ (2007b) (−)-Epicatechin stimulates ERK-dependent cyclic AMP response element activity and upregulates GLUR2 in cortical neurons. J Neurochem 101(6):1596–1606
Selkoe DJ (2001) Alzheimer's disease: genes, proteins, and therapy. Physiol Rev 81(2):741–766
Selkoe DJ (2005) Defining molecular targets to prevent Alzheimer disease. Arch Neurol 62(2): 192–195
Shahani N, Brandt R (2002) Functions and malfunctions of the tau proteins. Cell Mol Life Sci 39(10):1668–1680
Shen Y, Zhang J, Sheng R, Dong X, He Q, Yang B, Hu Y (2009) Synthesis and biological evaluation of novel flavonoid derivatives as dual binding acetylcholinesterase inhibitors. J Enzyme Inhib Med Chem 24(2):372–380
Sheng R, Lin X, Zhang J, Chol KS, Huang W, Yang B, He Q, Hu Y (2009) Design, synthesis and evaluation of flavonoid derivatives as potent AChE inhibitors. Bioorg Med Chem 17(18): 6692–6698
Shevtsova EF, Maltsev AV, Vinogradova DV, Shevtsov PN, Bachurin SO (2021) Mitochondria as a promising target for developing novel agents for treating Alzheimer's disease. Med Res Rev 41(2):803–827
Shevtzova EF, Kireeva EG, Bachurin SO (2001) Effect of beta-amyloid peptide fragment 25–35 on nonselective permeability of mitochondria. Bull Exp Biol Med 132(6):1173–1176
Shi D-H, Yan Z-Q, Zhang L-N, Wang Y-R, Jiang C-P, Wu J-H (2012) A novel 7-O-modified genistein derivative with acetylcholinesterase inhibitory effect, estrogenic activity and neuroprotective effect. Arch Pharm Res (Seoul) 35(9):1645–1654

Shoaib M, Shah SWA, Ali N, Shah I, Umar MN, Shafiullah, Ayaz M, Tahir MN, Akhtar S (2015) In vitro enzyme inhibition potentials and antioxidant activity of synthetic flavone derivatives. J Chem 2015:516878

Silva T, Reis J, Teixeira J, Borges F (2014) Alzheimer's disease, enzyme targets and drug discovery struggles: from natural products to drug prototypes. Ageing Res Rev 15:116–145

Šimić G, Leko MB, Wray S, Harrington C, Delalle I, Jovanov-Milošević N, Bažadona D, Buée L, de Silva R, Giovanni GD, Wischik C, Hof PR (2016) Intracellular amyloid-β in Alzheimer's disease. Biomolecules 6(1):6

Singh M, Silakari O (2016) Design, synthesis and biological evaluation of novel 2-phenyl-1-benzopyran-4-one derivatives as potential poly-functional anti-Alzheimer's agents. RSC Adv 6(110):108411–108422

Singh RP, Sharad S, Kapur S (2004) Free radicals and oxidative stress in neurodegenerative diseases: relevance of dietary antioxidants. J Indian Acad Clin Med 5(3):218–225

Singh M, Kaur M, Singh N, Silakari O (2017) Exploration of multi-target potential of chromen-4-one based compounds in Alzheimer's disease: design, synthesis and biological evaluations. Bioorg Med Chem 25(24):6273–6285

Solfrizzi V, Panza F, Frisardi V, Seripa D, Logroscino G, Imbimbo BP, Pilotto A (2011) Diet and Alzheimer's disease risk factors or prevention: the current evidence. Expert Rev Neurother 11(5):677–708

Sorond FA, Lipsitz LA, Hollenberg NK, Fisher NDL (2008) Cerebral blood flow response to flavanol-rich cocoa in healthy elderly humans. Neuropsychiatr Dis Treat 4(2):433–440

Sorond FA, Hurwitz S, Salat DH, Greve DN, Fisher NDL (2013) Neurovascular coupling, cerebral white matter integrity, and response to cocoa in older people. Neurology 81(10):904–909

Spencer JPE, Rice-Evans C, Williams RJ (2003) Modulation of pro-survival Akt/protein kinase B and ERK1/2 signaling cascades by quercetin and its in vivo metabolites underlie their action on neuronal viability. J Biol Chem 278(37):34783–34793

Spencer JP, Vafeiadou K, Williams RJ, Vauzour D (2012) Neuroinflammation: modulation by flavonoids and mechanisms of action. Mol Asp Med 33(1):83–97

Strooper BD, Karran E (2016) The cellular phase of Alzheimer's disease. Cell 164(4):603–615

Sun Y, Chen J, Chen X, Huang L, Li X (2013) Inhibition of cholinesterase and monoamine oxidase-B activity by Tacrine-Homoisoflavonoid hybrids. Bioorg Med Chem 21(23):7406–7417

Swerdlow RH (2018) Mitochondria and mitochondrial cascades in Alzheimer's disease. J Alzheimers Dis 62(3):1403–1416

Swerdlow RH, Burns JM, Khan SM (2014) The Alzheimer's disease mitochondrial cascade hypothesis: progress and perspectives. Biochim Biophys Acta 1842(8):1219–1231

Szekely CA, Thorne JE, Zandi PP, Ek M, Messias E, Breitner JCS, Goodman SN (2004) Nonsteroidal anti-inflammatory drugs for the prevention of Alzheimer's disease: a systematic review. Neuroepidemiology 23(4):159–169

Tanaka M, Saito S, Inoue T, Satoh-Asahara N, Ihara M (2019) Novel therapeutic potentials of Taxifolin for Amyloid-β-associated neurodegenerative diseases and other diseases: recent advances and future perspectives. Int J Mol Sci 20(9):2139

Tarkowski E, Liljeroth AM, Minthon L, Tarkowski A, Wallin A, Blennow K (2003) Cerebral pattern of pro- and anti-inflammatory cytokines in dementias. Brain Res Bull 61(3):255–260

Terry AV, Buccafusco JJ (2003) The cholinergic hypothesis of age and Alzheimer's disease-related cognitive deficits: recent challenges and their implications for novel drug development. J Pharmacol Exp Ther 306(3):821–827

Trollor JN, Smith E, Baune BT, Kochan NA, Campbell L, Samaras K, Crawford J, Brodaty H, Sachdev P (2010) Systemic inflammation is associated with MCI and its subtypes: the Sydney Memory and Aging Study. Dement Geriatr Cogn Disord 30(6):569–578

Troussard AA, Tan C, Yoganathan TN, Dedhar S (1999) Cell extracellular matrix interactions stimulate the AP-1 transcription factor in an integrin-linked kinase- and glycogen synthase kinase 3-dependent manner. Mol Cell Biol 19(11):7420–7427

Tseng J (2014) Discovery and characterisation of multi-target directed ligands as inhibitors of amyloid-β aggregation and regulators of Alzheimer disease. PhD thesis, University of South Carolina

Tully T, Bourtchouladze R, Scott R, Tallman J (2003) Targeting the CREB pathway for memory enhancers. Nat Rev Drug Discov 2(4):267–277

Ucar G, Gokhan N, Yesilada A, Bilgin AA (2005) 1-N-Substituted thiocarbamoyl-3-phenyl-5-thienyl-2-pyrazolines: a novel cholinesterase and selective monoamine oxidase B inhibitors for the treatment of Parkinson's and Alzheimer's diseases. Neurosci Lett 382(3):327–331

Urbanelli L, Emiliani C, Massini C, Persichetti E, Orlacchio A, Pelicci G, Sorbi S, Hasilik A, Bernardi G, Orlacchio A (2008) Cathepsin D expression is decreased in Alzheimer's disease fibroblasts. Neurobiol Aging 29(1):12–22

Vafeiadou K, Vauzour D, Lee HY, Rodriguez-Mateos A, Williams RJ, Spencer JPE (2009) The citrus flavanone naringenin inhibits inflammatory signalling in glial cells and protects against neuroinflammatory injury. Arch Biochem Biophys 484(1):100–109

Van Dyke MW (2014) Lysine deacetylase (KDAC) regulatory pathways: an alternative approach to selective modulation. ChemMedChem 9(3):511–522

Vance JE, Hayashi H, Karten B (2005) Cholesterol homeostasis in neurons and glial cells. Semin Cell Dev Biol 16(2):193–212

Vassar R, Kovacs DM, Yan R, Wong PC (2009) The β-secretase enzyme BACE in health and Alzheimer's disease: regulation, cell biology, function, and therapeutic potential. J Neurosci 29(41):12787–12794

Vauzour D (2012) Dietary polyphenols as modulators of brain functions: biological actions and molecular mechanisms underpinning their beneficial effects. Oxidative Med Cell Longev 2012: 914273

Vauzour D, Vafeiadou K, Rice-Evans C, Williams RJ, Spencer JPE (2007) Activation of pro-survival Akt and ERK1/2 signalling pathways underlie the anti-apoptotic effects of flavanones in cortical neurons. J Neurochem 103(4):1355–1367

Vepsäläinen S, Koivisto H, Pekkarinen E, Mäkinen P, Dobson G, McDougall GJ, Stewart D, Haapasalo A, Karjalainen RO, Tanila H, Hiltunen M (2013) Anthocyanin-enriched bilberry and blackcurrant extracts modulate amyloid precursor protein processing and alleviate behavioral abnormalities in the APP/PS1 mouse model of Alzheimer's disease. J Nutr Biochem 24(1): 360–370

Verbeek R, van Tol EA, van Noort JM (2005) Oral flavonoids delay recovery from experimental autoimmune encephalomyelitis in SJL mice. Biochem Pharmacol 70(2):220–228

Verma A, Waiker DK, Bhardwaj B, Saraf P, Shrivastava SK (2022) The molecular mechanism, targets, and novel molecules in the treatment of Alzheimer's disease. Bioorg Chem 119:105562

Visioli F, Bellomo G, Galli C (1998) Free radical-scavenging properties of olive oil polyphenols. Biochem Biophys Res Commun 247(1):60–64

Vos SJB, Gordon BA, Su Y, Visser PJ, Holtzman DM, Morris JC, Fagan AM, Benzinger TLS (2016) NIA-AA staging of preclinical Alzheimer disease: discordance and concordance of CSF and imaging biomarkers. Neurobiol Aging 44:1–8

Wallace T, Bertrand D (2013) Importance of the nicotinic acetylcholine receptor system in the prefrontal cortex. Biochem Pharmacol 85(12):1713–1720

Walsh DM, Selkoe DJ (2007) A beta oligomers—a decade of discovery. J Neurochem 101(5): 1172–1184

Waltereit R, Dammermann B, Wulff P, Scafidi J, Staubli U, Kauselmann G, Bundman M, Kuhl D (2001) Arg3.1/Arc mRNA induction by Ca^{2+} and cAMP requires protein kinase A and mitogen-activated protein kinase/extracellular regulated kinase activation. J Neurosci 21(15):5484–5493

Wang H-J, Murphy PA (1994) Isoflavone content in commercial soybean foods. J Agric Food Chem 42:1666–1673

Wang J-Z, Grundke-Iqbal I, Iqbal K (2007) Kinases and phosphatases and tau sites involved in Alzheimer neurofibrillary degeneration. Eur J Neurosci 25(1):59–68

Wang D-M, Li S-Q, Wu W-L, Zhu X-Y, Wang Y, Yuan H-Y (2014) Effects of long-term treatment with quercetin on cognition and mitochondrial function in a mouse model of Alzheimer's disease. Neurochem Res 39(8):1533–1543

Wang P, Chen H, Zhu Y, McBride J, Fu J, Sang S (2015) Oat avenanthramide-C (2c) is biotransformed by mice and the human microbiota into bioactive metabolites. J Nutr 145(2): 239–245

Wang Y, Wang H, Chen HZ (2016a) AChE Inhibition-based multi-target-directed ligands, a novel pharmacological approach for the symptomatic and disease modifying therapy of Alzheimer's disease. Curr Neuropharmacol 14(4):364–375

Wang Y, Sun Y, Guo Y, Wang Z, Huang L, Li X (2016b) Dual functional cholinesterase and MAO inhibitors for the treatment of Alzheimer's disease: pharmacological analysis and molecular modeling of homoisoflavonoid derivatives. J Enzyme Inhib Med Chem 31(3):389–397

Wei Y, Peng A-Y, Wang B, Ma L, Peng G, Du Y, Tang J (2014) Synthesis and biological evaluation of phosphorylated flavonoids as potent and selective inhibitors of cholesterol esterase. Eur J Med Chem 74:751–758

Weinreb O, Mandel S, Amit T, Youdim MBH (2004) Neurological mechanisms of green tea polyphenols in Alzheimer's and Parkinson's diseases. J Nutr Biochem 15(9):506–516

Weinreb O, Amit T, Mandel S, Youdim MBH (2009) Neuroprotective molecular mechanisms of (−)-epigallocatechin-3-gallate: a reflective outcome of its antioxidant, iron chelating and neuritogenic properties. Genes Nutr 4(4):283–296

Welsh GI, Proud CG (1993) Glycogen synthase kinase-3 is rapidly inactivated in response to insulin and phosphorylates eukaryotic initiation factor eIF-2B. Biochem J 294(Pt 3):625–629

Wilcock G, Esiri MM, Bowen DM, Smith CC (1982) Alzheimer's disease: correlation of cortical choline acetyltransferase activity with the severity of dementia and histological abnormalities. J Neurol Sci 57(2–3):407–417

Williams RJ, Spencer JPE (2012) Flavonoids, cognition, and dementia: actions, mechanisms, and potential therapeutic utility for Alzheimer disease. Free Radic Biol Med 52(1):35–45

Williams RJ, Spencer JPE, Rice-Evans C (2004) Flavonoids: antioxidants or signalling molecules? Free Radic Biol Med 36(7):838–849

Williams P, Sorribas A, Howes M-JR (2011) Natural products as a source of Alzheimer's drug leads. Nat Prod Rep 28(1):48–77

Wilquet V, De Strooper B (2004) Amyloid-beta precursor protein processing in neurodegeneration. Curr Opin Neurobiol 14(5):582–588

Winkler J, Suhr ST, Gage FH, Thal LJ, Fisher LJ (1995) Essential role of neocortical acetylcholine in spatial memory. Nature 375(6531):484–487

Wood JA, Wood PL, Ryan R, Graff-Radford NR, Pilapil C, Robitaille Y, Quirion R (1993) Cytokine indices in Alzheimer's temporal cortex: no changes in mature IL-1 beta or IL-1RA but increases in the associated acute phase proteins IL-6, alpha 2-macroglobulin and C-reactive protein. Brain Res 629(2):245–252

Worker A, Dima D, Combes A, Crum WR, Streffer J, Einstein S, Mehta MA, Barker GJ, Williams SCR, O'Daly O (2018) Test-retest reliability and longitudinal analysis of automated hippocampal subregion volumes in healthy ageing and Alzheimer's disease populations. Hum Brain Mapp 39(4):1743–1754

Xia M, Hyman BT (2002) GROα/KC, a chemokine receptor CXCR2 ligand, can be a potent trigger for neuronal ERK1/2 and PI-3 kinase pathways and for tau hyperphosphorylation—a role in Alzheimer's disease? J Neuroimmunol 122(1–2):55–64

Xie Y, Yang W, Tang F, Chen X, Ren L (2015) Antibacterial activities of flavonoids: structure-activity relationship and mechanism. Curr Med Chem 22(1):132–149

Yiannopoulou KG, Papageorgiou SG (2013) Current and future treatments for Alzheimer's disease. Ther Adv Neurol Disord 6(1):19–33

Zaciragic A, Lepara O, Valjevac A, Arslanagic S, Fajkic A, Hadzovic-Dzuvo A, Avdagic N, Alajbegovic A, Mehmedika-Suljic E, Coric G (2007) Elevated serum C-reactive protein concentration in Bosnian patients with probable Alzheimer's disease. J Alzheimers Dis 12(2): 151–156

Zhang L, Sheng S, Qin C (2013) The role of HDAC6 in Alzheimer's disease. J Alzheimers Dis 33(2):283–295

Zhang X, Wang J, Hong C, Luo W, Wang C (2015) Design, synthesis and evaluation of genistein-polyamine conjugates as multi-functional anti-Alzheimer agents. Acta Pharm Sin B 5(1):67–73

Zhang L, Wang Y, Li D, Ho C-T, Li J, Wan X (2016) The absorption, distribution, metabolism and excretion of procyanidins. Food Funct 7(3):1273–1281

Zhao C, Deng W, Gage FH (2008) Mechanisms and functional implications of adult neurogenesis. Cell 132(4):645–660

Zhao Z, Xu H, Gong W (2010) Histone deacetylase 6 (HDAC6) is an independent deacetylase for α-tubulin. Protein Pept Lett 17(5):555–558

Zipser BD, Johanson CE, Gonzalez L, Berzin TM, Tavares R, Hulette CM, Vitek MP, Hovanesian V, Stopa EG (2007) Microvascular injury and blood–brain barrier leakage in Alzheimer's disease. Neurobiol Aging 28(7):977–986

Zlokovic BV (2005) Neurovascular mechanisms of Alzheimer's neurodegeneration. Trends Neurosci 28(4):202–208

Zott B, Simon MM, Hong W, Unger F, Chen-Engerer H-J, Frosch MP, Sakmann B, Walsh DM, Konnerth A (2019) A vicious cycle of β amyloid-dependent neuronal hyperactivation. Science 365(6453):559–565

12 Vitamin-Based Derivatives for the Management of Alzheimer's Disease

Tanmaykumar Varma, Pradnya Kamble, Madhavi Kumari, Vineet Diwakar, and Prabha Garg

Abstract

Vitamins are organic compounds; they help in the regulation of many bodily functions like cell proliferation and differentiation, immunological response, and metabolism. A deficiency of these molecules can cause severe medical conditions, i.e., beriberi, xerophthalmia, scurvy, Crohn's disease, and others. Furthermore, studies have shown that vitamin deficiency might lead to many neuronal dysfunctions, even hampering the growth of neurons. This chapter explores the role of vitamins and their derivatives in the pathology of Alzheimer's disease. Alzheimer's disease is a multifactorial neurodegenerative disease, and the cause of this disease is still unknown. However, several hypotheses try to explain the aetiology of the disease, such as Aβ hypothesis, metal ion hypothesis, calcium homeostasis, cholinergic hypothesis, tau propagation, etc. Scientific literature reports several derivatives that show potential to treat Alzheimer's disease. Primarily these compounds act on nuclear receptors to activate ADAM10, inhibiting AChE or BuChE, neutralisation of ROS, inhibition of GSK-3, and amyloid-beta aggregation. Moreover, some can easily pass the BBB, which is crucial in targeting neurological disease. Vitamins and their derivatives show promising results in managing Alzheimer's disease, even several are in clinical trials.

Keywords

Vitamins · Alzheimer's disease · Vitamin-based derivatives

T. Varma · P. Kamble · M. Kumari · V. Diwakar · P. Garg (✉)
Department of Pharmacoinformatics, National Institute of Pharmaceutical Education and Research (NIPER), Mohali, Punjab, India
e-mail: prabhagarg@niper.ac.in

A. Sharma, G. P. Modi (eds.), *Natural Product-Based Synthetic Drug Molecules in Alzheimer's Disease*, https://doi.org/10.1007/978-981-99-6038-5_12

Abbreviations

AChE	Acetylcholinesterase
AD	Alzheimer's disease
ADAM10	A Disintegrin metalloproteinase 10
APL	Acute promyelocytic leukaemia
ApoE	Apolipoprotein E
APP	Amyloid precursor protein
ATP	Adenosine triphosphate
ATRA	All-*trans* RA
Aβ	Amyloid-beta
BACE1	Beta-site APP-cleaving enzyme 1
BBB	Blood-brain barrier
BEXA	Bexarotene
BFT	Benfotiamine
BHQ	Biotin-8-hydroxyquinoline
BuChE	Butyrylcholinesterase
CAS	Catalytic anionic site
CNS	Central nervous system
COVID-19	Coronavirus disease 2019
COX	Cyclooxygenase
DBT	Dibenzoylthiamine
DNA	Deoxyribonucleic acid
FDA	Food and Drug Administration
GLUT	Glucose transporters
GSH	Glutathione
GSK-3	Glycogen synthase kinase 3
HBHQ	8-Hydroxyquinolyl-biotin hydrazine
IL-6	Interleukin-6
LPS	Lipopolysaccharides
MAO-B	Monoamine oxidase B
MAPK	Mitogen-activated protein kinase
MARRS	Membrane-associated, rapid response steroid-binding
MEM	Memantine
NAD	Nicotinamide adenine dinucleotide
NADP	Nicotinamide adenine dinucleotide phosphate
NEP	Neprilysin
NMDA	*N*-Methyl-D-aspartate
PAS	Peripheral anionic site
RA	Retinoic acid
RAR-α	Retinoic acid receptor alpha
RNA	Ribonucleic acid
RNS	Reactive nitrogen species
ROS	Reactive oxygen species

RXR	Retinoid X receptor
SVCT1	Sodium-dependent vitamin C transporter-1
SVCT2	Sodium-dependent vitamin C transporter-2
ThT	Thioflavin T
TrkB	Tyrosine kinase receptor B
VDR	Vitamin D receptor
VK	Vitamin K
VK2	Vitamin K2
VK3	Vitamin K3
α-T3	Alpha-tocotrienol
α-TOC	Alpha-tocopherol
α-TQ	Alpha-tocopherolquinone
γ-T3	Gamma-tocotrienol

12.1 Introduction

In 1912, Casimir Funk coined the term vitamine, which is originated from the Latin word "vita" meaning "life" since these molecules are important part of normal functioning body, and "amine" since vitamins were assumed to include amino acids. Later, scientists uncovered the true nature of the substance, and vitamine was modified to vitamin (Semba 2012). Vitamins are organic compounds required for the body's development and normal functioning. They regulate various biological processes, including cell and tissue growth, differentiation, and mineral metabolism, and commonly act as antioxidants and enzymatic cofactors or precursors. Not all vitamins can be synthesised by the body. As a result, they must be obtained through dietary means. Vitamins can be classified into two main categories: (1) fat-soluble vitamins which includes vitamin A, D, E, K and (2) water-soluble vitamins which includes vitamin B complex and vitamin C. Generally, fat-soluble vitamins are stored in the body whereas water-soluble vitamins require a continual exogenous daily supply. Vitamin balance is necessary for biological processes and human body development.

Vitamin deficiency develops several disorders, such as night blindness, beriberi, pellagra, scurvy, megaloblastic anaemia, haemolytic anaemia, and osteomalacia. In recent years, vitamins are found to have advantageous effects in the management of numerous diseases such as peripheral vascular diseases (Collins et al. 2003), cardiovascular diseases (Czernichow and Hercberg 2001), coronavirus disease 2019 (COVID-19) (Abobaker et al. 2020), neurodegenerative diseases (Rai et al. 2021), and others. Among all neurodegenerative disorders, Alzheimer's disease (AD) is the most prevalent worldwide and has no cure (Dong et al. 2019).

Alzheimer's disease (AD) is named after the psychiatrist Alois Alzheimer who reported the first case of Alzheimer's disease. Major symptoms of AD are a decline in cognitive ability and progressive memory loss. Although the nature of this disorder's onset, progression, and severity is quite diverse, the types of symptoms, such as behavioural changes and the inability to perform everyday living tasks, are common. It is believed that the oxidative stress is a crucial factor that influences the initiation and progression of AD. Research shows that the antioxidative characteristics of vitamins are advantageous and can be utilised in the treatment and management of Alzheimer's disease (Singh et al. 2021). As a result, vitamins have been used as adjuvants in AD therapy (Kurutas 2016).

It is widely known that amyloid-beta (Aβ) induced oxidative stress, which aids in the formation of reactive oxygen species (ROS) that results into lipid peroxidation, tau hyperphosphorylation, protein oxidation and has deleterious effects on synapses and neurons. For these reasons, antioxidant therapy might be beneficial in AD patients. Vitamins A, C, and E have antioxidant activity; these vitamins and their derivatives enhance cognition and prevent dementia. Many studies have investigated the role of vitamins in the treatment of AD and other types of dementia. Such studies suggest that vitamin A can reduce the production of Aβ plaques, whereas vitamins like B, C, D, and E can intervene the progression of neurocognitive decline. Furthermore, vitamin D deficiency is associated with an increased risk of dementia and AD (Bhatti et al. 2016). Vitamin B supplements, such as pyridoxine (B6), folate (B9), and cyanocobalamin (B12), reduce the risk of dementia since they are involved in homocysteine metabolism (Bhatti et al. 2016; Littlejohns et al. 2014; Ford et al. 2010). Dietary vitamin A deficiency disrupts the retinoid signalling pathway, resulting in amyloid-beta (Aβ) deposition in the cerebral blood vessels through down-regulation of retinoic acid receptor alpha (RAR-α) in the forebrain neurons and loss of choline acetyltransferase (ChAT) expression. Administration of retinoic acid which is a vitamin A derivative can reverse these changes. Amyloid-β (Aβ) deposition and a similar RAR-α impairment were seen in the pathological samples from AD patients. The same has been reported for thiamine (vitamin B1) insufficiency and vitamin E (La Fata et al. 2014), which is linked to significant cognitive impairment and advancing dementia. Thiamine supplementation in those who were impacted helped with these symptoms (Lu'o'ng and Nguyen 2011). Cobalamin (vitamin B12) deficiency has also been linked to reversible dementia in older persons (Osimani et al. 2005).

A significant deal of interest in vitamins has been sparked by the development of the free radical theory, which explains the pathophysiology of brain ageing and neurodegenerative disorders like AD. The theory claims that oxidative and nitrosative modifications of several biological molecules are caused by reactive oxygen species (ROS) and reactive nitrogen species (RNS). Accumulation of free radical-mediated damage to neuronal components, combined with other age-related changes, is one of the leading causes of neurodegeneration. Therefore, preventing or reducing oxidative and nitrosative stress may counteract the molecular cause of pathological brain ageing and the subsequent neurodegeneration that results in brain damage, indicating that antioxidants vitamins may be an effective preventive and therapeutic approach (Bhatti et al. 2016; Singh et al. 2022).

The first acetylcholinesterase inhibitor approved for the symptomatic treatment of AD was tacrine although it is no longer used because of its hepatotoxicity (Joe and

Ringman 2019); Currently AD is managed by acetylcholinesterase inhibitors, i.e., galantamine, rivastigmine, and donepezil (Kumar et al. 2022) and antagonist of *N*-Methyl-D-aspartate (NMDA) receptor, i.e., memantine. Moreover, current pharmacological treatments for AD are expensive, sometimes accompanied by various side effects, and alleviate symptoms rather than halt the eventual decline in cognitive and behavioural function. Additionally, medication must be administered after a significant amount of cognitive impairment, which may be accompanied by irreversible functional deterioration. Antioxidant use might be an effective strategy that can assist in treating this illness. Apart from the excellent antioxidant properties of vitamins or their derivatives shown to have the inhibitory activities on AChE, BuChE, and GSK-3β. Several vitamin B derivatives can act as metal chelators and help to maintain the bio-metal homeostasis in the brain. Additionally, vitamins like ascorbic acid have special transporters which ensures the passage through BBB. Combining all these properties, vitamins and their derivatives can be explored to find new therapy for Alzheimer's disease. The current chapter explores the various vitamin-based derivatives and their ability to halt AD progression.

12.2 Vitamins and Their Derivatives

12.2.1 Vitamin A

Vitamin A is one of the fat-soluble and essential vitamins which is commonly known as retinol. The body cannot synthesise Vitamin A, so it must be taken in the diet. Retinol is metabolised in our body and converted to retinoic acid (RA), specifically all-*trans* RA (ATRA). Retinoic acid has several stereoisomers like all-*trans* RA, 13-*cis* RA, and 9-*cis* RA. Vitamin A is a vital nutrient for normal vision, growth, cell division, reproduction, and immunity. Vitamin A shows strong antioxidant properties and hence protects cells against ROS. Studies have shown that free radicals can cause heart diseases, cancer, and other diseases, including neurodegenerative diseases (like AD). Studies have shown that Alzheimer's disease (AD) patients have lower levels of Vitamin A, β-carotene, and Provitamin A in their plasma or cerebral fluid. Clinical studies have suggested that these vitamins may be effective in slowing the onset of dementia. In laboratory studies, Vitamin A (retinol, retinal, and retinoic acid) and β-carotene have been found to prevent the production, elongation, and are helpful in destabilisation of β-amyloid fibrils. Further, when transgenic animal models of AD were given, intraperitoneal infusions of Vitamin A, there was a reduction in Aβ deposition and tau phosphorylation, a slowdown in neuronal degeneration, and an enhancement of spatial learning and memory (Ding et al. 2008). Depriving mice of retinoic acid severely impairs their spatial learning and memory, demonstrating the importance of vitamin A in maintaining memory function. Retinoids have been found to suppress the expression of pro-inflammatory cytokines and chemokines in activated microglia and astrocytes in AD. Retinoid signalling has been linked to a variety of biological activities in the healthy brain, including cell proliferation, neurogenesis, dendritic spine

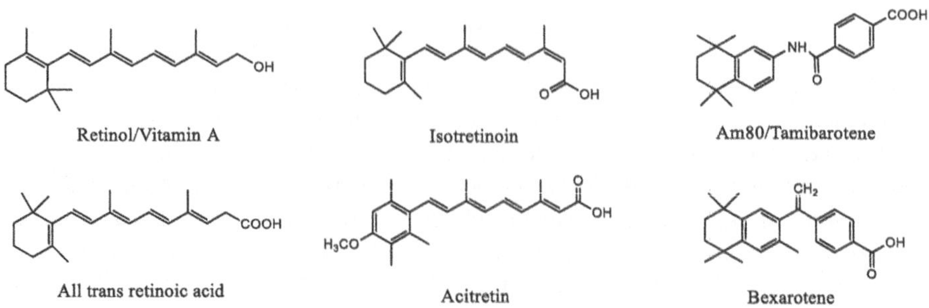

Fig. 12.1 Vitamin A and its derivatives

development, and immune system modulation (Hou et al. 2015). These findings suggest that vitamin A and β-carotene or their derivatives could be a promising treatment option for managing AD. Several derivatives of vitamin A were obtained either from natural sources or via synthesis and tested for its activity in AD. Compounds like ATRA, tamibarotene, bexarotene, isotretinoin, and acitretin showed beneficial effects (Fig. 12.1).

All-*trans*-retinoic acid (ATRA) is a bioactive derivative of vitamin A with established clinical indications. It is approved by the United States Food and Drug Administration (FDA) for treating acute promyelocytic leukaemia (APL) and has also been used to treat acne vulgaris. Being a RA isoform, ATRA interacts with retinoic acid receptors (RARs) and retinoid X receptors (RXRs). RARs and RXRs are nuclear receptors that each have three different subtypes (α, β, and γ); each subtypes have various isoforms. These isoforms have varying affinities towards the RA stereoisomers. These nuclear receptors can affect many cellular processes, including cellular differentiation, proliferation, and apoptosis. ATRA activates the manganese superoxide dismutase (MnSOD2) gene, which translates to the antioxidant enzyme in mitochondria (Kapoor et al. 2013).

ATRA is also involved in the regulation of arachidonic acid metabolites. Excessive generation of these metabolites can cause inflammation, oxidative stress, and neurodegeneration. ATRA might be passively involved in the regulation of the amyloid precursor protein (APP) processing. RAR-α and RAR-β can activate the disintegrin metalloproteinase 10 (ADAM10). Activation of ADAM10 promotes the production of a soluble form of Amyloid-β (APP-α) (Fahrenholz et al. 2010). It is also affected by other derivatives of vitamin A, like acitretin (Tippmann et al. 2009) and Am80.

Acitretin is an orally active drug and is currently used for the treatment of psoriasis. By activating the α-secretase ADAM10, the aromatic retinoid acitretin was discovered to have an anti-amyloidogenic impact in mice models and human patients. After being administered intravenously, it has previously been shown to traverse the blood–brain barrier and boost nest-building ability in the 5xFAD mouse model. It also shows immune-modulatory potential in the 5xFAD mice model. Administration of acitretin results in significant rise of interleukin-6 (IL-6) in both mice and humans, despite the fact that some serum analytes did not alter. This shows that, in addition to inducing α-secretase, acitretin also has an immune stimulatory effect that may help to improve the learning and memory in the mouse model (Endres et al. 2014). Acitretin improves non-amyloidogenic APP processing in a human and has a protective effect against neurodegeneration.

Am80 or Tamibarotene, a selective agonist for the retinoic acid receptor alpha/beta currently being used to treat leukaemia. Due to its transcriptional regulation of numerous genes implicated in the aetiology and pathogenesis of AD, tamibarotene is regarded as a potential candidate for the treatment of AD. Administration of Am80 to APP23 AD model mice showed reduced accumulation of insoluble amyloid-β. The retinoid-induced potentiation of either phagocytosis or α-secretase transcription by alternatively activated microglia may be the cause of this decrease in insoluble amyloid-β. Tamibarotene significantly reduced the inflammation of central nervous system cells caused by LPS-interferon (IFN). Micromolar doses of Am80 restored neuronal viability in organotypic midbrain slice cultures. An LPS-induced inflammatory model also revealed tamibarotene's neuroprotective effects on midbrain neurons. In cultured SH-SY5Y neuroblastoma cells, both RA and tamibarotene enhanced substantial neurite outgrowth and elevated expression of neurotrophic tyrosine kinase receptor B (TrkB), which encodes the cognate receptor for brain-derived neurotrophic factor (BDNF). Vitamin A derivatives, including tamibarotene, boost the gene transcription of ADAM10 (a member of the disintegrin and metalloprotease), as well as the activity of α-secretase, which cleaves the APP protein at a different site and releases the soluble neuroprotective protein sAPP (Fukasawa et al. 2012).

Bexarotene (BEXA), a retinoid X receptor agonist, is currently being used in the treatment of cutaneous T-cell lymphoma. The liver X receptor and retinoid X receptor (RXR) transcriptionally regulate the expression of apolipoprotein E (ApoE), which is essential for Aβ clearance. A possible treatment for AD that addresses amyloid pathology and memory loss might be bexarotene, which boosts ApoE expression and microglia phagocytosis (Tousi 2015). However, some investigations do not favour bexarotene as the new drug for AD due to several failures. A study has examined bexarotene's impact in vivo at various levels in TASTPM transgenic mice. These mice did not exhibit any discernible memory enhancement, plaque reduction, or increased microglial cell activation after receiving BEXA orally for 7 days. When analysed the microglial phagocytic state, no differences were discovered (Balducci et al. 2015). Because of these irregularities in results, bexarotene remains a topic of research and discussion for its use to treat AD. **Isotretinoin** is another vitamin A derived drug currently used for the treatment of acne. It acts on the retinoid receptors. Isotretinoin, patented by a pharmaceutical company Hexal AG, is in phase II clinical trials for the treatment of AD.

12.2.2 Vitamin B

Vitamin B is a set of eight water-soluble, structurally unrelated compounds that functions in highly similar ways inside the cells. These eight compounds include thiamine (B1), riboflavin (B2), niacin (B3), pantothenic acid (B5), pyridoxine (B6), biotin (B7), folic acid (B9), and cobalamin (B12). They participate in a variety of anabolic and catabolic enzymatic processes as co-enzymes and assist in synthesising neurochemicals and signalling molecules, DNA/RNA synthesis and repair, energy generation, and methylation of molecules. Most animals require vitamin B in their diets except vitamin B7, which the body can produce. Each of these vitamins performs a particular function and is essential to cellular processes. All the functions of vitamin B can be classified into two categories; one functions in catabolic metabolism, which are associated with energy generation, and second in anabolic metabolism, which are related to the formation and modification of biomolecules (Kennedy 2016).

Vitamin B is used by brain cells in a variety of enzymatic reactions that result in the production of energy and the creation of neurotransmitters. Lack of these substances may result in the failure of these processes, which can lead to brain dysfunction, such as disruption of glucose metabolism, which might lead to AD. In mammalian brains, the pentose phosphate pathway and the Krebs cycle are the two main mechanisms for glucose metabolism; thiamine is essential for both. Thiamine depletion and the disruption of thiamine-dependent pathways in glucose metabolism have long been recognised to be linked with AD. Thiamine deficiency causes pathophysiological changes that are strikingly similar to those of AD, including tau protein hyperphosphorylation, selective neuron loss, the formation of neurofibrillary tangles, increased secretion of the Aβ protein, and abnormal deposition in the hippocampus and nearby regions. Recent research has demonstrated that thiamine shortage increases the production and deposition of Aβ in the brains of animal models for AD. Unfortunately, due to its limited bioavailability, thiamine has not demonstrated a significant therapeutic effect in clinical studies for AD (Pan et al. 2010; Tapias et al. 2018).

Fursultiamine is a thiamine derivative (Fig. 12.2) containing a disulphide moiety; Yasuyo et al. studied the effect of fursultiamine on AD and concluded that it might have a modest therapeutic effect in patients with AD. Fursultiamine therapy not only showed improvement in emotional and mental symptoms but, in some cases, intellectual abilities were also enhanced. However, they also stated that a favourable effect was only evident in certain patients; people with severe impairment showed no beneficial effect (Mimori et al. 1996). Another study performed on the transgenic mice supports the above observation, although they reported that benfotiamine performs better than fursultiamine (Pan et al. 2010). **Benfotiamine** is *S*-acyl thiamine derivative (Fig. 12.2), whereas the majority of other thiamine derivatives, including fursultiamine, are disulphide derivatives. This specific alteration may impart distinctive pharmacological effects of benfotiamine. Benfotiamine (BFT) could reduce the production of amyloid plaques by modifying glycogen synthase kinase 3 (GSK-3) activities, it promotes the phosphorylation of Akt, which is the upstream kinase of GSK-3β (Pan et al. 2010, 2016). BFT therapy has been shown to have significant positive benefits in mice models of Alzheimer's and other brain diseases; however, to show its pharmacological action, it requires in large dosages (100–200 mg/kg per day). This would be equal to 10 g per day in people if

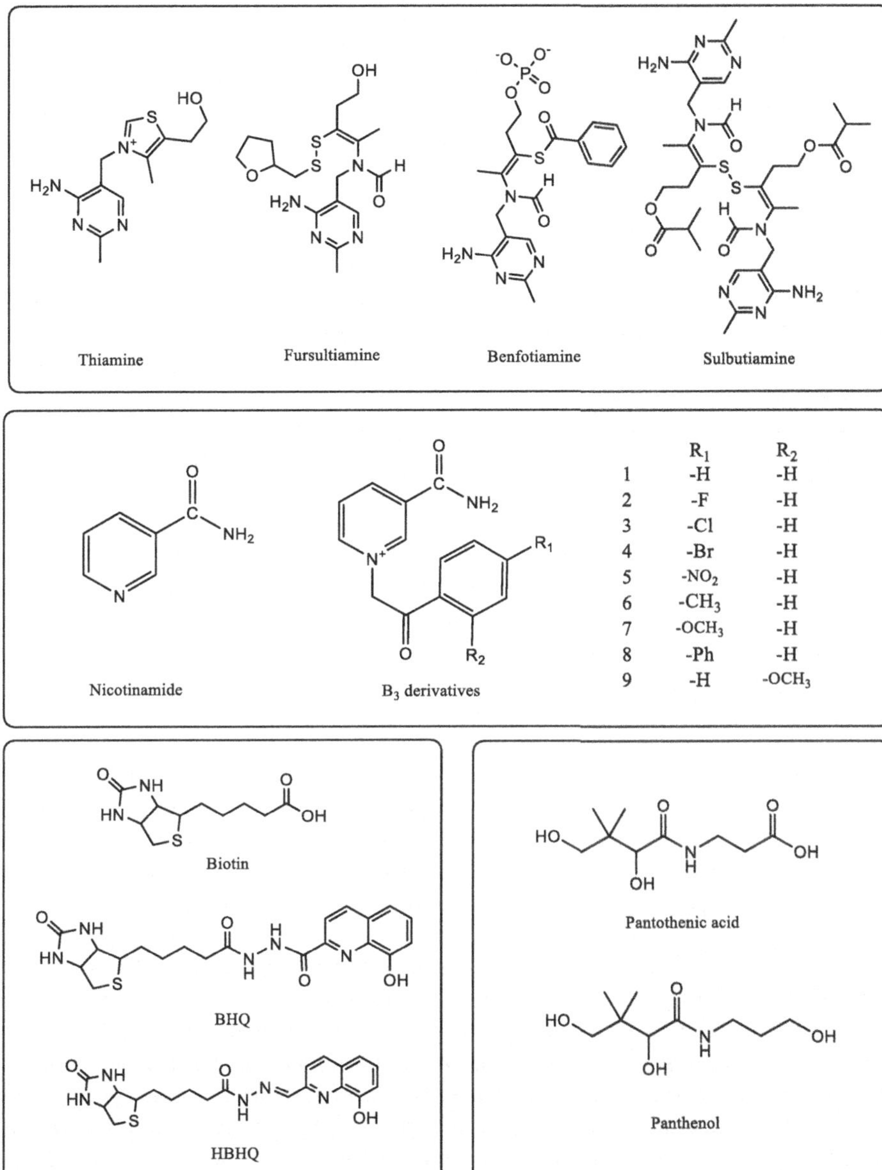

	R_1	R_2
1	-H	-H
2	-F	-H
3	-Cl	-H
4	-Br	-H
5	$-NO_2$	-H
6	$-CH_3$	-H
7	$-OCH_3$	-H
8	-Ph	-H
9	-H	$-OCH_3$

Fig. 12.2 Several vitamin B and their derivatives

equivalent. Though BFT has no noticeable adverse effects, it appears challenging to give patients significantly larger dosages of the drug. Thus, there is a need to identify precursors that work at lower dosages and have advantageous effects comparable to those of BFT. Recently, it was shown that dibenzoylthiamine (DBT) might fulfil these requirements. In Japan, DBT is legal as a food ingredient and has been known for a very long time. Even though no toxic or tumour-causing side effects have been documented, there is no data about its biological effects. It was more well tolerated than thiamine or BFT in salmon yearlings and increased thiamine retention over time. Moreover it can act as an anti-inflammatory agent, thiamine precursor, and antioxidant at very low concentration compared to BFT. DBT offers therapeutic promise for brain disorders connected to inflammation and oxidative stress (Sambon et al. 2020, 2021).

A thiamine derivative (Fig. 12.2) called **Sulbutiamine** was made in Japan in the middle of the 1960s as a beriberi therapeutic medication. Since then, other prospective uses have been proposed. Sulbutiamine, for instance, could have antioxidant, nootropic, and anti-fatigue benefits. Two thiamine molecules are fused to form sulbutiamine, after the respective thiazolium rings of each half-molecule are opened and esterification is done by an isobutyryl group to the primary alcohol of each molecule. When compared to thiamine, these changes make this derivative more lipophilic. In particular, sulbutiamine's thiol content, which can have neuroprotective benefits and can impact antioxidant status. Since it provides a large antioxidant pool made up of (1) sulphur bound in disulphide bridges, (2) thiols bound to proteins, and (3) free thiol primarily in the form of reduced glutathione (GSH), the global thiol concentration is well known to play a significant role in the regulation of cellular redox status. It's interesting to note that sulbutiamine and other thiol-containing substances have been demonstrated to increase GSH, which may decrease oxidative stress in neurons. Sulbutiamine administration has furthermore shown to be advantageous in AD (Starling-Soares et al. 2020).

Niacin-derived nucleotides such as nicotinamide adenine dinucleotide phosphate (NADP) and nicotinamide adenine dinucleotide (NAD) are essential for a wide range of enzymes and processes that are involved across every aspect of peripheral and nerve cell activity. These include oxidative reactions, DNA metabolism and repair, cellular signalling functions (through intracellular calcium), antioxidant protection, and folate to tetrahydrofolate conversion, in addition to energy generation (Kennedy 2016). Moreover, some **vitamin B3 derivatives** (Fig. 12.2) exhibit notable cholinesterase inhibitory action. In one study, the activity of the tested vitamin B3 derivatives on both cholinesterases was reversible and depended on the presence of a substituent on the side ring of the compound. The tested analogues were shown to be more selective acetylcholinesterase (AChE) inhibitors. According to this study, nicotinamide analogues 6 and 8 were the most effective AChE inhibitors. Derivative 6 shows selective inhibition of AChE which is 80 times higher than inhibition of BuChE. The nicotinamide derivative 8 was the most effective inhibitor of BuChE, indicating that the binding of both enzymes' active site is considerably influenced by the presence of an aromatic ring as a substituent. Four of the tested analogues, however, had a cytotoxic impact on both HEK293 and SH-SY5Y cells. Although the IC_{50} values did not place these molecules in the category of being very cytotoxic (Zandona et al. 2020).

Vitamin B5 (Pantothenic acid) is a water-soluble B complex that serves as a coenzyme A (CoA) synthesis substrate. Vitamin B5 deficiency in the cerebral region of the brain can cause dementia and neurodegeneration in AD, which may be prevented or perhaps reversible in its early stages with oral vitamin B5 doses adequate to normalise brain levels (Xu et al. 2020). Most of the pantothenic acid derivatives can biotransform into the CoA which leads to the numerous biological activity. Because acetyl-CoA is a key molecule in choline acetylation, the effects of pantothenic acid derivatives on CoA metabolism make them a promising tool for manipulating biosynthesis of acetylcholine. These compounds can potentially be employed in the treatment of AD in conjunction with glutamatergic drugs such as memantine (MEM). To modulate the cholinergic system, acetylcholinesterase inhibitors are typically employed in combination therapy; however, many of these medications are toxic and unsafe for daily use. In such cases, pantothenic acid derivatives may be advantageous because of their low cost, availability, and most importantly safety.

Pantothenic acid and its derivatives can potentially increase GSH concentrations in the brain. Tripeptide GSH plays a vital function as an antioxidant and in redox state maintenance in neurons and glial cells in the brain. Because of its interaction with glutamate receptors, GSH can operate as a neuromodulator. Deficits in the production and metabolism of GSH plays a pivotal role in neurodegenerative disorders. These deficiencies may lead to both neural plasticity and neuronal death. It has been suggested that GSH depletion occurs before neurodegeneration. Stepanichev et al. conducted behavioural study to examine the impact of calcium pantothenate (PAC) and panthenol (PL), which are derivatives of pantothenic acid, on the rat subjects, moreover they observed the CoA content and redox state of GSH in a scopolamine-induced amnesia model. They also investigated how these substances interacted with the anti-amnesic drug MEM. In order to regulate CoA metabolism in the brain, PAC and PL were used. Derivatives of pantothenic acid shielded different tissues including nerve tissue against various harmful elements. Most of these detrimental effects are brought on by oxidative stress. Pantothenic acid preincubation was found to dramatically boost the concentration of GSH in the cells. Pantothenic acid, PL, and PAC all have preventative antioxidant actions that are related to how they affect the GSH system (Turnaturi et al. 2016).

Transition metals are dangerous for the ageing brain and many disorders, including Alzheimer's disease (AD), despite being crucial for neuronal function. Cellular activity is hampered by oxidative stress and macromolecular damage brought on by the abnormal accumulation and distribution of reactive iron, copper, and zinc. A rear up body of research suggests that abnormal transition metal homeostasis plays a substantial role in the aetiology of AD, an age-related neurodegenerative disease that manifests massive accumulations of oxidative stress-induced damage (Turnaturi et al. 2016).

Metals' substantial significance in AD-type neurodegeneration is intriguing, and it opens up a possible opportunity for therapeutic intervention in these disorders. Graziella Vecchio and colleagues synthesised and characterised biotin derivatives with an 8-hydroxyquinoline moiety, such as **Biotin-8-hydroxyquinoline (BHQ)** and **8-hydroxyquinolyl-biotin hydrazone (HBHQ)**, a hydrazone conjugate of biotin (Fig. 12.2) with HQ and a series of transition metal complexes. Their metal complexes were also investigated as antioxidant agents. This research found that 8-hydroxyquinoline derivatives could be utilised to treat AD by regulating metal imbalances and reducing oxidative stress. Additionally, the inclusion of biotin may enhance their pharmacokinetics, such as improving absorption in the intestine and increasing bioavailability via the sodium-dependent multivitamin transporter (SMVT) (Turnaturi et al. 2016).

Pyridoxine (vitamin B6) was long thought to have only an enzyme cofactor role. Recently, it was discovered that it is a strong antioxidant. According to studies, pyridoxine prevents the synthesis of free radicals and acts as a single oxygen quencher. Additionally, epidemiological and clinical investigations revealed that an increased homocysteine (Hcy) level in the blood is a risk factor for developing AD. Pyridoxine deficiency causes Hcy levels to rise noticeably (Kennedy 2016; Yang et al. 2017). In addition, excessive Hcy concentrations may lead to elevated Aβ levels and are shown to accumulate in a transgenic mouse model of amyloidosis similar to AD. These advantages suggest that adding pyridoxine to a supplement may help with AD therapy.

The diverse spectrum of biological activities of resveratrol, a naturally occurring compound having a stilbene structure, has been studied intensively. Resveratrol seems to be an anti-AD agent with anti-inflammatory, antioxidant, and neuroprotective properties. Studies have also demonstrated that it has the capacity to inhibit monoamine oxidase, and these biological actions depend heavily on the stilbene structure (Yang et al. 2017). Additionally compounds with phenolic Mannich base moieties may show strong antioxidant, AChE inhibitory activity, and metal chelating characteristics. These findings point to vitamin B6 and resveratrol with Mannich base moieties as possible beginning points for the development of multifunctional medicines for the treatment of AD. **Pyridoxine-resveratrol hybrids** Mannich base compounds were synthesised and tested for their effect in AD. The majority of them demonstrated AChE and MAO-B-specific inhibition. In general, the inhibitory action against AChE was greatly boosted by the insertion of the Mannich base moiety to the 3′ or 4′ position of the benzene ring. However, it could drastically lessen MAO-B inhibitory action. Compounds 7d and 8b, having IC_{50} values of 2.11 μM and 1.56 μM, respectively, had the greatest efficacy for AChE inhibition among these derivatives. According to a kinetic analysis study, molecule 7d is bound to both the catalytic anionic site (CAS) and peripheral anionic site (PAS) of AChE, exhibiting a mixed-type inhibition. The majority of these molecules were specific MAO-B inhibitors for MAO. The strongest MAO-B inhibition was seen with compound 7e, having an IC_{50} value of 2.68 μM (Yang et al. 2017).

For the treatment of AD, another family of **pyridoxine-based triazole** compounds was identified as multi-target directed ligands. According to this study, 6 of the 17 produced pyridoxine-based triazoles exhibited high AChE inhibition and antioxidant activity. AChE activity was benefited by meta- and ortho-substitution on the aromatic ring. They discovered that 5i (EeAChE IC_{50} = 1.56 mM) is the best molecule in this new series through additional metal chelation studies (Pal et al. 2020).

Pyridoxine-carbamate-type derivatives of pyridoxine have metal chelating characteristics that can be used to modify bio-metals. Metal homeostasis in the brain is rather restrictive in normal physiological environments for healthy neuronal tissue functioning. Metal chelators may be employed to mitigate bio-metals' impairment. In several disorders, including Wilson's, thalassemia, multiple sclerosis, and AD, the preventive impact of metal chelators has been studied. In cases of copper, lead, mercury, and nickel poisoning, some medications, such as penicillamine, are utilised as chelators. The pyridoxine-carbamate series compound can be used to manage AD; these classes of molecules exhibited metal chelating and AChE binding capabilities in an experimental study (Pal et al. 2021). Pyridoxine and its derivatives are given in Fig. 12.3.

Folic acid and vitamin B12 are essential for the proper functioning of the central nervous system (CNS) at all stages of life, particularly in the conversion of homocysteine to methionine through the action of methionine synthase. This process is critical for nucleotide synthesis and methylation of both genomic and non-genomic material. Folic acid and vitamin B12 may also have a preventative role in the development of CNS abnormalities, mood disorders, and various forms of dementia, such as Alzheimer's disease and vascular dementia in the elderly (Reynolds 2006). Vitamin B12 and folate are risk factors for dementia based on their relations as cofactors in homocysteine metabolism, and homocysteine has been related to the risk of developing AD. Folic acid is also involved in the metabolic pathway for acetylcholine synthesis. Another possible biological mechanism of folate effects on dementia is folate deficiency, which may decrease acetylcholine, a neurotransmitter that helps send messages between nerve cells, thus which leads to AD. Furthermore, a lack of folate increases oxidative stress; in terms of dementia, it may contribute to cognitive impairment of the ageing brain, sometimes resulting in reversible dementia but also increasing the likelihood of AD and vascular dementia (Reynolds 2002).

12.2.3 Vitamin C

A class of water-soluble vitamins includes vitamin C, commonly known as ascorbic acid. Most mammalian organisms can produce ascorbic acid on their own. Unfortunately, some animals, including guinea pigs, fruit bats, humans, and other primates, lack this ability because they lack the l-gulono-1,4-lactone oxidase enzyme; this enzyme is part of the biochemical pathway that produces vitamin C from glucose. Consequently, gut bacteria do not generate vitamin C. The factors mentioned above make these species rely entirely on nutritional consumption. The general recommendation of everyday intake of vitamin C is 90 mg for men and 75 mg for women. This amount should be raised by 35 mg for smokers (Kocot et al. 2017).

Vitamin C has two forms inside the organism. One is ascorbic acid; at bodily pH, it occurs in its anion form, ascorbate. Another form is dehydroascorbic acid. Ascorbate has a wide range of actions in the central nervous system (CNS) and brain. It acts via the donation of a single electron. It can directly scavenge oxygen- or nitrogen-based reactive species produced during regular cellular metabolism in

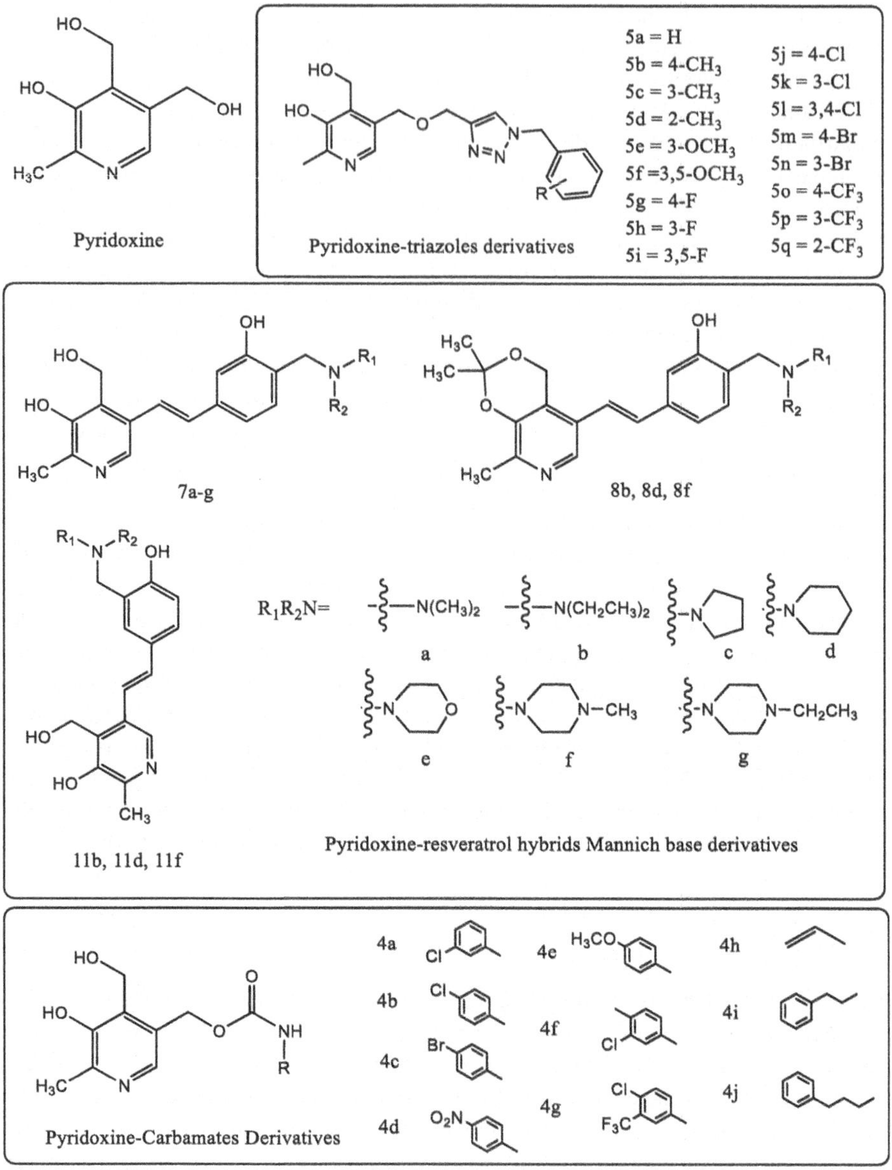

Fig. 12.3 Pyridoxine and its derivatives

terms of its antioxidant properties. Ascorbate will efficiently scavenge superoxide at the millimolar amounts found in neurons in vivo, a significant diffusible result of fast neuronal mitochondrial metabolism. According to research, ascorbic acid interacts with sodium-dependent vitamin C transporter-2 (SVCT2) to traverse the blood–brain barrier (BBB). Dehydroascorbic acid also penetrates the BBB through glucose transporters (GLUT) in its oxidised form. Vitamin C is a well-known antioxidant and can be helpful in AD (Harrison and May 2009).

Despite the advantages mentioned, ascorbic acid has significant pharmacological disadvantages, such as limited bioavailability. At neutral pH, ascorbic acid usually exists in the anionic form, which results in a slower diffusion through the plasma membrane. The bioavailability is observed to decline with increasing oral dosage, probably because of saturation of sodium-dependent vitamin C transporter-1 (SVCT1). Additionally, active transporters are necessary for distribution from the circulation to the intended organs. As a result, the concentration of ascorbic acid in plasma is 2.5 times greater than that in tissue. Applying a greater daily dose might not be able to get around this problem (Jiaranaikulwanitch et al. 2021).

Additionally, high consumption raises the risk of haemolysis in individuals with iron excess, paroxysmal nocturnal haemoglobinuria, oxalate kidney stone formation, and glucose 6-phosphate insufficiency. Moreover, vitamin C acts as a pro-oxidant because of its reducing nature. This phenomenon produces more ROS by reducing catalytic metals like Fe^{3+} or Cu^{2+} to Fe^{2+} and Cu^{+} (Jiaranaikulwanitch et al. 2021).

Ascorbic-triazole conjugates (Fig. 12.4) containing tryptoline and phenolic moieties were effective in AD. While phenolic moieties impact oxidative stress and amyloid aggregation, tryptoline is a well-known inhibitor of BACE1 (beta-site APP-cleaving enzyme 1) and choline esterase. Six ascorbic acid derivatives containing tryptoline and phenolic moieties connected by triazole linkers were described by Jutamas et al. 2c and 5c; two of these six compounds displayed encouraging results (Jiaranaikulwanitch et al. 2021). Both compounds demonstrated effectiveness against the amyloid cascade hypothesis, including inhibition of BACE1 (2c IC_{50} = 725.70 μM; 5c IC_{50} = 593.10 μM), suppression of amyloid aggregation (2c IC_{50} = 92.33 μM; 5c IC_{50} = 136.00 μM), and antioxidant (2c IC_{50} = 72.26 μM; 5c IC_{50} = 62.89 μM), anti-inflammatory, and neuroprotective action. Additionally, authors suggested that compounds 2c and 5c could work together to enhance molecules' anti-amyloid and BACE1 inhibitory effects. As these compounds interact with the SVCT2 transporter, in silico studies indicate their possible permeability through the BBB (Jiaranaikulwanitch et al. 2021).

Ascorbic acid can be conjugated with other molecules to treat various disorders linked with the brain since it can pass across the BBB. **Diclofenamic acid conjugates** of ascorbic acid (Fig. 12.4) were explored extensively for their potential use in treating Alzheimer's disease. Inhibition of cyclooxygenase (COX) might be beneficial in managing Alzheimer's (Manfredini et al. 2002). Axel et al. developed a natural product based on **palinurin-inspired derivatives** (Asc1, Asc2, and Asc3) of ascorbic acid (Fig. 12.4) that demonstrated activity against GSK-3β. IC_{50} values of Asc1, Asc2, and Asc3 were >50 μM, 6.1 μM and 5.8 μM, respectively, for inhibition of GSK-3β. A hydrophobic chain was attached to ascorbic acid; this modification improved the kinase inhibitory activity. GSK-3β is a well-known target for AD treatment. Moreover, the authors reported that all three derivatives (Asc1, Asc2, and Asc3) bind to the allosteric site and do not compete with ATP or peptide substrate (Bidon-Chanal et al. 2013).

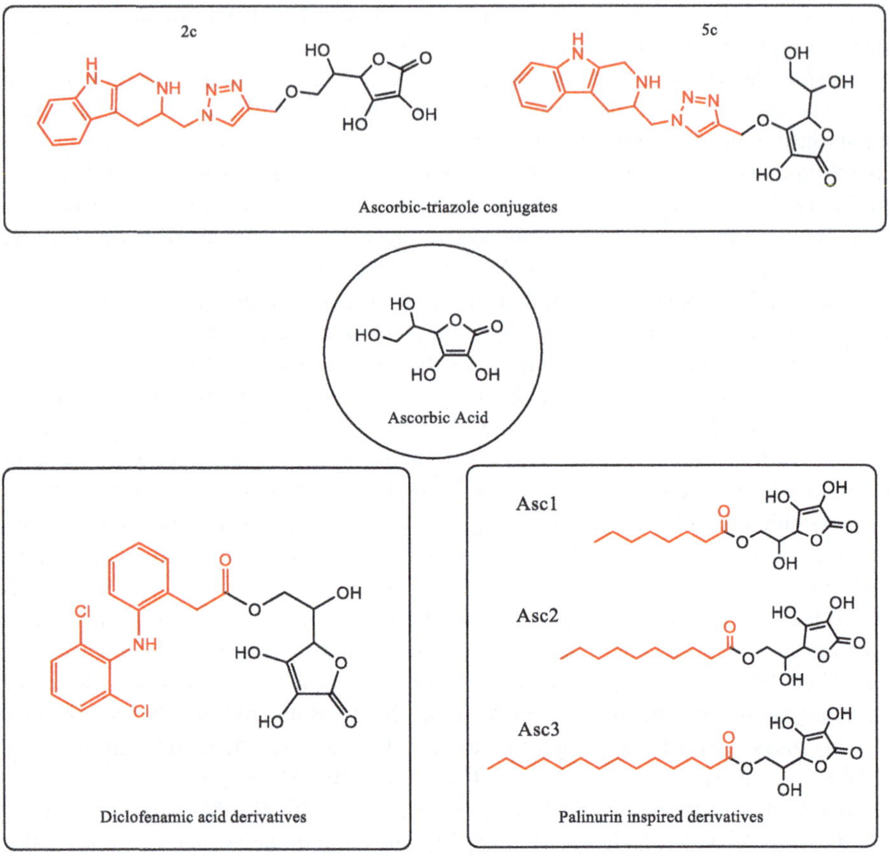

Fig. 12.4 Vitamin C and its derivatives

12.2.4 Vitamin D

The word "vitamin D" refers to a group of structurally similar and fat-soluble secosteroids such as ergocalciferol, cholecalciferol, 25-hydroxyvitamin D (calcidiol), and 1,25-dihydroxyvitamin D (calcitriol). Ergocalciferol (vitamin D2) and cholecalciferol (vitamin D3) are the two major compounds under vitamin D. Vitamin D3 (cholecalciferol) is generally synthesised by humans' epidermis (part of skin) from precursor molecule 7-dehydrocholesterol. Additionally it can be also obtained through the consumption of animal-based foods, while vitamin D2 (ergocalciferol) is mostly manufactured and added to food. Commercially produced vitamin D3 and vitamin D2 can be found in dietary supplements and foods that have been fortified. Only the side chain structure differentiates the D2 and D3 types. Both isoforms serve as prohormones; the variations do not alter metabolism (i.e., activation). When activated, it has been shown that the D2 and D3 forms show the same bodily reactions, and their effectiveness in treating vitamin D deficiency disease is equivalent. In addition, vitamin D2 has been found to be less harmful than vitamin D3 in experimental animal tests, although this has not been proven in humans (Jurutka et al. 2001; Gall and Szekely 2021).

Vitamin D is very crucial for humans and all other vertebrates. It plays a significant role in the maintenance of calcium and phosphate concentration in blood. It promotes the development of normal bone, optimum muscular contraction maintenance, and cellular functioning in many parts of the body. It is also involved in a wide range of biological processes, including cell proliferation and differentiation, immunological response, and cytokine control, in addition to their most well-known impact of regulating calcium absorption and stimulating bone mineralisation. According to published studies in the last two decades, the vitamin D signalling pathway has been implicated in several brain illnesses and brain functions (Gall and Szekely 2021). First, animal models' studies showed that it influences how the brain develops in infancy and supports neuroprotection, synaptic plasticity, neural connectivity, and dopaminergic system turnover in adulthood. Second, epidemiological evidence demonstrated a link between children's cognitive development abnormalities and low mother blood calcidiol levels during pregnancy. Third, serum calcidiol levels were lower in those with neurodegenerative disorders (such as Alzheimer's and Parkinson's disease), neuroinflammatory diseases, and neuropsychological ailments. Based on these findings, vitamin D supplementation was tried as a treatment for many neuropsychiatric conditions, such as dementia and cognitive impairment. However, there is no proof that higher vitamin D consumption increases vitamin D receptor (VDR) activation in the human brain (Gall and Szekely 2021).

Vitamin D can influence neurocognition via various pathways, including oxidative stress modulation, neuroprotection induction, inflammatory process suppression, and calcium homeostasis regulation (Bivona et al. 2019). The vitamin D receptor (VDR) goes through a conformational shift upon binding of vitamin D that enables the retinoid X receptor (RXR) association. Being a nuclear receptor RXR can bind to various sites on DNA and leads to many activities of vitamin D. It is hypothesised that the VDR-RXR complex actively interacts with the vitamin D response region found on the DNA. Vitamin D receptor as well as the CYP450 enzyme CYP27B1 that activates vitamin D both shows the expression in brain, additionally vitamin D and its metabolites are known to have blood–brain barriers permeability. Vitamin D decreases the breakdown of amyloid precursor protein (APP) by amyloidogenic β-secretase and increases the degradation of Aβ. In accordance with this, 25(OH) vitamin D therapy raised Aβ-degradation resulting in higher neprilysin (NEP) expression along with its activity, suggesting vitamin D derivatives as a rational approach for treatment of AD (Annweiler and Beauchet 2011; Grimm et al. 2017).

Most vitamin D analogues have side-chain modifications and have less calcaemic action than vitamin D3 which is naturally found, but they nonetheless have many of the same therapeutic qualities. For example, **Maxacalcitol**, a 1,25-hydroxylate vitamin D3 analogue which is used in the treatment of renal patients with secondary hyperparathyroidism. In contrast, **Calcipotriol**, a 1,24-hydroxylated vitamin D3

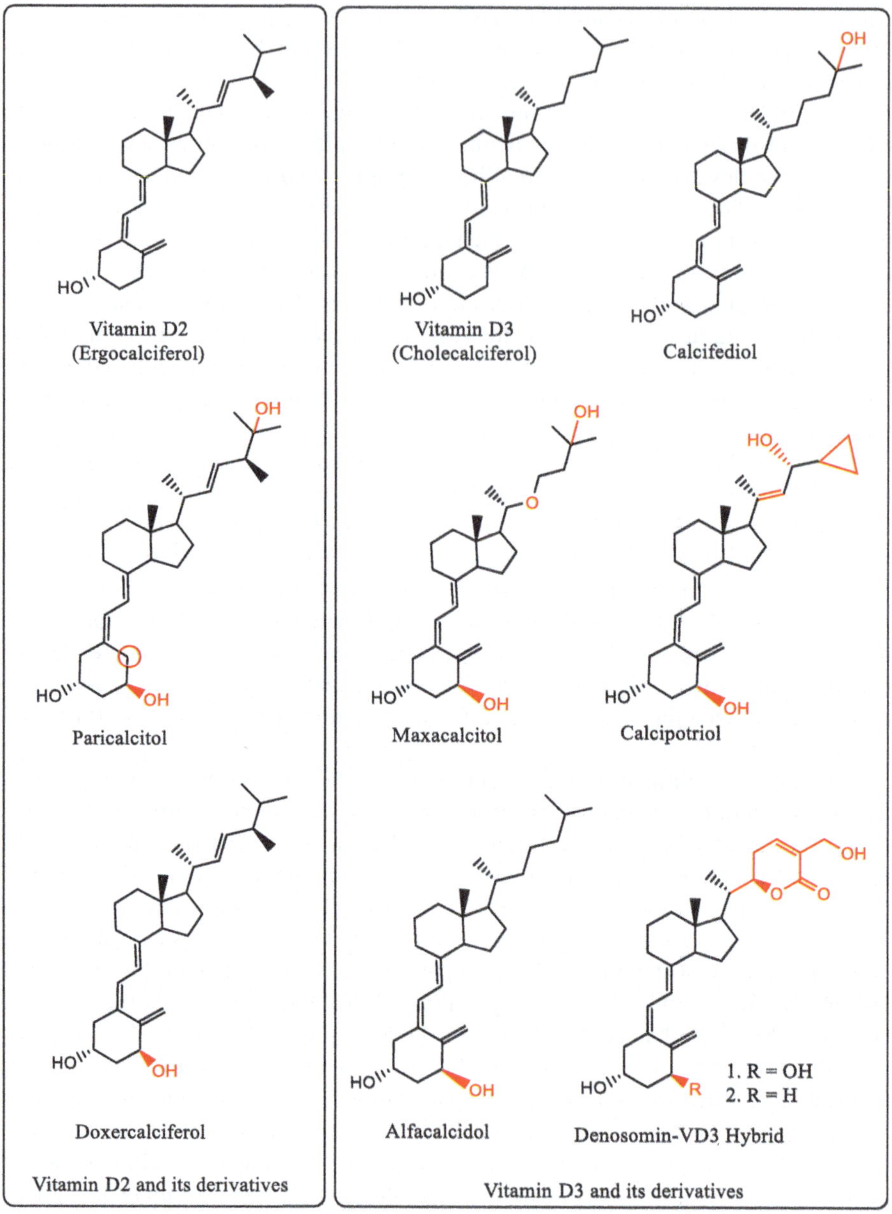

Fig. 12.5 Vitamin D and its derivatives

analogue which is used in the treatment of psoriasis, and **Alfacalcidol**, a 1-hydroxylated vitamin D3 analogue which is used to treat secondary hyperparathyroidism and osteoporosis (Grimm et al. 2017; Fan et al. 2019).

Paricalcitol and **Doxercalciferol** are vitamin D2 analogues. Paricalcitol has hydroxyl groups at C1 and C25, just like the vitamin D3 analogue maxacalcitol; however, it has a side chain of vitamin D2 instead of a vitamin D3 side chain. The vitamin D2 analogue 1-hydroxylated doxercalciferol is equivalent to alfacalcidol in terms of hydroxylation status. Paricalcitol and doxercalciferol treat secondary hyperparathyroidism, characterised by increased plasma parathyroid hormone levels (Grimm et al. 2017; Fan et al. 2019).

Vitamin D and all of the aforementioned D2 and D3 derivatives (Fig. 12.5) were investigated for their implications on Alzheimer's disease. Marcus et al. reported that vitamin D3, vitamin D2 and their derivatives could be beneficial in signalling pathways related to AD. They also noted that these molecules elevate A-β-degradation which indicates that vitamin D and its analogues play a crucial role in Aβ-homeostasis (Grimm et al. 2017). Their findings support vitamin D administration as a strategy for management and possible treatment of AD by increasing rate of Aβ catabolism, reducing Aβ anabolism, and lowering pro-inflammatory cytokines. Moreover, the studied vitamin D analogues have similar potencies for reducing Aβ levels. Still, they vary slightly in the effectiveness and exact mechanism, showing that different AD patients may benefit from vitamin D analogues to varying degrees. Vitamin D molecules like calcifediol and paricalcitol show increased non-amyloidogenic APP processing; therefore, AD patients having suppressed anti-amyloidogenic β-secretase activity may show better outcome from this treatment. As per their observations, vitamin D analogues with the major impacts on Aβ-degradation, such as calcipotriol and maxacalcitol, may be most beneficial for people with impaired Aβ-degradation (Grimm et al. 2017).

Almost all the vitamin D analogues show comparable effects for β- and γ-secretase processing, suggesting that those with higher amyloidogenic secretase activity shows beneficial effect from vitamin D derivatives in a similar way. It must be emphasised that there are only small differences in effect intensity and that all of the derivatives have been demonstrated to have similar positive effects on the Aβ level. Additionally, there were no discernible differences between vitamin D derivatives and vitamin D, indicates that vitamin D analogues as a treatment of AD had no therapeutic advantage over using calcifediol (or calcitriol) (Grimm et al. 2017).

Another important derivative of vitamin D3 is **Denosomin-VD3 hybrids** synthesised by Kenji et al. Denosomin (1-deoxy-24-norsominone) is a derivative of sominone, which is a natural steroid and found in Ashwagandha. Denosomin shows axonal extension activity on Aβ-damaged neurons. Kenji et al. reported that the denosomin-VD3 hybrid shows beneficial activity in Alzheimer's via a 1,25D3-MARRS pathway (Sugimoto et al. 2015).

12.2.5 Vitamin E

Fat-soluble vitamin E is an antioxidant that regulates the generation of RNS and ROS. The hydroxyl group of vitamin E has antioxidant activity. It also modulates

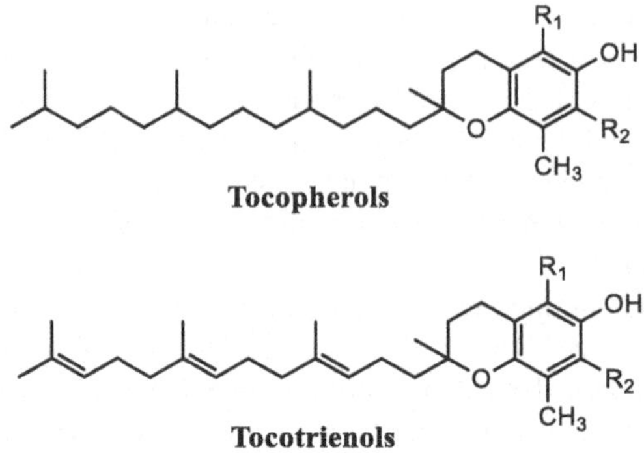

Types	R1	R2
α	CH3	CH3
β	CH3	CH3
γ	H	CH3
δ	H	H

Fig. 12.6 Vitamin E and its types

several cell signalling pathways and acts as a modulator of gene expression. Vitamin E has multiple forms depending upon their chemical structure, such as α-, β-, γ-, and δ-tocopherols and their related corresponding tocotrienols (Fig. 12.6), among which γ-tocopherol is the most prevalent in the diet. According to preclinical data, it is suggested that supplementing with vitamin E may be beneficial in AD since it not only reduces oxidative stress caused by Aβ but also improves memory and cognitive deficits. Tocotrienols exhibited promising results in AD animal models, exerting in some cases a more robust action than α-tocopherol, despite the fact that α-tocopherol is the vitamin E family member that has been the subject of most research. It shows synergistic antioxidant and anti-inflammatory activity when combining vitamin E with other compounds, thus which may help to treat AD (Gugliandolo et al. 2017).

Sang et al. designed several distinct **chalcone-Vitamin E-donepezil hybrids** (Fig. 12.7) multi-target active small molecules by combining vitamin E, donepezil, and chalcone. According to their studies, 17f is a good antioxidant, huMAO-B inhibitor, and a selective metal chelator as it inhibited AChE effectively (eeAChE IC_{50} 1.88 μM and ratAChE IC_{50} 0.41 μM). Also, the docking and kinetic experiments confirmed that 17f was an AChE inhibitor of mixed type. Additionally, it inhibited Cu^{2+}- and self-induced Aβ1–42 aggregation with notable percentage rates of 93.5% and 78.0% at 25 M, respectively, and disassembled self- and Cu^{2+}-induced aggregation of the accumulated Aβ1–42 fibrils with rates of 72.3% and 84.5%, respectively. Especially, 17f demonstrated good BBB permeability in vitro and neuroprotective impact on H_2O_2-induced PC12 cell damage, which indicates 17f is a potent multi-targeted ligand for treating AD (Sang et al. 2022).

An oxidative metabolite of α-tocopherol **Alpha-Tocopherolquinone** (α-TQ) (Fig. 12.7) has been shown to decrease Aβ induced cytotoxicity, oxidative stress, and inflammation by blocking Aβ42 fibril formation. It protects cells from oxidative stress by reversible two-electron redox cycling. Furthermore, it shows detox effects by eliminating reactive metabolites, suppressing ROS via quinine redox cycling, and generating the antioxidant hydroquinone (Crisostomo et al. 2007). α-TQ also inhibits Aβ aggregation, attenuated Aβ-induced cytotoxicity in SH-SY5Y neuroblastoma cells, and decreases inflammatory cytokine release as ROS and NO generation in BV-2 microglial cells in vitro (Yang et al. 2010). To investigate the effects and mechanisms of α-TQ on cognition and diseases in vivo, Wang et al. administered α-TQ orally to transgenic mice with AD. Their results demonstrated that α-TQ attenuated memory impairment, decreased Aβ oligomer levels, reduced lipid peroxidation, and restored superoxide dismutase activity in mice with AD. Furthermore, suppressing NF-κB signalling α-TQ decreased microglial activation and cytokine generation. These data imply that α-TQ has therapeutic promise in treating AD (Wang et al. 2016).

Tocotrienols and tocopherols vary structurally, as tocotrienols contain unsaturated isoprenoid side chains, whereas tocopherols comprise saturated side chains. It can address aspects of AD such as mitochondrial dysfunction, oxidative stress, and aberrant cholesterol production. Tocotrienol has been shown in preclinical investigations to lower oxidative stress by activating microbial activity and cellular repair, acting as a free radical scavenger; also, it protects cells from toxicity induced by glutamate. Human epidemiological research revealed a substantial inverse association between tocotrienol levels and AD development (Chin and Tay 2018). Tocotrienol is a potentially neuroprotective compound by lowering oxidative stress and promoting cellular repair, can halt the onset and progression of AD.

Ibrahim, Nor Faeizah, and colleagues studied the effects of vitamin E analogues such as α-tocotrienol (α-T3), γ-tocotrienol (γ-T3), and α-tocopherol (α-TOC) on Aβ oligomerization, aggregation, and disaggregation in vitro. They performed Thioflavin T (ThT) assay, which showed α-T3 reduced Aβ aggregation at 10 μM concentration. α-T3 and γ-T3 showed Aβ disaggregation, as evidenced by a decrease in ThT fluorescence α-TOC, on the other hand, had no impact. ThT assay results were confirmed with scanning electron microscopy imaging. Examination in the photoinduced cross-linking of unmodified protein assay revealed that γ-T3 reduced Aβ oligomerisation. Thus, tocotrienols have the potential role in developing therapeutic agents for AD (Ibrahim et al. 2021).

Trolox (Fig. 12.7) is the water-soluble and cell-permeable derivative of vitamin E that accumulates in the cytoplasm. It appeared to reduce Aβ plaque-induced oxidative stress and structural alterations in neurites and may neutralise free radical species, thus considered to be associated with ageing or AD. It can also prevent hydrogen peroxide and Aβ induced neurotoxicity and suppress GSK-3β (Munoz et al. 2002). As a result, various biological actions strengthen its neuroprotective capacity, making it a good lead for developing multifunctional drugs to treat AD. The increased tau hyperphosphorylation produced by Aβ depends on p38 mitogen-activated protein kinase (MAPK) and can be prevented by inhibiting MAPK. p38 is involved in the linkage of tau and Aβ. Giraldo et al. investigated the role of p38 signalling in AD and the protection provided by Trolox and found that incubation with Trolox inhibits Aβ-induced p38 activation (Giraldo et al. 2014).

Trolox has been used to develop **tacrine–trolox hybrids** to reduce metal-induced oxidative stress in AD. Teponnou et al. designed and synthesised tacrine-trolox hybrids having different linker chain lengths. Trolox moiety inhibited eqBuChE and TcAChE with IC_{50} of 3.16–128.82 nM, 17.37–2200 nM, respectively, and had free radical scavenging activities with IC_{50} in the range of 11.48–49.23 μM. Chains with

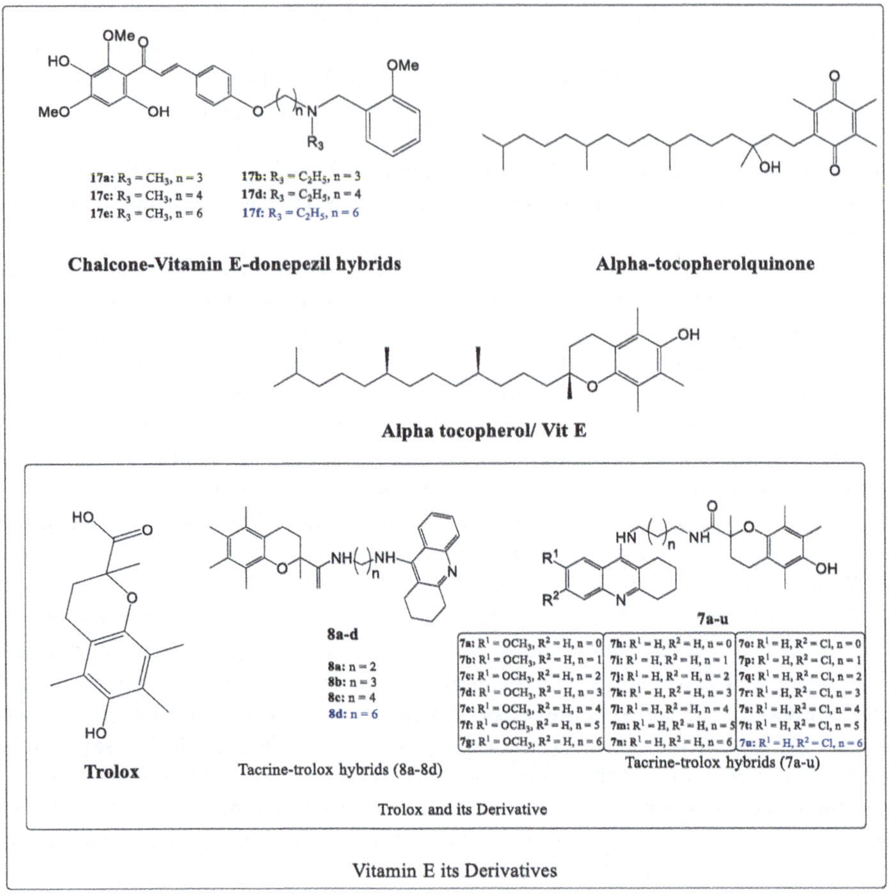

Fig. 12.7 Vitamin E and its derivatives

longer linker lengths hybrids improved ChE inhibitory activity. The docking experiments showed that compound 8d could bind to the PAS and CAS of TcAChE and eqBuChE, implying that it can inhibit ChE-induced Aβ aggregation. Out of all the derivatives, they found compound 8d the most potent and the best lead molecule showing an ability to inhibit AChE and BuChE and acting as a robust antioxidant. With these results, 8d can be used as a multifunctional agent for treating AD (Teponnou et al. 2017).

Nepovimova et al. designed and synthesised 21 tacrine–trolox hybrids (7a–u) as multifunctional candidates against AD which exerts antioxidant and anticholinesterase activities with less in vivo toxicity after intramuscular administration in rats and the possible to cross the BBB. Their results demonstrated that derivative 7u is the key structure (Nepovimova et al. 2015).

12.2.6 Vitamin K

Vitamin K (VK) refers to a class of fat-soluble chemicals known as naphthoquinones, which includes naturally present vitamins K1 and K2 as well as synthetic vitamin K3 (Fig. 12.8). Olive oil and other green vegetables are the main sources of vitamin K1, while vitamin K2 is found in trace amounts in eggs, butter, etc. VK is mainly found as menaquinone-4 in the CNS, which controls the action of proteins taking part in chemotaxis, mitogenesis, cell proliferation, myelination, and neuroprotection (DiNicolantonio et al. 2015; Ferland 2012). VK, long known for its involvement in bone development and blood clotting, has recently emerged as a critical nutrient for brain function. Its deficiency is associated with changes in sphingolipid homeostasis that seem to be detrimental to the mechanisms responsible for producing and absorbing Aβ peptides (Grimm et al. 2016).

Derivatives of vitamin K2 (VK2) by Josey et al. revealed that for neuroprotective effect, particular structural core is important; for example, naphthoquinone and amine substitution at the 2′ carbon considerably increased its protective activity. After adding a benzyl group at 2′ amine, the compound's safety index increased by eliminating neurotoxicity further; it also improved the compound's protective potency by chloro-substitution at the meta site of an aromatic ring. They derived a series of VK2 analogues as 1a–g, of which 2j and 2q are more effective even at lesser concentrations. Compounds 2j, 2q, and VK2 are most likely exert their effects through PARL regulatory mechanism (Josey et al. 2013).

Huy, Pham Dinh Quoc, and colleagues developed **analogues of vitamin K3** (VK3) that effectively protect neuroblastoma cells from Aβ-induced toxicity and prevent Aβ aggregation. They studied the VK3 analogue's effects on various Aβ amyloidogenic properties, viz. free radical formation, cell viability, and aggregation, using biological and spectroscopic methods. Using molecular dynamics modelling, they determined the binding mode and affinity of VK3 analogues binding to Aβ and found several VK3 analogues inhibited Aβ aggregation. The computed inhibition constants for VK3–6, VK3–9, and VK3–10 were in the μM range, which is close to the IC_{50} of curcumin. It also revealed that VK3–9 might efficiently reduce free radicals and protect against Aβ-induced cytotoxicity and inhibit Aβ aggregation. Modified analogues of VK3 can thus be designed as efficient anti-amyloidogenic drugs for treating AD (Huy et al. 2013).

12.3 Conclusion

In conclusion, vitamin-based derivatives may show to be a promising strategy for the treatment of Alzheimer's disease. Since it is a multifactorial disease, it needs therapy that can modify multiple cellular pathways. Vitamin-based derivatives exhibit anti-oxidant action, cholinesterase inhibitory activity, GSK-3β inhibitory activity, metal chelating activity, as well as some other properties that possibly slow the progression of Alzheimer's disease. This chapter discusses the several vitamin derivatives that

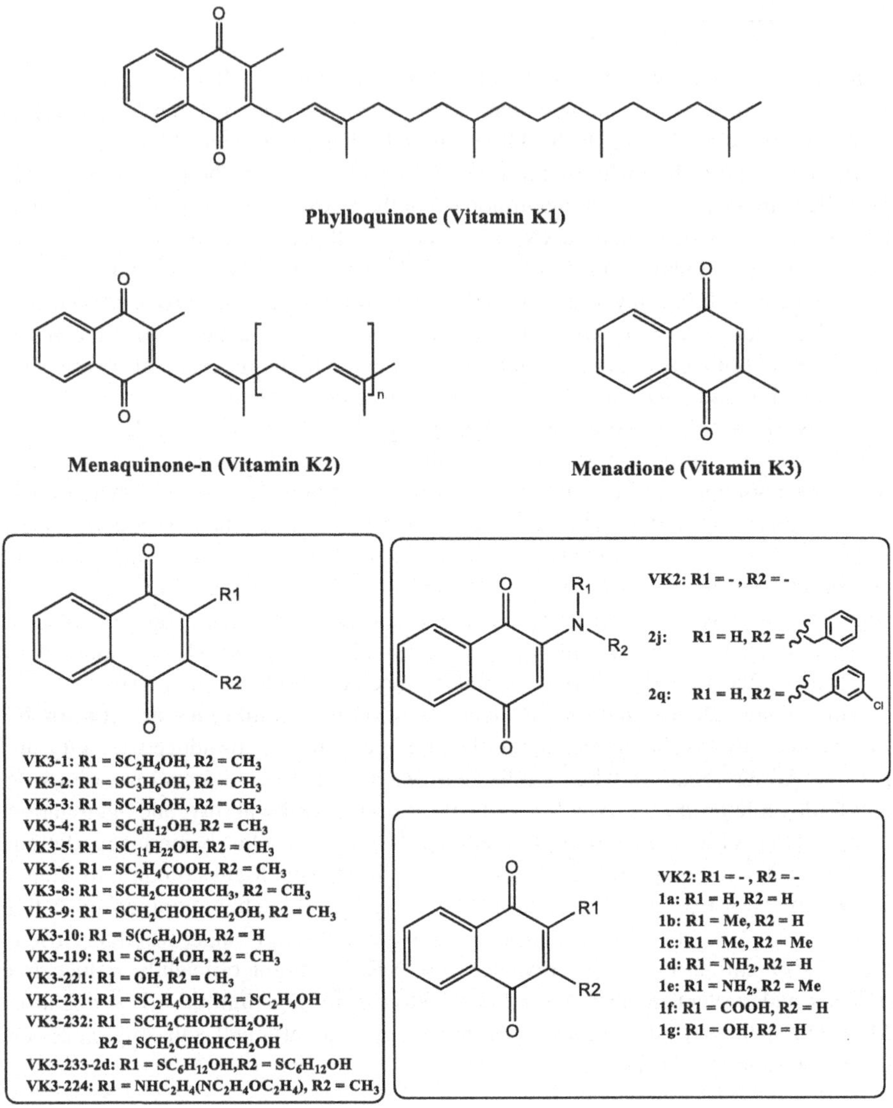

Fig. 12.8 Vitamin K and its derivatives

are developed and biologically evaluated for their effectiveness on Alzheimer's disease.

Conflict of Interest The authors declare no conflict of interest.

References

Abobaker A, Alzwi A, Alraied AHA (2020) Overview of the possible role of vitamin C in management of COVID-19. Pharmacol Rep 72(6):1517–1528

Annweiler C, Beauchet O (2011) Vitamin D-mentia: randomized clinical trials should be the next step. Neuroepidemiology 37(3–4):249–258

Balducci C, Paladini A, Micotti E, Tolomeo D, La Vitola P, Grigoli E, Richardson JC, Forloni G (2015) The continuing failure of bexarotene in Alzheimer's disease mice. J Alzheimers Dis 46(2):471–482

Bhatti AB, Usman M, Ali F, Satti SA (2016) Vitamin supplementation as an adjuvant treatment for Alzheimer's disease. J Clin Diagn Res 10(8):OE07–OE11

Bidon-Chanal A, Fuertes A, Alonso D, Perez DI, Martinez A, Luque FJ, Medina M (2013) Evidence for a new binding mode to GSK-3: allosteric regulation by the marine compound palinurin. Eur J Med Chem 60:479–489

Bivona G, Gambino CM, Iacolino G, Ciaccio M (2019) Vitamin D and the nervous system. Neurol Res 41(9):827–835

Chin KY, Tay SS (2018) A review on the relationship between tocotrienol and Alzheimer disease. Nutrients 10(7):881

Collins EG, Langbein WE, Orebaugh C, Bammert C, Hanson K, Reda D, Edwards LC, Littooy FN (2003) PoleStriding exercise and vitamin E for management of peripheral vascular disease. Med Sci Sports Exerc 35(3):384–393

Crisostomo AG, Moreno RB, Navaratnam S, Wilkinson JA, Bisby RH (2007) Generation of superoxide and singlet oxygen from alpha-tocopherolquinone and analogues. Free Radic Res 41(6):730–737

Czernichow S, Hercberg S (2001) Interventional studies concerning the role of antioxidant vitamins in cardiovascular diseases: a review. J Nutr Health Aging 5(3):188–195

Ding Y, Qiao A, Wang Z, Goodwin JS, Lee ES, Block ML, Allsbrook M, McDonald MP, Fan GH (2008) Retinoic acid attenuates beta-amyloid deposition and rescues memory deficits in an Alzheimer's disease transgenic mouse model. J Neurosci 28(45):11622–11634

DiNicolantonio JJ, Bhutani J, O'Keefe JH (2015) The health benefits of vitamin K. Open Heart 2(1):e000300

Dong R, Wang H, Ye J, Wang M, Bi Y (2019) Publication trends for Alzheimer's disease worldwide and in China: a 30-year bibliometric analysis. Front Hum Neurosci 13:259

Endres K, Fahrenholz F, Lotz J, Hiemke C, Teipel S, Lieb K, Tuscher O, Fellgiebel A (2014) Increased CSF APPs-alpha levels in patients with Alzheimer disease treated with acitretin. Neurology 83(21):1930–1935

Fahrenholz F, Tippmann F, Endres K (2010) Retinoids as a perspective in treatment of Alzheimer's disease. Neurodegener Dis 7(1–3):190–192

Fan YG, Guo T, Han XR, Liu JL, Cai YT, Xue H, Huang XS, Li YC, Wang ZY, Guo C (2019) Paricalcitol accelerates BACE1 lysosomal degradation and inhibits calpain-1 dependent neuronal loss in APP/PS1 transgenic mice. EBioMedicine 45:393–407

Ferland G (2012) Vitamin K and the nervous system: an overview of its actions. Adv Nutr 3(2): 204–212

Ford AH, Flicker L, Alfonso H, Thomas J, Clarnette R, Martins R, Almeida OP (2010) Vitamins B (12), B(6), and folic acid for cognition in older men. Neurology 75(17):1540–1547

Fukasawa H, Nakagomi M, Yamagata N, Katsuki H, Kawahara K, Kitaoka K, Miki T, Shudo K (2012) Tamibarotene: a candidate retinoid drug for Alzheimer's disease. Biol Pharm Bull 35(8): 1206–1212

Gall Z, Szekely O (2021) Role of vitamin D in cognitive dysfunction: new molecular concepts and discrepancies between animal and human findings. Nutrients 13(11):3672

Giraldo E, Lloret A, Fuchsberger T, Vina J (2014) Abeta and tau toxicities in Alzheimer's are linked via oxidative stress-induced p38 activation: protective role of vitamin E. Redox Biol 2:873–877

Grimm MO, Mett J, Hartmann T (2016) The impact of vitamin E and other fat-soluble vitamins on Alzheimer's disease. Int J Mol Sci 17(11):1785

Grimm MOW, Thiel A, Lauer AA, Winkler J, Lehmann J, Regner L, Nelke C, Janitschke D, Benoist C, Streidenberger O, Stotzel H, Endres K, Herr C, Beisswenger C, Grimm HS, Bals R, Lammert F, Hartmann T (2017) Vitamin D and its analogues decrease amyloid-beta (Abeta) formation and increase Abeta-degradation. Int J Mol Sci 18(12):2764

Gugliandolo A, Bramanti P, Mazzon E (2017) Role of vitamin E in the treatment of Alzheimer's disease: evidence from animal models. Int J Mol Sci 18(12):2504

Harrison FE, May JM (2009) Vitamin C function in the brain: vital role of the ascorbate transporter SVCT2. Free Radic Biol Med 46(6):719–730

Hou N, Ren L, Gong M, Bi Y, Gu Y, Dong Z, Liu Y, Chen J, Li T (2015) Vitamin A deficiency impairs spatial learning and memory: the mechanism of abnormal CBP-dependent histone acetylation regulated by retinoic acid receptor alpha. Mol Neurobiol 51(2):633–647

Huy PD, Yu YC, Ngo ST, Thao TV, Chen CP, Li MS, Chen YC (2013) In silico and in vitro characterization of anti-amyloidogenic activity of vitamin K3 analogues for Alzheimer's disease. Biochim Biophys Acta 1830(4):2960–2969

Ibrahim NF, Hamezah HS, Yanagisawa D, Tsuji M, Kiuchi Y, Ono K, Tooyama I (2021) The effect of alpha-tocopherol, alpha- and gamma-tocotrienols on amyloid-beta aggregation and disaggregation in vitro. Biochem Biophys Rep 28:101131

Jiaranaikulwanitch J, Pandith H, Tadtong S, Thammarat P, Jiranusornkul S, Chauthong N, Nilkosol S, Vajragupta O (2021) Novel multifunctional ascorbic triazole derivatives for amyloidogenic pathway inhibition, anti-inflammation, and neuroprotection. Molecules 26(6): 1562

Joe E, Ringman JM (2019) Cognitive symptoms of Alzheimer's disease: clinical management and prevention. BMJ 367:l6217

Josey BJ, Inks ES, Wen X, Chou CJ (2013) Structure-activity relationship study of vitamin K derivatives yields highly potent neuroprotective agents. J Med Chem 56(3):1007–1022

Jurutka PW, Whitfield GK, Hsieh JC, Thompson PD, Haussler CA, Haussler MR (2001) Molecular nature of the vitamin D receptor and its role in regulation of gene expression. Rev Endocr Metab Disord 2(2):203–216

Kapoor A, Wang BJ, Hsu WM, Chang MY, Liang SM, Liao YF (2013) Retinoic acid-elicited RARalpha/RXRalpha signaling attenuates Abeta production by directly inhibiting gamma-secretase-mediated cleavage of amyloid precursor protein. ACS Chem Neurosci 4(7): 1093–1100

Kennedy DO (2016) B Vitamins and the brain: mechanisms, dose and efficacy—a review. Nutrients 8(2):68

Kocot J, Luchowska-Kocot D, Kielczykowska M, Musik I, Kurzepa J (2017) Does vitamin C influence neurodegenerative diseases and psychiatric disorders? Nutrients 9(7):659

Kumar N, Gahlawat A, Kumar RN, Singh YP, Modi G, Garg P (2022) Drug repurposing for Alzheimer's disease: in silico and in vitro investigation of FDA-approved drugs as acetylcholinesterase inhibitors. J Biomol Struct Dyn 40(7):2878–2892

Kurutas EB (2016) The importance of antioxidants which play the role in cellular response against oxidative/nitrosative stress: current state. Nutr J 15(1):71

La Fata G, Weber P, Mohajeri MH (2014) Effects of vitamin E on cognitive performance during ageing and in Alzheimer's disease. Nutrients 6(12):5453–5472

Littlejohns TJ, Henley WE, Lang IA, Annweiler C, Beauchet O, Chaves PH, Fried L, Kestenbaum BR, Kuller LH, Langa KM, Lopez OL, Kos K, Soni M, Llewellyn DJ (2014) Vitamin D and the risk of dementia and Alzheimer disease. Neurology 83(10):920–928

Lu'o'ng K, Nguyen LT (2011) Role of thiamine in Alzheimer's disease. Am J Alzheimers Dis Other Dement 26(8):588–598

Manfredini S, Pavan B, Vertuani S, Scaglianti M, Compagnone D, Biondi C, Scatturin A, Tanganelli S, Ferraro L, Prasad P, Dalpiaz A (2002) Design, synthesis and activity of ascorbic

acid prodrugs of nipecotic, kynurenic and diclophenamic acids, liable to increase neurotropic activity. J Med Chem 45(3):559–562
Mimori Y, Katsuoka H, Nakamura S (1996) Thiamine therapy in Alzheimer's disease. Metab Brain Dis 11(1):89–94
Munoz FJ, Opazo C, Gil-Gomez G, Tapia G, Fernandez V, Valverde MA, Inestrosa NC (2002) Vitamin E but not 17beta-estradiol protects against vascular toxicity induced by beta-amyloid wild type and the Dutch amyloid variant. J Neurosci 22(8):3081–3089
Nepovimova E, Korabecny J, Dolezal R, Babkova K, Ondrejicek A, Jun D, Sepsova V, Horova A, Hrabinova M, Soukup O, Bukum N, Jost P, Muckova L, Kassa J, Malinak D, Andrs M, Kuca K (2015) Tacrine-trolox hybrids: a novel class of centrally active, nonhepatotoxic multi-target-directed ligands exerting anticholinesterase and antioxidant activities with low in vivo toxicity. J Med Chem 58(22):8985–9003
Osimani A, Berger A, Friedman J, Porat-Katz BS, Abarbanel JM (2005) Neuropsychology of vitamin B12 deficiency in elderly dementia patients and control subjects. J Geriatr Psychiatry Neurol 18(1):33–38
Pal T, Bhimaneni S, Sharma A, Flora SJS (2020) Design, synthesis, biological evaluation and molecular docking study of novel pyridoxine-triazoles as anti-Alzheimer's agents. RSC Adv 10(44):26006–26021
Pal T, Patil P, Sharma A (2021) Synthesis, molecular docking and spectroscopic studies of pyridoxine carbamates as metal chelator. J Mol Struct 1223:128837
Pan X, Gong N, Zhao J, Yu Z, Gu F, Chen J, Sun X, Zhao L, Yu M, Xu Z, Dong W, Qin Y, Fei G, Zhong C, Xu TL (2010) Powerful beneficial effects of benfotiamine on cognitive impairment and beta-amyloid deposition in amyloid precursor protein/presenilin-1 transgenic mice. Brain 133(Pt 5):1342–1351
Pan X, Chen Z, Fei G, Pan S, Bao W, Ren S, Guan Y, Zhong C (2016) Long-term cognitive improvement after benfotiamine administration in patients with Alzheimer's disease. Neurosci Bull 32(6):591–596
Rai SN, Singh P, Steinbusch HW, Vamanu E, Ashraf G, Singh MP (2021) The role of vitamins in neurodegenerative disease: an update. Biomedicine 9(10):1284
Reynolds EH (2002) Folic acid, ageing, depression, and dementia. BMJ (Clinical research ed) 324(7352):1512–1515
Reynolds E (2006) Vitamin B12, folic acid, and the nervous system. Lancet Neurol 5(11):949–960
Sambon M, Gorlova A, Demelenne A, Alhama-Riba J, Coumans B, Lakaye B, Wins P, Fillet M, Anthony DC, Strekalova T, Bettendorff L (2020) Dibenzoylthiamine has powerful antioxidant and anti-inflammatory properties in cultured cells and in mouse models of stress and neurodegeneration. Biomedicine 8(9):361
Sambon M, Wins P, Bettendorff L (2021) Neuroprotective effects of thiamine and precursors with higher bioavailability: focus on benfotiamine and dibenzoylthiamine. Int J Mol Sci 22(11):5418
Sang Z, Song Q, Cao Z, Deng Y, Zhang L (2022) Design, synthesis, and evaluation of chalcone-vitamin E-donepezil hybrids as multi-target-directed ligands for the treatment of Alzheimer's disease. J Enzyme Inhib Med Chem 37(1):69–85
Semba RD (2012) The discovery of the vitamins. Int J Vitam Nutr Res 82(5):310–315
Singh YP, Shankar G, Jahan S, Singh G, Kumar N, Barik A, Upadhyay P, Singh L, Kamble K, Singh GK, Tiwari S, Garg P, Gupta S, Modi G (2021) Further SAR studies on natural template based neuroprotective molecules for the treatment of Alzheimer's disease. Bioorg Med Chem 46:116385
Singh YP, Kumar N, Priya K, Chauhan BS, Shankar G, Kumar S, Singh GK, Srikrishna S, Garg P, Singh G, Rai G, Modi G (2022) Exploration of neuroprotective properties of a naturally inspired multifunctional molecule (F24) against oxidative stress and amyloid beta induced neurotoxicity in Alzheimer's disease models. ACS Chem Neurosci 13(1):27–42
Starling-Soares B, Carrera-Bastos P, Bettendorff L (2020) Role of the synthetic B1 vitamin sulbutiamine on health. J Nutr Metab 2020:9349063

Sugimoto K, Yajima H, Hayashi Y, Minato D, Terasaki S, Tohda C, Matsuya Y (2015) Synthesis of denosomin-vitamin D3 hybrids and evaluation of their anti-Alzheimer's disease activities. Org Lett 17(23):5910–5913

Tapias V, Jainuddin S, Ahuja M, Stack C, Elipenahli C, Vignisse J, Gerges M, Starkova N, Xu H, Starkov AA, Bettendorff L, Hushpulian DM, Smirnova NA, Gazaryan IG, Kaidery NA, Wakade S, Calingasan NY, Thomas B, Gibson GE, Dumont M, Beal MF (2018) Benfotiamine treatment activates the Nrf2/ARE pathway and is neuroprotective in a transgenic mouse model of tauopathy. Hum Mol Genet 27(16):2874–2892

Teponnou GAK, Joubert J, Malan SF (2017) Tacrine, trolox and tryptoline as lead compounds for the design and synthesis of multi-target agents for Alzheimer's disease therapy. Open Med Chem J 11:24–37

Tippmann F, Hundt J, Schneider A, Endres K, Fahrenholz F (2009) Up-regulation of the alpha-secretase ADAM10 by retinoic acid receptors and acitretin. FASEB J 23(6):1643–1654

Tousi B (2015) The emerging role of bexarotene in the treatment of Alzheimer's disease: current evidence. Neuropsychiatr Dis Treat 11:311–315

Turnaturi R, Oliveri V, Vecchio G (2016) Biotin-8-hydroxyquinoline conjugates and their metal complexes: exploring the chemical properties and the antioxidant activity. Polyhedron 110:254–260

Wang SW, Yang SG, Liu W, Zhang YX, Xu PX, Wang T, Ling TJ, Liu RT (2016) Alpha-tocopherol quinine ameliorates spatial memory deficits by reducing beta-amyloid oligomers, neuroinflammation and oxidative stress in transgenic mice with Alzheimer's disease. Behav Brain Res 296:109–117

Xu J, Patassini S, Begley P, Church S, Waldvogel HJ, Faull RLM, Unwin RD, Cooper GJS (2020) Cerebral deficiency of vitamin B5 (d-pantothenic acid; pantothenate) as a potentially-reversible cause of neurodegeneration and dementia in sporadic Alzheimer's disease. Biochem Biophys Res Commun 527(3):676–681

Yang SG, Wang WY, Ling TJ, Feng Y, Du XT, Zhang X, Sun XX, Zhao M, Xue D, Yang Y, Liu RT (2010) Alpha-tocopherol quinone inhibits beta-amyloid aggregation and cytotoxicity, disaggregates preformed fibrils and decreases the production of reactive oxygen species, NO and inflammatory cytokines. Neurochem Int 57(8):914–922

Yang X, Qiang X, Li Y, Luo L, Xu R, Zheng Y, Cao Z, Tan Z, Deng Y (2017) Pyridoxine-resveratrol hybrids Mannich base derivatives as novel dual inhibitors of AChE and MAO-B with antioxidant and metal-chelating properties for the treatment of Alzheimer's disease. Bioorg Chem 71:305–314

Zandona A, Lihtar G, Marakovic N, Mis K, Busic V, Gaso-Sokac D, Pirkmajer S, Katalinic M (2020) Vitamin B3-based biologically active compounds as inhibitors of human cholinesterases. Int J Mol Sci 21(21):8088

Part V

Metal Dyshomeostasis: A Hypothesis for Developing Drugs for the Treatment of Alzheimer's Disease

Metal Chelators as a Potential Therapeutic Agent for Alzheimer's Disease

13

Bhaskar Dewangan, Kunal Kumar, Amit Kumar,
Praveen Reddy Bodhe, Sunita Beni, and Bichismita Sahu

Abstract

Alzheimer's disease (AD) is a neurological condition that causes behavioural abnormalities and a progressive decline in intellectual abilities. In the brain, metal ions equilibria are essential for appropriate brain functioning, while dyshomeostasis of free metal ions is associated with Alzheimer's disease. In the brain, dysregulated metal ions trigger various responses, such as promoting $A\beta_{(1-42)}$ aggregation, tau hyperphosphorylation, oxidative stress, and disruption of organelles, which leads to autophagic failure. Approved drugs for AD only have palliative effects. Thus, potential molecules with contrasting features, such as chelating agents and metal-based medicines, have been proposed to act on various molecular targets to treat AD. This chapter aims to highlight the various applications and current advancements of metal-based medications, including metal complexes and metal-chelating agents, as potential therapeutic agents for treating AD.

Keywords

Alzheimer's disease · Biometals · Metal ions · Metal chelators · Metal homeostasis · Neurotoxicity

B. Dewangan · K. Kumar · A. Kumar · P. R. Bodhe · S. Beni · B. Sahu (✉)
Department of Medicinal Chemistry, National Institute of Pharmaceutical Education and Research, Ahmedabad, Gujarat, India
e-mail: bichismita@niperahm.res.in

A. Sharma, G. P. Modi (eds.), *Natural Product-Based Synthetic Drug Molecules in Alzheimer's Disease*, https://doi.org/10.1007/978-981-99-6038-5_13

13.1 Introduction to Alzheimer's Disease

Alzheimer's disease (AD) is the most common type of dementia. It results in a significant loss of nerve cells in the regions of the brain necessary for memory and learning. The amyloid-β (Aβ) plaques, tau-associated intraneuronal microtubules, neurofibrillary tangles (NFTs), and loss of synapses in the brain are the primary symptoms of this disease (Selkoe 2003; Selkoe 2002). The brain of AD patients has extracellular deposits of senile plaques surrounded by dystrophic neuritis and NFT inside the cells. The brain areas affected by plaques typically have fewer synapses and impaired neuritis that connect the plaques to other brain regions. Plaques and tangles are more common in brain parts like the hippocampus, entorhinal cortex, basal forebrain, and amygdala that help in learning, memory, and emotional behaviour (Selkoe and Hardy 2016; Grundke-Iqbal et al. 1984).

13.1.1 Discovered Therapies and Highlighted AD Targets

FDA had approved five drugs for AD treatments: tacrine, rivastigmine, donepezil, galantamine and memantine (Fig. 13.1). Except memantine, all four drugs are acetylcholinesterase enzyme (AChE) inhibitors, which are accountable for acetylcholine breakdown. The mechanism of AChE hydrolysis includes blocking acetylcholine metabolism, extending its action and increasing levels in the cerebral cortex (Table 13.1) (Santos et al. 2016).

Memantine, an NMDA receptor antagonist that reduces neuronal mortality and corrects the Aβ-induced Ca^{2+} imbalance (intracellular Ca^{2+} build up), is the only available non-cholinergic drug (Parsons et al. 2007). A combination of memantine and donepezil was also utilized in 2014. However, These FDA-approved medications give short-term (6–12 months) relief of cognitive and behavioural symptoms in moderate forms of AD only rather than cure the underlying illness or slow its development. Several hundred promising medicines have been developed and tried since the approval of memantine in 2003, but they have all failed

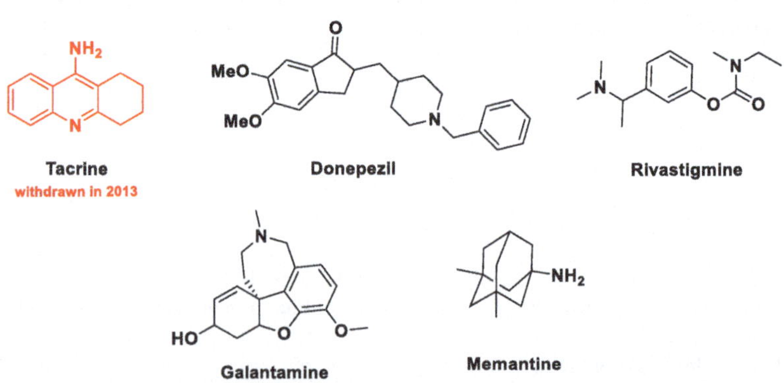

Fig. 13.1 Drugs for AD

Table 13.1 List of medications approved by USFDA for AD

Drug name	Brand name	Approved for	Approved in
Tacrine	Cognex	All stages/discontinued	1993
Donepezil	Aricept	All stages	1996
Rivastigmine	Exelon	All stages	2000
Galantamine	Razadyne	Mild to moderate	2001
Memantine	Namenda	Mild to moderate	2003

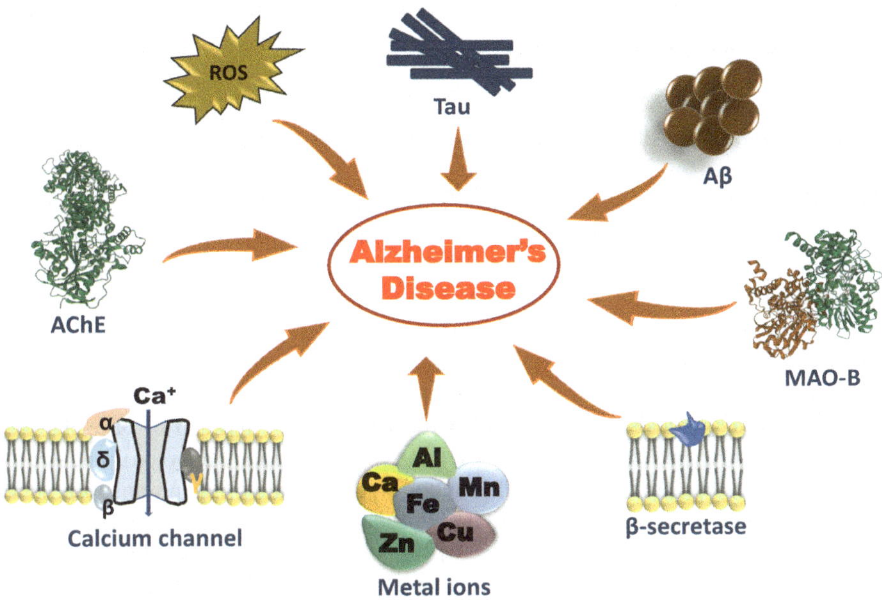

Fig. 13.2 Multiple pathological features of AD

(Cummings et al. 2014). Besides the cholinergic symptomatic medication relief, researchers have looked at additional AD targets, such as aggregated Aβ and tau proteins (Wischik et al. 2014), as well as oxidative stress dysregulation.

13.1.2 Molecular Targets for AD Treatment

The study of how the brain works normally and the factors that cause AD, has greatly increased. Numerous innovative treatments for AD that are now being developed aim to change the course of the disease by focusing on multiple brain changes in Alzheimer's patients (Dalvi et al. 2020). AD involves various pathological features (Fig. 13.2). Hence no cure has developed to date. According to research, Aβ protein plaques and metal dysregulation may be the focus of future pharmaceutical treatments (Selkoe 2001).

B-secretases and γ-secretases proteolytically cut Amyloid Precursor Protein (APP) to produce Aβ. AD brains are characterized by the manifestation of senile plaques formed when Aβ-oligomers clump together (Roland and Jacobsen 2009). These plaques are responsible for neuronal death and are connected to oxidative stress, loss of membrane coherence, mitochondrial dysfunction, abnormal calcium homeostasis, and triggered apoptosis (Crouch et al. 2008).

Studies have shown that AD patients have 3–7 times greater metal ion levels than healthy people (Huang et al. 2014). It is worthwhile to build specialized metal chelators with low toxicity that might re-establish metal homeostasis in AD brains, especially if the chelators have additional essential functional groups that could address additional disease targets (Santos et al. 2016; Grill and Cummings 2010).

13.2 Metals and Their Role in Central Nervous System

Metal ions are essential for sustaining plant, animal, and human life. Their importance in biological systems is well-known. They are crucial to survival because their absence can lead to developmental problems, severe malfunction, cancer, or even death. They play pivotal roles as macro- or micro-elements in various cellular structures and processes and are involved in a wide array of biochemical activities, and manifest in various morphological states. They also play a crucial role in cellular and intercellular communication, the preservation of electrical charges and osmotic pressure, photosynthesis and the transfer of electrons, the upkeep of base pairing, stacking stability, and the regulation of DNA transcription. Many biological processes rely on metals for their normal functioning, including oxygen transport, nerve and muscle cell function, brain and heart function, and many others, and it is impossible to conceive of the existence of life without them (Moustakas 2021).

The most common ionic forms of these metals are iron (Fe), cobalt (Co), nickel (Ni), calcium (Ca), zinc (Zn), copper (Cu), and chromium (Cr). Lack of these minerals can cause a variety of health problems, including anaemia (in the case of Fe and Co), brain and heart disease (in the case of Cu), stunted growth (in the case of Zn), skin changes (in the case of Ca), bone deterioration (in the case of Ca), and decreased glucose tolerance (in the case of Cr). As a result, modern medicinal chemistry strongly emphasises investigating and treating disorders associated with metal ion deficiencies and excesses at the molecular level (Gupta 2018).

13.2.1 Therapeutic Agents Containing Metals

In addition to its usage in the treatment of bladder, cervical, oesophageal, nasopharyngeal, and head and neck cancers, cisplatin is also effective in the treatment of germ-cell malignancies, prenatal trophoblastic tumours, epithelial ovarian cancer, and small-cell lung cancer. However, cisplatin resistance was discovered in some malignancies, and other unwanted adverse effects were discovered to accompany it. Titanocene (MTK4) and budotitane are two titanium complexes shown anticancer activity (INN). Titanocene had demonstrated anticancer potential in research labs and was the first non-platinum chemical to be evaluated as a chemotherapy drug. Budotitane (INN) is a highly sophisticated non-platinum compound. The most promising molecules have been made from ruthenium. However, there are currently no direct analogues in clinical use. Two ruthenium drugs, NAMI-A and KP1011, are tested for their anticancer properties. NAMI-A is effective against lung metastasis in vivo and tumour cell invasion in vitro, whereas KP1019 is useful for metastatic tumours and cisplatin-resistant malignancies. It displays strong cytotoxicity, when applied directly to primary tumours, especially those caused by colorectal cancer.

Compounds containing bismuth have been utilized to treat a wide range of gastrointestinal issues due to their antacid and astringent characteristics. Bismuth subsalicylate, sold under the brand name "Pepto-Bismol," is an antacid drug that temporarily relieve stomach and gastrointestinal discomforts like diarrhoea, indigestion, heartburn, and nausea. Several additional compounds are in the process of being developed. Researchers developed two gold complexes: aurothioglucose (also known as gold thioglucose) and auranofin, to alleviate the symptoms of rheumatoid arthritis. Joint discomfort, tenderness, swelling, and morning stiffness are all aided by auranofin treatment for arthritis. Although aurothioglucose and other gold complexes designed for the treatment of rheumatoid arthritis have been in use for over 70 years, they were recently removed from the US market, leaving just auranofin as the sole gold salt available there (Gupta 2018) (Table 13.2).

13.2.2 Role of Metals in Central Nervous System

The distribution of metals in every organ of the human body is distinct. Compared to other organs in the body, the human brain has the highest concentration of trace metals. It is widely acknowledged that growing older is the most significant risk factor for getting neurodegenerative disorders (Fig. 13.3) (Wang and Wang 2017). Iron (Fe), Zinc (Zn), Copper (Cu), and Manganese (Mn) are essential for a wide variety of human physiological and biochemical functions due to their roles in protein structure and regulation, cellular signalling pathways, oxygen transport, and other processes. Some other metals like Calcium (Ca), Aluminum (Al), and Magnesium (Mg) also play a crucial role in the neurological system.

Iron, or the heme iron, which is bound to haemoglobin in red blood cells, is widespread. Evidence shows that it can be found in every type of brain cell, including neurons, microglia, astrocytes, and oligodendrocytes. The brain requires a significant amount of both energy and iron to function properly. Iron is vital in ATP synthesis since it is a part of several enzymes and cellular structures involved in the process, including the electron transport chain's Fe-sulfur complexes, cytochrome oxidase, cytochromes a, b, and c. Apart from its function in oxygen transfer, iron is a cofactor for several enzymes involved in neurotransmitter production, particularly tyrosine hydroxylase and tryptophan hydroxylase (Yehuda et al. 1989). Iron also has a few additional tasks in the CNS, including regulating synaptic

Table 13.2 Metal containing therapeutic agents

S. No.	Complex type	Drug name	Structure	Uses
1.	Platinum complex	Cisplatin	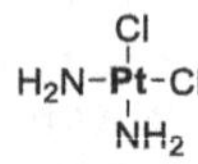	Anticancer
2.	Titanium complex	Titanocene (MTK4)	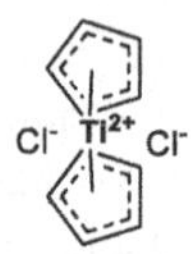	Anticancer
3.	Ruthenium complex	KP1019	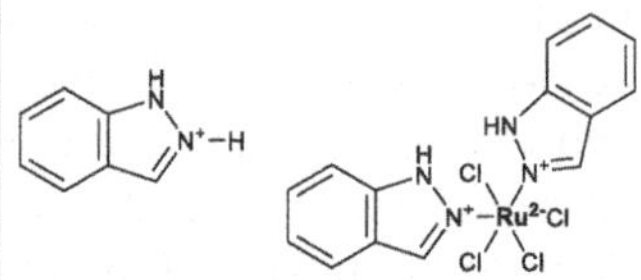	Anticancer
4.	Bismuth complex	Pepto-Bismol	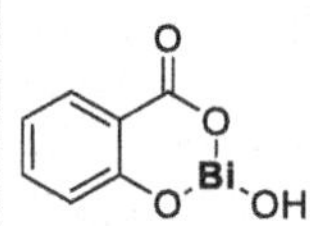	Antacid and astringent
5.	Gold complex	Auranofin	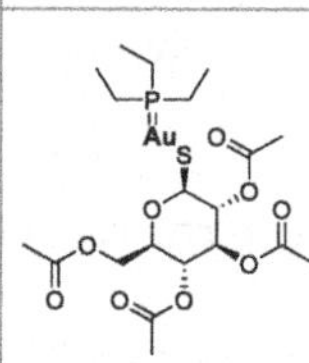	Anti-rheumatoid

plasticity, myelinating neuronal axons, and regulating the energy status of individual neurons (Clardy et al. 2006).

The second most prevalent trace element in the human brain is **zinc**. Like iron, zinc regulates synaptic plasticity. Being one of the brain's most prevalent divalent metal ions, it is a non-redox active important element vital to neurochemistry (James et al. 2017). It also influence neurotransmission, regulates neurogenesis, neuronal migration, and neuronal differentiation (Fukada et al. 2011). Over 300 enzymes use zinc as a cofactor, making it a vital structural and catalytic component of proteins. Zn-Importing Proteins (ZIPs), Zn Transporters (ZnTs), and buffering proteins such as metallothioneins are all present in large quantities (MTs) that bind cytosolic Zn are the critical regulators of Zn homeostasis in the brain (James et al. 2017). It also aids in maintaining the stability of several transcription regulators. Additionally, zinc protects against oxidative stress and has neuroprotective properties (Tapiero and Tew 2003).

Copper is the third most prevalent trace metal in the human brain. Like zinc, copper also functions as a cofactor for several enzymes and is necessary for producing neurotransmitters (Peña et al. 1999). Cu is an important micronutrient

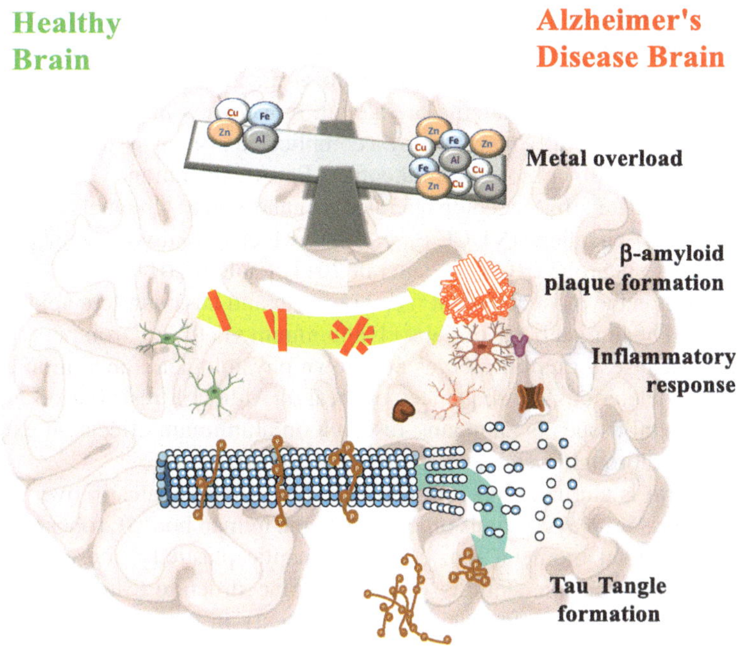

Fig. 13.3 Healthy brain vs. AD brain

and crucial for all aerobic species due to the prosthetic group's ability to accelerate electron transport during vital enzyme operations (Waggoner et al. 1999). The brain needs copper for healthy growth and function (Lutsenko et al. 2010), and an average adult's brain contains approximately 7.3% of the body's total copper (Lutsenko et al. 2010; Hung et al. 2009). The brain's primary cell compartments contain copper (nucleus, mitochondria, and cytosol). However, it is not evenly distributed throughout the CNS (Lutsenko et al. 2010). Cu is a metal that can exist in both reduced (Cu^{+}) and oxidized (Cu^{2+}) states. Both states can bind to various enzymes, including cytochrome c oxidase, ceruloplasmin, dopamine hydroxylase, and Cu, Zn dependent superoxide dismutase (Cu-ZnSOD or SOD 1) (Hung et al. 2009). Copper is needed as a cofactor by the enzyme Cu/Zn-superoxide dismutase, which is essential for scavenging ROS.

Manganese is needed as a cofactor by the enzymes pyruvate carboxylase, arginase, protein serine/threonine phosphatase 1, glutamine synthetase, and manganese superoxide dismutase (MnSOD). Additional Mn requirements include the ATM-p53 pathway and other Mn-sensitive pathways (Horning et al. 2015).

The intracellular **calcium** (Ca^{2+}) concentration is essential for normal neuronal function and has various other functions. Long-term potentiation (LTP) and long-term depression (LTD) are two types of synaptic plasticity that can be affected by modulating neurotransmitter release from presynaptic terminals. Ca^{2+} is a ubiquitous second messenger that has been shown to affect regulating gene expression, membrane excitability, dendritic development, synaptogenesis, and many other processes that contribute to neurons' basic activities, including information processing and memory storage (Kawamoto et al. 2012). The amplitude, spatial distribution, and temporal patterns of calcium movements inside neuronal cellular compartments are carefully regulated by a complicated protein system that allows for the specificity of distinct biological outcomes. Plasma membrane Ca^{2+} channels are principally regulated by receptor-operated (ligand-gated) channels and voltage-dependent Ca^{2+} channels (VDCCs), both of which are involved in controlling Ca^{2+} influx into neurons (ROCs) (Catterall 2011; Paoletti 2011).

Aluminum is the most abundant metal on earth and has countless applications in modern life. The human body can absorb aluminium via the environment, food, and medicine. Because aluminium serves no known physiological function in the body, it may have deleterious consequences on health. Although research on aluminum effects on neural tissues is abundant, research on aluminium effects on extraneural tissues is less thoroughly characterized (Nayak 2002).

Magnesium is a crucial macromineral because it participates in over 300 enzymatic reactions and plays various other roles in the human body (Gröber et al. 2015). Magnesium, a vital biometal ion, controls synaptic plasticity, muscular function, protein synthesis, and the stability of the ribosome structure (Slutsky et al. 2004). Magnesium's protective effect against excitotoxicity in the neurological system is only one of the many reasons why it is so crucial for proper nerve transmission and neuromuscular coordination (excessive excitement that kills cells) (Gröber et al. 2015). Magnesium's interaction with the *N*-methyl-D-aspartate (NMDA) receptor is essential for several neurological processes. Glutamatergic excitatory transmission requires magnesium's removal because it acts as a blocker of the NMDA receptor's calcium channel (Kirkland et al. 2018). Magnesium deficiency has been hypothesized to amplify glutamatergic neurotransmission, promote excitotoxicity, and ultimately result in oxidative stress and neuronal cell death (Castilho et al. 2001).

Trace metals are therefore necessary for the induction of enzymes, regulation of gene expression, catalysis, and defence against ROS. They are also essential for healthy brain development, growth, and operation. The blood–brain barrier (BBB) is actively crossed by various transport proteins, which maintain a constant level of metal ion concentration in the brain. Any disruption of the metal balance, including deficiencies in absorption, distribution, and excretion, gene mutations in metal transport or metal-binding proteins, competition between metals for binding sites (such as Zn and Cu), or other factors, can cause an imbalance of essential trace metals or an increase in non-essential trace elements (Sensi et al. 2018).

13.3 Metal Dyshomeostasis and Alzheimer's Disease

Many neurodegenerative illnesses, including Huntington's disease, Parkinson's disease, and AD (Fig. 13.4), contain the hallmarks of toxic metal accumulation and biometal dyshomeostasis. Among the neurotoxic consequences of metal imbalance are a decline in enzyme activity, an increase in protein aggregation, and

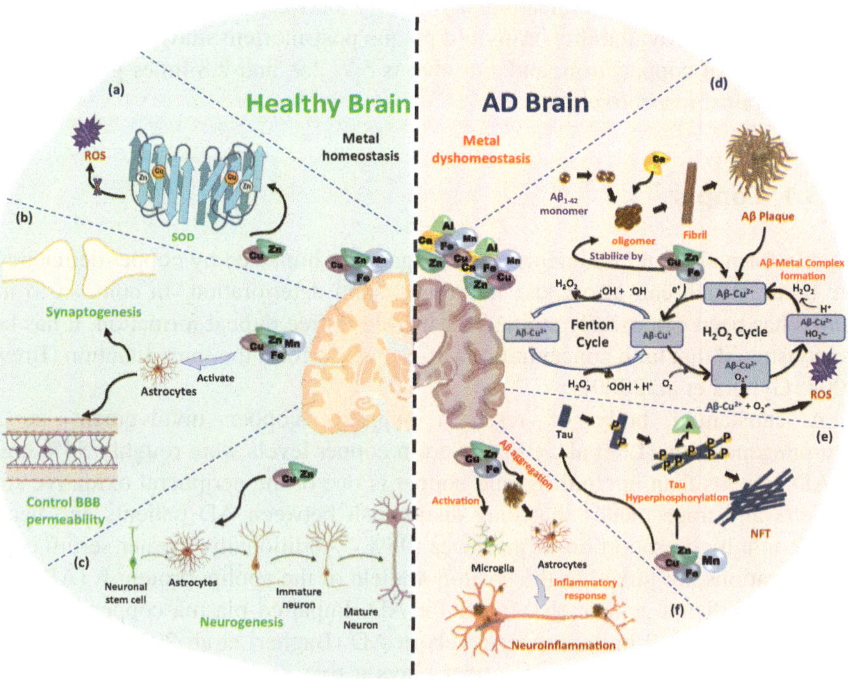

Fig. 13.4 Metal homeostasis vs. metal dyshomeostasis. (**a**) Cu/Zn-superoxide dismutase (SOD) requires Cu/Zn as a cofactor, essential for scavenging Reactive Oxygen Species (ROS). (**b**) Metal activates astrocytes which regulates synaptogenesis, controls BBB permeability (**c**) and helps in neural production. (**d**) Metal ions bind to amyloid beta to promote the formation of oligomers, stabilising $A\beta_{1-42}$, and by the Fenton pathway and the H_2O_2 cycle metal dyshomeostasis occurs which results in oxidative stress and excessive ROS production. (**e**) Free metals may directly bind to tau monomers, activate many kinases to phosphorylate tau proteins and promote tau aggregation and Neurofibrillary Tangles (NFT) formation. (**f**) Microglia, astrocytes, and neurons are all brain cell types, and when microglia become activated, they can emit free radicals and pro-inflammatory cytokines, which have been related to neuroinflammation and neurodegeneration

oxidative stress in the brain, which sets off a chain of events that results in cell death and neurodegeneration (Chin-Chan et al. 2015; Zhang et al. 2016). Although the connections between biometal imbalance and neurodegenerative diseases are still unclear, during the past few decades, research has focused more and more on a significant group of homeostatic proteins that facilitate metal transfer. The pathophysiology of AD and biometal imbalance have been linked to these proteins' aberrant expression (Li et al. 2017b).

Cu, Fe, and Zn have received the majority of the interest in this field of study as these transition metals are found prevalently in eukaryotes. Research on the brain has shown that amyloid plaques have higher concentrations of Cu, Fe, and Zn due to their binding to Aβ. However, their concentrations are reduced in the areas of the AD brain that are affected, indicating a disparity between these metals in AD and reducing their bioavailability. Amyloid plaque post-mortem study reveals an abnormal buildup of copper, iron, and zinc that is 5.7, 2.9, and 2.8 times greater than in healthy brains, respectively.

13.3.1 Copper

Myelin formation and maintenance are negatively impacted by copper deficiency in the brain, which can also cause nervous system deterioration. In contrast to iron, which has been extensively studied for its role in free radical formation, it has been demonstrated that high concentrations of copper perform the same function (Brewen 2008; Gybina et al. 2009).

A substantial body of research supports copper involvement in AD neurodegeneration (Li et al. 2017a). Serum copper levels were roughly 54% higher in AD patients than in controls, and copper is one of the peripheral oxidative stress markers and trace metals that may distinguish between AD patients and healthy people in a high proportion of instances (95%). Additionally, greater serum copper concentrations are linked to the epsilon 4 allele of the apolipoprotein E (APOE-ε4) gene, a significant genetic risk factor for AD. Impaired plasma copper levels may determine abnormal brain copper levels in AD (Bagheri et al. 2018; Hordyjewska et al. 2014). It is unclear how copper plays a role in AD pathophysiology. High copper concentrations have been discovered in senile plaques (Lovell et al. 1998). According to a recent statistical analysis, the total copper level in plasma and serum was greater in AD patients, some investigations showed low levels of overall copper in the AD brain (Klevay 2008; Bagheri et al. 2018). This variability can be rationally attributed to significant amounts of copper precipitating with AD-related senile plaques, causing copper deficiencies in other areas.

It is widely acknowledged that copper impacts tau and Aβ and worsens their pathogenic effects (Sparks and Schreurs 2003). Aβ and copper interact quite favourably and encourage the formation of its oligomers (Dalvi et al. 2021; Tõugu et al. 2008; Jin et al. 2011). Even though copper and Aβ can catalyse hydrogen peroxide in vitro, oxidative stress may play a role in the copper-mediated cytotoxicity of Aβ oligomers (Matlack et al. 2014). The copper-binding site is also present in APP and Aβ precursor-like protein 2 (APLP2) (Barnham et al. 2003). As APP or APLP2 mutant mice displayed increased cerebral cortex copper levels, it has been proposed that APP may have a role as a copper transporter (White et al. 1999a). The same author did, however, publish a different work in which cortical neurons with APP loss did not affect copper uptake (White et al. 1999b). There might be a possibility that APP is not a copper carrier but instead shows inappropriate interactions with copper, according to certain evidence. By increasing exocytosis and decreasing endocytosis, the cell membrane's redistribution of APP is aided by copper (Acevedo et al. 2011). Additionally, copper promotes the proteolytic cleavage of endogenous APP to produce Aβ by increasing its phosphorylation via GSK3B (Acevedo et al. 2014). It was discovered that copper can also connect to tau's microtubule-binding region. In vitro, Tau binds to copper, which causes it to aggregate (Su et al. 2007). Tau hyperphosphorylation and hydrogen peroxide generation were triggered by copper exposure in an in vivo AD animal model. The stimulation of the CDK5 and GSK3B pathways is suggested to be the mechanism of copper-mediated tau phosphorylation (Crouch et al. 2009a).

Other AD-related proteins, such as there have also been reports of interactions between copper and the beta-site APP-cleaving enzyme 1 (BACE1) (Vassar and Cole 2007). BACE1 is an aspartic protease that can break down APP during the beginning phases of the synthesis of Aβ. Therefore, experimental studies have demonstrated that the metalloproteins BACE1, APP, and Aβ are involved in copper homeostasis in the brain (Maynard et al. 2005; Kaden et al. 2011). Activated microglia surrounding Aβ plaques in the brains of TgCRND8 AD model mice have been shown to upregulate ATP7A, which may encourage Cu absorption by increasing CTR1 expression (Zheng et al. 2010). Therefore, the sequestration of copper by microglia in AD may act as a neuroprotective mechanism by lowering the quantity of free extracellular copper available for Aβ aggregation and plaque development (Choo et al. 2013). Contrarily, a lack of copper within cells brought on by downregulating APP, elevated ATP7A levels may prevent the development of Aβ, as evidenced by the ATP7A transfection in fibroblast cell lines, which led to decreased APP expression and copper loss in cells (Choo et al. 2013; Acevedo et al. 2011).

Copper's role in neurodegeneration caused to AD also involves oxidative stress. In vitro, APP or Aβ and Cu^{2+} interaction results in a decrease in Cu^{+} and increases the formation of neurotoxic H_2O_2. As a result, Cu^{2+} + $Aβ_{1–42}$ peptide complex regulates the redox cycles of oxidative stress, H_2O_2 production, Aβ oligomers, and plaque formation (Gu et al. 2018; Cheignon et al. 2018).

The APP's copper-binding domain can interact with Cu-Aβ interactions to produce Cu^{+} and ROS, which leading to neuronal malfunction or death (Cheignon et al. 2018; White et al. 1999a). Copper's site-directed redox activity has the potential to fragment APP and encourage amyloid peptide aggregation, both of which can lead to SP development. Additionally, copper has a direct impact on how peroxides are made. Although it is unclear whether clinical improvement or a slowdown of development may happen, substantial drops in peroxide concentrations have been observed in AD patients when bioavailable copper is reduced with D-penicillamine (Cheignon et al. 2018; Squitti et al. 2002). In neurons with insufficient GSH levels, Cu poisoning was found to be more pronounced. Both in vivo and in vitro alterations in glutathione peroxidase (GPx) were seen in AD. Glutathione reductase (GR) activity changed in the AD brain, and amyloidogenic AP exposure dramatically boosted GR activity in cultured neurons (White et al. 1999a; Wojsiat et al. 2018). These findings support the concept that altered GSH metabolism plays a role in the etiopathology of AD, with copper toxicity perhaps playing a part as well.

13.3.2 Iron

Iron deficiency causes several neurological abnormalities, particularly in infancy, which slows down brain development. Ageing increases brain iron levels, but too much iron triggers Fenton chemistry, producing too much ROS and eventually damaging cells (Dröge and Schipper 2007; Muhoberac and Vidal 2013). Furthermore, some research suggests that there may be a disturbance in iron homeostasis, a defining feature of various neurodegenerative diseases (Ward et al. 2014; Jiang et al. 2017).

Research has found that using various approaches the iron level is elevated in AD brains, particularly in the putamen and globus pallidus (Wang et al. 2015; Moon et al. 2016). However, statistical analyses have shown that AD patients' serum iron levels are either unchanged or lower than healthy persons (Tao et al. 2014; Wang et al. 2015). Scientific investigations into iron's impact on AD have revealed that having too much iron stimulates the Fenton reaction, which may lead to higher levels of oxidative stress in AD, even though the cause of this imbalance is still unknown. Iron chelators can counteract the harmful effects of oxidative damage to protein aggregates connected to AD, which can result in synaptic dysfunction and neurolysis (Salkovic-Petrisic et al. 2015). Senile plaques are loaded with high iron content, which increases the generation of Aβ by boosting the expression of APP. Because APP mRNA contains iron-responsive elements (IREs) in its 5′-untranslated portion, iron regulates APP transcription (Rogers et al. 2002). Iron regulatory proteins (IRPs) become detached from their IRE binding sites and lose their ability to repress APP mRNA, in contrast to lead (Pb), which prevents the IRE-mediated upregulation of APP translation (Rogers et al. 2016). Previous research has shown that Aβ-reduced ROS are present when neurodegeneration first begins. These ROS may benefit from reducing oxidative stress caused by binding iron (Jolivet-Gougeon and Bonnaure-Mallet 2018). However, an abundance of iron compromises the antioxidant system to become overworked, which over time damages cells and causes a vicious cycle in which Aβ production increases due to increased oxidative stress (Tamagno et al. 2007). Iron binds to tau in addition to Aβ, which encourages its aggregation in iron-enriched areas (Sayre et al. 2001). The activation of the glycogen synthase kinase 3b (GSK3b) and cyclin-dependent kinase 5 (CDK5) pathways has also been shown to trigger tau phosphorylation in response to iron (Lovell et al. 2005).

Transporter proteins are crucial for maintaining a dynamic equilibrium between iron influx and efflux for intracellular iron homeostasis. Deregulation of the iron importers transferrin (Tf), divalent metal transporter 1 (DMT1), lactoferrin (Lf), and melanotransferrin (MTf), as well as the iron exporter ferroportin, may be the source of iron accumulation in the damaged brain regions of AD patients (Fpn). Divalent cation transporter 1, or DMT1, is essential for the influx of Fe^{2+}. It is found on astrocytes, microglia, and neurons but not oligodendrocytes (Lovell et al. 2005). In the AD brain's plaques, two DMT1 isoforms, DMT1-IRE and DMT1CIRE, have been demonstrated to colocalize with Aβ and had considerably higher amounts of both DMT1 isoforms in the frontal brain and hippocampus in a transgenic mouse model of APP/PS1 (Zheng et al. 2009), which had an accompanying decrease in Fpn expression (Xian-hui et al. 2015), indicating that the iron-mediated neuropathogenesis of AD is significantly influenced by the deregulation of the proteins DMT1 and Fpn related to iron metabolism.

13.3.3 Zinc

After iron, the second most common metal in the body is zinc, an essential part of countless proteins and enzymes. Because it may function as a neurotransmitter, zinc is more extensively involved in cell signalling than other metals (Barr and Burdette 2017). By upregulating zinc transporters, it has been shown that zinc deficiency increases the overabundance of other metals, including nickel, copper, and possibly toxic metals (Antala and Dempski 2012). However, the mitochondria can hold onto zinc and cause the production of ROS by impeding complex III's ability to carry out the electron transport chain. This plays a crucial role in several disorders by causing mitochondrial dysfunction and neuronal death (Frazzini et al. 2006).

It is important to note how zinc promotes the formation of aggregates across the C-terminal histidine residues of Aβ. Aβ is bound to zinc with a stronger affinity than iron and copper at various pH levels (Yoshiike et al. 2001). For example, this binding hides the metalloproteases' proteolytic cleavage site where Aβ is degraded (Crouch et al. 2009b). However, binding causes a decrease in zinc bioavailability in the synaptic cleft, causing cognitive decline in AD and altering synaptic plasticity (Deshpande et al. 2009).

In neurons, three transporter classes, including the zinc transporters (ZnTs), iron- and zinc-regulated transporter-like proteins (ZIPs), and metallothioneins (MTs), are primarily responsible for maintaining zinc homeostasis. ZnTs promote zinc efflux from cells or mediate the accumulation of extra zinc in organelles and intracellular vesicles from the cytoplasm (Huang and Tepaamorndech 2013). ZIPs perform functions almost exactly the opposite of those of ZnTs, promoting the transport of zinc from intracellular vesicles to the cytoplasm or enabling the import of zinc into cells (Huang and Tepaamorndech 2013). MTs are proteins that regulate cellular levels of zinc and associated signalling pathways (Krężel et al. 2007). According to immunofluorescent research, ZnTs (ZnT1, 3, 4, 5, 6, and 7) were found in high concentrations in the cortex of human AD brains (Zhang et al. 2008). Of these, zinc-containing glutamatergic neurons' synaptic vesicles predominately contain ZnT3. Older people's levels of ZnT3 decline with age, especially in those with AD (Whitfield et al. 2014). In contrast, mice overexpressing APP but lacking ZnT3 exhibited reduced plaque load and synaptic zinc levels, suggesting that zinc from synapses plays a role in the development of amyloid plaques in AD. ZnT3 knockout animals showed an age-dependent cognitive loss. On the other hand, ZnT3 deletion increased intraneuronal zinc, which aggravated neuronal damage in AD (Adlard et al. 2008; Lee et al. 2002; Lovell et al. 1998).

13.4 Metal Chelators for AD

The Greek word "chele" is where the word "chelator" comes from, means "claw" and helps us understand how these molecules trap metal ions like a lobster claw. A chelator, or chelating agent, refers to a ligand that coordinates to a metal center by more than one point of attachment, thereby forming a ring with the metal atom. The resulting metal complex is referred to as a chelate. The name "chelator" does not discriminate between the number of attachment points to, a metal; hence, a chelator can be bidentate or multidentate. Multiple attachment sites boost the complex's stability due to the chelation effect. The name "chelate" indicates nothing else about the compound's characteristics, distinct from both chelates made of the same metal but various ligands and those made of the metal ion and ligand alone. Charge, solubility, lipophilicity, redox potential, shape, pH dependency, and the kinetics and thermodynamics of ligand and metal exchange are examples of chemical characteristics which might affect a chelator's biological effects (Haas and Franz 2009).

A metal-chelating agent must have a metal ion with high specificity to minimize unintended binding and depletion of other biometal ions. Traditional toxic chelators include polyaminocarboxylic acids like EDTA. To treat heavy metal toxicity, acids that bind calcium, magnesium, zinc, copper, and iron have been utilized. Bidentate or polydentate chelators can coordinate a metal ion at two or more sites. The metal complexes are stabilized by this chelation effect, which rises with the number of chelate rings and their size for the same sort of atoms, such as coordinating atoms and metal ions (and their oxidation state). The coordinating or donor atom a ligand uses to bind to a metal ion determines its selectivity (Fig. 13.5). Pearson's "hard and soft acids and bases" principle applies ligand-metal preference. For intermediate hard-soft copper(II), zinc(II), and iron(II) chelators, N- and O-donor atoms are needed, whereas S- and N-atoms are required for copper(I) chelators.

Hard iron chelators should have anionic O-donor atoms like phenoxides (III). The late first-row transition metal ions, such as Ni(II), Cu(II), and Zn(II), are stabilised mainly by ligands with two N-donor atoms, followed by N, O- and O, O-donor ligands, according to the Irving-Williams series, which categorizes metals based on the stability of metal–ligand complexes. Chelation affinity can be increased by adding an atom with an oxygen- or nitrogen-donor to bidentate scaffolds. There have been reports on tri- and tetradentate chelators for AD metal modulation.

Fig. 13.5 Chelating rings: metal-ligating atoms and number of ring atoms

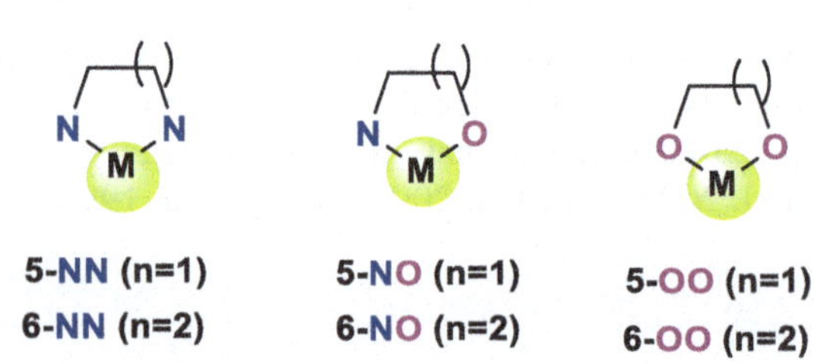

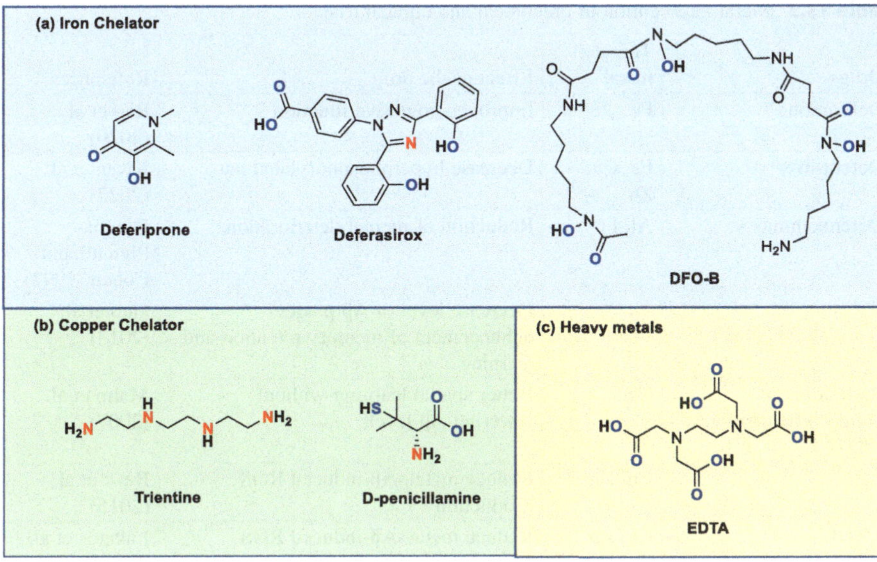

Fig. 13.6 Examples of chelating agents that have been approved by the FDA and used to manage copper overload (Wilson's disease), iron overload (beta thalassemia-related disorders), and heavy metal toxicity

Metal chelator (complex) materials should be polydentate (have several places of attachment for metal ions) for removing metal ions in overload circumstances. It also possess a high thermodynamic stability without relying on dilution effects. Consequently, fewer ligand molecules will be needed to achieve full coordination and prevent demetallization or transmetallation reactions. Fig. 13.6 contains Chelating agents structures that have been approved by the FDA for management of excess copper, zinc and iron.

For AD treatment, restoring metal ion balance is viewed as a significant difficulty. Metal chelator has not been explored significantly to date. It is probable that "classical chelators" and, by mistaken implication, all chelators are viewed as toxic by certain medical professionals. Therefore, there hasn't been much progress made with chelators. The truth is that toxic chelators, such as EDTA, are non-specific chelators that bind a wide variety of metal ions, such as iron, copper, zinc, calcium, magnesium, etc. However, the requirement for having chelators in order to reduce harmful side effects has received little attention. Table 13.3 enlists various representative chelators that were tested as anti-AD medications over the past decades, instances of biological data are cited. Very little information about in vivo effectiveness or safety is available for the majority of these ligands. The medications are advertised as metal-specific ligands; however, metal selectivity is infrequently and experimentally demonstrated.

Table 13.3 Metal intervention in preclinical and clinical trials

Drugs	Targeted metal	Effect of the drug	References
Deferiprone	Fe	Improved cognitive function	Rao et al. (2020)
Deferasirox	Fe, Cu, Zn	Decrease hyperphosphorylated tau	Kwan et al. (2022)
Deferoxamine	Al, Fe	Reduction of mental deterioration	Borgna-Pignatti and Cohen (1997)
		Decrease level of Aβ plaque, enhancement of memory retention and learning	Guo et al. (2013)
Pyrrolidine Dithiocarbamate (PDTC)	Cu	Better spatial learning without lowering Aβ levels	Malm et al. (2007)
L2b	Cu, Zn	Reduce metal–Aβ-induced ROS production	Beck et al. (2015)
ENDIP	Cu, Zn	Reduce metal–Aβ-induced ROS production	Lakatos et al. (2010)
L(a)	Cu	Rescuing antioxidant markers like SODs, malondialdehyde	Singh et al. (2013)
Bis (thiosemicarbazones) (BTSC)	Cu, Zn	Reduction of Aβ plaque	Donnelly et al. (2008b)
Clioquinol (CQ)	Fe, Cu, Zn	Reduction of plaque surface area, Aβ serum levels, and sedimentable Aβ by 50%, and rise in Aβ soluble fraction	Cherny et al. (2001)
		Preventing cognitive decline and lowering Aβ plasma levels	Ritchie et al. (2003)
PBT2	Fe, Cu, Zn	Improvements in learning and memory as well as a drop in soluble and insoluble Aβ levels and plaque burden in the MWM	Adlard et al. (2008)
		Lower Aβ plasma and CSF levels and improved executive function	Lannfelt et al. (2008), Faux et al. (2010)
Feralex-G	Al, Fe	Hyperphosphorylated tau-bound high Al/Fe elimination	Shin et al. (2003)
DP-109	Fe, Cu, Zn	Reduction of Aβ plaques and CAA	Lee et al. (2004), Petri et al. (2007)
Zn7MT3	Zn	Regulate metal Homeostasis, reduction of oxidative damage, neurodegeneration, and Aβ aggregation	Xu et al. (2017a), Carpenter et al. (2016)
XH1	Cu, Fe, Zn	Aβ plaque reduction	Dedeoglu et al. (2004)

13.4.1 Targeting Iron

Age-related or pathological disorders that affect iron homeostasis can cause neurodegeneration. Neurotoxicity may result from intracerebral heme release by microhemorrhages and an increase in the breakdown of heme enzymatically. Iron chelators are still being researched as a possible therapy for AD. It mainly consists of research ligands (deferasirox, deferiprone, and deferoxamine) or the already authorized conjugates of these drugs to treat chronic iron overload caused by blood transfusions given repeatedly to thalassemia patients (Guo et al. 2013).

Deferiprone (1). Based on deferiprone structure (**1**) (Fig. 13.7), other substituted 3-hydroxy-4-pyridinones, such as bis- and tris(3-hydroxy-4-pyridinone) derivatives, were investigated for iron chelation. To cross the BBB, which has many GLUT-1 transporters, glycoside derivatives that release 3-hydroxy-4-pyridinone upon enzymatic cleavage are used. The 3-hydroxy-4-pyridinone moiety, however, is not limited to Zn(II) and Cu(II), but also to iron chelation (III). Dendritic chelators with 3-hydroxy-2-methyl-4-pyridinone residues have been developed to improve copper-binding affinity. Researchers have looked on linking deferasirox (**2**) (Fig. 13.7) with lactoferrin or conjugating deferiprone with nanoparticles to enhance drug delivery to the brain (Ward et al. 2014; Liu et al. 2006).

Deferoxamine (DFO) (3). Intranasal DFO (**3**) (Fig. 13.7) therapy prevented the processing of amyloidogenic APP and restored Fe-induced memory deficits in APP/PS1 AD mice. DFO alters the expression of APP, iron regulatory protein-1, and hypoxia-induced factor, which may prevent the generation of ROS by suppressing the iron redox cycle (Guo et al. 2013).

13.4.2 Targeting Copper and Zinc

Dithiocarbamate of Pyrrolidine (PDTC) (4). PDTC (**4**) (Fig. 13.7) can deliver outside Cu(II) to a cell to reduce tau phosphorylation in the hippocampus of transgenic APP/PS1 mice and to shield hippocampal neurons from the toxicity of Aβ oligomers. Without changing the β-amyloid load, (**4**) was found to stop the deterioration of transgenic APP/PS1 mice's cognition. This was probably achieved by blocking Cu activation of the Akt pathway in both astrocytes and neurons (Malm et al. 2007).

Pyridine Derivatives. It was discovered that the 2-methylaminopyridine derivative (**5**), when chelated with Cu(II) and Zn(II), produced mixtures of 1/1 and 1/2 metal/ligand complexes with Kd values similar to those reported for Cu-Aβ and Zn-Aβ, (**5**) (Fig. 13.7) had a considerably stronger affinity for copper than zinc. In vitro, (**5**) was able to affect the structure of metal-induced Aβ-aggregates and govern Zn- or Cu-induced Aβ-aggregation, perhaps through a ternary (**5**)-Zn-Aβ complex. This compound demonstrated the usefulness of disassembling human Aβ aggregates and restoring the viability of human neuroblastoma cells treated

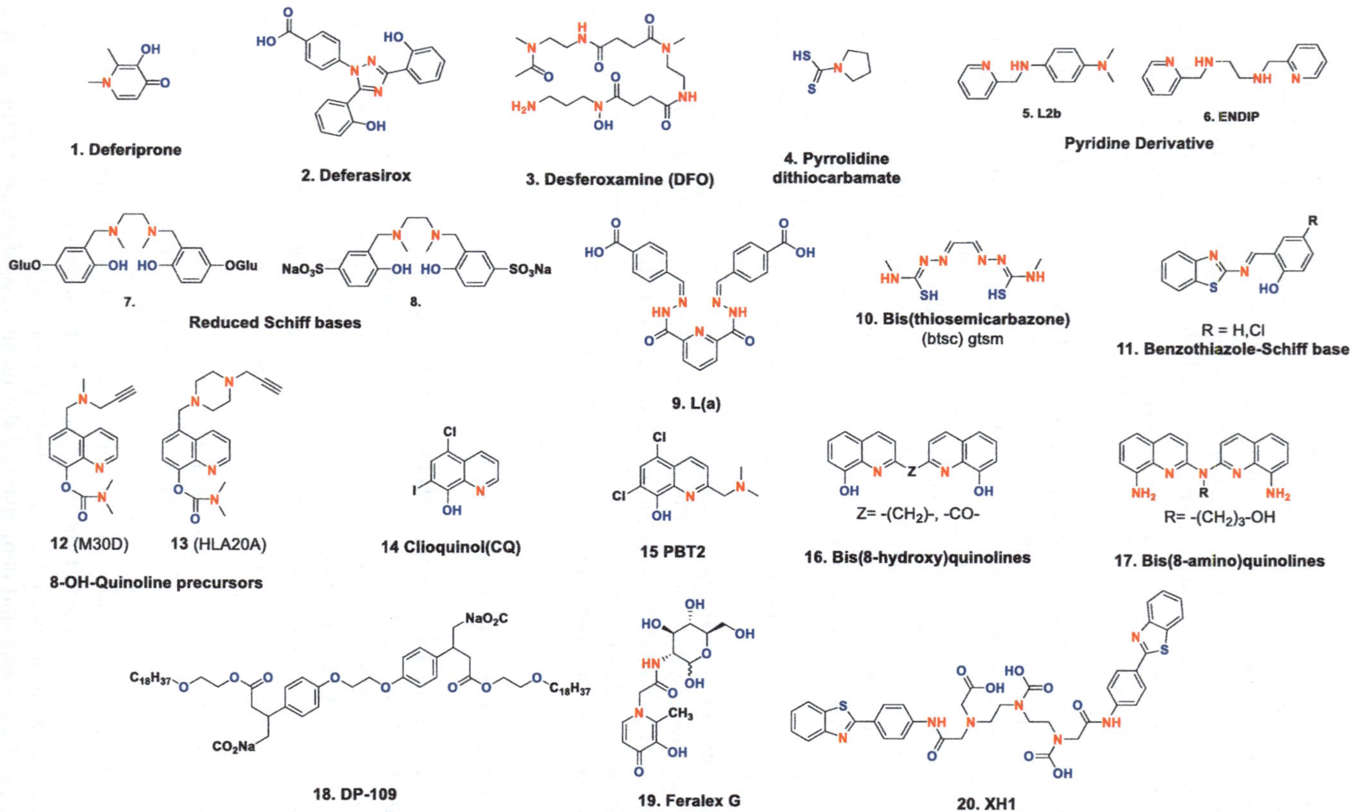

Fig. 13.7 Structures of representative metal chelators

with Aβ/Cu(II) or Aβ/Zn(II). Compound (**5**) was thought to specifically interact with metal-Aβ complexes and diminish Aβ aggregation and ROS generation after intraperitoneal given to 5XFAD AD animals. At physiological pH, the dipyridine derivative ENDIP (**6**) (Fig. 13.7) binds to Cu(II) and Zn(II) to produce a very stable tetradentate 1/1 complex with log K_{app} values of 16.6 and 11.4, respectively. By displacing Cu(II) from Aβ, compound (**6**) can resolubilize amyloid precipitates and stop metal-induced amyloid aggregation (Lakatos et al. 2010).

Reduced Schiff Bases (7) and (8). To remove Cu(II) from Cu(Aβ), prevent Aβ from aggregating, and serve as antioxidants, compounds (**7**) and (**8**) (Fig. 13.7) were proposed. In a typical square planar N_2O_2 environment, these water-soluble ligands join with copper(II) to create tetradentate complexes. Zinc and nickel complexes with a comparable structure were also produced for ligand (**7**). At physiological pH, it was determined that the log K_{app} values of Cu(II) and Zn (II) were around 9 and 6, respectively. At neutral pH, the log K_{app} values of 8 for Cu(II) and Zn(II) are 14.6 and 6.1, respectively, demonstrating a clear preference for copper for this ligand. These ligands claimed that biological functions have not yet been published (Noël et al. 2012).

L(a) (9). In a DMSO/water combination, this compound L(a) (**9**) (Fig. 13.7) forms a Cu/L complex with log $K_{aff} = 6$. Compound L(a) (**9**) decreases the Cu-induced retinal neurotoxicity in a transgenic Drosophila model of AD where Aβ is overexpressed (Singh et al. 2013).

Bis(thiosemicarbazones) (btsc) (10). Cu(II)(btsc) complexes have a square planar N_2S_2 shape, are stable (log $K_{aff} = 18$), and can pass through cell membranes. Intracellular reductants convert the series' leader, Cu(II)-**10**, to Cu(I), which subsequently separates from its ligand. Cu(II)-**10** (Fig. 13.7) therapy of neuron-like cells that overexpress APP led to intracellular copper release, a decrease in monomeric Aβ secretion, and an elevation of the phosphorylation of enzymes involved in apoptosis, including phosphoinositol 3-kinase and glycogen synthase kinase 3. A dose-dependent reduction of monomeric Aβ levels followed treatment induced by an increase in intracellular zinc levels with less stable Zn(II) (btsc) complexes (Donnelly et al. 2008a).

Clioquinol. Clioquinol, also known as 5-chloro-7-iodo-8-hydroxyquinoline (CQ, (**14**), Fig. 13.7), is a non-specific copper-zinc chelator that can enhance the learning and memory abilities of APP transgenic mice while reducing Aβ deposits. As a Zn-specific chelator, *N,N,N′,N′*-tetrakis (2-pyridylmethyl) ethylenediamine (TPEN) inhibits the neurotoxicity generated by soluble Aβ, demonstrating a strong link between Zn and Aβ neurotoxicity (Chen et al. 2021). In vitro, CQ also partially recovered Zn-containing Aβ amyloid's sensitivity to MMP2 destruction. Unfortunately, subacute myelo-optic neuropathy was caused by the antifungal and antiprotozoal drug clioquinol, which was taken off the market in 1983. Zinc chelation has been implicated in neurotoxicity. In actuality, 8-hydroxyquinolines are non-specific metal chelators, and CQ's affinity for Cu(II) is only slightly greater than that for Zn(II) (log $K = 10$ and 9, respectively). CQ, a bidentate ligand, interacts with Zn(II) ions and Cu(II) to create square planar complexes with a 2/1 (ligand/metal ratio) (Fig. 13.7). CQ is efficient in decreasing Aβ toxicity in yeast, but analogues without a nitrogen atom or hydroxyl group are ineffective, showing that bidentate metal binding is necessary for CQ to be effective. These findings also demonstrate a connection between Aβ's toxicity and its capacity to retain metal ions. In addition, amyloid is a competitor for the chelation of Zn and Cu. CQ raised the content of copper inside yeast cells (Matlack et al. 2014).

PBT2. Clioquinol served as a model for additional 8-hydroxyquinolines that could restrict metal-Aβ interactions and PBT2 (**15**) (Fig. 13.7), whose structure was just recently disclosed. PBT2 is a metal-protein attenuating substance (MPAC) that inhibits the harmful oligomerization of Aβ that is mediated by Cu^{2+} and Zn^{2+} in AD. Robust preclinical effectiveness results and early clinical safety studies preceded this phase IIa study. This drug candidate underwent a phase IIa, double-blind, placebo-controlled clinical trial, and it was well tolerated and showed some efficacy. PBT2 failed to achieve its primary goal lower the levels of statistically significantly amyloid plaques in the brains of patients with prodromal/mild AD (Adlard et al. 2008; Lannfelt et al. 2008; Faux et al. 2010).

Bis(8-hydroxyquinolines). Bis(8-hydroxyquinolines) (**16**) and (**17**) (Fig. 13.7) provide that one molecule has four N_2O_2 binding sites, as opposed to mono (8-hydroxyquinolines), leading to complexes with a ligand/metal ratio of 1/1 at low concentrations. These tetradentate ligands have a log K_{app} of 15.5–20.6 at physiological pH, which is 4 to 6 orders of magnitude higher than that of monovalent ligands for copper(II) (Log K_{app} for Zn(II) = 12.5–14.2). The range of copper's ligand selectivity toward zinc is 100–1000. These compounds are also very effective at dissolving Aβ peptides and inhibiting the formation of H_2O_2 by the CuAβ/ascorbic acid system (Nguyen et al. 2014).

13.4.3 Miscellaneous Compounds

DP-109. The large synthetic pro-drug known as hexadentate chelator, or DP-109 (**18**) (Fig. 13.7), is activated after its two long-chain esters cleavage. Over 3 months, daily oral gavage DP-109 treatment to female Tg2576 mice decreased the production and cerebral amyloid angiopathy deposition, amyloid plaques, and re-solubilized Aβ (Lee et al. 2004; Petri et al. 2007).

Feralex-G (19). Inhibition and dissociation of this Al(III)/Fe(III) binding linked to hyperphosphorylated tau (PHFtau) may slow or stop the neurofibrillary degeneration by tau in AD patients. Feralex-G and deferoxamine removal PHFtau-bound Al(III)/Fe(III) was compared and contrasted under reaction conditions at 37 °C. Both compounds successfully chelated Fe(III), with no apparent differences in their chelating capacities. In contrast, the two chelators in the current system each produced a distinct Al(III) chelation reaction. Deferoxamine failed to achieve noticeable Al(III) chelation during incubation at 37 °C, whereas Feralex-G (**19**) (Fig. 13.7) showed effective Al(III) chelation. Therefore, Feralex-G is a more effective chelator for Al(III) than deferoxamine when it comes to the competitive elimination of Al(III) from brain PHFtau (Shin et al. 2003).

Zn7MT3. APP/PS1 mice treated with prolonged pharmacological release of Zn7MT3 directly to the CNS and the role and molecular mechanism of Zn7MT3 to protect against AD mice were explored. Zn7MT3 reduced cognitive impairments, metal homeostasis, Aβ plaque burden, and oxidative stress. MT3 also penetrates AD mice's blood–brain barrier (Xu et al. 2017b; Carpenter et al. 2016).

XH1 (**20**). XH1 (Fig. 13.7) compound designed based on a "pharmacophore conjugation" concept lipophilic compound with amyloid-binding and metal-chelating amide linkages. Computational chemistry modelling suggests that it binds to Aβ peptide and reduces Zn(II)-induced Aβ aggregation *in vitro*. Pilot studies showed that XH1 has no appreciable neurotoxicity or acute animal toxicity at low micromolar dosages. XH1 reduced APP expression in human SH-SY5Y neuroblastoma cells and alleviated cerebral Aβ amyloid pathology in PS1/APP transgenic mice without toxicity or behaviour abnormalities (Dedeoglu et al. 2004).

13.5 Prospective and Conclusion

Long-term neuronal function necessitates a precise balance of transition metal activity. Errors inevitably happen in any physical system, and transition metals gradually occupy disproportionate positions in the brain, impairing cellular function. This aberration in metal homeostasis, at least in part, is the cause of AD, which exhibits a strong correlation with ageing. As a result, transition metal chelation offers a therapeutic approach that focuses on early disturbances in transition metal activity. Such sequestration would eventually aid in preventing ROS generation and accumulation that causes oxidative damage to neuronal tissue, as well as preventing Aβ peptide aggregation and deposition.

This book chapter examines the role played by various biometals in the primary pathological symptoms of AD. Due to increased APP expression, the aggregation of Aβ plaques, and tau hyperphosphorylation, biometals iron, zinc, and copper are known to be deposited in the brains of AD patients. Exposure to harmful metals like mercury, cadmium, lead, and aluminium may also begin an AD-specific pathology through mechanisms such as oxidative stress, neuroinflammation, and protein modification. As a result, any therapeutic approach aimed at specific drug delivery methods must be used when lowering trace metal levels in AD and consider how it will affect the concentration of a given metal.

As a result, any therapeutic approach aimed at reducing the levels of trace metals in AD must be precise regarding drug delivery and consider the impact on metal concentration. Even though they weren't entirely successful, preclinical and clinical research on manipulating trace metals was one of the most promising ways to treat AD and this could be achieved through the targeted distribution of trace metals to Aβ aggregates and NFTs. To achieve this, novel delivery platforms, novel metal chelators, and novel compounds will be required.

References

Acevedo KM, Hung YH, Dalziel AH, Li Q-X, Laughton K, Wikhe K, Rembach A, Roberts B, Masters CL, Bush AI, Camakaris J (2011) Copper promotes the trafficking of the amyloid precursor protein. J Biol Chem 286:8252–8262. https://doi.org/10.1074/JBC.M110.128512

Acevedo KM, Opazo CM, Norrish D, Challis LM, Li QX, White AR, Bush AI, Camakaris J (2014) Phosphorylation of amyloid precursor protein at threonine 668 is essential for its copper-responsive trafficking in SH-SY5Y neuroblastoma cells. J Biol Chem 289:11007–11019. https://doi.org/10.1074/JBC.M113.538710

Adlard PA, Cherny RA, Finkelstein DI, Gautier E, Robb E, Cortes M, Volitakis I, Liu X, Smith JP, Perez K, Laughton K, Li QX, Charman SA, Nicolazzo JA, Wilkins S, Deleva K, Lynch T, Kok G, Ritchie CW, Tanzi RE, Cappai R, Masters CL, Barnham KJ, Bush AI (2008) Rapid restoration of cognition in Alzheimer's transgenic mice with 8-hydroxy quinoline analogs is associated with decreased interstitial Aβ. Neuron 59:43–55. https://doi.org/10.1016/j.neuron.2008.06.018

Antala S, Dempski RE (2012) The human ZIP4 transporter has two distinct binding affinities and mediates transport of multiple transition metals. Biochemistry 51:963–973. https://doi.org/10.1021/BI201553P

Bagheri S, Squitti R, Haertlé T, Siotto M, Saboury AA (2018) Role of copper in the onset of Alzheimer's disease compared to other metals. Front Aging Neurosci 9:446. https://doi.org/10.3389/FNAGI.2017.00446/FULL

Barnham KJ, McKinstry WJ, Multhaup G, Galatis D, Morton CJ, Curtain CC, Williamson NA, White AR, Hinds MG, Norton RS, Beyreuther K, Masters CL, Parker MW, Cappai R (2003) Structure of the Alzheimer's disease amyloid precursor protein copper binding domain. A regulator of neuronal copper homeostasis. J Biol Chem 278:17401–17407. https://doi.org/10.1074/JBC.M300629200

Barr CA, Burdette SC (2017) The zinc paradigm for metalloneurochemistry. Essays Biochem 61: 225–235. https://doi.org/10.1042/EBC20160073

Beck MW, Oh SB, Kerr RA, Lee HJ, Kim SH, Kim S, Jang M, Ruotolo BT, Lee J-Y, Lim MH (2015) A rationally designed small molecule for identifying an in vivo link between metal-amyloid-β complexes and the pathogenesis of Alzheimer's disease. Chem Sci 6:1879–1886. https://doi.org/10.1039/c4sc03239j

Borgna-Pignatti C, Cohen A (1997) Evaluation of a new method of administration of the iron chelating agent deferoxamine. J Pediatr 130:86–88. https://doi.org/10.1016/S0022-3476(97)70314-7

Brewen GJ (2008) The risks of free copper in the body and the development of useful anticopper drugs. Curr Opin Clin Nutr Metab Care 11:727–732. https://doi.org/10.1097/MCO.0B013E328314B678

Carpenter MC, Shami Shah A, Desilva S, Gleaton A, Su A, Goundie B, Croteau ML, Stevenson MJ, Wilcox DE, Austin RN (2016) Thermodynamics of Pb(II) and Zn(II) binding to MT-3, a neurologically important metallothionein. Metallomics 8:605–617. https://doi.org/10.1039/c5mt00209e

Castilho RF, Ward MW, Nicholls DG (2001) Oxidative stress, mitochondrial function, and acute glutamate excitotoxicity in cultured cerebellar granule cells. J Neurochem 72:1394–1401. https://doi.org/10.1046/J.1471-4159.1999.721394.X

Catterall WA (2011) Voltage-gated calcium channels. Cold Spring Harb Perspect Biol 3:1–23. https://doi.org/10.1101/CSHPERSPECT.A003947

Cheignon C, Tomas M, Bonnefont-Rousselot D, Faller P, Hureau C, Collin F (2018) Oxidative stress and the amyloid beta peptide in Alzheimer's disease. Redox Biol 14:450–464. https://doi.org/10.1016/J.REDOX.2017.10.014

Chen W, Wang Y, Wang H, An D, Sun D, Li P, Zhang T, Lu W, Liu Y (2021) TPEN attenuates amyloid-β25–35-induced neuronal damage with changes in the electrophysiological properties of voltage-gated sodium and potassium channels. Mol Brain 14:124. https://doi.org/10.1186/S13041-021-00837-Z

Cherny RA, Atwood CS, Xilinas ME, Gray DN, Jones WD, McLean CA, Barnham KJ, Volitakis I, Fraser FW, Kim YS, Huang X, Goldstein LE, Moir RD, Lim JT, Beyreuther K, Zheng H, Tanzi RE, Masters CL, Bush AI (2001) Treatment with a copper-zinc chelator markedly and rapidly inhibits β-amyloid accumulation in Alzheimer's disease transgenic mice. Neuron 30:665–676. https://doi.org/10.1016/S0896-6273(01)00317-8

Chin-Chan M, Segovia J, Quintanar L, Arcos-López T, Hersh LB, Martin Chow K, Rodgers DW, Quintanilla-Vega B (2015) Mercury reduces the enzymatic activity of neprilysin in differentiated SH-SY5Y cells. Toxicol Sci 145:128–137. https://doi.org/10.1093/toxsci/kfv037

Choo XY, Alukaidey L, White AR, Grubman A (2013) Neuroinflammation and copper in Alzheimer's disease. Int J Alzheimers Dis 2013:1–12. https://doi.org/10.1155/2013/145345

Clardy SL, Wang X, Zhao W, Liu W, Chase GA, Beard JL, True Felt B, Connor JR (2006) Acute and chronic effects of developmental iron deficiency on mRNA expression patterns in the brain. J Neural Transm Suppl 71:173–196. https://doi.org/10.1007/978-3-211-33328-0_19

Crouch PJ, Harding SME, White AR, Camakaris J, Bush AI, Masters CL (2008) Mechanisms of Aβ mediated neurodegeneration in Alzheimer's disease. Int J Biochem Cell Biol 40:181–198. https://doi.org/10.1016/J.BIOCEL.2007.07.013

Crouch PJ, Lin WH, Adlard PA, Cortes M, Lal V, Filiz G, Perez KA, Nurjono M, Caragounis A, Du T, Laughton K, Volitakis I, Bush AI, Li QX, Masters CL, Cappai R, Cherny RA, Donnelly PS, White AR, Barnham KJ (2009a) Increasing Cu bioavailability inhibits Aβ oligomers and tau phosphorylation. Proc Natl Acad Sci U S A 106:381–386. https://doi.org/10.1073/PNAS.0809057106/SUPPL_FILE/0809057106SI.PDF

Crouch PJ, Tew DJ, Du T, Nguyen DN, Caragounis A, Filiz G, Blake RE, Trounce IA, Soon CPW, Laughton K, Perez KA, Li Q-X, Cherny RA, Masters CL, Barnham KJ, White AR (2009b) Restored degradation of the Alzheimer's amyloid-β peptide by targeting amyloid formation. J Neurochem 108:1198–1207. https://doi.org/10.1111/J.1471-4159.2009.05870.X

Cummings JL, Morstorf T, Zhong K (2014) Alzheimer's disease drug-development pipeline: few candidates, frequent failures. Alzheimers Res Ther 6:37. https://doi.org/10.1186/ALZRT269

Dalvi T, Dewangan B, Das R, Rani J, Shinde SD, Vhora N, Jain A, Sahu B (2020) Old drugs with new tricks: paradigm in drug development pipeline for Alzheimer's disease. Cent Nerv Syst Agents Med Chem 20:157–176. https://doi.org/10.2174/1871524920666201021164805

Dalvi T, Dewangan B, Agarwal G, Shinde Suchita D, Jain A, Srivastava A, Sahu B (2021) Design, synthesis and in-vitro evaluation of fluorinated triazoles as multi-target directed ligands for Alzheimer disease. Bioorg Med Chem Lett 42:127999. https://doi.org/10.1016/J.BMCL.2021.127999

Dedeoglu A, Cormier K, Payton S, Tseitlin KA, Kremsky JN, Lai L, Li X, Moir RD, Tanzi RE, Bush AI, Kowall NW, Rogers JT, Huang X (2004) Preliminary studies of a novel bifunctional metal chelator targeting Alzheimer's amyloidogenesis. Exp Gerontol 39:1641–1649. https://doi.org/10.1016/j.exger.2004.08.016

Deshpande A, Kawai H, Metherate R, Glabe CG, Busciglio J (2009) A role for synaptic zinc in activity-dependent Aβ oligomer formation and accumulation at excitatory synapses. J Neurosci 29:4004–4015. https://doi.org/10.1523/JNEUROSCI.5980-08.2009

Donnelly PS, Caragounis A, Du T, Laughton KM, Volitakis I, Cherny RA, Sharples RA, Hill AF, Li QX, Masters CL, Barnham KJ, White AR (2008a) Selective intracellular release of copper and zinc ions from bis(thiosemicarbazonato) complexes reduces levels of Alzheimer disease amyloid-beta peptide. J Biol Chem 283:4568–4577. https://doi.org/10.1074/JBC.M705957200

Donnelly PS, Caragounis A, Du T, Laughton KM, Volitakis I, Cherny RA, Sharples RA, Hill AF, Li QX, Masters CL, Barnham KJ, White AR (2008b) Selective intracellular release of copper and zinc ions from bis(thiosemicarbazonato) complexes reduces levels of Alzheimer disease amyloid-β peptide. J Biol Chem 283:4568–4577. https://doi.org/10.1074/jbc.M705957200

Dröge W, Schipper HM (2007) Oxidative stress and aberrant signaling in aging and cognitive decline. Aging Cell 6:361–370. https://doi.org/10.1111/J.1474-9726.2007.00294.X

Faux NG, Ritchie CW, Gunn A, Rembach A, Tsatsanis A, Bedo J, Harrison J, Lannfelt L, Blennow K, Zetterberg H, Ingelsson M, Masters CL, Tanzi RE, Cummings JL, Herd CM, Bush AI (2010) PBT2 rapidly improves cognition in Alzheimer's disease: additional phase II analyses. J Alzheimers Dis 20:509–516. https://doi.org/10.3233/JAD-2010-1390

Frazzini V, Rockabrand E, Mocchegiani E, Sensi SL (2006) Oxidative stress and brain aging: is zinc the link? Biogerontology 7:307–314. https://doi.org/10.1007/S10522-006-9045-7

Fukada T, Yamasaki S, Nishida K, Murakami M, Hirano T (2011) Zinc homeostasis and signaling in health and diseases. J Biol Inorg Chem 16:1123–1134. https://doi.org/10.1007/S00775-011-0797-4/FIGURES/7

Grill J, Cummings J (2010) Current therapeutic targets for the treatment of Alzheimer's disease. Expert Rev Neurother 10(5):711–728. https://doi.org/10.1586/ERN.10.29

Gröber U, Schmidt J, Kisters K (2015) Magnesium in prevention and therapy. Nutrients 7:8199–8226. https://doi.org/10.3390/NU7095388

Grundke-Iqbal I, Iqbal K, Tung YC, Wisniewski HM (1984) Alzheimer paired helical filaments: immunochemical identification of polypeptides. Acta Neuropathol 62:259–267. https://doi.org/10.1007/BF00687607

Gu M, Bode DC, Viles JH (2018) Copper redox cycling inhibits Aβ fibre formation and promotes fibre fragmentation, while generating a dityrosine Aβ dimer. Sci Rep 8:16190. https://doi.org/10.1038/s41598-018-33935-5

Guo C, Wang T, Zheng W, Shan ZY, Teng WP, Wang ZY (2013) Intranasal deferoxamine reverses iron-induced memory deficits and inhibits amyloidogenic APP processing in a transgenic mouse model of Alzheimer's disease. Neurobiol Aging 34:562–575. https://doi.org/10.1016/J.NEUROBIOLAGING.2012.05.009

Gupta SP (2018) Roles of metals in human health. MOJ Bioorg Org Chem 2(5):221–224. https://doi.org/10.15406/MOJBOC.2018.02.00085

Gybina AA, Tkac I, Prohaska JR (2009) Copper deficiency alters the neurochemical profile of developing rat brain. Nutr Neurosci 12:114–122. https://doi.org/10.1179/147683009X423265

Haas KL, Franz KJ (2009) Application of metal coordination chemistry to explore and manipulate cell biology. Chem Rev 109:4921–4960. https://doi.org/10.1021/cr900134a

Hordyjewska A, Popiołek Ł, Kocot J (2014) The many "faces" of copper in medicine and treatment. Biometals 27:611–621. https://doi.org/10.1007/S10534-014-9736-5

Horning KJ, Caito SW, Tipps KG, Bowman AB, Aschner M (2015) Manganese is essential for neuronal. Health 35:71–108. https://doi.org/10.1146/annurev-nutr-071714-034419

Huang L, Tepaamorndech S (2013) The SLC30 family of zinc transporters—a review of current understanding of their biological and pathophysiological roles. Mol Asp Med 34:548–560. https://doi.org/10.1016/J.MAM.2012.05.008

Huang W, Wei W, Shen Z (2014) Drug-like chelating agents: a potential lead for Alzheimer's disease. RSC Adv 4:52088–52099. https://doi.org/10.1039/C4RA09193K

Hung YH, Bush AI, Cherny RA (2009) Copper in the brain and Alzheimer's disease. J Biol Inorg Chem 15:1–15, 61–76. https://doi.org/10.1007/S00775-009-0600-Y

James SA, Churches QI, de Jonge MD, Birchall IE, Streltsov V, McColl G, Adlard PA, Hare DJ (2017) Iron, copper, and zinc concentration in Aβ plaques in the APP/PS1 mouse model of Alzheimer's disease correlates with metal levels in the surrounding neuropil. ACS Chem Neurosci 8:629–637. https://doi.org/10.1021/ACSCHEMNEURO.6B00362/ASSET/IMAGES/LARGE/CN-2016-003625_0005.JPEG

Jiang H, Wang J, Rogers J, Xie J (2017) Brain iron metabolism dysfunction in Parkinson's disease. Mol Neurobiol 54:3078–3101. https://doi.org/10.1007/S12035-016-9879-1

Jin L, Wu WH, Li QY, Zhao YF, Li YM (2011) Copper inducing Aβ42 rather than Aβ40 nanoscale oligomer formation is the key process for Aβ neurotoxicity. Nanoscale 3:4746–4751. https://doi.org/10.1039/C1NR11029B

Jolivet-Gougeon A, Bonnaure-Mallet M (2018) Treponema, iron and neurodegeneration. Curr Alzheimer Res 15:716–722. https://doi.org/10.2174/1567205013666161122093404

Kaden D, Bush AI, Danzeisen R, Bayer TA, Multhaup G (2011) Disturbed copper bioavailability in Alzheimer's disease. Int J Alzheimers Dis 2011:345614. https://doi.org/10.4061/2011/345614
Kawamoto EM, Vivar C, Camandola S (2012) Physiology and pathology of calcium signaling in the brain. Front Pharmacol 3:61. https://doi.org/10.3389/FPHAR.2012.00061/BIBTEX
Kirkland AE, Sarlo GL, Holton KF (2018) The role of magnesium in neurological disorders. Nutrients 10:730. https://doi.org/10.3390/NU10060730
Klevay LM (2008) Alzheimer's disease as copper deficiency. Med Hypotheses 70:802–807. https://doi.org/10.1016/j.mehy.2007.04.051
Krężel A, Hao Q, Maret W (2007) The zinc/thiolate redox biochemistry of metallothionein and the control of zinc ion fluctuations in cell signaling. Arch Biochem Biophys 463:188–200. https://doi.org/10.1016/J.ABB.2007.02.017
Kwan P, Ho A, Baum L (2022) Effects of deferasirox in Alzheimer's disease and tauopathy animal models. Biomol Ther 12:365. https://doi.org/10.3390/BIOM12030365/S1
Lakatos A, Zsigó É, Hollender D, Nagy NV, Fülöp L, Simon D, Bozsó Z, Kiss T (2010) Two pyridine derivatives as potential Cu(II) and Zn(II) chelators in therapy for Alzheimer's disease. Dalton Trans 39:1302–1315. https://doi.org/10.1039/B916366B
Lannfelt L, Blennow K, Zetterberg H, Batsman S, Ames D, Harrison J, Masters CL, Targum S, Bush AI, Murdoch R, Wilson J, Ritchie CW (2008) Safety, efficacy, and biomarker findings of PBT2 in targeting Aβ as a modifying therapy for Alzheimer's disease: a phase IIa, double-blind, randomised, placebo-controlled trial. Lancet Neurol 7:779–786. https://doi.org/10.1016/S1474-4422(08)70167-4
Lee JY, Cole TB, Palmiter RD, Suh SW, Koh JY (2002) Contribution by synaptic zinc to the gender-disparate plaque formation in human Swedish mutant APP transgenic mice. Proc Natl Acad Sci U S A 99:7705–7710. https://doi.org/10.1073/pnas.092034699
Lee JY, Friedman JE, Angel I, Kozak A, Koh JY (2004) The lipophilic metal chelator DP-109 reduces amyloid pathology in brains of human β-amyloid precursor protein transgenic mice. Neurobiol Aging 25:1315–1321. https://doi.org/10.1016/j.neurobiolaging.2004.01.005
Li Y, Jiao Q, Xu H, Du X, Shi L, Jia F, Jiang H (2017a) Biometal dyshomeostasis and toxic metal accumulations in the development of Alzheimer's disease. Front Mol Neurosci 10:339. https://doi.org/10.3389/FNMOL.2017.00339/BIBTEX
Li DD, Zhang W, Wang ZY, Zhao P (2017b) Serum copper, zinc, and iron levels in patients with Alzheimer's disease: a meta-analysis of case-control studies. Front Aging Neurosci 9:300. https://doi.org/10.3389/FNAGI.2017.00300/BIBTEX
Liu G, Men P, Harris PLR, Rolston RK, Perry G, Smith MA (2006) Nanoparticle iron chelators: a new therapeutic approach in Alzheimer disease and other neurologic disorders associated with trace metal imbalance. Neurosci Lett 406:189–193. https://doi.org/10.1016/J.NEULET.2006.07.020
Lovell MA, Robertson JD, Teesdale WJ, Campbell JL, Markesbery WR (1998) Copper, iron and zinc in Alzheimer's disease senile plaques. J Neurol Sci 158:47–52. https://doi.org/10.1016/S0022-510X(98)00092-6
Lovell MA, Xiong S, Xie C, Davies P, Markesbery WR (2005) Induction of hyperphosphorylated tau in primary rat cortical neuron cultures mediated by oxidative stress and glycogen synthase kinase-3. J Alzheimers Dis 6:659–671. https://doi.org/10.3233/JAD-2004-6610
Lutsenko S, Bhattacharjee A, Hubbard AL (2010) Copper handling machinery of the brain. Metallomics 2:596–608. https://doi.org/10.1039/C0MT00006J
Malm TM, Iivonen H, Goldsteins G, Keksa-Goldsteine V, Ahtoniemi T, Kanninen K, Salminen A, Auriola S, van Groen T, Tanila H, Koistinaho J (2007) Pyrrolidine dithiocarbamate activates Akt and improves spatial learning in APP/PS1 mice without affecting β-amyloid burden. J Neurosci 27:3712–3721. https://doi.org/10.1523/JNEUROSCI.0059-07.2007
Matlack KES, Tardiff DF, Narayan P, Hamamichi S, Caldwell KA, Caldwell GA, Lindquist S (2014) Clioquinol promotes the degradation of metal-dependent amyloid-β (Aβ) oligomers to restore endocytosis and ameliorate Aβ toxicity. Proc Natl Acad Sci U S A 111:4013–4018. https://doi.org/10.1073/pnas.1402228111

Maynard CJ, Bush AI, Masters CL, Cappai R, Li QX (2005) Metals and amyloid-β in Alzheimer's disease. Int J Exp Pathol 86:147–159. https://doi.org/10.1111/J.0959-9673.2005.00434.X

Moon Y, Han SH, Moon WJ (2016) Patterns of brain iron accumulation in vascular dementia and Alzheimer's dementia using quantitative susceptibility mapping imaging. J Alzheimers Dis 51: 737–745. https://doi.org/10.3233/JAD-151037

Moustakas M (2021) The role of metal ions in biology, biochemistry and medicine. Materials 14:1–4. https://doi.org/10.3390/MA14030549

Muhoberac BB, Vidal R (2013) Abnormal iron homeostasis and neurodegeneration. Front Aging Neurosci 5:32. https://doi.org/10.3389/FNAGI.2013.00032

Nayak P (2002) Aluminum: impacts and disease. Environ Res 89:101–115. https://doi.org/10.1006/ENRS.2002.4352

Nguyen M, Robert A, Sournia-Saquet A, Vendier L, Meunier B (2014) Characterization of new specific copper chelators as potential drugs for the treatment of Alzheimer's disease. Chem Eur J 20:6771–6785. https://doi.org/10.1002/CHEM.201402143

Noël S, Perez F, Pedersen JT, Alies B, Ladeira S, Sayen S, Guillon E, Gras E, Hureau C (2012) A new water-soluble Cu(II) chelator that retrieves Cu from Cu(amyloid-β) species, stops associated ROS production and prevents Cu(II)-induced Aβ aggregation. J Inorg Biochem 117:322–325. https://doi.org/10.1016/J.JINORGBIO.2012.05.016

Paoletti P (2011) Molecular basis of NMDA receptor functional diversity. Eur J Neurosci 33:1351–1365. https://doi.org/10.1111/J.1460-9568.2011.07628.X

Parsons CG, Stöffler A, Danysz W (2007) Memantine: a NMDA receptor antagonist that improves memory by restoration of homeostasis in the glutamatergic system—too little activation is bad, too much is even worse. Neuropharmacology 53:699–723. https://doi.org/10.1016/j.neuropharm.2007.07.013

Peña MMO, Lee J, Thiele DJ (1999) A delicate balance: homeostatic control of copper uptake and distribution. J Nutr 129:1251–1260. https://doi.org/10.1093/JN/129.7.1251

Petri S, Calingasan NY, Alsaied OA, Wille E, Kiaei M, Friedman JE, Baranova O, Chavez JC, Beal MF (2007) The lipophilic metal chelators DP-109 and DP-460 are neuroprotective in a transgenic mouse model of amyotrophic lateral sclerosis. J Neurochem 102:991–1000. https://doi.org/10.1111/J.1471-4159.2007.04604.X

Rao SS, Portbury SD, Lago L, Bush AI, Adlard PA (2020) The iron chelator deferiprone improves the phenotype in a mouse model of tauopathy. J Alzheimers Dis 77:753–771. https://doi.org/10.3233/JAD-200551

Ritchie CW, Bush AI, Mackinnon A, Macfarlane S, Mastwyk M, MacGregor L, Kiers L, Cherny R, Li QX, Tammer A, Carrington D, Mavros C, Volitakis I, Xilinas M, Ames D, Davis S, Beyreuther K, Tanzi RE, Masters CL (2003) Metal-protein attenuation with iodochlorhydroxyquin (clioquinol) targeting Aβ amyloid deposition and toxicity in Alzheimer disease: a pilot phase 2 clinical trial. Arch Neurol 60:1685–1691. https://doi.org/10.1001/ARCHNEUR.60.12.1685

Rogers JT, Randall JD, Cahill CM, Eder PS, Huang X, Gunshin H, Leiter L, McPhee J, Sarang SS, Utsuki T, Greig NH, Lahiri DK, Tanzi RE, Bush AI, Giordano T, Gullans SR (2002) An iron-responsive element type II in the 5′-untranslated region of the Alzheimer's amyloid precursor protein transcript. J Biol Chem 277:45518–45528. https://doi.org/10.1074/JBC.M207435200

Rogers JT, Venkataramani V, Washburn C, Liu Y, Tummala V, Jiang H, Smith A, Cahill CM (2016) A role for amyloid precursor protein translation to restore iron homeostasis and ameliorate lead (Pb) neurotoxicity. J Neurochem 138:479–494. https://doi.org/10.1111/JNC.13671

Roland JR, Jacobsen H (2009) Alzheimer's disease: from pathology to therapeutic approaches. Angew Chem Int Ed 48:3030–3059. https://doi.org/10.1002/ANIE.200802808

Salkovic-Petrisic M, Knezovic A, Osmanovic-Barilar J, Smailovic U, Trkulja V, Riederer P, Amit T, Mandel S, Youdim MBH (2015) Multi-target iron-chelators improve memory loss in a rat model of sporadic Alzheimer's disease. Life Sci 136:108–119. https://doi.org/10.1016/J.LFS.2015.06.026

Santos MA, Chand K, Chaves S (2016) Recent progress in multifunctional metal chelators as potential drugs for Alzheimer's disease. Coord Chem Rev 327–328:287–303. https://doi.org/10.1016/J.CCR.2016.04.013

Sayre LM, Perry G, Harris PLR, Liu Y, Schubert KA, Smith MA (2001) In situ oxidative catalysis by neurofibrillary tangles and senile plaques in Alzheimer's disease. J Neurochem 74:270–279. https://doi.org/10.1046/J.1471-4159.2000.0740270.X

Selkoe DJ (2001) Alzheimer's disease: genes, proteins, and therapy. Physiol Rev 81:741–766. https://doi.org/10.1152/PHYSREV.2001.81.2.741

Selkoe DJ (2002) Alzheimer's disease is a synaptic failure. Science 298:789–791. https://doi.org/10.1126/SCIENCE.1074069

Selkoe DJ (2003) Folding proteins in fatal ways. Nature 426:900–904. https://doi.org/10.1038/nature02264

Selkoe DJ, Hardy J (2016) The amyloid hypothesis of Alzheimer's disease at 25 years. EMBO Mol Med 8:595–608. https://doi.org/10.15252/EMMM.201606210

Sensi SL, Granzotto A, Siotto M, Squitti R (2018) Copper and zinc dysregulation in Alzheimer's disease. Trends Pharmacol Sci 39:1049–1063. https://doi.org/10.1016/J.TIPS.2018.10.001

Shin RW, Kruck TPA, Murayama H, Kitamoto T (2003) A novel trivalent cation chelator Feralex dissociates binding of aluminum and iron associated with hyperphosphorylated τ of Alzheimer's disease. Brain Res 961:139–146. https://doi.org/10.1016/S0006-8993(02)03893-3

Singh SK, Sinha P, Mishra L, Srikrishna S (2013) Neuroprotective role of a novel copper chelator against Aβ 42 induced neurotoxicity. Int J Alzheimers Dis 2013:567128. https://doi.org/10.1155/2013/567128

Slutsky I, Sadeghpour S, Li B, Liu G (2004) Enhancement of synaptic plasticity through chronically reduced Ca^{2+} flux during uncorrelated activity. Neuron 44:835–849. https://doi.org/10.1016/j.neuron.2004.11.013

Sparks DL, Schreurs BG (2003) Trace amounts of copper in water induce β-amyloid plaques and learning deficits in a rabbit model of Alzheimer's disease. Proc Natl Acad Sci U S A 100:11065–11069. https://doi.org/10.1073/pnas.1832769100

Squitti R, Rossini PM, Cassetta E, Moffa F, Pasqualetti P, Cortesi M, Colloca A, Rossi L, Finazzi-Agró A (2002) D-penicillamine reduces serum oxidative stress in Alzheimer's disease patients. Eur J Clin Investig 32:51–59. https://doi.org/10.1046/J.1365-2362.2002.00933.X

Su X-Y, Wu W-H, Huang Z-P, Hu J, Lei P, Yu C-H, Zhao Y-F, Li Y-M (2007) Hydrogen peroxide can be generated by tau in the presence of Cu(II). Biochem Biophys Res Commun 358:661–665. https://doi.org/10.1016/J.BBRC.2007.04.191

Tamagno E, Guglielmotto M, Aragno M, Borghi R, Autelli R, Giliberto L, Muraca G, Danni O, Zhu X, Smith MA, Perry G, Jo D-G, Mattson MP, Tabaton M (2007) Oxidative stress activates a positive feedback between the γ- and β-secretase cleavages of the β-amyloid precursor protein. J Neurochem 2007:071115163316002. https://doi.org/10.1111/J.1471-4159.2007.05072.X

Tao Y, Wang Y, Rogers JT, Wang F (2014) Perturbed iron distribution in Alzheimer's disease serum, cerebrospinal fluid, and selected brain regions: a systematic review and meta-analysis. J Alzheimers Dis 42:679–690. https://doi.org/10.3233/JAD-140396

Tapiero H, Tew KD (2003) Trace elements in human physiology and pathology: zinc and metallothioneins. Biomed Pharmacother 57:399–411. https://doi.org/10.1016/S0753-3322(03)00081-7

Tõugu V, Karafin A, Palumaa P (2008) Binding of zinc(II) and copper(II) to the full-length Alzheimer's amyloid-β peptide. J Neurochem 104:1249–1259. https://doi.org/10.1111/j.1471-4159.2007.05061.x

Vassar R, Cole S (2007) The basic biology of BACE1: a key therapeutic target for Alzheimer's disease. Curr Genomics 8:509–530. https://doi.org/10.2174/138920207783769512

Waggoner DJ, Bartnikas TB, Gitlin JD (1999) The role of copper in neurodegenerative disease. Neurobiol Dis 6:221–230. https://doi.org/10.1006/nbdi.1999.0250

Wang P, Wang ZY (2017) Metal ions influx is a double edged sword for the pathogenesis of Alzheimer's disease. Ageing Res Rev 35:265–290. https://doi.org/10.1016/j.arr.2016.10.003

Wang ZX, Tan L, Wang HF, Ma J, Liu J, Tan MS, Sun JH, Zhu XC, Jiang T, Yu JT (2015) Serum iron, zinc, and copper levels in patients with Alzheimer's disease: a replication study and meta-analyses. J Alzheimers Dis 47:565–581. https://doi.org/10.3233/JAD-143108

Ward RJ, Zucca FA, Duyn JH, Crichton RR, Zecca L (2014) The role of iron in brain ageing and neurodegenerative disorders. Lancet Neurol 13:1045–1060. https://doi.org/10.1016/S1474-4422(14)70117-6

White AR, Multhaup G, Maher F, Bellingham S, Camakaris J, Zheng H, Bush AI, Beyreuther K, Masters CL, Cappai R (1999a) The Alzheimer's disease amyloid precursor protein modulates copper-induced toxicity and oxidative stress in primary neuronal cultures. J Neurosci 19:9170–9179. https://doi.org/10.1523/JNEUROSCI.19-21-09170.1999

White AR, Reyes R, Mercer JFB, Camakaris J, Zheng H, Bush AI, Multhaup G, Beyreuther K, Masters CL, Cappai R (1999b) Copper levels are increased in the cerebral cortex and liver of APP and APLP2 knockout mice. Brain Res 842:439–444. https://doi.org/10.1016/S0006-8993(99)01861-2

Whitfield DR, Vallortigara J, Alghamdi A, Howlett D, Hortobágyi T, Johnson M, Attems J, Newhouse S, Ballard C, Thomas AJ, O'Brien JT, Aarsland D, Francis PT (2014) Assessment of ZnT3 and PSD95 protein levels in Lewy body dementias and Alzheimer's disease: association with cognitive impairment. Neurobiol Aging 35:2836–2844. https://doi.org/10.1016/J.NEUROBIOLAGING.2014.06.015

Wischik CM, Harrington CR, Storey JMD (2014) Tau-aggregation inhibitor therapy for Alzheimer's disease. Biochem Pharmacol 88:529–539. https://doi.org/10.1016/J.BCP.2013.12.008

Wojsiat J, Zoltowska KM, Laskowska-Kaszub K, Wojda U (2018) Oxidant/antioxidant imbalance in Alzheimer's disease: therapeutic and diagnostic prospects. Oxidative Med Cell Longev. https://doi.org/10.1155/2018/6435861

Xian-hui D, Wei-juan G, Tie-mei S, Hong-lin X, Jiang-tao B, Jing-yi Z, Xi-qing C (2015) Age-related changes of brain iron load changes in the frontal cortex in APPswe/PS1ΔE9 transgenic mouse model of Alzheimer's disease. J Trace Elem Med Biol 30:118–123. https://doi.org/10.1016/J.JTEMB.2014.11.009

Xu W, Xu Q, Cheng H, Tan X (2017a) The efficacy and pharmacological mechanism of Zn7MT3 to protect against Alzheimer's disease. Sci Rep 7:13763. https://doi.org/10.1038/s41598-017-12800-x

Xu W, Xu Q, Cheng H, Tan X (2017b) The efficacy and pharmacological mechanism of Zn7MT3 to protect against Alzheimer's disease. Sci Rep 7:1–7. https://doi.org/10.1038/s41598-017-12800-x

Yehuda S, Youdim MBH, Hill JM, Levitsky DA (1989) Brain iron: a lesson from animal models. Am J Clin Nutr 50:618–629. https://doi.org/10.1093/AJCN/50.3.618

Yoshiike Y, Tanemura K, Murayama O, Akagi T, Murayama M, Sato S, Sun X, Tanaka N, Takashima A (2001) New insights on how metals disrupt amyloid β-aggregation and their effects on amyloid-β cytotoxicity. J Biol Chem 276:32293–32299. https://doi.org/10.1074/JBC.M010706200

Zhang L-H, Wang X, Stoltenberg M, Danscher G, Huang L, Wang Z-Y (2008) Abundant expression of zinc transporters in the amyloid plaques of Alzheimer's disease brain. Brain Res Bull 77:55–60. https://doi.org/10.1016/J.BRAINRESBULL.2008.03.014

Zhang Z, Miah M, Culbreth M, Aschner M (2016) Autophagy in neurodegenerative diseases and metal neurotoxicity. Neurochem Res 41:409–422. https://doi.org/10.1007/s11064-016-1844-x

Zheng W, Xin N, Chi Z-H, Zhao B-L, Zhang J, Li J-Y, Wang Z-Y (2009) Divalent metal transporter 1 is involved in amyloid precursor protein processing and Aβ generation. FASEB J 23:4207–4217. https://doi.org/10.1096/FJ.09-135749

Zheng Z, White C, Lee J, Peterson TS, Bush AI, Sun GY, Weisman GA, Petris MJ (2010) Altered microglial copper homeostasis in a mouse model of Alzheimer's disease. J Neurochem 114:1630–1638. https://doi.org/10.1111/J.1471-4159.2010.06888.X

Ferroptosis Modulators: A Potential Therapeutic Target in Alzheimer's Disease

14

Gourav Singh, Nishant Kumar Rana, Indubhusan Mishra, and Gyan Prakash Modi

Abstract

The most commonly known dementia, i.e., Alzheimer's disease (AD) occurs in older age people and it is an irreversible neurodegenerative disorder of the brain that is primarily regarded as a decline in thinking and independence in daily activities. It is well documented that the disease development starts the years before the warning signs are evidenced and at that point most treatments become ineffective. Over a long time, vascular risk factors have been connected with cognitive decline and higher risk of AD through increased tau and cerebral amyloid β (Aβ) aggregates. Unlike programmed cell fate to death, ferroptosis is oxidative stress or aggregation of lipid peroxide and iron-reliant cell death associated with AD. The exact mechanism of action of ferroptosis in AD, related pathways, and differentially expressed genes need to be widely investigated. Ferroptosis hallmarks such as high levels of iron level, oxidative stress, and accumulation of lipid peroxides were investigated in the AD brain. Further, it was noticed that the development of neurofibrillary tangles and Aβ plaques are associated with iron burden in AD mice models. Here, we summarize the recent development of a ferroptosis-related molecular mechanisms, and chemical modulators of ferroptosis in order to control the progression of AD progression. The details of the existing potential targets for therapeutics in order to avert the ferroptosis in AD have been discussed. Ferroptosis-related studies for AD treatment in the future might be based on creating novel techniques and therapeutic

G. Singh · N. K. Rana · G. P. Modi (✉)
Department of Pharmaceutical Engineering & Technology, Indian Institute of Technology (Banaras Hindu University), Varanasi, Uttar Pradesh, India
e-mail: gpmodi.phe@itbhu.ac.in

I. Mishra
Department of Medicine, Heritage Hospital, Varanasi, Uttar Pradesh, India

A. Sharma, G. P. Modi (eds.), *Natural Product-Based Synthetic Drug Molecules in Alzheimer's Disease*, https://doi.org/10.1007/978-981-99-6038-5_14

agents to modulate ferroptosis in AD brain and construct animal models of AD in order to identify ferroptosis modulators to stop or reverse the neurodegeneration in AD.

Keywords

Ferroptosis · Alzheimer's disease · Ferrostatin-1 · Lipoxygenase · Glutathione · Lipid peroxidation · Liproxstatin-1

14.1 Introduction

Alzheimer's disease (AD) is a multifactorial progressive neurodegenerative disorder that normally occurs in elderly people. AD is one of the major causes of dementia globally where irreversible neuronal loss leads to loss of memory and reduced ability to carry out their daily routine independently. The first case of AD was reported by Alois Alzheimer, a German physician in 1907 who described particular changes in the cortical cell clusters after brain biopsy and accredited the behavioral changes in the patient due to these lesions. Alzheimer portrayed nerve fibrillary tangles and plaques which were identified as amyloid β (Aβ) plaques and NFTs (neurofibrillary tangles) in 1980. The diagnostic basis for AD was standardized in 1984 in the USA. These criteria were revised in 2011 and 2018 to distinguish the role of biomarkers in diagnosing AD and to create distinct diagnoses for the preclinical stages, cognitive impairment, and different phases of dementia in AD. This disease is usually diagnosed at a late stage when significant neurodegeneration has occurred. Diagnosis is based on clinical tests as no robust biomarkers have been identified so far.

14.1.1 Statistics of AD

The occurrence of dementia globally is approximately over 45 million where AD is the main reason of dementia and is accountable for approximately more than 70% of cases (Crous-Bou et al. 2017). It is estimated that globally approximately 35.6 million people are affected by AD. It is projected to rise up to 65.7 million by 2030 and by 2050 it will be around 115.4 million (Mcgirr et al. 2020). Whereas in India deaths due to AD were highest in three decades in 2019. AD-related deaths increased almost five times in India between 1990 and 2019. According to India Report 2010, 3.7 million people were living with AD in India which is now increased to 4 million. The actual figure of patients affected by dementia in India is expected to double by 2050 (Perianayagam et al. 2022).

14.1.2 Current Treatment for AD

Currently, available therapy for AD includes NMDA and cholinesterase inhibitors. Cholinesterase inhibitors and NMDA receptor antagonists are the two classes of medications used to treat AD. The acetylcholine neurotransmitter plays a crucial role in learning and memory, and cholinesterase enzyme hydrolyzes acetylcholine and converts it into acetic acid and choline whereas cholinesterase inhibitors like donepezil, rivastigmine, and galantamine prevent the hydrolysis of acetylcholine (Cai et al. 2018). By increasing the amount of acetylcholine in the brain, these medications can improve cognitive function and manage the symptoms of AD, including memory loss, confusion, and difficulties with language and thinking (Colović et al. 2013). Cholinesterase inhibitors are primarily used to treat mild to moderate AD, and they are generally well-tolerated. NMDA receptor antagonists, such as memantine, work by regulating the neurotransmitter glutamate, which is involved in learning and memory. In AD, excessive glutamate release can lead to neuronal damage and death. By blocking the overactivity of glutamate receptors, NMDA receptor antagonists can protect neurons from this damage and improve cognitive function (Liu et al. 2019). Memantine is primarily used to treat moderate to severe AD, although its side effects can include dizziness, confusion, and headaches (Khalaf et al. 2019). While cholinesterase inhibitors and NMDA receptor antagonists can help manage the symptoms of AD, they do not cure the disease, and their effectiveness varies from person to person. Moreover, they do not address the underlying causes of AD, which are complex and not fully understood. Ongoing research is focused on identifying new targets for drug development and developing more effective treatments that can slow or halt the progression of the disease.

The most recent drug approved for the treatment of Alzheimer's disease (AD) is Aducanumab, which was approved by the "US Food and Drug Administration (FDA)" in June 2021 (Vaz et al. 2022). Aducanumab is a monoclonal antibody that targets amyloid beta, a protein that accumulates in AD patients' brains and is believed to contribute in the development of the disease (Beshir and Aadithsoorya 2022). The approval of aducanumab was based on data from clinical trials that showed a reduction in the amount of amyloid beta in the brains of people with early-stage AD who received the drug (Tampi et al. 2021). However, the efficacy of aducanumab in slowing down cognitive decline and improving the overall function of people with AD remains controversial, with some experts questioning the clinical relevance of the trial data (Tampi et al. 2021).

Despite the approval of aducanumab, there is still a need for more effective treatments for AD, as the disease remains a significant public health challenge. Ongoing research is focused on identifying new targets for drug development, including tau protein and inflammation, as well as developing more personalized and targeted treatments that can address the complex nature of the disease. Moreover, other treatment options consist of escalating cognitive reserve and offering a nutritional approach to slow down and prevent the disease progression. AD progression starts years prior to the symptoms get visible and at this stage the available therapies become ineffective. The major proteins occupied through JAK2/STAT3 pathway in the hippocampus, like p-STAT3-Tyr705 and p-JAK2-Tyr1007 were overexpressed in several AD experimental models. It is well documented that glial cells like astrocytes, play a major part in the succession of AD. Devoid of showing a substantial result on amyloid pathologies and tau protein, the above pathway has shown a considerable impact on animal models of AD via tau and Aβ pathologies. Moreover, in AD, cholinergic atrophy that is necessary to the maintenance or survival of BFCN (basal forebrain cholinergic neurons) was outlined as a trophic failure during the NGF (nerve growth factor) metabolic pathway. In the case of AD, a modification in the alteration of the proNGF—NGF as well as augment in degradation of mature NGF occurs. Therefore, in the AD experimental studies, the function of exogenous mature NGF was exposed to recover the atrophic BFCN. In addition to this, it is also reported that the microRNA-107 mediated signaling pathway like FGF7/FGFR2/PI3K/Akt signaling pathway is also involved in the pathophysiology of AD. Several studies have suggested that dysregulation of miRNA expression could be associated with various human pathological conditions, including neurological disorders. Studies on miRNA therapies in neurodegenerative diseases are still in infant stage. Moreover, it is coming into the knowledge that, vascular dysfunction is connected to the reasonable turn down and higher threat of AD. Vascular risk factors were linked with elevated Aβ and tau protein burden, whereas collegially performing with Aβ to persuade reasonable turndown. For example, one of the main vascular risk factors like, apolipoprotein E4 polymorphism is genetically associated risk factor of AD. Various mechanisms of cell death were studied in AD pathology.

In recent decades, several reports suggest that besides apoptosis and necrosis, ferroptosis appears to be associated with AD (Dixon 2017; Hambright et al. 2017). Ferroptosis, which is increasing with aging, is genetically different from another form of cell death even though its biochemical and morphological properties are different (Dixon et al. 2012; Zhou et al. 2020).

14.1.3 Ferroptosis

It is a type of nonapoptotic cell death reliant on intracellular iron. The process of ferroptosis is one of the newest types of synchronized cell death associated with the overloading of iron species and lipid-based ROS (reactive oxygen species) accumulation, reduced glutathione levels, and inhibition of GPX4, resulting in disturbance in iron homeostasis and cell death (Stockwell et al. 2017; Dixon and Stockwell 2019). In 2012, the term ferroptosis was introduced by Dixon and his group, in order to describe the erastin-driven cell death, which mainly act by inhibiting the system Xc. The process ferroptosis can be incited by inhibiting system Xc, glutathione depletion, targeting excess iron, and directly inhibiting GPX4 enzyme. At the same time, ferroptosis inhibition is achieved by blocking excessive lipid peroxidation by targeting different pathways. These mechanisms are often related to the pathophysiology and development of diseases like cancer and AD. It is fundamentally different from other cell deaths like necrosis and apoptosis in biochemistry, functions, and morphology. The mechanism by which ferroptosis acts as a regulatory factor in many diseases remains elusive. The activation and inhibition of ferroptosis to alleviate different disease progression is a topic of great interest and has become a hotspot for etiological research and treatment. Here, we focus on systemically summarizing the various mechanisms implicated in the inhibition and initiation of ferroptosis. We have also discussed in depth various ferroptosis inducers and inhibitors.

The overload of iron molecules may persuade ferroptosis in the AD mice model, and it could be exclusively prohibited by means of an iron chelator (Dixon et al. 2012; Wang et al. 2017). Rising evidences have recommended that ferroptosis might be revealed in cell death connected to neurons in several neurological diseases, like PD, Huntington's and stroke escorted by mitochondrial dysfunction, lipid peroxidation, and GPX4 reduction (Alim et al. 2019; Wu et al. 2018; Tuo et al. 2017; Skouta et al. 2014). Moreover, ferroptotic inhibitors have been revealed as neuron protectors which also recover empirical function in stroke disease animal models (Alim et al. 2019; Tuo et al. 2017). GPX4 knockout mice directly showed significant neuronal loss and age-dependent neurodegenerative changes (Hambright et al. 2017). Taking into consideration that the iron burden in AD significantly generates elevated ROS level in the brain somehow strengthen the point that ferroptosis is implicated in the cognitive impairment and loss of neurons. However, the evidences are not enough to make a direct connection to it (Praticò et al. 2001; Praticò and Sung 2004; Castellani et al. 2007; Derry et al. 2020). Ferroptosis hallmarks, such as augmented oxidative stress and iron levels have been extensively noted in the AD brain. It is reported that formation of NFTs and Aβ plaques is connected with iron overload in AD mice model (Yamamoto et al. 2002; Peters et al. 2018). Moreover, iron level completely associated with cognitive decline in glutathione peroxidase (GPX4) and human subjects is protective in mice model of AD (Yoo et al. 2010; Ayton et al. 2017). After various efforts made by researcher globally on tau protein hyperphosphorylation and Aβ deposition, still an efficient and effective therapeutic option to decline the neurodegeneration in case of AD is missing. However, besides these rising evidences advocate that ferroptosis contributes to neurodegeneration. Therefore the involvement of ferroptosis and the mechanism connected with a classical pathway of ferroptosis during progression of AD became one of the major hurdles to deal with novel pathophysiological studies and therapeutics.

14.2 Mechanism Concerned to the Fundamental Ferroptotic Pathway

On the basis of available literature till date mechanism of ferroptosis can be separated into three fractions: (1) iron homeostasis, (2) GSH (glutathione) metabolism, and (3) lipid peroxidation or oxidative stress. Disturbance in any of these mechanisms either alone or in combination can induce ferroptosis.

14.2.1 Iron Homeostasis

It was reported that transferrin 1 receptor facilitates the entry of iron inside the cells where it get reduced from Fe^{3+} to Fe^{2+} in the endosome through

STEAP3 (six-transmembrane epithelial antigen of the prostate 3) metalloreductase and plays a crucial role in ferroptosis (Zhang et al. 2012; Yan and Zhang 2020). In 2019, Gao et al. reported that Fe^{2+} accumulated in ferritin and ferroportin mediates its exportation from the cell. Further ferritin deprivation through NCOA4 (nuclear receptor coactivator 4) strengthen the ferroptotic process by accumulating the intracellular iron. In addition to this Fe^{2+} accumulation further leads to production of oxidative stress and ROS (Hou et al. 2016; Ward et al. 2014). Production of ROS such as O_2^- (superoxide anion) and hydroxyl free radical ($^{\cdot}OH$) due to excessive free intracellular iron accumulation results into oxidative damage to membrane or lipid enclosed molecules and lipid peroxidation which further lead to harm at cellular level and eventually to ferroptotic cell death (Aprioku 2013).

14.2.2 Glutathione Metabolism

Blocking of the glutamate/cystine antiporter can cause a decrease in intracellular GSH levels that ultimately results in the deactivation of GPX4, a critical enzyme involved in antioxidant defense and the regulation of ferroptosis (Seibt et al. 2019). These processes ultimately lead to ferroptosis cell death via ROS accumulation and enhanced lipid peroxidation (Wang et al. 2020). Further, it was reported that GPX4 reduces PL-PUFA(PE)-OOH to PL-PUFA(PE)-OH, where GPX4, also referred as GS-SG (glutathione disulfide), changes GSH into its oxidized form (Seibt et al. 2019; Cozza et al. 2017). Moreover, under oxidative stress despite Nrf2, xCT mRNA positively coordinated by ATF4 activation (transcription factor 4), whereas negative regulation of xCT mRNA via P53 leads to cysteine deprivation and susceptibility toward ferroptosis (Cai et al. 2018; Sato et al. 2004; Jiang et al. 2015; Habib et al. 2015).

14.2.3 Lipid Peroxidation and Oxidative Stress

It is well known that ROS-mediated lipid peroxidation that leads to oxidative stress is the main key to ferroptosis (Kuang et al. 2020). A decrease in GSH levels and GPX4 inhibition leads to 12/15 LOX (12/15-lipoxygenase) activation. Further the association of Fe^{2+} with 12/15-LOX results in the oxygenation of PUFA (polyunsaturated fatty-acids) and triggers lipid peroxidation- mediated ferroptosis (Kagan et al. 2017). Among the six isoforms of LOX, 12/15-LOX are most the abundant and considered as key ferroptotic cell death regulators (Kagan et al. 2017; Yang et al. 2016). Besides accumulated free intracellular iron-induced ROS generation, mitochondria also facilitate ROS production. Electron leakage from mitochondrial complexes I and III during the electron transport process gives rise to the production of ROS like superoxide (O_2^-) and hydrogen peroxide (H_2O_2). Consequently, this can lead to a reduction in mitochondrial membrane potential (Gao et al. 2019). Recently in 2020, Kuang et al. reported that reduction in mitochondrial membrane potential associated with ferroptosis and engaged regulatory mechanism different

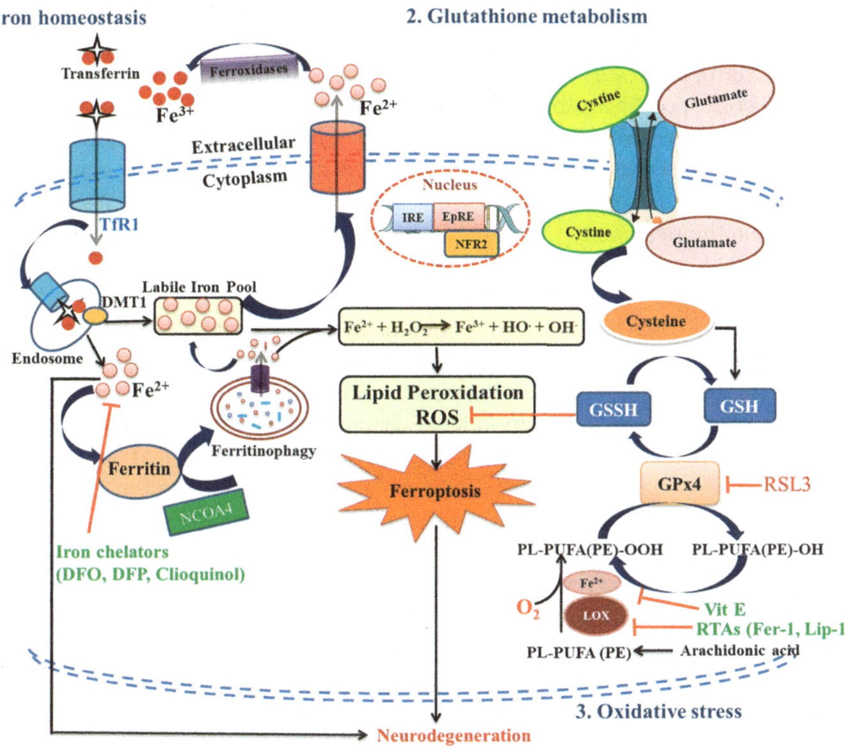

Fig. 14.1 Molecular mechanisms involved in cell death due to ferroptosis. Ferroptosis pathway can be separated into three modes (1) iron metabolism, (2) glutathione metabolism, and (3) polyunsaturated fatty acid metabolism. Well-recognized ferroptosis inhibitors and their method of action are indicated in green

from apoptosis. Activation of 12/15-LOX due to GSH depletion can enhance cytosolic calcium ion (Ca^{2+}) via importation from the mitochondrial extracellular compartment and by releasing from endoplasmic reticulum and mitochondria (Maher et al. 2018). GSH levels depletion may also induce the dysregulation of in and out Ca^{2+} transport from mitochondria via VDAC (voltage dependent anion channels) and MCU (mitochondrial Ca^{2+} uniporter) (Zorov et al. 2014; DeHart et al. 2018). Due to overload of Ca^{2+} and loss of mitochondrial function induce the Ca^{2+} dependent proteases (DeHart et al. 2018). Subsequently, ROS-mediated BID transactivation and induction of AIF (apoptosis-inducing factor) translocation to nucleus from mitochondria lead to cell death (Neitemeier et al. 2017). Reduction of MAMs (mitochondria-associated endoplasmic reticulum membranes) communication and activation of calcium-activated potassium channel have the potential to defence against ferroptosis dependent cell death (Krabbendam et al. 2020). The various pathways associated with ferroptosis are shown in Fig. 14.1.

14.3 Role of Ferroptosis During AD Progression

In 2012, ferroptosis was first illustrated and stated that it differs from normal apoptosis corresponding to biochemical reaction and structural morphology, further ferroptosis is characterized by the accumulation of lipid peroxidation induced by free intracellular iron (Dixon et al. 2012; Stockwell et al. 2017). Erastin is considered as one of the major ferroptosis inducer that inhibits import of cystine from cystine-glutamate antiporter system Xc− and further leads to glutathione reduction (Dixon et al. 2012). A well-known selenoenzyme GPX4 uses glutathione as substrate in order to convert a potent toxic lipid hydroperoxides to non-toxic lipid alcohols (Yang et al. 2014; Ursini et al. 1982; Friedmann Angeli et al. 2014). Several reports suggest that tiny mitochondria with diminishing mitochondrial crista, solid mitochondrial membrane, transparent nuclei, and ruptured outer membrane of mitochondria are the morphology-related characteristic quality of ferroptosis (Mou et al. 2019; Miyake et al. 2020). Recently, it was reported that ferritinophagy facilitates ferroptosis regulation (Tang et al. 2018). It is reported that when ferroptosis occurs due to cysteine depletion, ferritinophagy comes into play and releases the free intracellular iron with the help of nuclear receptor coactivator 4 (Tang et al. 2018). A well-known modulator of ferritinophagy, i.e., HER2C regulates the FBXL5 turnover, which emerges to play a key function in the metabolism of iron and is basically a part of complex that aims proteasomal degradation via iron regulatory protein 2 (Tang et al. 2018; Moroishi et al. 2014). Further it is reported that HER2C deficiency may trigger neurodegeneration via the release of free iron, a damaging response to enhanced iron levels that may lead to ferritinophagy and damage of neurons (Tang et al. 2018). Nrf2 (Erythroid 2-related factor 2) is a well-known transcriptional regulator which control genes related to the mitochondrial function and involved in the lipid, glutathione, and iron metabolism (Abdalkader et al. 2018). Further Nrf2 can assist to defend against ferroptosis, and hence increasing Nrf2 signal may offer neuroprotective effects (Abdalkader et al. 2018). In recent decades, researchers demonstrate that ferroptosis may affect AD prognosis or pathogenesis, and elevated lipid peroxides level in brain is one of the characteristic quality of AD (Adibhatla and Hatcher 2010). GPX4 deletion in mice model of AD displays degeneration of hippocampus and cognitive impairment similar to ferroptosis-mediated neurodegeneration which includes neuroinflammation, lipid hydroperoxides, neuroinflammation, and ERK establishment which is improved by ferroptotic inhibitors (Hambright et al. 2017). Ferroptosis may intensify AD via heme oxygenase-1 pathways and targeting the Hif-1α (Tang et al. 2018). Reports suggest that it is altered by the interruption in lipid restoration involving GPX4 and glutathione (Stockwell et al. 2017). Cellular death via ferroptosis can be obscured by lipid peroxidation inhibitors, iron chelators, reduction of PUFAs, and lipophilic antioxidants (Stockwell et al. 2017). Therapeutic studies in AD models with potent anti-ferroptosis activity and experimentation with vitamin E supplements to treat AD have produced contradictory results or prevent cognitive decline (Lloret et al. 2019). Hybrids of donepezil-butylated hydroxytoluene used as potential anti-AD therapies (Cai et al. 2018). Moreover, cognitive improvement in AD has been achieved by idebenone, but irrefutable evidence for a clinically significant benefit is lacking (Thal et al. 2003; Senin et al. 1992; Weyer et al. 1997). Further vildagliptin showed significant biochemical changes, declined tau phosphorylation, memory improvement, amyloid-β attenuation, and enhanced expression of synaptic plasticity proteins in the hippocampus of AD rat models (Khalaf et al. 2019; Ma et al. 2018; Kosaraju et al. 2013). Linagliptin alleviated tau phosphorylation, amyloid-β attenuation, cognitive deficits, and neuroinflammation in rodent model of 3xTg-AD (Kosaraju et al. 2017). It is shown that linagliptin improved amyloid-β-activated intracellular ROS and dysfunction of mitochondria in cultured human neuronal cells. It was shown that baicalein prevents tau phosphorylation, declined amyloid-β and improved cognition in AD mouse models (Zhang et al. 2013; Gu et al. 2016). Certain reports demonstrated that in the case of AD, PD-146176 decreases tau pathology and amyloid-β, diminished autophagy and preserved memory in a 3xTg mouse model (Di Meco et al. 2017; Chu et al. 2015). It is reported that zileuton compound improves memory and lessens tau and Aβ pathophysiology in 3xTg AD mice model (Di Meco et al. 2014; Chu et al. 2013). A controlled and randomized use of sodium selenate reduce the raised concentration of selenium in CSF sample. This decrease in CSF selenium concentration correlated with changes in cognitive function as assessed by the Mini-Mental Status Examination (MMSE) (Cardoso et al. 2019).

14.4 Ferroptosis Inhibitors

14.4.1 Radical Trapping Antioxidants (RTAs)

These agents act by inhibiting the autoxidation of lipids, which is a crucial step in lipid peroxidation. The RTAs are also known as chain-breaking antioxidants that act by reacting with the chain that carries peroxyl radical, due to which the process of autoxidation of lipids does not occur (Ingold 1961; Ingold and Pratt 2014). The petroleum industry has widely used the RTAs for many years. It prevents the autoxidation of oil, fuels, rubber, and plastics; due to this property, they are widely used as excipients in petroleum (Burton and Ingold 1986). The role of RTA in suppressing lipid peroxidation was revealed after the characterization of α-tocopherol (Fig. 14.7), the highest biologically active form of vitamin E. It is one of the most potent RTA reported till date. The research on tocopherol has concluded that antioxidants have essential roles in treating cardiovascular and Alzheimer's disease (Bowry and Stocker 1993). One of the primary disadvantages of antioxidants is that they frequently fail in in vivo tests. The activity that occurs in vitro does not cross over to the in vivo research. There are many misconceptions about antioxidants' beneficial effects in disease prevention, and as a result, very little research has been done in this field (Angeli et al. 2017). Vitamin E deficiency leads to premature neurodegeneration onset, which is mainly related to ferroptosis-linked disease (Ulatowski and Manor 2015). So it is suggested that antioxidants carry beneficial therapeutic effects. Still, presently we have very little understanding regarding antioxidants, due to which fewer compounds have come into existence (Harding et al. 1982; Doba et al. 1985).

14.4.2 Vitamin E Analogs

After the identification of α-tocopherol, it has come to notice that vitamin analogs are one of the promising candidates that act as antioxidants. It is one of the naturally occurring lipid-soluble RTA which acts on the chain carrying peroxyl radical (Angeli et al. 2017). Much work has been done to enhance the reactivity of these analogs with peroxyl radicals. Many researchers have introduced different groups in naturally occurring vitamin E analogs. After synthesizing a large number of compounds and studying their SAR, the compound tetrahydronaphthyridinols (THNs) is identified as (Yang 2016) the promising lead with enhanced activity toward peroxyl radical. The SAR studies have suggested that the introduction of nitrogen group on the 3 and 5 positions in the ring relative to a hydroxyl group of classical phenolic RTA promisingly stabilizes electron-rich phenol against autoxidation. The substitution of *N,N*-dialkylamino (strongly electron-donating group) increased the rate of reaction toward the free radicals, mainly peroxyl radicals, by weakening the O–H bond and not compromising their resistance toward autoxidation (Wijtmans et al. 2003; Pratt et al. 2001). Considering these facts, the compound tetrahydronaphthyridinols THNs (Aza-Phenols) comes into existence, which possesses the best combination of stability and reactivity. The THNs showed 30-folds higher activity toward peroxyl radicals in organic solution and liposomes compared with α-tocopherol (Nam et al. 2007). After recognizing that oxidative lipid degradation is necessary for the execution of ferroptosis, the excellent activity of RTA toward the inhibition of lipid peroxidation makes them one of the most privileged candidates to inhibit ferroptosis, which is one of the key processes in many diseases like cardiac diseases, ischemia, and neurodegenerative disorders. Zilka et al. have shown that when THNs with different chain lengths were treated with alkyl chain ranging from 22 to 15, they exhibited excellent inhibitory properties toward ferroptosis and even showed more significant activity than ferrostatin and liproxstatin-1, well-known ferroptosis inhibitors (Zilka et al. 2017). THNs have also demonstrated higher tissue distribution because they efficiently bind with α-tocopherol transfer protein (αTPP) and protect themselves from liver metabolism (Li et al. 2013). These properties make THNs the essential candidates in the form of ferroptosis inhibitors.

14.4.3 Lipoxygenase Inhibitor

Different studies have suggested that lipoxygenase is one of the key enzymes that plays a crucial role in the execution of ferroptosis by accelerating the process of lipid peroxidation (Fan and Fan 2018; Feng and Stockwell 2018). Lipoxygenase enzyme can activate the nuclear factor-κB (NF-κB) pathway, which expresses the transcription of genes like inducible nitric oxide synthase (iNOS), which is recognized as one of the critical factors in inflammatory responses (Mendes et al. 2017; Guo et al. 2019; Vallance and Leiper 2002). The iNOS accelerates the generation of NO radicals, which is an important parameter in many physiological processes.

Excessive production of NO leads to the generation of reactive nitrogen species (RNOS), and excessive accumulation of RNOS shows a toxic effect and causes cell death and tissue damage (Blaise et al. 2005). The main drawback of lipoxygenase as a target is that it lacks specificity, as lipoxygenase exists in six isoforms in humans. So many researchers have studied the different isoforms to design isoform-specific inhibitors, which were of great medicinal effect (Gilbert et al. 2011). The other main obstacle regarding lipoxygenase was the lack of 3D structure information of human lipoxygenase. There has been no conclusive evidence that a particular isoform is entirely responsible for ferroptosis. One genetic study has suggested that the deficiency of Alox 15, the isoform of lipoxygenase, is capable of oxidizing esterified PUFAs and further suggested that it is responsible for cell death (Brütsch et al. 2015). Alox15 is an iron-containing enzyme without heme, which generates lipid peroxides from polyunsaturated acids like arachidonic acid and linoleic acid (LA) (Guo et al. 2019). Alox15 acts by oxidizing either AA or LA and converts them into their respective peroxides, which are 15-hydroxyeicosatetraenoic acid and 13-hydroperoxy-octadecadienoic acid (13-HpODE), respectively. Alox15 also generates lipoxins, which act as important inflammatory mediators (Solomon et al. 1997; Ivanov et al. 2010). Excessive generation of these lipid peroxides has toxic effects on the cell and causes death by ferroptotic cells. Although the cellular antioxidant system balances lipid peroxides, an imbalance of this process leads to ferroptotic cell death (Yang and Stockwell 2016). Several types of research have shown that the development of new lipoxygenase inhibitors can prevent ferroptosis. Based on their mechanism, the lipoxygenase inhibitors are divided into five groups: (1) redox inhibitors maintain the concentration of active site iron atom into its inactive form, which was reported as reduced form; (2) iron chelators remove the iron from the active sites due to which the enzyme losts its activity; (3) competitive inhibitors, (4) suicide substrates have a property to irreversibly bind with enzyme and inactivate the enzyme; and (5) allosteric inhibitors have a property to bind with a specific site which is responsible for the enzyme activity and inactivate that specific site (Haeggström and Funk 2011). Alox15 has been involved in a wide variety of illnesses, including asthma, diabetes, stroke, atherogenesis, cancer, Alzheimer's disease, and Parkinson disease. This exerts the focus of many researchers to design Alox15 inhibitors for drug discovery. Rather than targeting a single lipoxygenase, several studies show that focusing on multiple lipoxygenase inhibition yields superior outcomes with greater ferroptosis suppression (Kagan et al. 2017; Friedmann Angeli et al. 2014). It remains elusive whether to develop specific lipoxygenase or pan-lipoxygenase inhibitors as novel drug which efficiently inhibits the execution of the ferroptosis process. Some compounds are reported as lipoxygenase inhibitors, as shown in Fig. 14.2.

14.4.4 Ferrostatin

Ferrostatin is an alkylamine derivative that came into existence by high-throughput screening of a large library of compounds targeting ferroptosis (Skouta et al. 2014;

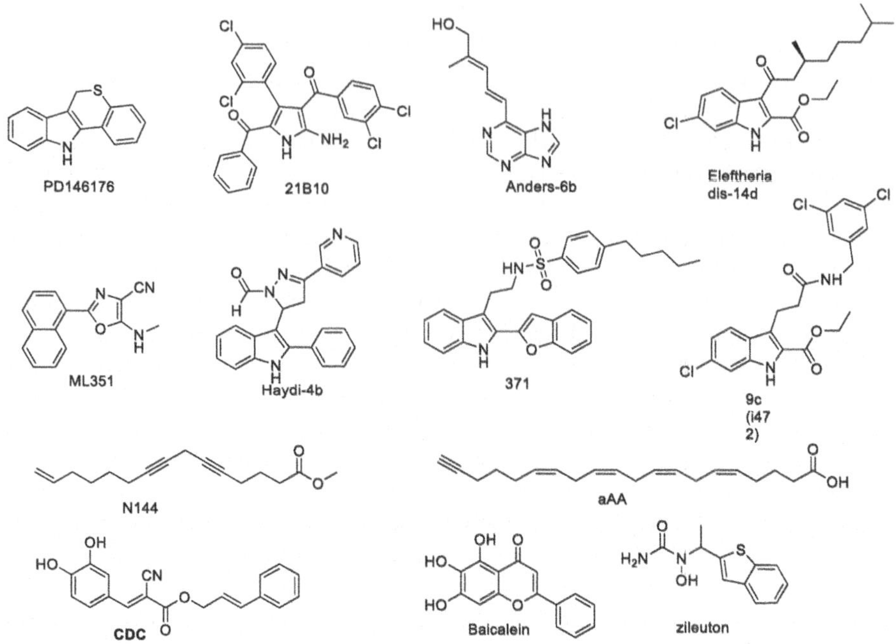

Fig. 14.2 The chemical structures of various lipoxygenase inhibitors

Hofmans et al. 2016). Ferrostatin mainly acts by inhibiting lipid peroxidation, but it does not inhibit the accumulation of mitochondrial ROS and does not inhibit lysosomal membrane permeability (Skouta et al. 2014). It has previously been reported that diarylamines and hindered dialkylamines effectively reduce agents and antioxidants. They are already employed as a preservative in the food industry to slow the oxidation process and prevent food spoilage. These antioxidants are widely used in the material sector to prevent oxidative deterioration of polymers, plastics, and other materials and in the chemical industry to prevent the autoxidation of rubber and plastics (DeHart et al. 2018). The diarylamines are exciting molecules and are under investigation as possible antioxidants that work as excellent free radical scavengers and have the therapeutic potential. The exploration of diarylamines has led to the discovery of ferrostatin-1, the first ferroptosis inhibitor. Ferrostatin-1 suppresses ferroptosis caused by erastin, although the specific site where ferrostatin-1 reacts is unclear. It is hypothesized that it works by preserving the levels of PUFA and their derivatives rather than changing the cysteine and glutathione levels. PUFAs are prone to oxidation. PUFAs are highly prone to oxidation, and it gets oxidized by both enzymatic (lipoxygenase mediated) and non-enzymatic (Fenton process mediated) processes (Loscalzo 2008). Ferrostain-1 acts by preventing the degradation of oxidized PUFA and their fragments by inhibiting their oxidative destruction. The ferrostatin-1 traps the free lipid peroxyl radical either by transferring the hydrogen atoms or by directly reducing lipid

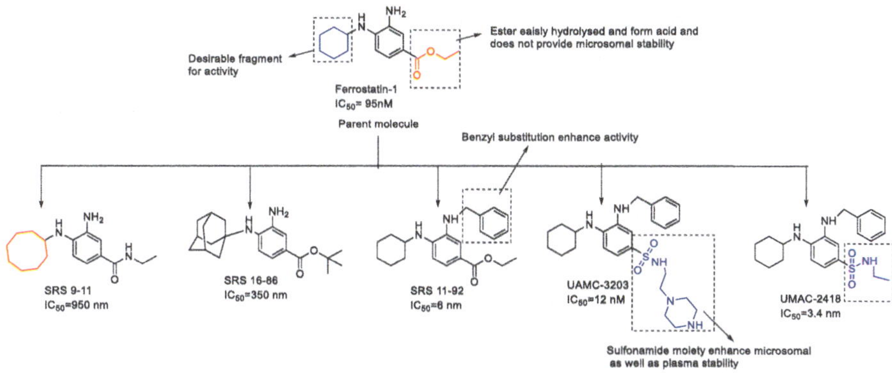

Fig. 14.3 Structures of ferroptosis inhibitors derived from ferrostatin-1

peroxide (Skouta et al. 2014). Ferrostatin-1 suffers from poor efficacy and metabolic stability. The ester group in this gets hydrolyzed into the inactive carboxylic acid group. It also exhibits weak microsomal activity and gets trapped into the microsomes due to its acidic nature (Hofmans et al. 2016; Devisscher et al. 2018).

Stockwell et al. established structure–activity relationship and suggested that the replacement of the amine group (-NH_2) of ferrostatin with a nitro group (NO_2) leads to the drastically reduction of ferroptosis. The amine group acts as a radical trapper and helps to quench the free radical, which was an essential factor as an excellent radical trapping agent. The amino group acts as the tremendous hydrogen source and transfers one of the hydrogen atoms into free radicals to retard or inhibit the free radical chain reactions. Similarly, in phenolic-based RTA, the OH group of phenol acts as a hydrogen atom source. The N-substituted cycloalkyl (cyclohexyl) present in the ferrostatin is necessary for the activity. They also vary the number of carbon atom of the N-substituted cycloalkyl moiety and synthesized ten analogs. After studying these analogs, the conclusion comes that the compound with a large lipophilic bulky group, cycloalkyl moiety shows greater potency (Dixon et al. 2012; Dixon and Stockwell 2014). Skouta et al. also performed an SAR study of ferrostatin-1. They suggested that the introduction of heteroatoms in the N-substituted cycloalkyl moiety can reduce the potency and also studied the role of hydrophobicity in the interaction with the lipophilic membrane environment. The substitution of aryl group with electron withdrawing or donating group or containing heteroatom on central primary amine can ameliorate the potency of ferrostatin. The tertiary amines turned out to be less potent compared to secondary or primary amines. The replacement of ester with the amide group can decrease the activity by ten folds (SRS9–11, Fig. 14.3, $IC_{50} = 950$ nM) when compared to ferrostatin-1 ($IC_{50} = 95$ nM) (Skouta et al. 2014). The substitution of the benzyl group on the central amine leads to a potent compound (~6-fold) with higher activity (SRS11–92, Fig. 14.3, $IC_{50} = 6$ nm) when compared to ferrostain-1. These studies are mainly focused on ferrostatin-1 to improve its existing potency (Hofmans et al. 2016). Linkermann et al. designed a ferrostatin-1-derived compound with improved plasma and microsomal activity. In this compound, ethyl ester was replaced with the tertiary-butyl ester to increase microsomal stability. Also, the imine group was introduced to the central amine group to generate SRS16–84, with higher plasma stability ($t_{1/2} \geq 120$ min), and to show more excellent metabolic stability toward mouse microsomes. Intriguingly, SRS16–84 turned out to be of poor potency ($IC_{50} = 350$ nm) when compared with ferrostatin-1. Further, the compound UMAC-2418 was designed by substituting the benzyl group on central amine and replacing ethyl ester with sulfonamide, which could inhibit erastin-induced ferroptotic cell death. In human and mouse microsomes, the compound UMAC-2418 showed better potency with IC50 = 4 nM and *t*/2 of 90 min and 180 min. With complete plasma stability up to 6 h (Hofmans et al. 2016). Further, the same research group reported UAMC-3203 (Jiang et al. 2015) with slightly less potency (12 nM) over UMAC-2418 with improved microsomal stability (with $t_{1/2}$ (Human) = 20.5 h, $t_{1/2}$ (Rat) = 16.5 h and $t_{1/2}$ (Mouse) = 3.5 h), enhanced plasma stability up to 6 h and with improved water solubility. The compound UAMC-3203 (Fig. 14.3) also showed rapid redistribution into various tissues. It is worth mention that UAMC-3203 was found to be efficacious in the acute iron-induced poisoning injury in mice (Devisscher et al. 2018).

14.4.5 Liproxstatin

Liproxstatin is an anti-ferroptotic molecule discovered in high-throughput screening of a vast library of small molecules utilizing the cell-based assays (Zilka et al. 2017). The ferroptotic cell death was induced primarily through the deletion of the glutathione peroxidase-4 (GPX4) gene or by inhibiting the System Xc antiporter that allows the exchange of intracellular glutamate and extracellular cysteine, which is necessary for the generation or synthesis of glutathione (GSH). The liproxstatin suppresses the ferroptosis activity by inhibiting the accumulation of lipid hydroperoxide (LOOH). Liproxstatin was evaluated in an inducible GPX4 KO mouse cell line, and it inhibited the cell death in this model (Friedmann Angeli et al. 2014). The liproxstatin is a highly potent active spiroquinoxalinamine derivative that acts by scavenging the lipid peroxide by transferring hydrogen atom from the amine group (-NH) to the free radical and acts as radical trapping antioxidant, which differs from a well-known phenolic RTA, where phenolic OH group acts as the trapping free radicals (Sheng et al. 2017). Several efforts are underway further to improve the anti-ferroptosis activity of the ferroptosis inhibitors to optimize these inhibitors as better drug candidates for AD (Reichert et al. 2020), ischemia (Magtanong and Dixon 2018), and heart diseases (Xie et al. 2016). The underlying molecular mechanism of action of liproxstatin is still illusive (Sheng et al. 2017). It inhibits the cell death induced by ferroptosis by inactivating the 15-lipoxygenase (15-ALOX) enzyme activity (Kagan et al. 2017). In another study, Pratt et al. reported inhibition of autoxidation of lipid or lipid peroxidation process by liproxstatin (Zilka et al. 2017). The study also suggested that liproxstatin also acts as an excellent radical trapping antioxidant, which scavenges lipid peroxide by

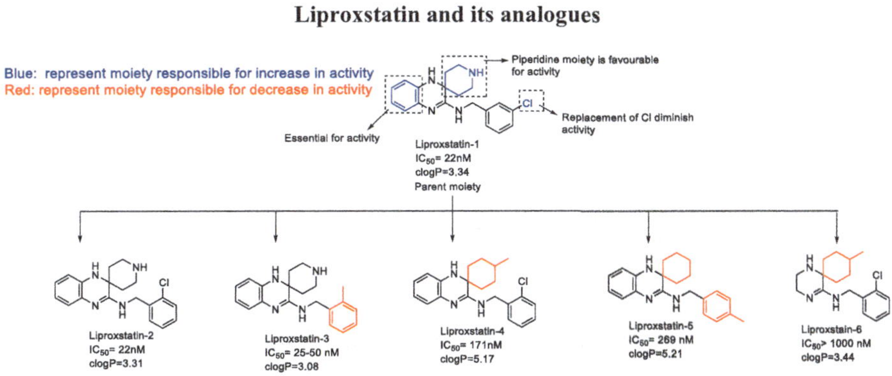

Fig. 14.4 Chemical structures of liproxstatin-1 derived ferroptosis inhibitors

trapping free radicals. The computational studies revealed the intrinsic mechanism of liproxstatin activity using density functional theory (DFT). They used the methyl peroxyl radical (CH_3OO^-) model to carry this experiment to know how liproxstatin traps the free radicals (Tejero et al. 2007).

Further, a detailed SAR study on liproxstatin suggested the presence of the benzene group to be essential for the activity and its replacement, completely abolished activity (Friedmann Angeli et al. 2014). Similarly, amine (NH_2) also plays a vital role in the activity because the amine group is essential for antioxidant properties for scavenging free radicals. The process of radicalization occurs due to the transfer of an electron from the HOMO of the liproxstatin to the LUMO of methyl peroxy radical (CH_3OO^-). The compounds with a higher value of HOMO energy exhibit potent antioxidant properties. Therefore, liproxstatin-1, -2, and -3 showed promising antioxidant properties with the HOMO value of −5.44, −5.36, and −5.34, respectively (Fig. 14.4). The compounds liproxstatin-4 and liproxstatin-5 are found to be weak antioxidants due to the lack of one of the amine groups in the structure of liproxstatin-4, and the substitution of methyl with chlorine further decreased the antioxidant property of liproxstatin-5 with HOMO values −5.50 eV, and 5.49 eV, respectively. Liproxstatin-6 showed less bioactivity with the lowest HOMO value of −6.17 eV. The main factor in reducing the HOMO value for liproxstatin-6 is the absence of the phenyl group, which plays a crucial role in the HOMO energy (Fig.14.4). This study further suggests that the aromatic amine group plays an important role and is an essential requirement for better activity. The aromatic amine is also necessary for the resonance effect. So, after studying a SAR by substituting and withdrawing different groups in the parent molecule (liproxstatin-1), we notice that the aromatic amine is crucial for an antioxidant property; by increasing the HOMO value, we can further improve the antioxidant property of the compounds (Sheng et al. 2017).

The SAR study by Sheng et al. has suggested that the amine group is essential for the activity; however, out of the three amines, the amine group participating in the radical trapping by donating a hydrogen atom to the free radicals is still illusive. All

Fig. 14.5 Structure of liproxstatin-1 with their amine numbering

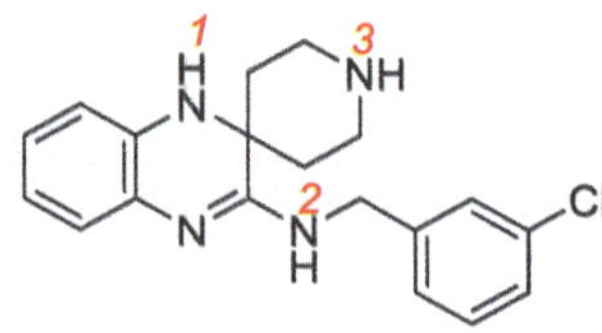

three -NH groups act as attacking sites for radicals carrying the radicalization process. Therefore, Sheng et al. carried out a study to find out the most reactive -NH site by using methyl peroxyl (CH_3COO^-) radical. Thus, the energy required to cross the barrier was calculated and found to be 13.45, 18.28, and 20.41 kcal mol^{-1} for 1-NH, 2-NH, and 3-NH, respectively. So, the smaller the energy required to cross the barrier, the better the reactive site. Comparing all three -NH sites, 1-NH and 2-NH (Fig. 14.5) sites required less energy to cross the energy barrier. Still, the 3-NH group showed a high energy barrier, so it might be possible that it does not show any antioxidant property. Therefore, the reactivity arrangement for -NH group in liproxstatin is 1-NH > 2-NH > 3-NH, based on the reaction energy barrier.

14.4.5.1 Role of Ubiquinol (UQH_2) or Coenzyme Q10 (Antioxidant Synthesized by Body) in the Activity of Liproxtatin-1

UQH_2 (Fig. 14.7) is one of the crucial components which acts as an antioxidant agent synthesized by the body. It is mainly a 2,3 dimethoxy-5-methyl-6-poly prenyl-1,4-benzoquinol, in which poly prenyl side chain contains 10 units with lipid-soluble properties (Dhanasekaran and Ren 2005; Saini 2011). CoQ 10 is converted into ubiquinol with the help of ferroptosis, thereby suppressing protein 1 (FSP1). During the radical trapping mechanism, liproxstatin itself gets oxidized into liproxstatin-1 radical, which acts as a new radical and causes damage to the bio-membrane and the DNA (Niki et al. 2005). So here, the role of UQH2 comes into existence, as it acts like an antioxidant system of the living organism and can trap the free radical generated by liproxstatin-1 and donates one hydrogen atom to the radical by which the radical again comes into its active form, the process is reported as regeneration of liproxstatin-1 and acts as a key factor to study the antioxidant property of liproxstatin in vivo (Wang and Hekimi 2013; Bowry and Stocker 1993). So, ubiquinol is the critical parameter in studying the regeneration process of liproxstatin-1. Therefore, ubiquinol has a phenolic OH group, which acts as a phenolic radical trapping antioxidant and transfers one hydrogen atom to the liproxstatin-1 radical. The energy required by the UQH2 hydrogen atom to cross the energy barrier is reported to be 15.65 kcal mol^{-1}, and the process is exothermic (4.49 kcal mol^{-1}) in nature (Sheng et al. 2017). The UQH2 radical is further reduced to its active form with the other redox reactions (Nohl et al. 2001). This free radical of tocopherol is recycled back to its active form by UQH_2. The essential energy barrier required to transfer hydrogen atoms from UQH2 is reported as 12.1 kcal mol^{-1}, which has been proved in an experiment (Ouchi et al. 2010). This study suggested that liproxstatin-1 acts as a good candidate as an antioxidant and inhibits the process of lipid peroxidation to a great extent by trapping a free radical (LOO) generated during the lipid peroxidation process. The free radical generated by the oxidation of liproxstatin-1 is recycled back by other antioxidants present in the body, like ubiquinol (Sheng et al. 2017). So ubiquinol shows a synergistic effect with liproxstatin and increases its antioxidant efficacy.

14.4.6 ACSL4 Inhibitors

ACSL4 plays a crucial role in the execution of ferroptosis. During ferroptosis, it accelerates the esterification of polyunsaturated fatty acids (PUFA) such as arachidonic acid and adrenic acid to generate oxidized phosphatidylethanolamines (PE). It also accelerates the synthesis of AA-COA or AdA-CoA, which is required to form phospholipid hydroperoxide (PL-OOH), a key component in the lipid peroxidation process (Ivanov et al. 2010). The rosiglitazone, a peroxisome proliferator-activated receptor-γ (PPAR-γ) activator, inhibits ACSL4 activity and suppresses ferroptotic death (Askari et al. 2007). Rosiglitazone was used in in vivo studies to inhibit ACSL4 activity from finding its impact on ferroptosis during lung ischemia-reperfusion, it reduced the MDA production in ischemia reperfusion-injured lung tissues and attenuated ferroptosis (Xu et al. 2020). In the genetic model of ferroptosis, rosiglitazone (Fig. 14.7) can also alleviate and reduce the rate of mortality due to acute renal failure (Angeli et al. 2017; Doll 2017). Similarly, thiazolidinedione also inhibited ACSL4 and suppressed ferroptosis (Angeli et al. 2017). The above findings suggest that ACSL4 may be a potential therapeutic target. The pharmacological inhibition of ACSL4 by its inhibitors acts as the new paradigm in inhibiting ferroptosis and other lipid peroxidation-derived processes by limiting substrate availability. The ACSL4 inhibitors may offer opportunities for novel interventions for the treatment of neurodegenerative diseases caused due to lipid peroxidation.

14.4.7 Deuterated Phospholipids

The oxidative damage contributes to ferroptosis's execution. The PUFA is more prone to oxidative damage because it contains bis-allylic protons, which are easily abstracted and produce alkyl radicals. These alkyl-radicals react with free oxygen to form peroxyl radicals, which then react with additional PUFAs to form a lipid peroxidation chain reaction. Malondialdehydes (MDA) and 4-hydroxynonenal, the two end products of lipid peroxidation, cause cell damage (Yang 2016). In 2007 Shchepinov et al. substituted linoleic acid (LA) with deuterated linoleic acid (D-LA). The 11,11-d2-LA (D-LA) experienced a propagation step of autoxidation, a 13-fold lesser than LA (Shchepinov 2007). Based on the findings, deuterated LA supplementation of cell medium was investigated and found to be beneficial in many cell models of lipid peroxidation linked to neurological illnesses such as Parkinson's disease and Friedreich's ataxia (Shchepinov et al. 2011; Cotticelli et al. 2013). In 2011, Shchepinov et al. did a study to find the role of PUFAs in ferroptosis. When

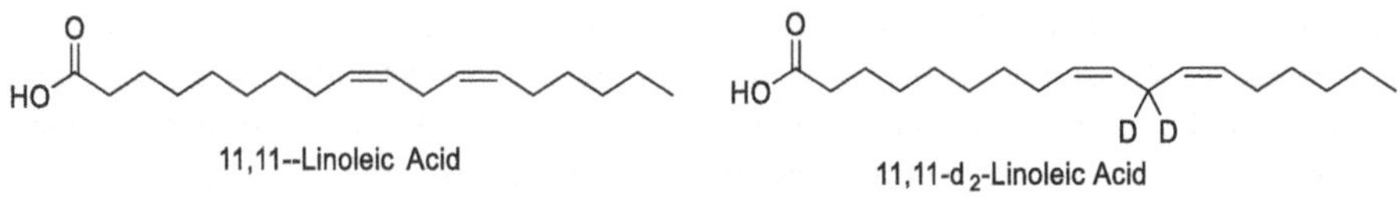

Fig. 14.6 Structure of 11,11-Linoleic Acid and 11,11-d_2-Linoleic Acid

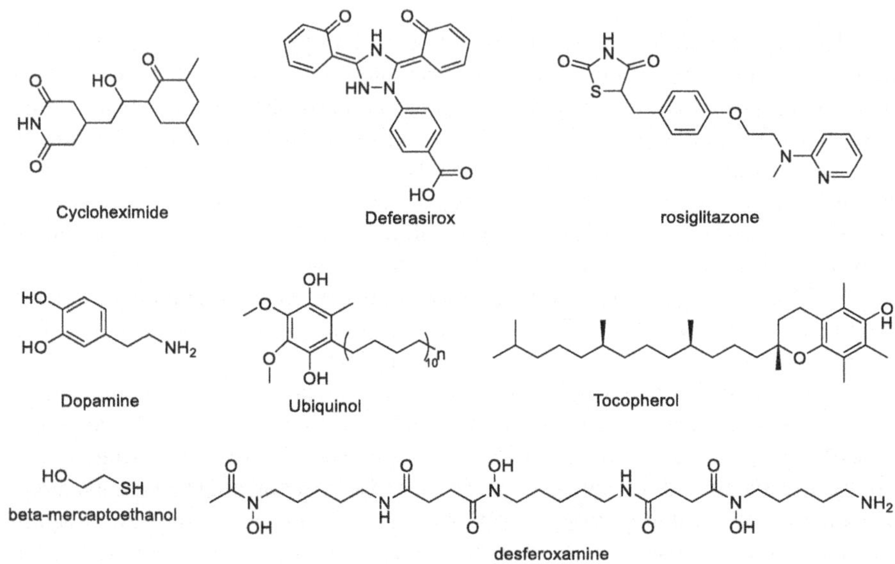

Fig. 14.7 Chemical structures of some different classes of ferroptosis inhibitors

H-PUFA is replaced with D-PUFA, which has deuterium in place of hydrogen at all bis-allylic positions, the initiation of deuterium abstraction and radical production gets decelerated as compared to H-PUFA (Shchepinov et al. 2011). Similarly in 2016, Yang et al. treated G-401 cell with D-linoleate (D-Lin), H-linoleate (H-lin), and vehicle (0.1% ethanol) for overnight and then exposed these cells to erastin and RSL3 to evaluate the impact of D-PUFA in the execution of ferroptosis. The D-PUFA showed a potential protective effect against erastin and RSL3 treated conditions and suppressed ferroptosis by inhibiting lipid peroxides generation which was measured by C11-BODIPY (Yang et al. 2016). In 2018, Shah et al. conducted a study to determine the role of LOX enzyme. In this study, they used deuterated arachidonic acid. All bis-allylic hydrogen atoms were replaced with their heavy isotope deuterium. The LOX did not abstract hydrogen to initiate lipid peroxidation and suppressed RSL3 induced cell death (Shah et al. 2018). The above finding suggested that deuterated phospholipids give important contribution to suppress ferroptosis by inhibiting lipid peroxidation process and can therefore be used as ferroptosis inhibitors to treat various neurodegenerative diseases (Figs. 14.6 and 14.7; Table 14.1).

Table 14.1 Ferroptosis inhibitor along with their target and mechanism followed to inhibit ferroptosis

S. No.	Ferroptosis inhibitor	Target	Mechanism	Reference
1	Ferrostatin	Lipid peroxidation	It suppresses lipid peroxidation	Skouta et al. (2014)
2	Liproxstatin	Lipid peroxidation	It suppresses lipid peroxidation	Zilka et al. (2017)
3	Deferoxamine	Iron	It acts as an iron chelator and obstructs an iron-dependent lipid peroxidation	Kontoghiorghe and Kontoghiorghes (2016)
4	CDC, PD-146176, AA-861, baicalein	Lipoxygenase	Inhibits lipoxygenase-mediated lipid peroxidation	Yang (2016)
5	Cycloheximide	Protein synthesis	It represses ferroptosis persuaded by system Xc inhibitors	Kwon et al. (2015)
6	Dopamine	Neurotransmitter	It inhibits GPX4 degradation	Ayton et al. (2013)
7	Beta-mercaptoethanol	Reducing agent	It reduces extracellular cystine into cysteine, which bypasses system Xc	Shimada and Stockwell (2015), Cao and Dixon (2016)
8	Vitamin E and its analog	Lipid peroxidation	It inhibits the proliferation of lipid peroxidation. It may inhibit lipoxygenases	Brigelius-Flohé (2009)
9	Deferiprone	Iron	It attenuates iron and acts as an iron chelator	Kontoghiorghe and Kontoghiorghes (2016)
10	Selenium	Selenoprotein	It increases the abundance of selenoprotein	Arbiser et al. (2018)
11	D-PUFAs	Lipid peroxidation	It suppresses the initiation and proliferation of lipid peroxidation	Probst et al. (2017)
12	Ubiquinol	Lipid peroxidation	It is a natural antioxidant released in the body that functions as a free radical trapper and inhibits lipid peroxidation	Ouchi et al. (2010)
13	Thiazolidinediones (TZNs)	ASCL4	It inhibits the ASCL4 enzyme and oxidation of AA and Adrenic acid (AdA)	Doll et al. (2017)

14.5 Conclusion

The prognosis and pathogenesis of AD are multifactorial, and further continual discovery of novel signaling pathways on ferroptosis driven AD imitates the complexity of the disease. These complexity should be addressed in management of AD in patients for better therapeutic outcomes, and anti-ferroptotic drugs like iron chelators or inhibitors of better fit should be included as AD treatment options. This chapter sum up the evidences following the role of ferroptosis in AD pathogenesis and signifies what is known about the therapeutic targets for inhibition of AD progression. In addition to this ferroptosis-related differentially expressed genes in AD supported that the inhibition of ferroptosis might slow down the progression of disease and declination of memory, however this area is still uncovered and need to be widely investigated. Further development of new models for AD is needed to understand how ferroptosis affects cell-to-cell interaction and time duration of development of ferroptosis in AD. In future, studies must be focused on developing detection tools of ferroptosis and systematizing huge and randomized clinical trials of drugs related to ferroptosis inhibition in early or later stage of disease progression in AD models.

Acknowledgement GM is thankful to Indian Council of Medical Research (ICMR) for providing research grant sanctioned under ICMR 5/4-5/3/24/Neuro/2022-NCD-I (ICMR-EMR-2021-10363).

References

Abdalkader M, Lampinen R, Kanninen KM, Malm TM, Liddell JR (2018) Targeting Nrf2 to suppress ferroptosis and mitochondrial dysfunction in neurodegeneration. Front Neurosci 12: 466

Adibhatla RM, Hatcher JF (2010) Lipid oxidation and peroxidation in CNS health and disease: from molecular mechanisms to therapeutic opportunities. Antioxid Redox Signal 12(1): 125–169

Alim I, Caulfield JT, Chen Y, Swarup V, Geschwind DH, Ivanova E et al (2019) Selenium drives a transcriptional adaptive program to block ferroptosis and treat stroke. Cell 177(5):1262–79.e25

Angeli JPF, Shah R, Pratt DA, Conrad M (2017) Ferroptosis inhibition: mechanisms and opportunities. Trends Pharmacol Sci 38(5):489–498

Aprioku JS (2013) Pharmacology of free radicals and the impact of reactive oxygen species on the testis. J Reprod Infertil 14(4):158–172

Arbiser JL, Bonner MY, Ward N, Elsey J, Rao S (2018) Selenium unmasks protective iron armor: a possible defense against cutaneous inflammation and cancer. Biochim Biophys Acta Gen Subj 1862(11):2518–2527

Askari B, Kanter JE, Sherrid AM, Golej DL, Bender AT, Liu J et al (2007) Rosiglitazone inhibits acyl-CoA synthetase activity and fatty acid partitioning to diacylglycerol and triacylglycerol via a peroxisome proliferator-activated receptor-gamma-independent mechanism in human arterial smooth muscle cells and macrophages. Diabetes 56(4):1143–1152

Ayton S, Lei P, Duce JA, Wong BX, Sedjahtera A, Adlard PA et al (2013) Ceruloplasmin dysfunction and therapeutic potential for Parkinson disease. Ann Neurol 73(4):554–559

Ayton S, Fazlollahi A, Bourgeat P, Raniga P, Ng A, Lim YY et al (2017) Cerebral quantitative susceptibility mapping predicts amyloid-β-related cognitive decline. Brain 140(8):2112–2119

Beshir SA, Aadithsoorya AM (2022) Aducanumab therapy to treat Alzheimer's disease: a narrative review. Int J Alzheimers Dis 2022:9343514

Blaise GA, Gauvin D, Gangal M, Authier S (2005) Nitric oxide, cell signaling and cell death. Toxicology 208(2):177–192

Bowry VW, Stocker R (1993) Tocopherol-mediated peroxidation. The prooxidant effect of vitamin E on the radical-initiated oxidation of human low-density lipoprotein. J Am Chem Soc 115(14): 6029–6044

Brigelius-Flohé R (2009) Vitamin E: the shrew waiting to be tamed. Free Radic Biol Med 46(5): 543–554

Brütsch SH, Wang CC, Li L, Stender H, Neziroglu N, Richter C et al (2015) Expression of inactive glutathione peroxidase 4 leads to embryonic lethality, and inactivation of the Alox15 gene does not rescue such knock-in mice. Antioxid Redox Signal 22(4):281–293

Burton GW, Ingold KU (1986) Vitamin-E—application of the principles of physical organic-chemistry to the exploration of its structure and function. Acc Chem Res 19:194

Cai P, Fang SQ, Yang HL, Yang XL, Liu QH, Kong LY et al (2018) Donepezil-butylated hydroxytoluene (BHT) hybrids as anti-Alzheimer's disease agents with cholinergic, antioxidant, and neuroprotective properties. Eur J Med Chem 157:161–176

Cao JY, Dixon SJ (2016) Mechanisms of ferroptosis. Cell Mol Life Sci 73(11–12):2195–2209

Cardoso BR, Roberts BR, Malpas CB, Vivash L, Genc S, Saling MM et al (2019) Supranutritional sodium selenate supplementation delivers selenium to the central nervous system: results from a randomized controlled pilot trial in Alzheimer's disease. Neurotherapeutics 16(1):192–202

Castellani RJ, Moreira PI, Liu G, Dobson J, Perry G, Smith MA et al (2007) Iron: the Redox-active center of oxidative stress in Alzheimer disease. Neurochem Res 32(10):1640–1645

Chu J, Li JG, Praticò D (2013) Zileuton improves memory deficits, amyloid and tau pathology in a mouse model of Alzheimer's disease with plaques and tangles. PLoS One 8(8):e70991

Chu J, Li JG, Giannopoulos PF, Blass BE, Childers W, Abou-Gharbia M et al (2015) Pharmacologic blockade of 12/15-lipoxygenase ameliorates memory deficits, Aβ and tau neuropathology in the triple-transgenic mice. Mol Psychiatry 20(11):1329–1338

Colović MB, Krstić DZ, Lazarević-Pašti TD, Bondžić AM, Vasić VM (2013) Acetylcholinesterase inhibitors: pharmacology and toxicology. Curr Neuropharmacol 11(3):315–335

Cotticelli MG, Crabbe AM, Wilson RB, Shchepinov MS (2013) Insights into the role of oxidative stress in the pathology of Friedreich ataxia using peroxidation resistant polyunsaturated fatty acids. Redox Biol 1(1):398–404

Cozza G, Rossetto M, Bosello-Travain V, Maiorino M, Roveri A, Toppo S et al (2017) Glutathione peroxidase 4-catalyzed reduction of lipid hydroperoxides in membranes: the polar head of membrane phospholipids binds the enzyme and addresses the fatty acid hydroperoxide group toward the redox center. Free Radic Biol Med 112:1–11

Crous-Bou M, Minguillón C, Gramunt N, Molinuevo JL (2017) Alzheimer's disease prevention: from risk factors to early intervention. Alzheimers Res Ther 9(1):71

DeHart DN, Fang D, Heslop K, Li L, Lemasters JJ, Maldonado EN (2018) Opening of voltage dependent anion channels promotes reactive oxygen species generation, mitochondrial dysfunction and cell death in cancer cells. Biochem Pharmacol 148:155–162

Derry PJ, Hegde ML, Jackson GR, Kayed R, Tour JM, Tsai AL et al (2020) Revisiting the intersection of amyloid, pathologically modified tau and iron in Alzheimer's disease from a ferroptosis perspective. Prog Neurobiol 184:101716

Devisscher L, Van Coillie S, Hofmans S, Van Rompaey D, Goossens K, Meul E et al (2018) Discovery of novel, drug-like ferroptosis inhibitors with in vivo efficacy. J Med Chem 61(22): 10126–10140

Dhanasekaran M, Ren J (2005) The emerging role of coenzyme Q-10 in aging, neurodegeneration, cardiovascular disease, cancer and diabetes mellitus. Curr Neurovasc Res 2(5):447–459

Di Meco A, Lauretti E, Vagnozzi AN, Praticò D (2014) Zileuton restores memory impairments and reverses amyloid and tau pathology in aged Alzheimer's disease mice. Neurobiol Aging 35(11): 2458–2464

Di Meco A, Li JG, Blass BE, Abou-Gharbia M, Lauretti E, Praticò D (2017) 12/15-Lipoxygenase inhibition reverses cognitive impairment, brain amyloidosis, and tau pathology by stimulating autophagy in aged triple transgenic mice. Biol Psychiatry 81(2):92–100

Dixon SJ (2017) Ferroptosis: bug or feature? Immunol Rev 277(1):150–157

Dixon SJ, Stockwell BR (2014) The role of iron and reactive oxygen species in cell death. Nat Chem Biol 10(1):9–17

Dixon SJ, Stockwell BR (2019) The hallmarks of ferroptosis. Annu Rev Cancer Biol 3(1):35–54

Dixon SJ, Lemberg KM, Lamprecht MR, Skouta R, Zaitsev EM, Gleason CE et al (2012) Ferroptosis: an iron-dependent form of nonapoptotic cell death. Cell 149(5):1060–1072

Doba T, Burton GW, Ingold KU (1985) Antioxidant and co-antioxidant activity of vitamin C. The effect of vitamin C, either alone or in the presence of vitamin E or a water-soluble vitamin E analogue, upon the peroxidation of aqueous multilamellar phospholipid liposomes. Biochim Biophys Acta Lipids Lipid Metab 835:298

Doll S (2017) ACSL4 dictates ferroptosis sensitivity by shaping cellular lipid composition. Nat Chem Biol 13:91

Doll S, Proneth B, Tyurina YY, Panzilius E, Kobayashi S, Ingold I et al (2017) ACSL4 dictates ferroptosis sensitivity by shaping cellular lipid composition. Nat Chem Biol 13(1):91–98

Fan EKY, Fan J (2018) Regulation of alveolar macrophage death in acute lung inflammation. Respir Res 19(1):50

Feng H, Stockwell BR (2018) Unsolved mysteries: how does lipid peroxidation cause ferroptosis? PLoS Biol 16(5):e2006203

Friedmann Angeli JP, Schneider M, Proneth B, Tyurina YY, Tyurin VA, Hammond VJ et al (2014) Inactivation of the ferroptosis regulator Gpx4 triggers acute renal failure in mice. Nat Cell Biol 16(12):1180–1191

Gao M, Yi J, Zhu J, Minikes AM, Monian P, Thompson CB et al (2019) Role of mitochondria in ferroptosis. Mol Cell 73(2):354–63.e3

Gilbert NC, Bartlett SG, Waight MT, Neau DB, Boeglin WE, Brash AR et al (2011) The structure of human 5-lipoxygenase. Science (New York, NY) 331(6014):217–219

Gu XH, Xu LJ, Liu ZQ, Wei B, Yang YJ, Xu GG et al (2016) The flavonoid baicalein rescues synaptic plasticity and memory deficits in a mouse model of Alzheimer's disease. Behav Brain Res 311:309–321

Guo H, Verhoek IC, Prins GGH, van der Vlag R, van der Wouden PE, van Merkerk R et al (2019) Novel 15-lipoxygenase-1 inhibitor protects macrophages from lipopolysaccharide-induced cytotoxicity. J Med Chem 62(9):4624–4637

Habib E, Linher-Melville K, Lin HX, Singh G (2015) Expression of xCT and activity of system xc (−) are regulated by NRF2 in human breast cancer cells in response to oxidative stress. Redox Biol 5:33–42

Haeggström JZ, Funk CD (2011) Lipoxygenase and leukotriene pathways: biochemistry, biology, and roles in disease. Chem Rev 111(10):5866–5898

Hambright WS, Fonseca RS, Chen L, Na R, Ran Q (2017) Ablation of ferroptosis regulator glutathione peroxidase 4 in forebrain neurons promotes cognitive impairment and neurodegeneration. Redox Biol 12:8–17

Harding AE, Muller DP, Thomas PK, Willison HJ (1982) Spinocerebellar degeneration secondary to chronic intestinal malabsorption: a vitamin E deficiency syndrome. Ann Neurol 12(5): 419–424

Hofmans S, Vanden Berghe T, Devisscher L, Hassannia B, Lyssens S, Joossens J et al (2016) Novel ferroptosis inhibitors with improved potency and ADME properties. J Med Chem 59(5): 2041–2053

Hou W, Xie Y, Song X, Sun X, Lotze MT, Zeh HJ III et al (2016) Autophagy promotes ferroptosis by degradation of ferritin. Autophagy 12(8):1425–1428

Ingold KU (1961) Inhibition of the autoxidation of organic substances in the liquid phase. Chem Rev 61:563

Ingold KU, Pratt DA (2014) Advances in radical-trapping antioxidant chemistry in the 21st century: a kinetics and mechanisms perspective. Chem Rev 114:9022

Ivanov I, Heydeck D, Hofheinz K, Roffeis J, O'Donnell VB, Kuhn H et al (2010) Molecular enzymology of lipoxygenases. Arch Biochem Biophys 503(2):161–174

Jiang L, Kon N, Li T, Wang SJ, Su T, Hibshoosh H et al (2015) Ferroptosis as a p53-mediated activity during tumour suppression. Nature 520(7545):57–62

Kagan VE, Mao G, Qu F, Angeli JP, Doll S, Croix CS et al (2017) Oxidized arachidonic and adrenic PEs navigate cells to ferroptosis. Nat Chem Biol 13(1):81–90

Khalaf SS, Hafez MM, Mehanna ET, Mesbah NM, Abo-Elmatty DM (2019) Combined vildagliptin and memantine treatment downregulates expression of amyloid precursor protein, and total and phosphorylated tau in a rat model of combined Alzheimer's disease and type 2 diabetes. Naunyn Schmiedebergs Arch Pharmacol 392(6):685–695

Kontoghiorghe CN, Kontoghiorghes GJ (2016) Efficacy and safety of iron-chelation therapy with deferoxamine, deferiprone, and deferasirox for the treatment of iron-loaded patients with non-transfusion-dependent thalassemia syndromes. Drug Des Dev Ther 10:465–481

Kosaraju J, Murthy V, Khatwal RB, Dubala A, Chinni S, Muthureddy Nataraj SK et al (2013) Vildagliptin: an anti-diabetes agent ameliorates cognitive deficits and pathology observed in streptozotocin-induced Alzheimer's disease. J Pharm Pharmacol 65(12):1773–1784

Kosaraju J, Holsinger RMD, Guo L, Tam KY (2017) Linagliptin, a dipeptidyl peptidase-4 inhibitor, mitigates cognitive deficits and pathology in the 3xTg-AD mouse model of Alzheimer's disease. Mol Neurobiol 54(8):6074–6084

Krabbendam IE, Honrath B, Dilberger B, Iannetti EF (2020) SK channel-mediated metabolic escape to glycolysis inhibits ferroptosis and supports stress resistance in C. elegans. Cell Death Dis 11(4):263

Kuang F, Liu J, Tang D, Kang R (2020) Oxidative damage and antioxidant defense in ferroptosis. Front Cell Dev Biol 8:586578

Kwon M-Y, Park E, Lee S-J, Chung SW (2015) Heme oxygenase-1 accelerates erastin-induced ferroptotic cell death. Oncotarget 6(27):24393–24403

Li B, Harjani JR, Cormier NS, Madarati H, Atkinson J, Cosa G et al (2013) Besting vitamin E: sidechain substitution is key to the reactivity of naphthyridinol antioxidants in lipid bilayers. J Am Chem Soc 135(4):1394–1405

Liu J, Chang L, Song Y, Li H, Wu Y (2019) The role of NMDA receptors in Alzheimer's disease. Front Neurosci 13:43

Lloret A, Esteve D, Monllor P, Cervera-Ferri A, Lloret A (2019) The effectiveness of vitamin E treatment in Alzheimer's disease. Int J Mol Sci 20(4):879

Loscalzo J (2008) Membrane redox state and apoptosis: death by peroxide. Cell Metab 8(3): 182–183

Ma QH, Jiang LF, Mao JL, Xu WX, Huang M (2018) Vildagliptin prevents cognitive deficits and neuronal apoptosis in a rat model of Alzheimer's disease. Mol Med Rep 17(3):4113–4119

Magtanong L, Dixon SJ (2018) Ferroptosis and brain injury. Dev Neurosci 40(5–6):382–395

Maher P, van Leyen K, Dey PN, Honrath B, Dolga A, Methner A (2018) The role of Ca(2+) in cell death caused by oxidative glutamate toxicity and ferroptosis. Cell Calcium 70:47–55

Mcgirr S, Venegas C, Swaminathan A (2020) Alzheimers disease: a brief review. J Exp Neurol 1(3):89–98

Mendes KL, Lelis DF, Santos SHS (2017) Nuclear sirtuins and inflammatory signaling pathways. Cytokine Growth Factor Rev 38:98–105

Miyake S, Murai S, Kakuta S, Uchiyama Y, Nakano H (2020) Identification of the hallmarks of necroptosis and ferroptosis by transmission electron microscopy. Biochem Biophys Res Commun 527(3):839–844

Moroishi T, Yamauchi T, Nishiyama M, Nakayama KI (2014) HERC2 targets the iron regulator FBXL5 for degradation and modulates iron metabolism. J Biol Chem 289(23):16430–16441

Mou Y, Wang J, Wu J, He D, Zhang C, Duan C et al (2019) Ferroptosis, a new form of cell death: opportunities and challenges in cancer. J Hematol Oncol 12(1):34

Nam TG, Rector CL, Kim HY, Sonnen AF, Meyer R, Nau WM et al (2007) Tetrahydro-1,8-naphthyridinol analogues of alpha-tocopherol as antioxidants in lipid membranes and low-density lipoproteins. J Am Chem Soc 129(33):10211–10219

Neitemeier S, Jelinek A, Laino V, Hoffmann L, Eisenbach I, Eying R et al (2017) BID links ferroptosis to mitochondrial cell death pathways. Redox Biol 12:558–570

Niki E, Yoshida Y, Saito Y, Noguchi N (2005) Lipid peroxidation: mechanisms, inhibition, and biological effects. Biochem Biophys Res Commun 338(1):668–676

Nohl H, Kozlov AV, Staniek K, Gille L (2001) The multiple functions of coenzyme Q. Bioorg Chem 29(1):1–13

Ouchi A, Nagaoka S-i, Mukai K (2010) Tunneling effect in regeneration reaction of vitamin E by ubiquinol. J Phys Chem B 114(19):6601–6607

Perianayagam A, Bloom D, Lee J, Parasuraman S, Sekher TV, Mohanty SK et al (2022) Cohort profile: the longitudinal ageing study in India (LASI). Int J Epidemiol 51(4):e167–e176

Peters DG, Pollack AN, Cheng KC, Sun D, Saido T, Haaf MP et al (2018) Dietary lipophilic iron alters amyloidogenesis and microglial morphology in Alzheimer's disease knock-in APP mice. Metallomics 10(3):426–443

Praticò D, Sung S (2004) Lipid peroxidation and oxidative imbalance: early functional events in Alzheimer's disease. J Alzheimers Dis 6(2):171–175

Praticò D, Uryu K, Leight S, Trojanoswki JQ, Lee VM (2001) Increased lipid peroxidation precedes amyloid plaque formation in an animal model of Alzheimer amyloidosis. J Neurosci 21(12): 4183–4187

Pratt DA, DiLabio GA, Brigati G, Pedulli GF, Valgimigli L (2001) 5-Pyrimidinols: novel chain-breaking antioxidants more effective than phenols. J Am Chem Soc 123(19):4625–4626

Probst L, Dächert J, Schenk B, Fulda S (2017) Lipoxygenase inhibitors protect acute lymphoblastic leukemia cells from ferroptotic cell death. Biochem Pharmacol 140:41–52

Reichert CO, de Freitas FA, Sampaio-Silva J, Rokita-Rosa L, Barros PL, Levy D et al (2020) Ferroptosis mechanisms involved in neurodegenerative diseases. Int J Mol Sci 21(22):8765

Saini R (2011) Coenzyme Q10: the essential nutrient. J Pharm Bioallied Sci 3(3):466–467

Sato H, Nomura S, Maebara K, Sato K, Tamba M, Bannai S (2004) Transcriptional control of cystine/glutamate transporter gene by amino acid deprivation. Biochem Biophys Res Commun 325(1):109–116

Seibt TM, Proneth B, Conrad M (2019) Role of GPX4 in ferroptosis and its pharmacological implication. Free Radic Biol Med 133:144–152

Senin U, Parnetti L, Barbagallo-Sangiorgi G, Bartorelli L, Bocola V, Capurso A et al (1992) Idebenone in senile dementia of Alzheimer type: a multicentre study. Arch Gerontol Geriatr 15(3):249–260

Shah R, Shchepinov MS, Pratt DA (2018) Resolving the role of lipoxygenases in the initiation and execution of ferroptosis. ACS Cent Sci 4(3):387–396

Shchepinov MS (2007) Reactive oxygen species, isotope effect, essential nutrients, and enhanced longevity. Rejuvenation Res 10(1):47–59

Shchepinov MS, Chou VP, Pollock E, Langston JW, Cantor CR, Molinari RJ et al (2011) Isotopic reinforcement of essential polyunsaturated fatty acids diminishes nigrostriatal degeneration in a mouse model of Parkinson's disease. Toxicol Lett 207(2):97–103

Sheng X, Shan C, Liu J, Yang J, Sun B, Chen D (2017) Theoretical insights into the mechanism of ferroptosis suppression via inactivation of a lipid peroxide radical by liproxstatin-1. Phys Chem Chem Phys 19(20):13153–13159

Shimada K, Stockwell BR (2015) tRNA synthase suppression activates de novo cysteine synthesis to compensate for cystine and glutathione deprivation during ferroptosis. Mol Cell Oncol 3(2): e1091059

Skouta R, Dixon SJ, Wang J, Dunn DE, Orman M, Shimada K et al (2014) Ferrostatins inhibit oxidative lipid damage and cell death in diverse disease models. J Am Chem Soc 136(12): 4551–4556

Solomon EI, Zhou J, Neese F, Pavel EG (1997) New insights from spectroscopy into the structure/function relationships of lipoxygenases. Chem Biol 4(11):795–808

Stockwell BR, Friedmann Angeli JP, Bayir H, Bush AI, Conrad M, Dixon SJ et al (2017) Ferroptosis: a regulated cell death nexus linking metabolism, redox biology, and disease. Cell 171(2):273–285

Tampi RR, Forester BP, Agronin M (2021) Aducanumab: evidence from clinical trial data and controversies. Drugs Context 10:2021-7-3

Tang M, Chen Z, Wu D, Chen L (2018) Ferritinophagy/ferroptosis: iron-related newcomers in human diseases. J Cell Physiol 233(12):9179–9190

Tejero I, González-García N, González-Lafont À, Lluch JM (2007) Tunneling in green tea: understanding the antioxidant activity of catechol-containing compounds. A variational transition-state theory study. J Am Chem Soc 129(18):5846–5854

Thal LJ, Grundman M, Berg J, Ernstrom K, Margolin R, Pfeiffer E et al (2003) Idebenone treatment fails to slow cognitive decline in Alzheimer's disease. Neurology 61(11):1498–1502

Tuo QZ, Lei P, Jackman KA, Li XL, Xiong H, Li XL et al (2017) Tau-mediated iron export prevents ferroptotic damage after ischemic stroke. Mol Psychiatry 22(11):1520–1530

Ulatowski LM, Manor D (2015) Vitamin E and neurodegeneration. Neurobiol Dis 84:78–83

Ursini F, Maiorino M, Valente M, Ferri L, Gregolin C (1982) Purification from pig liver of a protein which protects liposomes and biomembranes from peroxidative degradation and exhibits glutathione peroxidase activity on phosphatidylcholine hydroperoxides. Biochim Biophys Acta 710(2):197–211

Vallance P, Leiper J (2002) Blocking NO synthesis: how, where and why? Nat Rev Drug Discov 1(12):939–950

Vaz M, Silva V, Monteiro C (2022) Role of aducanumab in the treatment of Alzheimer's disease: challenges and opportunities. Clin Interv Aging 17:797–810

Wang Y, Hekimi S (2013) Molecular genetics of ubiquinone biosynthesis in animals. Crit Rev Biochem Mol Biol 48(1):69–88

Wang H, An P, Xie E, Wu Q, Fang X, Gao H et al (2017) Characterization of ferroptosis in murine models of hemochromatosis. Hepatology 66(2):449–465

Wang L, Liu Y, Du T, Yang H, Lei L, Guo M et al (2020) ATF3 promotes erastin-induced ferroptosis by suppressing system Xc. Cell Death Differ 27(2):662–675

Ward RJ, Zucca FA, Duyn JH, Crichton RR, Zecca L (2014) The role of iron in brain ageing and neurodegenerative disorders. Lancet Neurol 13(10):1045–1060

Weyer G, Babej-Dölle RM, Hadler D, Hofmann S, Herrmann WM (1997) A controlled study of 2 doses of idebenone in the treatment of Alzheimer's disease. Neuropsychobiology 36(2):73–82

Wijtmans M, Pratt DA, Valgimigli L, DiLabio GA, Pedulli GF, Porter NA (2003) 6-Amino-3-pyridinols: towards diffusion-controlled chain-breaking antioxidants. Angew Chem Int Ed Engl 42(36):4370–4373

Wu JR, Tuo QZ, Lei P (2018) Ferroptosis, a recent defined form of critical cell death in neurological disorders. J Mol Neurosci 66(2):197–206

Xie Y, Hou W, Song X, Yu Y, Huang J, Sun X et al (2016) Ferroptosis: process and function. Cell Death Differ 23:369

Xu Y, Li X, Cheng Y, Yang M, Wang R (2020) Inhibition of ACSL4 attenuates ferroptotic damage after pulmonary ischemia-reperfusion. FASEB J 34(12):16262–16275

Yamamoto A, Shin RW, Hasegawa K, Naiki H, Sato H, Yoshimasu F et al (2002) Iron (III) induces aggregation of hyperphosphorylated tau and its reduction to iron (II) reverses the aggregation: implications in the formation of neurofibrillary tangles of Alzheimer's disease. J Neurochem 82(5):1137–1147

Yan N, Zhang J (2020) Iron metabolism, ferroptosis, and the links with Alzheimer's disease. Front Neurosci 13:1443

Yang WS (2016) Peroxidation of polyunsaturated fatty acids by lipoxygenases drives ferroptosis. Proc Natl Acad Sci U S A 113:E4966

Yang WS, Stockwell BR (2016) Ferroptosis: death by lipid peroxidation. Trends Cell Biol 26(3): 165–176

Yang WS, SriRamaratnam R, Welsch ME, Shimada K, Skouta R, Viswanathan VS et al (2014) Regulation of ferroptotic cancer cell death by GPX4. Cell 156(1–2):317–331

Yang WS, Kim KJ, Gaschler MM, Patel M, Shchepinov MS, Stockwell BR (2016) Peroxidation of polyunsaturated fatty acids by lipoxygenases drives ferroptosis. Proc Natl Acad Sci U S A 113(34):E4966–E4975

Yoo MH, Gu X, Xu XM, Kim JY, Carlson BA, Patterson AD et al (2010) Delineating the role of glutathione peroxidase 4 in protecting cells against lipid hydroperoxide damage and in Alzheimer's disease. Antioxid Redox Signal 12(7):819–827

Zhang F, Tao Y, Zhang Z, Guo X, An P, Shen Y et al (2012) Metalloreductase Steap3 coordinates the regulation of iron homeostasis and inflammatory responses. Haematologica 97(12): 1826–1835

Zhang SQ, Obregon D, Ehrhart J, Deng J, Tian J, Hou H et al (2013) Baicalein reduces β-amyloid and promotes nonamyloidogenic amyloid precursor protein processing in an Alzheimer's disease transgenic mouse model. J Neurosci Res 91(9):1239–1246

Zhou RP, Chen Y, Wei X, Yu B, Xiong ZG, Lu C et al (2020) Novel insights into ferroptosis: implications for age-related diseases. Theranostics 10(26):11976–11997

Zilka O, Shah R, Li B, Friedmann Angeli JP, Griesser M, Conrad M et al (2017) On the mechanism of cytoprotection by ferrostatin-1 and liproxstatin-1 and the role of lipid peroxidation in ferroptotic cell death. ACS Cent Sci 3(3):232–243

Zorov DB, Juhaszova M, Sollott SJ (2014) Mitochondrial reactive oxygen species (ROS) and ROS-induced ROS release. Physiol Rev 94(3):909–950

Part VI

Natural Products/Synthetic Molecules as Epigenetic Modulators in Alzheimer's Disease

Sirtuin Modulator: Design, Synthesis, and Biological Evaluation

15

Shivam Kumar Kori, Satyamshyam Vishwakarma, Sandhya Pathak, Varsha Kashaw, Vandana Soni, Arun K. Iyer, and Sushil Kumar Kashaw

Abstract

A family of signalling proteins called sirtuins is involved in the control of metabolism. The sirtuin family of NAD^+-dependent protein lysine deacylases controls a range of physiological processes, including stress reactions and energy metabolism. For ageing-related illnesses such type 2 diabetes, inflammatory diseases, gene repression, metabolic regulation, apoptosis and cell survival, DNA repair, and neurodegenerative disorders, the human sirtuin isoforms (1–7) are thought to be promising therapeutic targets. The search for small compounds that alter the activity of sirtuins is becoming more and more popular since it may have positive implications on treating human ailments. Here, we discussed the sirtuin synthesis, biological importance, potent and specific pharmacological sirtuin activators and inhibitors, isoforms, and the current status of sirtuin-targeted therapeutic research. The rationale behind continued medication development is based on the progressive understanding of the sirtuin modulation processes by such compounds.

S. K. Kori · S. Vishwakarma · V. Soni · S. K. Kashaw (✉)
Integrated Drug Discovery Research Laboratory, Department of Pharmaceutical Sciences, Dr. Hasrisingh Gour University (A Central University), Sagar, Madhya Pradesh, India

S. Pathak
Department of Chemistry, Dr. Harisingh Gour University (A Central University), Sagar, Madhya Pradesh, India

V. Kashaw
Sagar Institute of Pharmaceutical Sciences, Sagar, Madhya Pradesh, India

A. K. Iyer
Use-Inspired Biomaterials & Integrated Nano Delivery (U-BiND) Systems Laboratory, Department of Pharmaceutical Sciences, Wayne State University, Detroit, MI, USA

Molecular Imaging Program, Karmanos Cancer Institute, Detroit, MI, USA

A. Sharma, G. P. Modi (eds.), *Natural Product-Based Synthetic Drug Molecules in Alzheimer's Disease*, https://doi.org/10.1007/978-981-99-6038-5_15

Keywords

Sirtuins · Cancer therapy · Alzheimer's disease · Neurodegenerative disorders · Sirtuin modulation

15.1 Introduction

Sirtuins are existing in both prokaryotes and eukaryotes, which are NAD^+-dependent mono-ADP ribosylases and protein lysine deacylases. The sirtuin family in living things consists of seven isoforms, each with a unique subcellular location and biological functions (Alvarez et al. 2011; Fiorentino et al. 2022b). Sirtuins have established cumulative consideration in the past couple of decades specified their vital roles in a range of biological processes, including cytodifferentiation, transcriptional control, cell cycle progression, apoptosis, swelling, breakdown, brain-related and heart physiology, and cancer. Thus, it has been noted that sirtuin activity regulation represents a prospective therapeutic approach for several illness (Fiorentino et al. 2022a).

Since 2000, sirtuins (SIRT1–7) have drawn increasing interest for their associations with an extensive variety of biological functions, including the control of cellular metabolism, neuroprotection, apoptosis, inflammation, and the growth of cancer (Mellini et al. 2015). In particular, SIRT1 has received the greatest research attention, both for its involvement in calorie restriction and as an eventual path for the therapy of illnesses associated with aging (Yeong et al. 2020). SIRT suppression can have advantageous effects on aging-related conditions including metabolic, cardiovascular, and neurodegenerative illnesses, whereas SIRT inhibitors may be beneficial for the treatment of muscular disorders, HIV infection, or cancer (Mellini et al. 2015; Valente et al. 2016).

The sirtuin route has drawn superior interest since it is linked to the advantages of calorie restriction for anti-aging (Mohamad Nasir et al. 2018). In aging-related laboratory models, pharmacological or genetically overexpression of the sirtuin system has also revealed encouraging outcomes. Mammalian sirtuins, which are related of the yeast Sir2 family, consist of seven members (SIRT1–SIRT7). ADP-ribosyltransferase and a deacetylase activity are two distinct actions that sirtuins have been discovered to exhibit (Liszt et al. 2005; Landry et al. 2000). Histones, transcription factors, and apoptosis modulators are just a few of the targets that sirtuins use to convey out their deacetylase activity (Youcef et al. 2007; Dai et al. 2018).

Similar to this, the revelation which are catalytic action requires NAD^+, as opposed to Zn^{2+} in the situations of other groups of deacetylases, suggests their potential significance in balancing management in the cell's breakdown and its functional state. Seven sirtuin iso-forms regulate numerous metabolic, anxiety, and aging processes in living things. For illnesses associated with metabolism and aging, such as metabolic syndrome and neurodegenerative disorders, sirtuins are supposed to be promising therapeutic targets. Sirtuins are thought to be prospective therapeutic targets for disorders of metabolism and aging, such as neurodegenerative illnesses and metabolic disorders.

Few powerful and selective compounds have been produced despite intensive attempts to generate minor-molecule sirtuin inhibitors and activators, in part because of the limitations of currently available assays and a lack of mechanistic characterisation of discovered compounds (Schutkowski et al. 2014).

15.2 Types of Sirtuin

See Table 15.1.

15.3 Pharmacological Sirtuin Modulation

Sirtuins have fascinated attention as possible therapeutic targets, therefore much work has been put into developing specialised sirtuin agonist and inhibitors, both are utilised in research for understanding sirtuin function and as potential anti-ageing medications. Recent growths in sirtuin biochemistry, analytical test, and crystal structures of sirtuin/modulator composite, which disclose a challenging relationship between numerous chemicals and the structure and function of the enzyme, are currently assisting in the discovery of pharmacological sirtuin modulators (Dai et al. 2018).

15.3.1 Sirtuin Activators

Since more than 75 years ago, it has been recognised that calorie restriction (CR) increases mammalian longevity and promotes better health. When sirtuins were eliminated, it had been assumed that food restriction did not increase lifespan (Rogina and Helfand 2004; Lin et al. 2000) and that calorie restriction might lengthen mammalian lifetime by bringing SIRT1 expression (Cohen et al. 2004) led researchers to study sirtuins and to discover and develop molecules that could stimulate them.

Despite the fact that the function of sirtuins in enhancing lifespan is currently being called into doubt, based on the beneficial effects of calorie restriction on mammalian health and the consequent rises in SIRT1 (Cohen et al. 2004), sirtuins are gaining a lot of attention and compounds that can activate them due to the identification and formation of medicines that activate SIRT1.

15.3.1.1 Resveratrol

Two phenyl rings of the polyphenol resveratrol (RSV), which are connected by a methylene bridge (Fig. 15.1a), was the primary substance to be found that can imitate calorie restriction via activating sirtuins (Howitz et al. 2003; Wood et al. 2004). The primary substance identified that may imitate calorie restriction through

Table 15.1 Aspects of sirtuin's biochemistry and functionality (Anekonda and Reddy 2006; Baur et al. 2012; Bonda et al. 2011; Outeiro et al. 2008; Wątroba and Szukiewicz 2016)

Types of sirtuins	Molecular mass (kDa)	Sub-cellular site	High tissue expression	Core substrates	Roles	Enzyme function
Sirtuin 1	81.7	Cytosolic and nuclear	Uterus, skeletal muscle, heart, kidney, brain	Histones H1, H3 and H4 Transcription factors p53, FOXO family, Ku70, p300, NF-κB, PGC-1α, PPAR-γ, UCP2 Acetyl-CoA synthetase 1	Cell survival, lifespan regulation, metabolism regulation, inflammation, oxidative stress response	NAD^+-dependent deacetylase
Sirtuin 2	43.2	Nuclear and cytosol	Brain	α-Tubulin	Cell cycle regulation, nervous system development	NAD^+-dependent deacetylase
Sirtuin 3	43.6	Nuclear, mitochondrial and cytosol	Brain, heart, liver, kidney, and brown adipose tissue	Acetyl-CoA synthetase 2 Isocitrate dehydrogenase 2 Ku70, FOXO 3a, MnSOD, Mitochondrial ribosomal protein L10 Long-chain acyl-CoA dehydrogenase 3-Hydroxy-3-Methylglutaryl CoA synthase 2 Succinate dehydrogenase, NADH: quinone oxidoreductase	Regulation of mitochondrial metabolism	NAD^+-dependent deacetylase
Sirtuin 4	35.2	Mitochondria	Pancreatic β-cells, brain, liver, kidney, and heart	Glutamate dehydrogenase	Regulation of mitochondrial metabolism	ADP-ribosyl transferase
Sirtuin 5	33.9	Mitochondria			Apoptosis	NAD^+-dependent

			Brain, testis, heart muscle, and lymphoblast	Cytochrome *c* Carbamoyl phosphate synthetase 1		deacetylase Desuccinylase Demalonylase
Sirtuin 6	39.1	Nucleus	Bone cells (absent in bone marrow), heart, muscle and ovary	Histone H3 TNF-α PAPR1	DNA repair and genomic stability	NAD^+-dependent deacetylase ADP-ribosyl transferase
Sirtuin 7	44.8	Nucleus	Peripheral blood cells, $CD33^+$ myeloid bone marrow precursor cells	RNA polymerase- I p53	Regulation of rRNA transcription, cell cycle regulation	NAD^+-dependent deacetylase

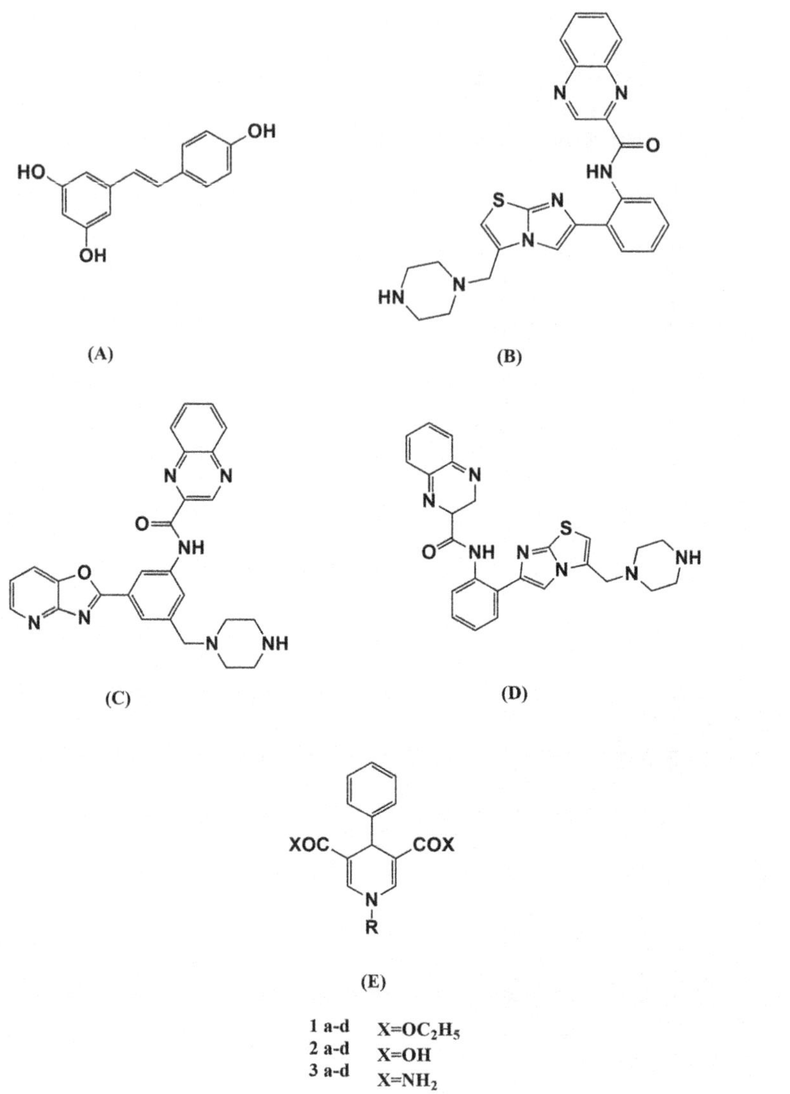

Fig. 15.1 Sirtuin activators. (**a**) Resveratrol (3,5,4′-Trihydroxy-*trans*-stilbene) (**b**) SRT1720; (**c**) Oxazolo[4,5-b]pyridines derivative; (**d**) Imidazo[1,2-b]thiazole derivatives; (**e**) 1,4-Dihydropyridine (DHP) derivatives are a few examples

activating sirtuins is resveratrol, a polyphenol comprising two phenyl rings split apart by a methylene bridge (Fig. 15.1a) (Howitz et al. 2003; Wood et al. 2004).

RSV significantly reduced ageing symptoms without altering the pattern of expression of any sirtuin genetic factor and mimicked CR-induced gene expression patterns in several organs (Barger et al. 2008; Pearson et al. 2008). RSV's bioavailability and pharmacokinetics have been updated and reported in the past (Alcain and Villalba 2009; Vang et al. 2011). It has been shown that RSV, in in vivo testing to significantly slow down the ageing process in mice fed a diet loaded with calories (Baur et al. 2006). The occurrences of cataracts and albuminuria are reduced, vascular endothelium inflammation and apoptosis are decreased, aortic flexibility is increased, motor coordination is enhanced, and bone mineral density is maintained (Pearson et al. 2008). RSV's limited bioavailability has led to modifications to boost bioavailability. A nutraceutical product called resVida, which contains 150 mg of resveratrol daily, has demonstrated success in lowering intrahepatic lipid levels, moving glucose, fatty acids, alanine-aminotransferase, or signs of inflammation in healthy obese males. These outcomes are identical to CRs (Timmers et al. 2011).

Severe dose-sensitive improvements in vasodilation mediated by the endothelium were also seen by oral RSV supplementation, and these improvements were associated with higher plasma RSV concentrations (Wong et al. 2011). After 3 months of treatment, RSV administered when nutritional preparation Longevinex® increased dilatation mediated by flow, but 3 months later Longevinex® was stopped. This option went back to its default value. The medication had no effect on inflammatory markers, lipid profiles, blood pressure, insulin resistance, or any of these parameters (Fujitaka et al. 2011).

15.3.1.2 Sirtuin Activators Structurally Unrelated to Resveratrol

The hunt for new compounds that may activate sirtuins more efficiently than RSV has gained attention because sirtuins are important for maintaining metabolic activities and disorders linked to ageing. Milne et al., in 2007 found small molecule SIRT1 activators that are structurally different from RSV but 100 times more powerful.

The most efficient substance was SRT1720, which at 10 M increased SIRT1 activity by 750%. Due to additive SIRT1 activation caused by medication combination, it was shown that SRT1720 (Fig. 15.1b) binds to and activates the enzyme at the same molecular site as RSV. The key domain's nitrogen-terminal protein sequences 183–225 were significant in identifying the chemical attaching region.

In vivo and in vitro, SRT1720 boosted the deacetylation of SIRT1 substrates such p53, the transcriptional co-activator PGC-1α, and Foxo1a. Innately overweight mice (Lep ob/ob), DIO mice, and the Zucker fa/fa rat were used as in vivo disease models to examine the healing potential of SIRT1720 to cure insulin resistance and type-2 diabetes (Smith et al. 2009).

SRT1720 treatment significantly decreased fasting blood glucose in mice on a high-fat diet to levels close to normal, moderately controlling raised insulin amount, and protected mice from DIO and insulin resistance by improving oxidative metabolism in skeletal muscle, the liver, and brown adipose tissue through a general metabolic change that mimicked low energy levels (Feige et al. 2008).

Mice fed a diet rich in fat received SRT1720, which significantly dropped their fasting blood glucose levels to levels that were almost normal, partially normalised their increased insulin levels, and reduced their feeding glucose levels (Messa et al.

2020). By increasing oxidative metabolism in skeletal muscle, the liver, and brown adipose tissue through a general metabolic adaptation imitating low energy levels, these results effectively protected mice from DIO and insulin resistance (Feige et al. 2008).

Novel small molecule SIRT1 stimulants that are distinct from RSV have been identified, including a series of **Oxazolo[4,5-b]pyridines** (Fig. 15.1c). Additionally, a series of **imidazo(1,2-b)thiazole** derivatives bearing an oxazolopyridine core have been synthesised, and they may represent novel healing aims for the treatment of several illnesses.

Compound 29 (Fig. 15.1d), the least effective homologue in this generation, demonstrated oral antidiabetic activity for three distinct forms of type-2 diabetes and had a significant oral bioavailability in mouse and rat models of type-2 diabetes. In the DIO model after 2 weeks of dosage, in the genetic Zucker fa/fa rat model after 3 weeks, and in the ob/ob deficient mice after just 1 week of therapy, an intake of 100 mg/kg of compound 29 administered per day resulted in a considerable drop in fasting blood glucose content. Compound 29 was capable to control a noticeably minor amount of administered insulin and glucose in the DIO mice even after 10 weeks of dosage without having any effects on body weight, overall clinical chemistry, or haematology (Vu et al. 2009) (Fig. 15.2).

15.3.2 Sirtuin Blockers

Entirely sirtuins comprise the co-factor nicotinamide adenine dinucleotide (NAD^+) (Landry et al. 2000). The acetyl-peptide remains bound by ADP-ribose, which causes the creation of a 0-alkylamidate intermediate, in the first stage of the suggested reaction process, which involves the cleavage of nicotinamide (NIC) from NAD^+.

NIC is a powerful process product inhibitor due to its potential to re-bind the enzyme's intermediate form the 0-alkylamidate and target the intermediate (Sauve and Schramm 2003). The NIC promotes megakaryocyte maturation and ploidy, in part, by inhibiting SIRT1 and by enhancing p53 binding to the NIC consensus DNA binding sequence (Giammona et al. 2009).

Due to up-regulated SIRT1 being reported in cancer cell lines, sirtuin inhibitors may also be effective as therapeutic drugs. That opens up the possibility of sirtuin inhibition can restrict the tumorigenesis. In various animal models, increased SIRT1 activity was discovered to be advantageous, which stimulated the creation of pharmacological sirtuin activators, perhaps as CR mimics. Sirtuin blockers are also suggested for the therapy for other condition like Parkinson's disease, leishmaniosis, or HIV in along with treatment of cancer (Pagans et al. 2005). It is also crucial to take into account that SIRT3, SIRT4, and SIRT5 are found in the mitochondria, wherever many mitochondrial proteins undergo run of acetylation/deacetylation (Verdin et al. 2010) processes crucial for energy utilisation and apoptotic the beginning, as well as in the situation of certain illnesses such as cancer and the metabolic disorder (Pereira et al. 2012).

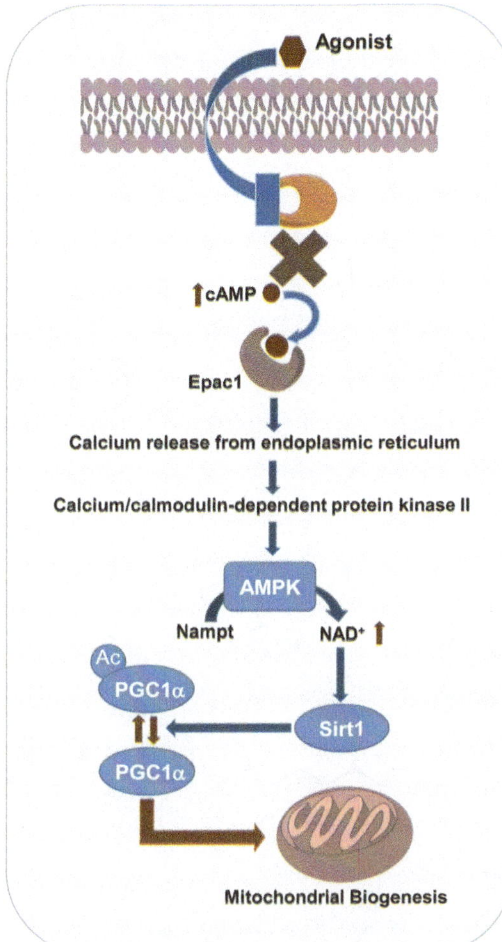

Fig. 15.2 Sirt1 activation pathway by resveratrol

High-throughput and computational analysis screens identified sirtuin inhibitors for SIRT1, SIRT2, SIRT3, and SIRT5, in contrast to the activators sector, where only SIRT1 activators have been discovered (Villalba and Alcaín 2012). The majority of sirtuin inhibitors that have been studied till now, only SIRT1 and/or SIRT2 have been inhibited; however, a few of them have lesser affinity inhibitory effects on SIRT3 and SIRT5.

15.3.2.1 Splitomicin and Its Derivatives

Bedalov et al., discovered a substance called splitomicin (Fig. 15.3a) that produces a restricted phenocopy of a Sir2 deletions variant in *Saccharomyces cerevisiae*, provide a fresh method to study the vital function of Sir2 in vivo. IC50 for splitomicin's inhibition of Sir2 is 60 μM (Bedalov et al. 2001). Splitomicin's effects on human SIRT1 were, however, only moderately inhibited.

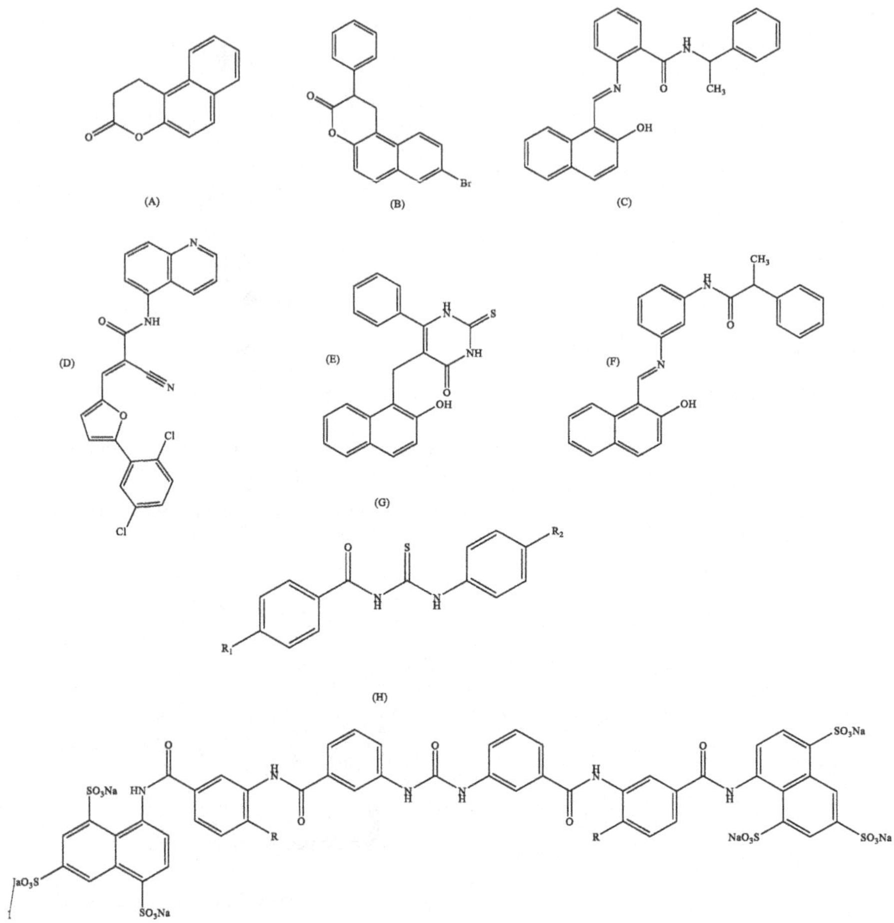

Fig. 15.3 Sirtuin inhibitors. (**a**) Splitomicin, (**b**) HR73, (**c**) sirtinol, (**d**) AGK2, (**e**) cambinol, (**f**) salermide, (**g**) tenovin, (**h**) Suramin

One of the two β-phenylsplitomicin stereoisomers was clearly preferred, as shown by docking and free energy computing, and a connection between enhanced enzyme inhibition and antiproliferative action in MCF-7 breast cancer cells was identified (Neugebauer et al. 2008).

By heterochromatinizing the regulator and silencing the gene, SIRT1 contributes significantly to the amplification of the CGG·CCG-repeat tract that causes Fragile X syndrome, one of maximum prevalent genetic form of psychological illness. Its interesting to note that the Fragile X intellectual disabilities syndrome gene silence is reduced by splitomicin-mediated regulation of SIRT1 activity (Biacsi et al. 2008).

The HIV Tat protein is controlled by process of acetylation and deacetylation. Through its acetyl group and the bromodomain of PCAF, acetylated Tat that is linked to the expanding polymerase attracts the transcriptional coactivator PCAF. Tat's separation from the expanding polymerase and PCAF might result from SIRT1's deacetylation of the protein. A splitomicin analogue known as HR73 (Fig. 15.3b) was shown by Pagans et al. to decrease SIRT1 functioning in vitro with an IC50 value less than 5 μM, confirming SIRT1 as a new therapeutic aim for HIV infection (Pagans et al. 2005).

The Tat protein of the HIV virus is controlled by processes of acetylation and deacetylation (Chen et al. 2020). Through the polymerases elongating domain and the bromodomain of PCAF, acetylated Tat attracts the transcriptional coactivator PCAF. Tat's separation from the elongating polymerase and PCAF could occur if SIRT1 deacetylates Tat. A splitomicin analogue (Villalba and Alcaín 2012) identified as HR73 (Fig. 15.3b) was discovered by Pagans et al. (2005) and it suppressed SIRT1 action in vitro with an IC50 value of less than 5 μM, establishing SIRT1 as a new HIV infection treatment target (Pagans et al. 2005).

15.3.2.2 Sirtinol

An additional sirtuin antagonist, sirtinol (2-[(2-hydroxy-naphthalen-1-ylmethylene)-amino]-*N*-(1-phenyl-ethyl)-benzamide) (Fig. 15.3c), was also found by Grozinger et al. (2001), in a cell-based screen. This substance decreased both yeast Sir2 and human SIRT2 action in vitro. It was proven that the 2-hydroxyl-1-napthol portion was enough to block (Grozinger et al. 2001). Sirtinol's two derivatives, *m*- and *p*-sirtinol, were two- to tenfold highly effective against human SIRT1 and SIRT-2 than sirtinol (Mai et al. 2005). The activation of SIRT1 may be crucial in fostering cell development. As a result of decrease in apoptosis in consequence of diverse genotoxic stimuli caused by deacetylation of p53, SIRT1 inhibitors like sirtinol have potential to treat cancer (Mirzayans et al. 2017). Sirtinol also led to senescence-like growth arrest in human breast cancer MCF-7 and lung cancer H1299 cells, in addition to raising chemosensitivity to camptothecin and cisplatin in PC3, DU145, and HeLa cells (Carafa et al. 2016). The result of an increase in programmed cell death, this led to a large decrease in viable cells (Kojima et al. 2008; Peck et al. 2010).

It has also been discovered that sirtinol, nicotinamide, and SIRT3 down-regulation reduced cell proliferation and expansion in oral squamous cell carcinoma (OSCC) cell lines in vitro and in vivo, whereas SIRT3 is abundantly expressed comparison to other sirtuins, and triggered apoptosis. Additionally, SIRT3 downregulation increased OSCC cells susceptible to radiotherapy and cis-platin's cytotoxic effects (Alhazzazi et al. 2011).

Oculopharyngeal muscular dystrophy is modeled using nematodes, a condition brought on via polyalanine increase in the nuclear protein PABPN1, sirtinol therapy proved protective, by boosting the dose of Sir2/SIRT1, increased muscle prognosis (Pasco et al. 2010). Furthermore, sirtinol treatment decreased pain and swelling in human superficial microvascular endothelial cells, modulated the appearance of binding molecules and monocyte sticking in main human cutaneous microvascular endothelial cells and induced cell death in Leishmania infantum, greatly reducing this axenic amastigote's in vitro proliferation (Orecchia et al. 2011).

15.3.2.3 AGK2

According to reports, neuroprotection is provided by inhibiting SIRT2 activity. The protein α-synuclein (α-syn) is found in Lewy bodies which are the most prevalent histological characteristic of Parkinson disease (PD). The inhibition of SIRT2 with short interfering RNA or including AGK2 (Fig. 15.3d) prevented the impairment of dopaminergic nerve cells caused on by α-syn toxicity in a Drosophila PD model and lessened the neurotoxicity generated by mutant α-syn in rat primary dopamine-positive neurons (Outeiro et al. 2007).

Treatment with AGK2 increased the amounts of acetylated tubulin, but it also made PC12 cells more susceptible to necrosis without changing autophagy and caused C-6 glioma cells to undergo caspase-3-dependent apoptosis (He et al. 2012; Nie et al. 2011). Additionally, through down-regulated the RNAs necessary for sterol production. In both animal and cell-based models of Huntington's disease (HD), sirt2 inhibition produced neuroprotection (Luthi-Carter et al. 2010). In contrast to the neuroprotective effects of SIRT2 suppression in PD and HD models, pharmacological inhibition of SIRT1/2 by nicotinamide, AGK2, or cambinol increased multiplication in cultivated megakaryocytic cells, boosting acetylation of nucleosomes and p53 (Giammona et al. 2009).

15.3.2.4 Cambinol

The chemically stable molecule cambinol (Fig. 15.3e), which suppresses both SIRT1 and SIRT2 in vitro with IC50 values of 56 and 59 μM, consequently, is known as β-naphtol. It shares a pharmacophore with sirtinol and splitomicin. Cambinol has individual marginal suppression effect (42% suppression at 300 μM) against SIRT5 (Heltweg et al. 2006). Mice showed good tolerance to cambinol and prevented the spread of Burkitt lymphoma xenografts by triggering apoptosis through hyperacetylation of the BCL6 oncoprotein and p53 (Heltweg et al. 2006).

Although the efficacy against SIRT2 improved in vitro when the substituent at the N1-position was utilised or this was not the case for SIRT1, changes to the phenyl ring of cambinol improved activity and increased selectivity for SIRT1, leading to the identification of a variety of SIRT2 selective analogues (Medda et al. 2009).

15.3.2.5 Suramin

Suramin, which has an IC50 value of 22 M, is a more powerful blocker of SIRT5 NAD^+-dependent deacetylase activity than cambinol (Fig. 15.3h). Suramin has a strong inhibitory effect on SIRT1 and SIRT2 (IC50 = 0.297 μM and 1.15 μM, respectively) (Trapp et al. 2007). In addition to its antiproliferative and antiviral properties, the polyanionic naphthylurea known as suramin was first used to treat trypanosomiasis (Perabo and Müller 2005). A stronger inhibition of SIRT5 NAD^+-dependent deacetylase activity than cambinol is suramin (Fig. 15.3h), with an IC50 value of 22 μM. SIRT1 and SIRT2 are both significantly inhibited by suramin (IC50 values of 0.297 μM and 1.15 μM, respectively) (Trapp et al. 2007). Originally used to treat trypanosomiasis, suramin is a polyanionic naphthylurea that also has antiproliferative and antiviral properties (Perabo and Müller 2005).

15.3.2.6 Tenovin

Two SIRT1 inhibitors were discovered by Lain et al. (2008) looking for tiny molecules that might stimulate p53 and stop cancer development utilising a cell-based screening method: tenovin-1 and its more water-soluble counterpart, tenovin-6 (Fig. 15.3g). Both substances inhibited tumour development in vivo at one-digit micromolar doses in vitro, all without producing appreciable over-all toxicity. In chronic myelogenous leukaemia (CML), SIRT1 activation enhances cell survival, and proliferation was linked to the deacetylation of many SIRT1 substrates, including FOXO1, p53, and KU70. Tenovin-6, an inhibitor of SIRT1, was administered to mice to stop the course of the illness (Yuan et al. 2012).

Tenovin-1 and tenovin-6 have recently undergone a series of more water-soluble counterparts that, generally, kept the required biological activity. In the presence of a solution, tenovin-1 analogues take on a preferred shape that includes an intramolecular hydrogen bond necessary for SIRT1 binding. Additionally, When the 4-*tert*-butyl substituent in tenovin-6 was replaced with shorter alkyl chains (4-propyl or 4-iso-propyl substituent), analogues with longer *n*-alkyl chains (4-*n*-butyl or 4-*n*-pentyl substituent) showed hazardous or ineffective in cells (Mccarthy et al. 2012).

15.3.2.7 Salermide

A reverse amide with a strong in vitro inhibitory effect on SIRT1 and SIRT2 is salermide is (*N*-{3-[(2-hydroxy-1-naphthalenylmethylene)-amino]-phenyl2}-phenyl-propionamide) (Fig. 15.3f). At dosages up to 100 μM, salermide was tolerated effectively by mice and induced p53-independent apoptosis in cancer cells but not in healthy cells. This was accomplished by regenerating proapoptotic genes that SIRT1 had been epigenetically suppressed in cancer cells (Lara et al. 2009). A further SIRT1 and SIRT2 inhibitor called sirtinol requires p53, as well as salermide-induced apoptosis, according to a different research utilising breast cancer cell lines and p53-insufficient mice fibroblasts (Peck et al. 2010). The salermide impact in human non-small cell lung cancer cells could be mediated by an increase in death receptor 5 expression (Liu et al. 2012).

15.3.2.8 Other Inhibitors of Human Sirtuins

There are many of other SIRT1 and SIRT2 inhibitors that have been found and thoroughly explored. Tripeptide analogues based on lysine, *N*-thioacetyl lysine found in non-peptides, thiobarbiturates, indole derivatives (EX-527), and other inhibitors with various structural cores are among them.

15.4 Natural Sirtuin Inhibitors and Modulators Beneficial Effects on Health

See Table 15.2.

Table 15.2 Sirtuin inhibitors and modulators

Compound	Medical advantages	Mode of action	References
Fisetin	Against cardiovascular disease and anti-aging	• Senotherapeutic action is demonstrated in mice and human tissue • Lowers the concentrations of p25, the p35 cleavage product of the cyclin-dependent kinase 5 (Cdk5) activator, in the brains of patients with Alzheimer's disease and control subjects • Increases the p25/p35 ratio, which raises p25 concentrations and leads to a dysregulation of Cdk5 activity, which culminates in neuroinflammation and neurodegeneration	Heltweg et al. (2006), Trapp et al. (2007), Perabo and Müller (2005)
	Chemopreventive/ chemotherapeutic agent	Activates caspases • Boosts Bak synthesis and causes its oligomerization in the mitochondria • Akt/mTOR signalling blockers	Lain et al. (2008), Yuan et al. (2012)
	Antioxidant agent	Unknown	Mccarthy et al. (2012)
	Antidiabetic	Reduces glycation of methylglyoxal-dependent proteins	Lara et al. (2009)
Orientin	Anti-inflammatory	• Reduces cytokine production and myeloperoxidase (MPO) activity in rats • Suppresses the translocation of NF-κB p65, the activity of NF-κB-luciferase, and the expression of NF-κB target genes • By blocking TLR4 and deactivating the NF-κB and MAPK pathways, it lessens the severity of experimental inflammatory bowel disease (IBD) in rats	Peck et al. (2010)
	Antioxidant and antiaging	Decreases the β-galactosidase activity generated by H_2O_2	Liu et al. (2012)

(continued)

Table 15.2 (continued)

Compound	Medical advantages	Mode of action	References
	Antiviral and antibacterial agent	• Shows a moderate to strong antiviral response to the para 3 virus • On Hep-2 cells, it was demonstrated that the flavonoid combination, which contains orientin, rutin, quercetin, and kaempferol, completely inhibited Herpes Simplex Virus Type 2 (HSV-2) of various viral titres (1, 10, and 100 TCID50)	Yousefzadeh et al. (2018), Currais et al. (2014), Zhu et al. (2017)
	Anti-inflammatory agent	• HMGB1-mediated cytoskeletal rearrangements as well as lipopolysaccharide (LPS)-induced elevation of the protein level of HMGB1 are both prevented in umbilical vein endothelial cells (HUVECs) • Suppresses the effects of LPS on membrane rupturing, monocyte movement, cell adhesion molecule (CAM) expression, and EPCR detachment	Syed et al. (2013), Lall et al. (2016), Tripathi et al. (2011)
	Anticancer effects	• Controls the expression of the p53 and bcl-2 genes that are involved to apoptosis • Oesophageal cancer	Khan et al. (2013)
	Weight loss	• Prevents the formation of intracellular triglycerides (TG) in mouse adipocyte 3T3-L1 cells • Reduces the mRNA levels of the genes involved in adipogenesis, lipogenesis, lipolysis, and the generation of TG, inhibits glycerol from being released, and inhibits the release of glycerol • Reduces the expression of adipogenic	Maher et al. (2011)

(continued)

Table 15.2 (continued)

Compound	Medical advantages	Mode of action	References
		master transcription factors like C/EBP and PPAR in the early stages of adipogenesis by downregulating the CCAAT/enhancer binding protein (C/EBP) gene	
	Protect bone marrow	Decreases bone marrow cells with chromosomal aberrations	Sun et al. (2016), Nayak and Uma (2006)
	Other (e.g., vasodilatation, cardioprotective, radioprotective, neuroprotective, antidepressant-like, antiadipogenesis, antinociceptive)	• Blocks the production of PPAR γ and C/EBPα proteins in proteins • Reduces the writhing and discomfort that capsaicin and glutamate in mice cause • Acetylsalicylic acid (commonly known as aspirin), a popular analgesic, and indomethacin, a common anti-inflammatory medicine, were shown to be 20 times more powerful and 3.5 times more dynamic, respectively, than orientin	Lin et al. (2004)
Piceatannol	Anticancer effects	• Prevents prostate cancer cells from migrating and invading, potentially through reduced interleukin-6 signalling • Inhibits CSN-associated kinase and COX-1/2	Li et al. (2002), Boominathan et al. (2014)
	Metabolic diseases	Inhibits adipogenesis • Restricts the growth of clonal mitotic • Non-competitive binding to the insulin receptor slows down insulin signalling • Reduces lipid buildup during the last phases of differentiation	Yoo et al. (2014)

(continued)

Table 15.2 (continued)

Compound	Medical advantages	Mode of action	References
	Cardiovascular diseases	• Peroxisome proliferator-activated receptor alpha (PPAR-) isoform is activated on rat hepatoma (H4IIEC3 cells) in vitro • Reduces cholesterol and lipoprotein levels	Bae (2015), An et al. (2015)
Quercetin	Cancer treatment	Tyrosine kinase inhibition	Nagai et al. (2018), Nayak and Devi (2005)
	Colitis and gastric ulcer therapy	Unknown	Uma Devi et al. (1999)
	Treatment of respiratory tract infection	Unknown	Lam et al. (2016)
	Treatment of type 2 diabetes	Increase the antioxidant level of type 2 diabetic individuals	Syed et al. (2013), Kwon et al. (2012)
	Treatment of high blood pressure	Unknown	Kershaw and Kim (2017)
	Treatment of oral lichen planus	• Control of cytokines including IL12, INFγ, INFα, IL8, cyclooxygenase 2, and prostaglandin E	Tang and Chan (2014)
	Other health benefits	Unknown	Rimando et al. (2005)
Resveratrol	Obesity treatment	Modulation of sirtuin	Yuan et al. (2006)
	Colon cancer prevention	Inhibits signaling pathway involved in colon cancer initiation	Ferry et al. (1996)
	Cardioprotection	Reduce low-density lipoprotein (LDL) oxidation	Hamdy and Ibrahem (2010)
Trans-(−)-ε-Viniferin	Anti-inflammatory, antioxidant, platelet antiaggregatory, and anticarcinogenic properties	Monoamine oxidase activity (MOA)	Heinz et al. (2010)
	Rotaviral diarrhoea	Intestine's calcium-activated chloride channel is blocked	Mazloom et al. (2014)
	Alzheimer's disease	Induces the disaggregation of amyloid β (Aβ) peptide	Zahedi et al. (2013)

(continued)

Table 15.2 (continued)

Compound	Medical advantages	Mode of action	References
	Diabetes	• Maintenance of Ca^{2+} and preservation of mitochondrial membrane potential (MMP) to prevent high glucose-induced apoptosis	Rezvan et al. (2017)
	Anticancer effects	• Controls the expression of the apoptosis-related genes p53 and bcl-2 • Potential medicinal agents for oesophageal cancer therapy • Specifically targeting U937 cell apoptosis	Serban et al. (2016), Amirchaghmaghi et al. (2015)
	Anti-oxidant effects	Effects of antioxidants reduces the action of the antioxidant enzymes in cells and the sulfhydryl in the protein of the red cell membrane, as well as oxygen free radicals	Miles et al. (2014), Poulsen et al. (2013)
Vitexin	Anti-inflammatory effects	• Prevent IL-1β, IL-6, IL-8, IL-17, and IL-33 • Prevent tumour necrosis factor-α (TNF-α) secretion • Prevent COX-2 • Prevent NF-κB activation • Prevent iNOS (inducible nitric oxide synthase) • Prevent NO, PGE2, monocyte chemoattractant protein-1 (MCP-1), and neutrophil influx • Increase in IL-10 and reduce the expression of p-p38, p-ERK and p-JNK	Nguyen et al. (2009), Magyar et al. (2012), Yáñez et al. (2006)
	Anti-neoplastic effects	Promotes autophagy	Yu et al. (2018)
	Anti-microbial and anti-viral effects	• Activity against anti-*H. pylori* • Anti-phytoviral activity against Tobacco mosaic virus	Vion et al. (2018), Zhao et al. (2016)

15.5 Conclusions

Sirtuins are a significant group of histone deacetylases that take part in NAD^+-dependent deacetylation processes. Activation, inhibition, and regulation of sirtuins are engaged in a variety of key metabolic processes that are connected to conditions including type 2 diabetes, ageing, and inflammation. As a result, they have anti-cancer, anti-oxidant, anti-microbial, and anti-viral effects and are a significant drug target class. Natural products are renowned for being significant sources of lead compounds. In this study, we've made an effort to give a quick rundown of the most current findings about natural products that have been found to interact with sirtuins.

Natural sirtuin modulators and inhibitors may have positive effect on health in addition to well-recognised mechanisms of action and individuals being researched in clinical studies. Alkaloids, bichalcones, resveratrol, xanthone, tanikolide, and flavonoids are a few examples of natural substances and chemical types that have been demonstrated to have actions against sirtuins. The next generation sirtuins modulators are anticipated to be natural product derivates beginning from the discovered lead compounds.

References

Alcain FJ, Villalba JM (2009) Sirtuin activators. Expert Opin Ther Pat 19(4):403–414

Alhazzazi TY, Kamarajan P, Joo N, Huang JY, Verdin E, D'Silva NJ, Kapila YL (2011) Sirtuin-3 (SIRT3), a novel potential therapeutic target for oral cancer. Cancer 117(8):1670–1678

Alvarez R, Altucci L, Gronemeyer H, de Lera AR (2011) Epigenetic multiple modulators. Curr Top Med Chem 11(22):2749–2787

Amirchaghmaghi M, Delavarian Z, Iranshahi M, Shakeri MT, Mosannen Mozafari P, Mohammadpour AH, Farazi F, Iranshahy M (2015) A randomized placebo-controlled double blind clinical trial of quercetin for treatment of oral lichen planus. J Dent Res Dent Clin Dent Prospects 9:23–28

An F, Wang S, Tian Q, Zhu D (2015) Effects of orientin and vitexin from Trollius chinensis on the growth and apoptosis of esophageal cancer EC-109 cells. Oncol Lett 10:2627–2633

Anekonda TS, Reddy PH (2006) Neuronal protection by sirtuins in Alzheimer's disease. J Neurochem 96(2):305–313

Bae JS (2015) Inhibitory effect of orientin on secretory group IIA phospholipase A2. Inflammation 38:1631–1638

Barger JL, Kayo T, Pugh TD, Prolla TA, Weindruch R (2008) Short-term consumption of a resveratrol-containing nutraceutical mixture mimics gene expression of long-term caloric restriction in mouse heart. Exp Gerontol 43(9):859–866

Baur JA, Pearson KJ, Price NL, Jamieson HA, Lerin C, Kalra A, Prabhu VV, Allard JS, Lopez-Lluch G, Lewis K, Pistell PJ (2006) Resveratrol improves health and survival of mice on a high-calorie diet. Nature 444(7117):337–342

Baur JA, Ungvari Z, Minor RK, Le Couteur DG, De Cabo R (2012) Are sirtuins viable targets for improving healthspan and lifespan? Nat Rev Drug Discov 11(6):443–461

Bedalov A, Gatbonton T, Irvine WP, Gottschling DE, Simon JA (2001) Identification of a small molecule inhibitor of Sir2p. Proc Natl Acad Sci U S A 98(26):15113–15118

Biacsi R, Kumari D, Usdin K (2008) SIRT1 inhibition alleviates gene silencing in Fragile X mental retardation syndrome. PLoS Genet 4(3):e1000017

Bonda DJ, Lee HG, Camins A, Pallàs M, Casadesus G, Smith MA, Zhu X (2011) The sirtuin pathway in ageing and Alzheimer disease: mechanistic and therapeutic considerations. Lancet Neurol 10(3):275–279

Boominathan SP, Sarangan G, Srikakulapu S, Rajesh S, Duraipandian C, Srikanth P, Jo L (2014) Antiviral activity of bioassay guided fractionation of Plumbago zeylanica roots against herpes simplex virus type 2. World J Pharm Sci 3:1003–1017

Carafa V, Rotili D, Forgione M, Cuomo F, Serretiello E, Hailu GS, Jarho E, Lahtela-Kakkonen M, Mai A, Altucci L (2016) Sirtuin functions and modulation: from chemistry to the clinic. Clin Epigenetics 8:1–21

Chen J, Liu Q, Zeng L, Huang X (2020) Protein acetylation/deacetylation: a potential strategy for fungal infection control. Front Microbiol 11:574736

Cohen HY, Miller C, Bitterman KJ, Wall NR, Hekking B, Kessler B, Howitz KT, Gorospe M, de Cabo R, Sinclair DA (2004) Calorie restriction promotes mammalian cell survival by inducing the SIRT1 deacetylase. Science 305(5682):390–392

Currais A, Prior M, Dargusch R, Armando A, Ehren J, Schubert D, Quehenberger O, Maher P (2014) Modulation of p25 and inflammatory pathways by fisetin maintains cognitive function in Alzheimer's disease transgenic mice. Aging Cell 13:379–390

Dai H, Sinclair DA, Ellis JL, Steegborn C (2018) Sirtuin activators and inhibitors: promises, achievements, and challenges. Pharmacol Ther 188:140–154

Dali-Youcef N, Lagouge M, Froelich S, Koehl C, Schoonjans K, Auwerx J (2007) Sirtuins: the 'magnificent seven', function, metabolism and longevity. Ann Med 39(5):335–345. https://doi.org/10.1080/07853890701408194

Feige JN, Lagouge M, Canto C, Strehle A, Houten SM, Milne JC, Lambert PD, Mataki C, Elliott PJ, Auwerx J (2008) Specific SIRT1 activation mimics low energy levels and protects against diet-induced metabolic disorders by enhancing fat oxidation. Cell Metab 8(5):347–358

Ferry DR, Smith A, Malkhandi J, Fyfe DW, deTakats PG, Anderson D, Baker J, Kerr DJ (1996) Phase I clinical trial of the flavonoid quercetin: pharmacokinetics and evidence for in vivo tyrosine kinase inhibition. Clin Cancer Res 2:659–668

Fiorentino F, Castiello C, Mai A, Rotili D (2022a) Therapeutic potential and activity modulation of the protein lysine deacylase sirtuin 5. J Med Chem 65(14):9580–9606

Fiorentino F, Mautone N, Menna M, D'Acunzo F, Mai A, Rotili D (2022b) Sirtuin modulators: past, present, and future perspectives. Future Med Chem 14(12):915–939

Fujitaka K, Otani H, Jo F, Jo H, Nomura E, Iwasaki M, Nishikawa M, Iwasaka T, Das DK (2011) Modified resveratrol Longevinex improves endothelial function in adults with metabolic syndrome receiving standard treatment. Nutr Res 31(11):842–847

Giammona LM, Panuganti S, Kemper JM, Apostolidis PA, Lindsey S, Papoutsakis ET, Miller WM (2009) Mechanistic studies on the effects of nicotinamide on megakaryocytic polyploidization and the roles of NAD+ levels and SIRT inhibition. Exp Hematol 37(11):1340–1352

Grozinger CM, Chao ED, Blackwell HE, Moazed D, Schreiber SL (2001) Identification of a class of small molecule inhibitors of the sirtuin family of NAD-dependent deacetylases by phenotypic screening. J Biol Chem 276(42):38837–38843

Hamdy AA, Ibrahem MA (2010) Management of aphthous ulceration with topical quercetin: a randomized clinical trial. J Contemp Dent Pract 11:E009–E016

He X, Nie H, Hong Y, Sheng C, Xia W, Ying W (2012) SIRT2 activity is required for the survival of C6 glioma cells. Biochem Biophys Res Commun 417(1):468–472

Heinz SA, Henson DA, Austin MD, Jin F, Nieman DC (2010) Quercetin supplementation and upper respiratory tract infection: a randomized community clinical trial. Pharm Res 62:237–242

Heltweg B, Gatbonton T, Schuler AD, Posakony J, Li H, Goehle S, Kollipara R, DePinho RA, Gu Y, Simon JA, Bedalov A (2006) Antitumor activity of a small-molecule inhibitor of human silent information regulator 2 enzymes. Cancer Res 66(8):4368–4377

Howitz KT, Bitterman KJ, Cohen HY, Lamming DW, Lavu S, Wood JG, Zipkin RE, Chung P, Kisielewski A, Zhang LL, Scherer B (2003) Small molecule activators of sirtuins extend Saccharomyces cerevisiae lifespan. Nature 425(6954):191–196

Kershaw J, Kim K-H (2017) The therapeutic potential of piceatannol, a natural stilbene, in metabolic diseases: a review. J Med Food 20:427–438

Khan N, Syed DN, Ahmad N, Mukhtar H (2013) Fisetin: a dietary antioxidant for health promotion. Antioxid Redox Signal 19:151–162

Kojima K, Ohhashi R, Fujita Y, Hamada N, Akao Y, Nozawa Y, Deguchi T, Ito M (2008) A role for SIRT1 in cell growth and chemoresistance in prostate cancer PC3 and DU145 cells. Biochem Biophys Res Commun 373(3):423–428

Kwon GT, Jung JI, Song HR, Woo EY, Jun JG, Kim JK, Her S, Park JH (2012) Piceatannol inhibits migration and invasion of prostate cancer cells: possible mediation by decreased interleukin-6 signaling. J Nutr Biochem 23:228–238

Lain S, Hollick JJ, Campbell J, Staples OD, Higgins M, Aoubala M, McCarthy A, Appleyard V, Murray KE, Baker L, Thompson A (2008) Discovery, in vivo activity, and mechanism of action of a small-molecule p53 activator. Cancer Cell 13(5):454–463

Lall RK, Adhami VM, Mukhtar H (2016) Dietary flavonoid fisetin for cancer prevention and treatment. Mol Nutr Food Res 60:1396–1405

Lam KY, Ling AP, Koh RY, Wong YP, Say YH (2016) A review on medicinal properties of orientin. Adv Pharm Sci 2016:4104595

Landry J, Sutton A, Tafrov ST, Heller RC, Stebbins J, Pillus L, Sternglanz R (2000) The silencing protein SIR2 and its homologs are NAD-dependent protein deacetylases. Proc Natl Acad Sci U S A 97(11):5807–5811. https://doi.org/10.1073/pnas.110148297

Lara E, Mai A, Calvanese V, Altucci L, Lopez-Nieva P, Martinez-Chantar ML, Varela-Rey M, Rotili D, Nebbioso A, Ropero S, Montoya G (2009) Salermide, a Sirtuin inhibitor with a strong cancer-specific proapoptotic effect. Oncogene 28(6):781–791

Li YL, Ma SC, Yang YT, Ye SM, But PP (2002) Antiviral activities of flavonoids and organic acid from Trollius chinensis Bunge. J Ethnopharmacol 79:365–368

Lin SJ, Defossez PA, Guarente L (2000) Requirement of NAD and SIR2 for life-span extension by calorie restriction in Saccharomyces cerevisiae. Science 289(5487):2126–2128

Lin Q, Feng S, Cen Y, Yang Y, Wang L (2004) Study on the antibacterial and antiviral activity compositions of Trollium chinensis Bunge. J Zhejiang Univ (Sci Ed) 31:412–415

Liszt G, Ford E, Kurtev M, Guarente L (2005) Mouse Sir2 homolog SIRT6 is a nuclear ADP-ribosyltransferase. J Biol Chem 280(22):21313–21320. https://doi.org/10.1074/jbc.m413296200

Liu G, Su L, Hao X, Zhong N, Zhong D, Singhal S, Liu X (2012) Salermide up-regulates death receptor 5 expression through the ATF4-ATF3-CHOP axis and leads to apoptosis in human cancer cells. J Cell Mol Med 16(7):1618–1628

Luthi-Carter R, Taylor DM, Pallos J, Lambert E, Amore A, Parker A, Moffitt H, Smith DL, Runne H, Gokce O, Kuhn A (2010) SIRT2 inhibition achieves neuroprotection by decreasing sterol biosynthesis. Proc Natl Acad Sci U S A 107(17):7927–7932

Magyar K, Halmosi R, Palfi A, Feher G, Czopf L, Fulop A, Battyany I, Sumegi B, Toth K, Szabados E (2012) Cardioprotection by resveratrol: a human clinical trial in patients with stable coronary artery disease. Clin Hemorheol Microcirc 50:179–187

Maher P, Dargusch R, Ehren JL, Okada S, Sharma K, Schubert D (2011) Fisetin lowers methylglyoxal dependent protein glycation and limits the complications of diabetes. PLoS One 6:e21226

Mai A, Massa S, Lavu S, Pezzi R, Simeoni S, Ragno R, Mariotti FR, Chiani F, Camilloni G, Sinclair DA (2005) Design, synthesis, and biological evaluation of sirtinol analogues as class III histone/protein deacetylase (Sirtuin) inhibitors. J Med Chem 48(24):7789–7795

Mazloom Z, Abdollahzadeh SM, Dabbaghmanesh MH, Rezaianzadeh A (2014) The effect of quercetin supplementation on oxidative stress, glycemic control, lipid profile and insulin resistance in type 2 diabetes: a randomized clinical trial. J Health Sci Surveill Syst 2:8–14

Mccarthy AR, Pirrie L, Hollick JJ, Ronseaux S et al (2012) Synthesis and biological characterisation of sirtuin inhibitors based on the tenovins. Bioorg Med Chem 20:1779–1793

Medda F, Russell RJ, Higgins M, McCarthy AR, Campbell J, Slawin AM, Lane DP, Lain S, Westwood NJ (2009) Novel cambinol analogs as sirtuin inhibitors: synthesis, biological evaluation, and rationalization of activity. J Med Chem 52(9):2673–2682

Mellini P, Valente S, Mai A (2015) Sirtuin modulators: an updated patent review (2012–2014). Expert Opin Ther Pat 25(1):5–15

Messa GA, Piasecki M, Hurst J, Hill C, Tallis J, Degens H (2020) The impact of a high-fat diet in mice is dependent on duration and age, and differs between muscles. J Exp Biol 223(6): jeb217117

Miles SL, McFarland M, Niles RM (2014) Molecular and physiological actions of quercetin: need for clinical trials to assess its benefits in human disease. Nutr Rev 72:720–734

Milne JC, Lambert PD, Schenk S, Carney DP, Smith JJ, Gagne DJ, Jin L, Boss O, Perni RB, Vu CB, Bemis JE (2007) Small molecule activators of SIRT1 as therapeutics for the treatment of type 2 diabetes. Nature 450(7170):712–716

Mirzayans R, Andrais B, Kumar P, Murray D (2017) Significance of wild-type p53 signaling in suppressing apoptosis in response to chemical genotoxic agents: impact on chemotherapy outcome. Int J Mol Sci 18(5):928

Mohamad Nasir NF, Zainuddin A, Shamsuddin S (2018) Emerging roles of Sirtuin 6 in Alzheimer's disease. J Mol Neurosci 64:157–161

Nagai S, Matsumoto C, Shibano M, Fujimori K (2018) Suppression of fatty acid and triglyceride synthesis by the flavonoid orientin through decrease of C/EBPδ expression and inhibition of PI3K/Akt-FOXO1 signaling in adipocytes. Nutrients 10:130

Nayak V, Devi PU (2005) Protection of mouse bone marrow against radiation-induced chromosome damage and stem cell death by the ocimum flavonoids orientin and vicenin. Radiat Res 163:165–171

Nayak V, Uma P (2006) Antioxidant and radioprotective effects of ocimum flavonoids orientin and vicenin in Escherichia coli. Def Sci J 56:179

Neugebauer RC, Uchiechowska U, Meier R, Hruby H, Valkov V, Verdin E, Sippl W, Jung M (2008) Structure-activity studies on splitomicin derivatives as sirtuin inhibitors and computational prediction of binding mode. J Med Chem 51(5):1203–1213

Nguyen AV, Martinez M, Stamos MJ, Moyer MP, Planutis K, Hope C, Holcombe RF (2009) Results of a phase I pilot clinical trial examining the effect of plant-derived resveratrol and grape powder on Wnt pathway target gene expression in colonic mucosa and colon cancer. Cancer Manag Res 1:25–37

Nie H, Chen H, Han J, Hong Y, Ma Y, Xia W, Ying W (2011) Silencing of SIRT2 induces cell death and a decrease in the intracellular ATP level of PC12 cells. Int J Physiol Pathophysiol Pharmacol 3(1):65

Orecchia A, Scarponi C, Di Felice F, Cesarini E, Avitabile S, Mai A, Mauro ML, Sirri V, Zambruno G, Albanesi C, Camilloni G (2011) Sirtinol treatment reduces inflammation in human dermal microvascular endothelial cells. PLoS One 6(9):e24307

Outeiro TF, Kontopoulos E, Altmann SM, Kufareva I et al (2007) Sirtuin 2 inhibitors rescue alpha-synuclein-mediated toxicity in models of Parkinson's disease. Science 317:516–519

Outeiro TF, Marques O, Kazantsev A (2008) Therapeutic role of sirtuins in neurodegenerative disease. Biochim Biophys Acta Mol Basis Dis 1782(6):363–369

Pagans S, Pedal A, North BJ, Kaehlcke K, Marshall BL, Dorr A, Hetzer-Egger C, Henklein P, Frye R, McBurney MW, Hruby H (2005) SIRT1 regulates HIV transcription via Tat deacetylation. PLoS Biol 3(2):e41

Pasco MY, Rotili D, Altucci L, Farina F, Rouleau GA, Mai A, Neri C (2010) Characterization of sirtuin inhibitors in nematodes expressing a muscular dystrophy protein reveals muscle cell and behavioral protection by specific sirtinol analogues. J Med Chem 53(3):1407–1411

Pearson KJ, Baur JA, Lewis KN, Peshkin L, Price NL, Labinskyy N, Swindell WR, Kamara D, Minor RK, Perez E, Jamieson HA (2008) Resveratrol delays age-related deterioration and mimics transcriptional aspects of dietary restriction without extending life span. Cell Metab 8(2):157–168

Peck B, Chen CY, Ho KK, Di Fruscia P, Myatt SS, Coombes RC, Fuchter MJ, Hsiao CD, Lam EW (2010) SIRT inhibitors induce cell death and p53 acetylation through targeting both SIRT1 and SIRT2SIRT inhibitors target SIRT1/2 to activate p53. Mol Cancer Ther 9(4):844–855
Perabo FG, Müller SC (2005) New agents in intravesical chemotherapy of superficial bladder cancer. Scand J Urol Nephrol 39(2):108–116
Pereira CV, Lebiedzinska M, Wieckowski MR, Oliveira PJ (2012) Regulation and protection of mitochondrial physiology by sirtuins. Mitochondrion 12(1):66–76
Poulsen MM, Vestergaard PF, Clasen BF, Radko Y, Christensen LP, Stødkilde-Jørgensen H, Møller N, Jessen N, Pedersen SB, Jørgensen JO (2013) High-dose resveratrol supplementation in obese men: an investigator-initiated, randomized, placebo-controlled clinical trial of substrate metabolism, insulin sensitivity, and body composition. Diabetes 62:1186–1195
Rezvan N, Moini A, Janani L, Mohammad K, Saedisomeolia A, Nourbakhsh M, Gorgani-Firuzjaee S, Mazaherioun M, Hosseinzadeh-Attar MJ (2017) Effects of quercetin on adiponectin-mediated insulin sensitivity in polycystic ovary syndrome: a randomized placebo-controlled double-blind clinical trial. Horm Metab Res 49:115–121
Rimando AM, Nagmani R, Feller DR, Yokoyama W (2005) Pterostilbene, a new agonist for the peroxisome proliferator-activated receptor alpha-isoform, lowers plasma lipoproteins and cholesterol in hypercholesterolemic hamsters. J Agric Food Chem 53:3403–3407
Rogina B, Helfand SL (2004) Sir2 mediates longevity in the fly through a pathway related to calorie restriction. Proc Natl Acad Sci U S A 101(45):15998–16003
Sauve AA, Schramm VL (2003) Sir2 regulation by nicotinamide results from switching between base exchange and deacetylation chemistry. Biochemistry 42(31):9249–9256
Schutkowski M, Fischer F, Roessler C, Steegborn C (2014) New assays and approaches for discovery and design of Sirtuin modulators. In: Expert opinion on drug discovery, vol 9(2). Informa UK Limited, pp 183–199. https://doi.org/10.1517/17460441.2014.875526
Serban M-C, Sahebkar A, Zanchetti A, Mikhailidis DP, Howard G, Antal D, Andrica F, Ahmed A, Aronow WS, Muntner P et al (2016) Effects of quercetin on blood pressure: a systematic review and meta-analysis of randomized controlled trials. J Am Heart Assoc 5:e002713
Smith JJ, Kenney RD, Gagne DJ, Frushour BP, Ladd W, Galonek HL, Israelian K, Song J, Razvadauskaite G, Lynch AV, Carney DP (2009) Small molecule activators of SIRT1 replicate signaling pathways triggered by calorie restriction in vivo. BMC Syst Biol 3(1):1–4
Sun A, Ren G, Deng C, Zhang J, Luo X, Wu X, Mani S, Dou W, Wang Z (2016) C-glycosyl flavonoidorientin improves chemically induced inflammatory bowel disease in mice. J Funct Foods 21:418–430
Syed DN, Adhami VM, Khan MI, Mukhtar H (2013) Inhibition of Akt/mTOR signaling by the dietary flavonoid fisetin. Anti Cancer Agents Med Chem 13:995–1001
Tang Y-L, Chan S-W (2014) A review of the pharmacological effects of piceatannol on cardiovascular diseases. Phytother Res 28:1581–1588
Timmers S, Konings E, Bilet L, Houtkooper RH, van de Weijer T, Goossens GH, Hoeks J, van der Krieken S, Ryu D, Kersten S, Moonen-Kornips E (2011) Calorie restriction-like effects of 30 days of resveratrol supplementation on energy metabolism and metabolic profile in obese humans. Cell Metab 14(5):612–622
Trapp J, Meier R, Hongwiset D, Kassack MU, Sippl W, Jung M (2007) Structure–activity studies on suramin analogues as inhibitors of NAD+-dependent histone deacetylases (sirtuins). ChemMedChem 2(10):1419–1431
Tripathi R, Samadder T, Gupta S, Surolia A, Shaha C (2011) Anticancer activity of a combination of cisplatin and fisetin in embryonal carcinoma cells and xenograft tumors. Mol Cancer 10:255–268
Uma Devi P, Ganasoundari A, Rao BS, Srinivasan KK (1999) In vivo radioprotection by ocimum flavonoids: survival of mice. Radiat Res 151:74–78
Valente S, Mellini P, Spallotta F, Carafa V, Nebbioso A, Polletta L, Carnevale I, Saladini S, Trisciuoglio D, Gabellini C, Tardugno M (2016) 1,4-Dihydropyridines active on the SIRT1/

AMPK pathway ameliorate skin repair and mitochondrial function and exhibit inhibition of proliferation in cancer cells. J Med Chem 59(4):1471–1491
Vang O, Ahmad N, Baile CA, Baur JA, Brown K, Csiszar A, Das DK, Delmas D, Gottfried C, Lin HY, Ma QY (2011) What is new for an old molecule? Systematic review and recommendations on the use of resveratrol. PLoS One 6(6):e19881
Verdin E, Hirschey MD, Finley LW, Haigis MC (2010) Sirtuin regulation of mitochondria: energy production, apoptosis, and signaling. Trends Biochem Sci 35(12):669–675
Villalba JM, Alcaín FJ (2012) Sirtuin activators and inhibitors. Biofactors 38(5):349–359
Vion E, Page G, Bourdeaud E, Paccalin M, Guillard J, Rioux Bilan A (2018) Trans ε-viniferin is an amyloid-β disaggregating and anti-inflammatory drug in a mouse primary cellular model of Alzheimer's disease. Mol Cell Neurosci 88:1–6
Vu CB, Bemis JE, Disch JS, Ng PY, Nunes JJ, Milne JC, Carney DP, Lynch AV, Smith JJ, Lavu S, Lambert PD (2009) Discovery of imidazo [1, 2-b] thiazole derivatives as novel SIRT1 activators. J Med Chem 52(5):1275–1283
Wątroba M, Szukiewicz D (2016) The role of sirtuins in aging and age-related diseases. Adv Med Sci 61(1):52–62
Wong RH, Howe PR, Buckley JD, Coates AM, Kunz I, Berry NM (2011) Acute resveratrol supplementation improves flow-mediated dilatation in overweight/obese individuals with mildly elevated blood pressure. Nutr Metab Cardiovasc Dis 21(11):851–856
Wood JG, Rogina B, Lavu S, Howitz K, Helfand SL, Tatar M, Sinclair D (2004) Sirtuin activators mimic caloric restriction and delay ageing in metazoans. Nature 430(7000):686–689
Yáñez M, Fraiz N, Cano E, Orallo F (2006) (−)-Trans-epsilon-viniferin, a polyphenol present in wines, is an inhibitor of noradrenaline and 5-hydroxytryptamine uptake and of monoamine oxidase activity. Eur J Pharmacol 542:54–60
Yeong KY, Berdigaliyev N, Chang Y (2020) Sirtuins and their implications in neurodegenerative diseases from a drug discovery perspective. ACS Chem Neurosci 11(24):4073–4091
Yoo H, Ku SK, Lee T, Bae JS (2014) Orientin inhibits HMGB1-induced inflammatory responses in HUVECs and in murine polymicrobial sepsis. Inflammation 37:1705–1717
Yousefzadeh MJ, Zhu YI, McGowan SJ, Angelini L, Fuhrmann-Stroissnigg H, Xu M, Ling YY, Melos KI, Pirtskhalava T, Inman CL, McGuckian C (2018) Fisetin is a senotherapeutic that extends health and lifespan. EBioMedicine 36:18–28
Yu B, Jiang Y, Zhang B, Yang H, Ma T (2018) Resveratrol dimer trans-ε-viniferin prevents rotaviral diarrhea in mice by inhibition of the intestinal calcium-activated chloride channel. Pharm Res 129:453–461
Yuan Z-P, Chen L-J, Fan L-Y, Tang M-H, Yang G-L, Yang H-S, Du X-B, Wang G-Q, Yao W-X, Zhao Q-M et al (2006) Liposomal quercetin efficiently suppresses growth of solid tumors in murine models. Clin Cancer Res 12:3193–3199
Yuan H, Wang Z, Li L, Zhang H, Modi H, Horne D, Bhatia R, Chen W (2012) Activation of stress response gene SIRT1 by BCR-ABL promotes leukemogenesis. Blood 119(8):1904–1914
Zahedi M, Ghiasvand R, Feizi A, Asgari G, Darvish L (2013) Does quercetin improve cardiovascular risk factors and inflammatory biomarkers in women with type 2 diabetes: a double-blind randomized controlled clinical trial. Int J Prev Med 4:777–785
Zhao H, Ma T, Fan B, Yang L, Han C, Luo J, Kong L (2016) Protective effect of trans-δ-viniferin against high glucose-induced oxidative stress in human umbilical vein endothelial cells through the SIRT1 pathway. Free Radic Res 50:68–83
Zhu Y, Doornebal EJ, Pirtskhalava T, Giorgadze N, Wentworth M, Fuhrmann-Stroissnigg H, Niedernhofer LJ, Robbins PD, Tchkonia T, Kirkland JL (2017) New agents that target senescent cells: the flavone, fisetin, and the BCL-X(L) inhibitors, A1331852 and A1155463. Aging (Albany NY) 9:955–963

Histone Deacetylase Inhibitors: Design, Synthesis, and Biological Evaluation

16

Sravani Pulya, Abhiram Ganorkar, Swati Biswas, and Balaram Ghosh

Abstract

Alzheimer's disease (AD), a neurodegenerative disease, is a chronic form of dementia. AD is mostly prevalent in the elderly and is characterised by the impairment of learning and loss of memory. AD is pathologically characterised by neurofibrillary tangles, amyloid-β (Aβ) plaques with eventual loss of neurons. The abnormal Aβ plaques accumulation leads to the apoptosis of the neurons resulting in the pathogenesis of AD. Though AD is caused by multiple factors, its pathogenesis hasn't been completely studied. Recent studies in this direction found that epigenetic regulation might offer a new way of understanding AD and its pathogenesis. Among various epigenetic processes, targeting histone acetylation was found to be one of the promising treatment strategies of AD. Histone deacetylases (HDACs) regulate the levels of histone acetylation and deacetylation, thus affecting the gene expression and underlying pathways. It is believed that the targeting of HDACs might offer more potential treatment regimens for AD. The epigenetic research reported that numerous HDAC inhibitors (HDAC*is*) were found to be promising in improving cognition and memory through several in vitro cellular and in vivo animal mouse models of AD. Though several HDAC*is* were reported and studied, the identification of isoform-specific HDAC*is* was found to be critical in understanding the action of specific HDAC isoforms in the pathogenesis of AD and its cure. In this direction, HDAC*is* either derived from natural products or their analogues were designed, synthesised, and biologically evaluated towards enhancing the therapeutic

S. Pulya · A. Ganorkar · S. Biswas · B. Ghosh (✉)
Epigenetic Research Laboratory, Department of Pharmacy, Birla Institute of Technology and Science-Pilani, Hyderabad, India
e-mail: balaram@hyderabad.bits-pilani.ac.in

A. Sharma, G. P. Modi (eds.), *Natural Product-Based Synthetic Drug Molecules in Alzheimer's Disease*, https://doi.org/10.1007/978-981-99-6038-5_16

specificity and efficacy in AD models. However, extensive research in this area is needed to address the clinical limitations of the HDAC*is* in the treatment of AD.

Keywords

Alzheimer's disease · HDAC inhibitor · Epigenetics · Natural product · Drug discovery

Abbreviations

AD	Alzheimer's disease
APP	Amyloid precursor protein
Aβ	Amyloid-beta
Bcl 2	B-cell lymphoma 2 protein
BDNF	Brain derived neurotrophic factor
CBP	CREB binding protein
CNS	Central nervous system
CREB	cAMP response element-binding protein
ERK	Extracellular regulated kinase
Gap 43	Growth associated protein 43
GRIP1	Glutamate receptor-interacting protein
Grn	Granulin
GSK 3β	Glycogen synthase kinase 3 β
HAT	Histone acetyltransferase
HDAC	Histone deacetylases
IL 1β	Interleukin
IL 6	Interleukin
Keap1/Nrf2	Kelch like ECH associated protein 1/nuclear factor erythroid 2 related factor 2
LTM	Long-term memory
MAP 2	Microtubule-associated protein 2
MMP9	Matrix metalloproteinase 9
N2a	Neuro 2a cells
NEP	Neprilysin
NMDA	*N*-methyl-D-aspartate receptor
NOR	Novel object recognition
Nr4a2	Nuclear receptor 4A2
PI3K/Akt	Phosphoinositide 3 kinase protein kinase B/Ak strain transforming
PKC	Protein kinase C
PS1	Presenilin 1
PSD 95	Postsynaptic density protein 95
PTM	Post translational modifications
SAP 102	Synapse-associated protein 102
SYP	Synaptophysin
TNF	Tumour necrosis factor

16.1 Introduction

Alzheimer's disease (AD) is a neurodegenerative disorder, that is mostly seen in the elderly, after 65 years of age and is the cause of premature death in many (Liang et al. 2021). Nearly, >47 million population is suffering from AD in the world today that is impacting their daily activities (Lane et al. 2018). AD is characterised by the loss of memory and the impairment of cognitive abilities. Two types of AD are known either genetic (familiar) or sporadic. Sporadic form of AD is more frequent in the patients with the late onset of the disease and also with no known family history of brain disorders (Vickers et al. 2016). Even though AD is caused due to various factors, its cause and pathogenesis are not completely understood. The major cause of AD pathogenesis is due to the synaptic and neuronal loss caused by the extracellular Aβ plaques produced by the abnormal processing of amyloid precursor protein (APP), abnormal phosphorylation of tau proteins leading to intracellular neurofibrillary tangles (NFT), and generation of reactive oxygen species (ROS) that may lead to protein oxidation, degradation, mitochondrial damage, and eventually neuronal cell death (Wenk 2003; Hardy and Higgins 1992; Misrani et al. 2021). The aggregation of Aβ neuritic plaques is neurotoxic, and their deposition leads to pathophysiological changes in cerebral cortex and eventually nerve cell death (Chu and Liu 2019). Another major cause of AD might also be due to genetic and environmental factors including life style, which indicates that epigenetic mechanisms and their intervention might have a major role in the pathogenesis and disease progression of AD (Chouliaras et al. 2010).

Epigenetics is one of the new and most promising approach for investigating AD pathogenesis. Epigenetics involves the gene expression regulation, chromatin remodelling, and reorganisation and micro-RNA (miRNA) regulation with no change in the underlying DNA sequence both transcriptionally and post-transcriptionally (Wang et al. 2013a). Among the different epigenetic mechanisms, DNA methylation and histone post-translational modifications (PTMs) are most widely studied in the implication of brain disorders in particular, AD (Xiao et al. 2020). The role of histone PTMs has been widely documented over the past decade in the therapeutic intervention of AD and among these, histone acetylation is found to be very critical in rescuing the cognitive decline, learning and memory (Li et al. 2022). Two enzymes, histone acetyl transferases (HATs) and histone deacetylases (HDACs), catalyse and regulate the acetylation and deacetylation processes of histones, and the dysregulation of these enzymes leads to abnormal gene expression and their further implications (Fig. 16.1) (Turner 2000). Histone acetylation upregulates gene expression and activates transcription, whereas histone deacetylation represses gene transcription (Haggarty and Tsai 2011). Abnormal histone acetylation is reported to contribute to the disease progression in AD, and HDACs were found to be implicating several mechanisms or pathways leading to the pathogenesis of the disease in various animal models (Haggarty and Tsai 2011; Levenson and Sweatt 2005; McQuown et al. 2011). Histone acetylation is known to occur in the hippocampus region of the brain in response to the contextual fear conditioning and learning. Moreover, literature reports suggested that the reduction

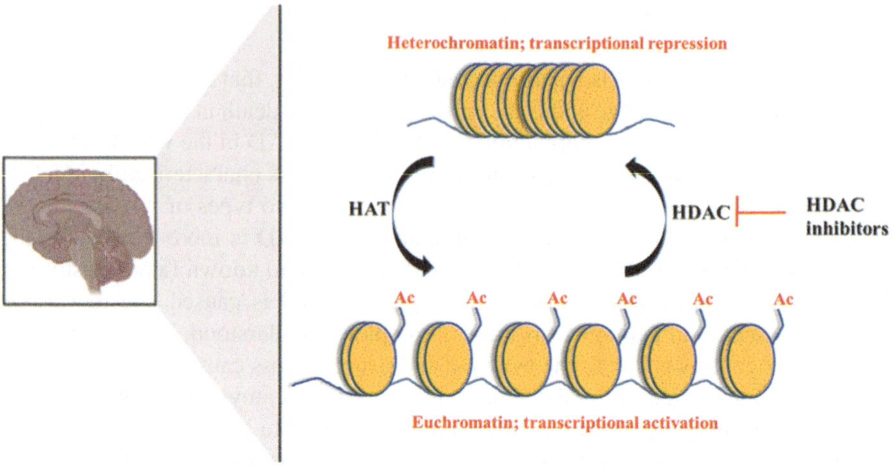

Fig. 16.1 Histone modifications and transcriptional regulation by HATs and HDACs

of histone acetylation by inhibiting HAT promoted amnesia and also disrupted the consolidated learning and memory in the hippocampus (Levenson et al. 2004; Hemstedt et al. 2017; Vecsey et al. 2007). Moreover, HDAC inhibition led to increased acetylation with enhanced gene transcription, thus promoting hippocampus-dependent memory formation (Levenson et al. 2004; Vecsey et al. 2007; Lattal et al. 2007; McQuown et al. 2011; Stefanko et al. 2009; Fischer et al. 2007). Thus, targeting HDACs might be considered as one of the most promising therapeutic strategies in the treatment as well as in understanding the underlying mechanisms in the AD pathogenesis. Therefore, HDAC inhibitors (HDAC*is*) have been widely reported as an attractive and alternative treatment method in case of Alzheimer's disease.

HDAC*is* were known to exhibit both neurotoxic and neuroprotective effects in different in vivo animal models of AD and thus the design and development of HDAC*is* are very crucial in the drug discovery of AD (Haggarty and Tsai 2011). Several HDAC*is* known so far are *pan*-HDAC*is* and very few are isoform-specific HDAC*is*. These *pan*-HDAC*is* are non-selective and thus lead to toxic side effects and also poor bioavailability. Hence, the development of isoform selective HDAC*is* is very essential in understanding the role of specific HDAC isoform in the disease mechanisms, pathogenesis, and progression in case of AD. Few isoform-specific HDAC*is* were also reported until recently, and much more research needs to be carried out to obtain a more suitable therapeutic molecule towards the treatment and studying the pathology of Alzheimer's disease.

This book chapter details the role of different HDAC isoforms and their selective inhibitors with respect to the treatment and the pathogenesis of Alzheimer's disease. Further, the literature reports of the existing HDAC*is*, both from natural sources and their derivatives including *pan*-HDAC*is* and isoform selective HDAC*is* will be discussed. The details of their design and synthesis of the molecules along with their biological evaluation in vitro and in vivo animal models of AD are presented in this review. The various pathways that are implicated in the disease progression as well as pathology will be mentioned across the reported HDAC inhibitors.

16.2 Histone Deacetylation and Its Implication in Alzheimer's Disease

Mammalian histone deacetylases constitute a family of 18 HDAC isoforms reported till date. These are further classified based on their mechanism of action, their structural homology with yeast protein, and cellular localisation into four classes. The four different classes of HDAC isoforms are distinct in terms of their structural features, cellular localisation and their expression patterns (de Ruijter et al. 2003). Based on their mechanism of action, they are also further classified into zinc dependent HDACs where Zn^{2+} acts as a co-factor includes HDACs class I, II, and IV and nicotinamide adenine dinucleotide (NAD^+) dependent HDACs where NAD^+ acts as a co-factor for their catalytic activity including class III HDACs called as sirtuins (Park and Kim 2020). Among the 18 HDAC isoforms, class I HDACs include isoforms HDAC1–3 and HDAC8 primarily located in the nucleus with histones as their major substrate. Class II HDACs, divided into class IIa that includes HDAC4, HDAC5, HDAC7, and HDAC9 and class IIb that includes HDAC6 and HDAC10 based on their sequence homology and catalytic active sites. Class II HDACs were found in both nucleus and cytoplasm with class IIb HDACs being primarily cytoplasmic in its localisation and these act through various non-histone proteins as their substrates. Another Zn^{2+} dependent HDAC is isoform HDAC11 which belongs to class IV HDAC, that is known to be similar to HDAC3 and HDAC8 in its characterisation. Class III HDACs, known as sirtuins include SirT 1–SirT 7, are the only class of NAD^+ dependent HDACs known to share homologous sequence with Sir2 yeast protein (Xu et al. 2011; Pulya et al. 2020).

The dysregulation of HDAC activity that is associated with chromatin remodelling and transcription factors is involved in many neurological disorders, particularly in AD. This imbalance leads to the alteration of transcriptional regulation of various genes involved in the AD pathogenesis. Therefore, inhibiting HDACs indirectly increases the acetylated histone levels, which further promotes the transcription of memory genes and proteins such as Aβ, GSK-3β, and tau that are known to restore cognitive functions of the brain (Fig. 16.2).

One of the first reports suggested that class I HDACs have a major role in affecting learning and memory, thus leading to cognitive decline in the elderly AD patients (Fischer et al. 2010; Gräff et al. 2012; Penney and Tsai 2014). Reports demonstrated the increase of class I HDACs in the hippocampal region and their decrease in the cortex of the brain (Kilgore et al. 2010; Rumbaugh et al. 2015). A very interesting study by Pascoal et al. reported the use of [^{11}C] Martinostat, a novel PET imaging agent that selectively binds to class I HDACs and is known to cause

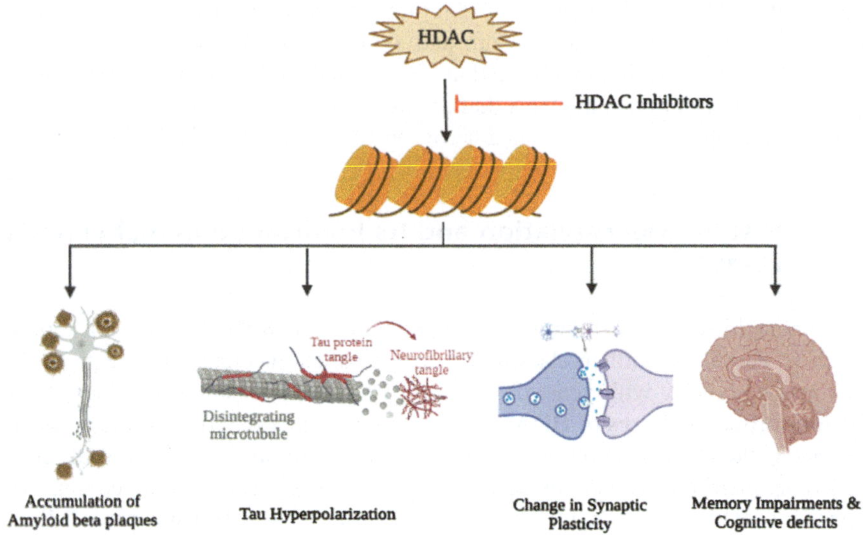

Fig. 16.2 Role of HDACs and HDAC*is* in Alzheimer's disease

changes in the genes associated with synaptic plasticity and neurodegeneration (Pascoal et al. 2022). Among the class I HDACs, HDAC2 isoform overexpression in CK-p25 AD mice resulted in reduced synaptic plasticity and eventually cognitive impairment and later the treatment with vorinostat, a *pan*-HDAC*i* is found to restore the cognitive function of the brain. Further, HDAC2 but not HDAC1 or HDAC3 has increased in the hippocampal brain region of the AD patients (Gräff et al. 2012). HDAC2 is also known to be negatively regulating learning and memory functions (Guan et al. 2009). Only HDAC2 is known to significantly contribute to the hippocampus-dependent age-related cognitive decline (Akhtar et al. 2009; Guan et al. 2009). The overexpression of HDAC3 in the brain known to cause neuronal cell death. HDAC3 inhibition caused significant enhancement of the long-term memory formation in HDAC3-flox-modified mice or in the mice treated with HDAC3*i* RGFP136 (McQuown et al. 2011; McQuown and Wood 2011). The first evidence provided by Krishna et al. demonstrated that inhibiting HDAC3 by RGFP966 in either single or a population of neurons could prevent Aβ oligomer induced synaptic plasticity (Krishna et al. 2016). Another study by Janczura et al. has reported that selective HDAC3 inhibition or silencing has increased *Bdnf* expression, decreased tau phosphorylation, acetylation, and β-secretase cleavage of APP at the diseased sites, thus decreasing $A\beta_{1-42}$ accumulation (Janczura et al. 2018). Furthermore, the overexpression of HDAC3 increased Aβ levels and reduced dendritic spine density in APP/PS2 mice of 6 month old, thus indicating the role of HDAC3 as a negative regulator of spatial memory (Zhu et al. 2017). HDAC4 was known to be highly expressed in the brain tissue and thus is involved in the synaptic plasticity and memory (Sando et al. 2012). In another report, loss of HDAC5 has led to memory impairment suggesting its role in the implication of AD (Agis-Balboa et al. 2013). Moreover, several other recent reports have identified and indicated the role of HDAC4 and HDAC5 in AD disease pathogenesis (Sando et al. 2012; Parra 2015; Mielcarek et al. 2015; Colussi et al. 2023). A study involving the use of a novel radiotracer, [18F]TFAHA reported an increase in the HDAC4 expression in response to Aβ deposition in a triple transgenic AD mouse model, and when treated with a HDAC4 selective inhibitor, the upregulation of the genes related to synaptic plasticity was reported, further establishing HDAC4 as another promising therapeutic target of AD (Chen et al. 2021). HDAC6, a class IIb HDAC isoform, is one of the most essential isoforms that is attributed to Alzheimer's disease pathology. Higher levels of HDAC6 were found in the hippocampus and cortex brain regions of AD patients when compared to normal brains (Simões-Pires et al. 2013). HDAC6 is also known to facilitate hyperphosphorylation of tau and its aggregation which further leads to microtubule instability and eventually apoptotic neuronal cell death (Selenica et al. 2014; Tran et al. 2007). Inhibiting HDAC6 is well known to restore cognitive abilities in the mouse model of AD (Govindarajan et al. 2013). Several mechanisms were described for the implication of HDAC6 in AD, and selective inhibition of HDAC6 led to its neurogenesis and neuroprotective activity of cortical neurons (Pulya et al. 2020; Govindarajan et al. 2013; Rivieccio et al. 2009; Chen et al. 2010).

Altogether, the role of various HDAC isoforms particularly, HDAC1–3 of class I and HDAC6 of class IIb, has been largely implicated in the pathogenesis of AD. Furthermore, studies have also demonstrated the use of HDAC inhibitors as possible therapeutic strategies that restore and improve cognitive abilities in the in vivo animal models of AD.

16.3 HDAC Inhibitors in Alzheimer's Disease

A number of HDAC*is* are reported till date as therapeutics against life-threatening diseases such as cancer and neurodegenerative disorders. These HDAC*is* are further classified into five different categories based on their zinc binding groups. They are hydroxamates, benzamides, cyclic tetrapeptides, and short chain fatty acids, and the most recent one is hydrazides. Most of the reported HDAC*is* are *pan*-HDAC*is*, and few selectively inhibit one or the other HDAC isoform. Though most of the clinically approved HDAC*is* are reported for their anti-tumour therapy, studies have also reported the activity of HDACs in neurodegenerative disorders. Therefore, the attention towards HDAC*is* as promising therapeutics for neurodegenerative disorders, namely Alzheimer's disease and other CNS disorders has increased (Hahnen et al. 2008).

Several literature reports have been established since then by the usage of several HDAC*is* as cognition enhancing agents using different animal models. The animal

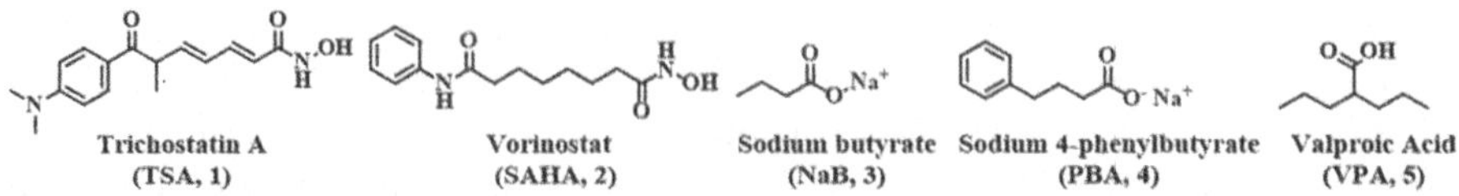

Fig. 16.3 Structures of reported *pan*-HDAC inhibitors

models used different experiments such as behavioural paradigms, contextual fear conditioning, formation of learning and memory in rat and mice models. In this direction, studies proved that upregulation of histone acetylation by HDAC inhibition was found to improve memory and learning in Alzheimer's disease (AD) (Xu et al. 2011; Kazantsev and Thompson 2008; Fischer 2014). In this exploration, most widely used *pan*-HDAC*is* (Fig. 16.3) are trichostatin A (**1**, TSA), vorinostat (**2**, SAHA). sodium butyrate (**3**, NaB), phenyl butyrate (**4**, PBA), and valproic acid (**5**, VPA) and further, research also developed their analogues and studied their activity in various animal models against AD.

Fischer et al. for the first time tested HDAC*is* in CK-p25 mice of AD and demonstrated that the inducible prolonged overexpression of p25 in hippocampus caused amyloid and tau aggregates that leads to the neurodegeneration and cognitive impairment (Fischer 2014). Another study by employing the use of NaB via intraperitoneal (i.p.) route of administration for a period of 4 weeks was found to restore learning and memory in CK-p25 mice suggesting the neuroprotective and neurogenesis properties of HDAC*is* (Fischer et al. 2007). One of the first and most widely studied HDAC*i*, Trichostatin A (TSA), a hydroxamate class of HDAC*i*, naturally derived from *Streptomyces hygroscopicus* bacterial strain, is a highly potent *pan*-HDAC*i* with reported activity at nanomolar range and was found to inhibit mammalian histones both in vitro and in vivo (Yoshida et al. 1990). Several studies reported the neuroprotective effects of TSA in the in vivo animal models. In a study conducted by Francis et al., the administration of TSA intraperitoneally in the mouse model of AD demonstrated the improved contextual fear conditioning as well as hippocampal long-term potentiation along with the upregulation of acetylated histone H4 in APP/PS1 mice (Francis et al. 2009). In another study conducted by Yang et al., TSA treatment was found to increase plasma gelsolin (a protein known for the clearance of amyloid-beta protein) levels in the transgenic mouse model of AD leading to the clearance of Aβ protein from the brain tissues (Yang et al. 2014a). Same group reported the increased gelsolin expression in the hippocampus and cortex regions of the brain tissue and also further increased activity of γ- and β-secretase, with no change in the amyloid load in the brain of the APP/PS1 transgenic mice. Interestingly, this study also demonstrated the prevention of the formation of new amyloid plaques (Yang et al. 2014b). Recent studies reported the various mechanistic pathways of TSA activity in AD mouse models. In one of the studies, the TSA when treated in SH-SY5Y cells was found to increase cell viability and its antioxidant capacity through Keap1-Nrf2 pathway, thus inhibiting Aβ peptide mediated cell death (Li et al. 2019). Recent report by Su et al. demonstrated the TSA treatment improved cognition abilities that is attributed to the increased Aβ oligomer clearance further reducing the aggregation of Aβ plaques in the brain of APP/PS1 transgenic mice. The in vitro study demonstrated the upregulation of albumin expression led to the inhibition of Aβ aggregation and also promoted the phagocytosis of Aβ oligomers (Su et al. 2021). Another natural *pan*-HDAC*i*, sodium butyrate (NaB), produced by gut bacteria, *Cytophaga* and *Flavobacterium* group in the colon, is well known to exert neuroprotective effects, and several reports have demonstrated that NaB has attenuated memory deficits in vivo (Tan et al. 2014). The studies involving the supplementation of NaB in diet to 5×FAD mice demonstrated significant effects on Aβ levels further on the learning and cognitive functions when studied in behavioural testing experiments at an early disease stage (Fernando et al. 2020). Another important in vitro study in N2a cells reported that NaB regulated the expression of *APP*, *Nep*, *Bdnf*, AD-related genes by acting on GPR109A receptor, thus protecting N2a cells from injury (Sun et al. 2020). Jiang et al. reported that at the early stage of the disease progression, the administration of NaB leads to the increased synaptic plasticity in 5×FAD mice through increasing PSD 95, SYP, NR2B, synapse-associated proteins and also by decreasing TNF α, IL 6, IL 1β (Jiang et al. 2021). Recently, Wang et al. reported that NaB promoted the differentiation of astrocytes. In this study, the oral administration of NaB has increased the astrocyte-neuron lactate shuttle (ANLS) by promoting the mitochondrial function of astrocytes, that is known to be important in synaptic plasticity and LTM formation (Wang et al. 2022). All these findings further reinstate that NaB, at the early stages of the disease, could be potentially enhancing the cognitive deficits caused by $A\beta_{25-35}$ in mice. Another *pan*-HDAC*i*, 4-phenyl butyrate sodium (Na salt of PBA) was found to reinstate fear learning and memory in Tg2576 mouse model irrespective of the disease stage. PBA was found to restore dendritic spine densities of hippocampal CA1 neurons and also effect NR2B and SAP102 expression, a synaptic scaffold responsible for memory and learning (Ricobaraza et al. 2012). Several mechanistic studies involved the use of PBA in the enhancement of spatial learning and memory associated with improved synaptic function through various pathways such as PKC-cAMP, CREB activation thus activating the neutrophil synthesis in the astrocytes in in vivo mouse model of AD (Corbett et al. 2013; Ricobaraza et al. 2009; Ricobaraza et al. 2011; Corbett et al. 2017). Valproic acid (VPA), a *pan*-HDAC*i*, though initially approved as an anticonvulsant and a mood stabilising agent, has been extensively studied in the in vivo mouse models of AD for its neuroprotective and neurogenesis properties. It was found to reduce Aβ production and plaque formation of neurites, thus enhancing spatial learning and memory in the APP23 and APP23/PS45 mice (Qing et al. 2008). VPA was widely studied in the pathways leading to neurogenesis and neuroprotective activities in various transgenic mouse models of AD. VPA is known to inhibit GSK-3β by activating Wnt/β-catenin pathways leading to neurogenesis and PI3K/Akt pathway, thus promoting synaptogenesis and plasticity as well as through regulating mitochondria mediated apoptotic pathway (Zeng et al. 2019; Zhao et al. 2018a; Zhang et al. 2010; Singh et al. 2021). VPA is reported to alleviate spatial learning and memory deficits, also reducing Aβ deposits via activating ERK pathways, thus upregulating GAP 43 and *Bcl 2* expression and

Table 16.1 Summary of the HDAC*is* tested so far

Class of HDAC	HDAC isoform	Name of inhibitor/cpd no.[a]
Class I	HDAC1	Trichostatin A Sodium butyrate Vorinostat Valproic acid Crebinostat Sodium phenylbutyrate Entinostat, 10
	HDAC2	RGFP968 RGFP963
	HDAC3	RGFP966 RGFP136 BG 45 PT3, 17
	HDAC8	xx
Class IIa	HDAC4	9, 10, 26, 27
	HDAC5	xx
	HDAC7	xx
	HDAC9	xx
Class IIb	HDAC6	Tubastatin A ACY-1215 MPT0G211 ACY-738 9, 10, 12, 23, 26, 27, 29 T-518
	HDAC 10	xx

[a] Compound number mentioned in the table corresponds to the number assigned in Figures

also the downregulation of NEP in the APP/PS1 mouse model (Xuan et al. 2015; Wang et al. 2013b). Vorinostat (SAHA) is another *pan*-HDAC*i* that was found to have a role in cognition and memory. SAHA was found to selectively enhance postsynaptic excitatory synapses in its in vitro treatment. In case of in vivo Tg2576 mouse model of AD, SAHA treatment did not rescue the memory impairment (Hanson et al. 2013). On the other hand, the intra-hippocampal administration of SAHA demonstrated to rescue age-related cognitive impairment (Peleg et al. 2010). But the recent reports demonstrate the use of combination of SAHA, curcumin, and silibinin reversed Aβ-induced neuronal toxicity through the activation of Akt-MDM2-p53 pathway, thus improving therapeutic selectivity and neuroprotective abilities (Meng et al. 2019). Thus, all these studies suggested the decreased histone acetylation levels in the AD models which was then upregulated via HDAC inhibition, thus rescuing the cognitive deficits in the AD mice (Table 16.1).

Apart from the *pan*-HDAC*is*, several isoform selective HDAC*is* (Figs. 16.4 and 16.5) were also reported to be promising drug candidates for Alzheimer's disease. Entinostat (MS 275, **6**), a class I selective HDAC*i* with an IC_{50} (HDAC1 = 163 nM) is shown to prevent neuronal cell death and also inhibits neuroinflammation

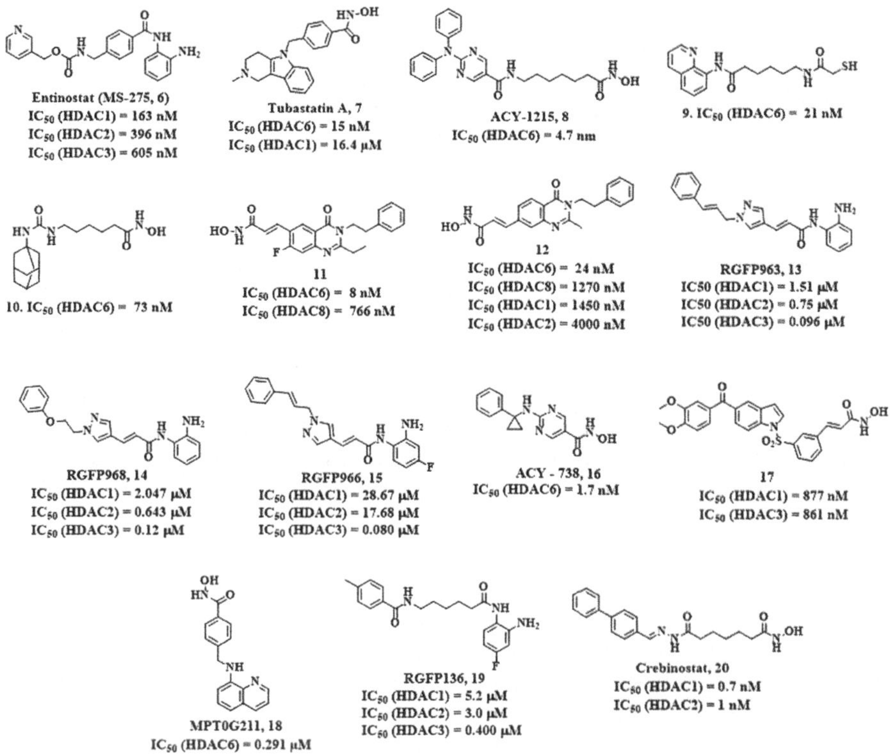

Fig. 16.4 Structures of reported isoform selective HDAC inhibitors (6–20) with their HDAC isoform inhibition values (IC50)

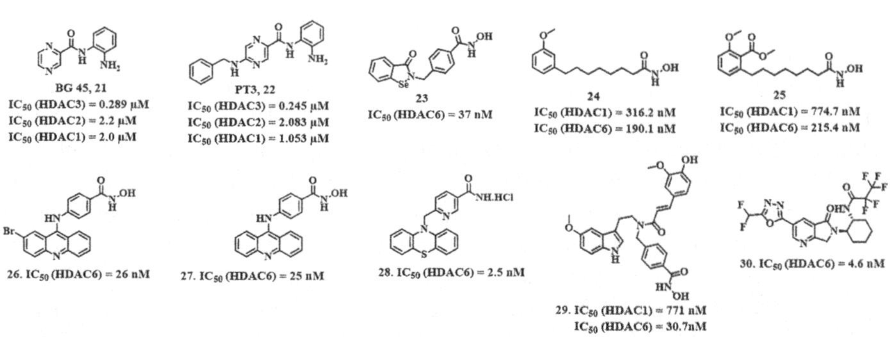

Fig. 16.5 Structures of reported isoform selective and multi-target-directed inhibitors (21–30) with their HDAC isoform inhibition values

activities in transgenic mice (Zhang et al. 2014). MS275 when evaluated in the in vivo transgenic mice, demonstrated increased H4 acetylation in the cortex and amygdala leading to decreased Aβ plaque deposition in the hippocampal and cortical regions, thus enhancing the consolidation of fear memory extinction (Fischer 2014; Whittle et al. 2016; Zhang and Schluesener 2013). Another study reported that MS275 treatment enhanced the activation of *Bdnf*/TrkB pathway leading to the object recognition memory consolidation in the cortex (Whittle et al. 2016). Tubastatin A (Tub A, **7**), one of the very first HDAC6 selective inhibitor identified from homology modelling of structure-based drug discovery method, when tested in primary cortical neuronal cultures demonstrated the dose-dependent neuro protective activity against the oxidative stress due to the glutathione depletion with no neuronal toxicity (Butler et al. 2010). Zhang et al. reported that the use of Tub A and ACY1215 (**8**) in the transgenic mouse model of AD reduced tau hyperphosphorylation, promoted Aβ autophagic clearance, improvement of microtubule stability leading to improved cognitive abilities (Zhang et al. 2014). Two HDAC*is* compound **9** and **10** were reported by Sung et al. and studied their effects on Aβ levels and cognition (Sung et al. 2013). Compound **9** is a mercaptoacetamide based class II HDAC*i*, whereas compound **10** is a hydroxamic acid-based class I and II HDAC*i*. Compound **9** was found to decrease $A\beta_{40}$ and $A\beta_{42}$ in vitro, whereas compound **10** was found to decrease only $A\beta_{40}$ but not $A\beta_{42}$. These HDAC*is* were found to improve learning and memory functions when studied in the morris water maze (MWM) test and also were found to decrease tau phosphorylation. They also reduced Aβ levels by affecting Aβ synthesis and also reduced the γ-secretase enzyme expression and further increased the Aβ degradation enzyme, MMP2 (Sung et al. 2013). Yu et al. reported a series of potent HDAC6*is* with hydroxamates as ZBG and quinazoline-4-one as a cap group against AD. Among the series, compound **11**, (*E*)-3-(2-Ethyl-7-fluoro-4-oxo-3-phenethyl-3,4-dihydroquinazolin-6-yl)-*N*-hydroxyacrylamide, was the most potent with of 8 nM (HDAC6 IC_{50}) potency. The in vitro studies of these compounds enhanced neurite outgrowth via GAP 43 expression and were also found to enhance synaptic plasticity in PC12 and SH-SY5Y cells with no neuronal toxicity. Another compound **12**, *N*-Hydroxy-3-(2-methyl-4-oxo-3-phenethyl-3,4-dihydroquinazolin-7-yl)-acrylamide, most promising HDAC6 inhibitor with a IC_{50} of 24 nM reported to improve learning abilities of mice with Aβ induced hippocampal lesions (Yu et al. 2013). Rumbaugh et al. reported the class I selective HDAC*is* such as RGFP963 (**13**), RGFP968 (**14**), and RGFP966 (**15**) to stimulate synaptogenesis in the dorsal hippocampus of mature mice (Rumbaugh et al. 2015). Similarly, oral administration of ACY-738 (**16**), a HDAC6*i* in the amyloid mouse model was reported to improve the in vivo axonal transport causing improved hyperactivity and contextual learning and memory. ACY-738 treatment was also found to improve amyloid pathology in transgenic mice by decreasing the levels of insoluble $A\beta_{1-42}$ (Majid et al. 2015). Recent study involving the use of RGFP966, a selective HDAC3*i*, increased *Bdnf* expression, decreased tau acetylation and phosphorylation, and further reduced β-secretase cleavage of APP along with decreasing $A\beta_{1-42}$ accumulation in vitro. Moreover, this compound when treated in vivo in the APP-PS1 triple transgenic mouse model of AD revealed the increased levels of Aβ degrading enzyme, NEP, thus reducing the $A\beta_{1-42}$ levels in the brain, leading to improved spatial learning and memory (Janczura et al. 2018). A series of aroyl-indole hydroxamate derivatives as HDAC6*is* were reported by Lee et al. and compound **17**, the most potent among them with 3.92 nM, HDAC6 IC_{50}. Compound **17** displayed high selectivity towards HDAC6 with >50-fold selectivity over other HDAC isoforms. This compound upon treatment in vivo was found to downregulate phosphorylated tau proteins and also Aβ in the hippocampal CA1 region of mice. Compound **17** was also found to exert its neuroprotective effects through ubiquitination, thus ameliorating AD phenotypes (Lee et al. 2018). MPT0G211 (**18**), an orally administrated, blood–brain barrier (BBB) penetrant, highly selective HDAC6*i*, improved learning and memory deficits in the in vivo mouse model of AD by reducing tau phosphorylation and aggregation (Fan et al. 2018). MPT0G211 induced the acetylation of Hsp90, due to the decreased binding of HDAC6–Hsp90, further reduced tau phosphorylation followed by ubiquitination of tau proteins and their degradation. Furthermore, MPT0G211 treatment also reduced the GSK 3β activity that is associated with phosphor-tau Ser396 through Akt phosphorylation. Oral treatment of MPT0G211 demonstrated significant reduction of phosphorylated tau in the hippocampal CA1 regions in the mouse models of AD (Fan et al. 2018). HDAC3 was found to be highly expressed in brain tissue and is found to negatively regulate the learning and memory abilities when treated with HDAC3 selective inhibitors in various in vivo mouse models. RGFP136 (**19**), a selective HDAC3*i* was found to regulate the CBP-dependent transcription mechanism, therefore enhancing LTM formation. Crebinostat (**20**), a class I selective HDAC*i*, upregulated the expression of the CREB target Erg 1. When treated in the in vivo mouse model by systemic administration, Crebinostat was found to enhance the learning, memory, and synaptic plasticity by upregulating *Bdnf*, *Grn* and downregulating tau gene expression in the hippocampus-dependent, contextual fear conditioning study, thus demonstrating its wide implication in the age-related cognitive decline and cognitive disorders (Fass et al. 2013). Further, studies involving the use of class I selective HDAC*i*s were found to enhance long-term memory formation by upregulating the memory associated genes, thus modulating neuroplasticity in the hippocampal tissue of treated mice (Fass et al. 2013; Zhao et al. 2018b; Pulya et al. 2021). BG 45 (**21**), a class I HDAC*i* when administered to an Aβ-treated cells, in vitro and in vivo has upregulated the synapse-related genes *grik2*, *scn3b*, *synpr* via HDAC inhibition (Han et al. 2021). Much recently, Ma et al. and group investigated BG 45 in the APP-transfected cells and also in APP/PS1 transgenic mouse model of AD. BG 45 further upregulated SYP, PSD 95, spinophilin and also repaired cytoskeletal damage. When treated in vivo, BG 45 reduced the Aβ plaque deposition and tau phosphorylation further reducing the hippocampal neuronal loss by apoptosis. Therefore, the application of BG 45 by inhibiting class I HDACs in particular HDAC1, HDAC2, and HDAC3 rescued the synaptic damage by enhancing synapse-related gene expression and modulating AMPA receptor transduction via GRIP1/AMPA pathway, thus increasing the function of excitatory synapses in the early stage of in AD disease progression (Han et al. 2022). PT3 (**22**), a selective HDAC3*i* was found to enhance long-term memory in the novel object recognition experiment. PT3 has a good BBB permeability and is known to upregulate CREB1, *Bdnf*, TRKB, *Nr4a2*, *c-fos*, *PKA*, GAP 43, PSD 95, and

MMP9 gene expression both in vitro and in vivo of C57/BL6 mice (Pulya et al. 2021).

Apart from the selective HDAC*is*, recently multi-target agents including HDAC*is* have gathered attention towards the treatment of AD. A set of multi-target agents were synthesised incorporating the pharmacophores of SAHA and the antioxidant ebselen and were tested against AD. These compounds were found to exert HDAC inhibitory activity along with the pharmacological effect of ebselen. Among the series, compound **23** was reported to be the most potent with the HDAC6 IC_{50} of 37 nM. These compounds also exhibited free oxygen radical scavenging ability along with the neuroprotective effects against peroxidase damage in PC12 cells in vitro (Hu et al. 2018). One of the recent interesting reports described the first in class HDAC*is* derived from the natural source, cashew-nut shell liquid (CNSL), compounds **24** and **25** displaying similar HDAC inhibitory profile and increased safety compared to that of SAHA. These hydroxamate derivatives were found to regulate inflammatory responses via glial cell activation thus suggesting their further development as sustainable drug candidates for AD (Uliassi et al. 2021; Soares Romeiro et al. 2019). In a report published by Tseng et al., hybrid compounds that incorporated HDAC inhibitor core into the acridine-derived moiety were studied as HDAC-Ab-AChE inhibitors and they were found to enhance neurite outgrowth with no neurotoxicity effects. Among the compounds in the series, compounds **26** and compound **27** were found to be strongly inhibiting Aβ-aggregation and also disrupted Aβ oligomer along with inhibiting acetylcholinesterase enzyme (Tseng et al. 2020). Novel multifunctional HDAC6 inhibitors incorporated with phenothiazine-, memantine-, and 1,2,3,4-tetrahydro-γ-carboline-cap groups with different linkers have been reported. In that series, compound **28**, a HDAC6*i* with phenothiazine based hydroxamic acid containing pyridyl moiety as the linker was found to reduce the hyperphosphorylated tau at Ser396. It was also found to inhibit Cu^{2+}-induced $A\beta_{1-42}$ aggregation and disrupted Cu^{2+}-induced $A\beta_{1-42}$ oligomers. The in vitro studies revealed that the mRNA expression of GAP 43, *N-myc*, and MAP 2 were enhanced, thus promoting the neurite outgrowth differentiation (Wang et al. 2021).

A multi-target-directed inhibitor series was designed and developed as the hybrid of melatonin and ferulic acid structures into the amide-based HDAC6*is* that is known to have antioxidant effects along with potent HDAC6 inhibition. From the series, compound **29** has exhibited an HDAC6 IC_{50} of 30.7 nM with >25-fold HDAC6 selectivity over other HDAC isoforms. This compound exhibited comparable antioxidant activity to melatonin with no neurotoxicity on HT-22 cells when used in vitro. Further, compound **29**, when treated in vivo displayed significant efficacy in vivo by preventing $A\beta_{25-35}$ induced spatial learning and memory at a very low oral dose of 0.3 mg/kg as against ACY-1215 alone or the combination of ACY-1215, melatonin and ferulic acid together (He et al. 2021). Recently, a novel HDAC6*i*, T-518 (**30**) with oxadiazole as ZBG reported as a potential drug candidate against AD and tauopathy. T-518, an orally available, highly potent and selective HDAC6*i* has displayed 4.6 nM IC_{50} with good CNS penetration ability. Short-term treatment with T-518 was found to restore damaged axonal transport and NOR, whereas the long-term treatment reduced tau accumulation in the in vivo AD model (He et al. 2021).

16.4 Summary

In conclusion, various studies in both in vitro and in vivo animal models of AD indicated the role of specific HDAC isoforms by inhibiting HDAC enzyme activity using either *pan*-HDAC*i*s or isoform selective inhibitors with respect to their neuroprotective as well as neurogenesis activity. Most of the clinically approved anticancer HDAC*i*s are *pan*-HDAC*i*s and their use led to unwanted side effects and toxicities due to their non-specific nature. Further, multi-targeted agents that include HDAC inhibitor pharmacophore incorporated into a hybrid structure compound, thus addressing several other targets of AD has also been extensively studied recently. Altogether, the major challenge ahead lies in identifying and developing HDAC isoform-specific inhibitors, and to further study in detail their implications in various underlying mechanisms and pathways in relation to the pathogenesis of Alzheimer's disease and of course finding a potential therapeutic candidate for AD treatment.

Acknowledgements The research in the epigenetic research laboratory of Dr. Balaram Ghosh has been supported by the Council of Scientific and Industrial Research (**File no: CSIR-37(1722)/19/EMR-II**). Authors sincerely acknowledge the Department of Pharmacy, Birla Institute of Technology and Science-Pilani, Hyderabad Campus, India.

References

Agis-Balboa RC, Pavelka Z, Kerimoglu C, Fischer A (2013) Loss of HDAC5 impairs memory function: implications for Alzheimer's disease. J Alzheimers Dis 33(1):35–44. https://doi.org/10.3233/JAD-2012-121009

Akhtar MW, Raingo J, Nelson ED, Montgomery RL, Olson EN, Kavalali ET, Monteggia LM (2009) Histone deacetylases 1 and 2 form a developmental switch that controls excitatory synapse maturation and function. J Neurosci 29(25):8288–8297. https://doi.org/10.1523/JNEUROSCI.0097-09.2009

Butler KV, Kalin J, Brochier C, Vistoli G, Langley B, Kozikowski AP (2010) Rational design and simple chemistry yield a superior, neuroprotective HDAC6 inhibitor, Tubastatin A. J Am Chem Soc 132(31):10842–10846. https://doi.org/10.1021/ja102758v

Chen S, Owens GC, Makarenkova H, Edelman DB (2010) HDAC6 regulates mitochondrial transport in hippocampal neurons. PLoS One 5(5):e10848. https://doi.org/10.1371/journal.pone.0010848

Chen Y-A, Lu C-H, Ke C-C, Chiu S-J, Chang C-W, Yang B-H, Gelovani JG, Liu R-S (2021) Evaluation of class IIa histone deacetylases expression and in vivo epigenetic imaging in a transgenic mouse model of Alzheimer's disease. Int J Mol Sci 22(16):8633. https://doi.org/10.3390/ijms22168633

Chouliaras L, Rutten BPF, Kenis G, Peerbooms O, Visser PJ, Verhey F, van Os J, Steinbusch HWM, van den Hove DLA (2010) Epigenetic regulation in the pathophysiology of Alzheimer's disease. Prog Neurobiol 90(4):498–510. https://doi.org/10.1016/j.pneurobio.2010.01.002

Chu D, Liu F (2019) Pathological changes of tau related to Alzheimer's disease. ACS Chem Neurosci 10(2):931–944. https://doi.org/10.1021/acschemneuro.8b00457

Colussi C, Aceto G, Ripoli C, Bertozzi A, Puma DDL, Paccosi E, Dascenzo M, Grassi C (2023) Cytoplasmic HDAC4 recovers synaptic function in the 3×tg mouse model of Alzheimer's disease. Neuropathol Appl Neurobiol 49(1):e12861. https://doi.org/10.1111/nan.12861

Corbett GT, Roy A, Pahan K (2013) Sodium phenylbutyrate enhances astrocytic neurotrophin synthesis via protein kinase C (PKC)-mediated activation of CAMP-response element-binding protein (CREB): implications for Alzheimer disease therapy*. J Biol Chem 288(12):8299–8312. https://doi.org/10.1074/jbc.M112.426536

Corbett BF, You JC, Zhang X, Pyfer MS, Tosi U, Iascone DM, Petrof I, Hazra A, Fu C-H, Stephens GS, Ashok AA, Aschmies S, Zhao L, Nestler EJ, Chin J (2017) ΔFosB regulates gene expression and cognitive dysfunction in a mouse model of Alzheimer's disease. Cell Rep 20(2):344–355. https://doi.org/10.1016/j.celrep.2017.06.040

de Ruijter AJM, van Gennip AH, Caron HN, Kemp S, van Kuilenburg ABP (2003) Histone deacetylases (HDACs): characterization of the classical HDAC family. Biochem J 370(3): 737–749. https://doi.org/10.1042/bj20021321

Fan S-J, Huang F-I, Liou J-P, Yang C-R (2018) The novel histone de acetylase 6 inhibitor, MPT0G211, ameliorates tau phosphorylation and cognitive deficits in an Alzheimer's disease model. Cell Death Dis 9(6):1–14. https://doi.org/10.1038/s41419-018-0688-5

Fass DM, Reis SA, Ghosh B, Hennig KM, Joseph NF, Zhao W-N, Nieland TJF, Guan J-S, Kuhnle CEG, Tang W, Barker DD, Mazitschek R, Schreiber SL, Tsai L-H, Haggarty SJ (2013) Crebinostat: a novel cognitive enhancer that inhibits histone deacetylase activity and modulates chromatin-mediated neuroplasticity. Neuropharmacology 64:81–96. https://doi.org/10.1016/j.neuropharm.2012.06.043

Fernando WMADB, Martins IJ, Morici M, Bharadwaj P, Rainey-Smith SR, Lim WLF, Martins RN (2020) Sodium butyrate reduces brain amyloid-β levels and improves cognitive memory performance in an Alzheimer's disease transgenic mouse model at an early disease stage. J Alzheimers Dis 74(1):91–99. https://doi.org/10.3233/JAD-190120

Fischer A (2014) Targeting histone-modifications in Alzheimer's disease. What is the evidence that this is a promising therapeutic avenue? Neuropharmacology 80:95–102. https://doi.org/10.1016/j.neuropharm.2014.01.038

Fischer A, Sananbenesi F, Wang X, Dobbin M, Tsai L-H (2007) Recovery of learning and memory is associated with chromatin remodelling. Nature 447(7141):178–182. https://doi.org/10.1038/nature05772

Fischer A, Sananbenesi F, Mungenast A, Tsai L-H (2010) Targeting the correct HDAC(s) to treat cognitive disorders. Trends Pharmacol Sci 31(12):605–617. https://doi.org/10.1016/j.tips.2010.09.003

Francis YI, Fà M, Ashraf H, Zhang H, Staniszewski A, Latchman DS, Arancio O (2009) Dysregulation of histone acetylation in the APP/PS1 mouse model of Alzheimer's disease. J Alzheimers Dis 18(1):131–139. https://doi.org/10.3233/jad-2009-1134

Govindarajan N, Rao P, Burkhardt S, Sananbenesi F, Schlüter OM, Bradke F, Lu J, Fischer A (2013) Reducing HDAC6 ameliorates cognitive deficits in a mouse model for Alzheimer's disease. EMBO Mol Med 5(1):52–63. https://doi.org/10.1002/emmm.201201923

Gräff J, Rei D, Guan J-S, Wang W-Y, Seo J, Hennig KM, Nieland TJF, Fass DM, Kao PF, Kahn M, Su SC, Samiei A, Joseph N, Haggarty SJ, Delalle I, Tsai L-H (2012) An epigenetic blockade of cognitive functions in the neurodegenerating brain. Nature 483(7388):222–226. https://doi.org/10.1038/nature10849

Guan J-S, Haggarty SJ, Giacometti E, Dannenberg J-H, Joseph N, Gao J, Nieland TJF, Zhou Y, Wang X, Mazitschek R, Bradner JE, DePinho RA, Jaenisch R, Tsai L-H (2009) HDAC2 negatively regulates memory formation and synaptic plasticity. Nature 459(7243):55–60. https://doi.org/10.1038/nature07925

Haggarty SJ, Tsai L-H (2011) Probing the role of HDACs and mechanisms of chromatin-mediated neuroplasticity. Neurobiol Learn Mem 96(1):41–52. https://doi.org/10.1016/j.nlm.2011.04.009

Hahnen E, Hauke J, Tränkle C, Eyüpoglu IY, Wirth B, Blümcke I (2008) Histone deacetylase inhibitors: possible implications for neurodegenerative disorders. Expert Opin Investig Drugs 17(2):169–184. https://doi.org/10.1517/13543784.17.2.169

Han Y, Chen L, Guo Y, Wang C, Zhang C, Kong L, Ma H (2021) Class I HDAC inhibitor improves synaptic proteins and repairs cytoskeleton through regulating synapse-related genes in vitro and in vivo. Front Aging Neurosci 12:619866. https://doi.org/10.3389/fnagi.2020.619866

Han Y, Chen L, Liu J, Chen J, Wang C, Guo Y, Yu X, Zhang C, Chu H, Ma H (2022) A class I HDAC inhibitor rescues synaptic damage and neuron loss in APP-transfected cells and APP/PS1 mice through the GRIP1/AMPA pathway. Molecules 27(13):4160. https://doi.org/10.3390/molecules27134160

Hanson JE, La H, Plise E, Chen Y-H, Ding X, Hanania T, Sabath EV, Alexandrov V, Brunner D, Leahy E, Steiner P, Liu L, Scearce-Levie K, Zhou Q (2013) SAHA enhances synaptic function and plasticity in vitro but has limited brain availability in vivo and does not impact cognition. PLoS One 8(7):e69964. https://doi.org/10.1371/journal.pone.0069964

Hardy JA, Higgins GA (1992) Alzheimer's disease: the amyloid cascade hypothesis. Science 256(5054):184–185. https://doi.org/10.1126/science.1566067

He F, Chou CJ, Scheiner M, Poeta E, Yuan Chen N, Gunesch S, Hoffmann M, Sotriffer C, Monti B, Maurice T, Decker M (2021) Melatonin- and ferulic acid-based HDAC6 selective inhibitors exhibit pronounced immunomodulatory effects in vitro and neuroprotective effects in a pharmacological Alzheimer's disease mouse model. J Med Chem 64(7):3794–3812. https://doi.org/10.1021/acs.jmedchem.0c01940

Hemstedt TJ, Lattal KM, Wood MA (2017) Reconsolidation and extinction: using epigenetic signatures to challenge conventional wisdom. Neurobiol Learn Mem 142:55–65. https://doi.org/10.1016/j.nlm.2017.01.007

Hu J, An B, Pan T, Li Z, Huang L, Li X (2018) Design, synthesis, and biological evaluation of histone deacetylase inhibitors possessing glutathione peroxidase-like and antioxidant activities against Alzheimer's disease. Bioorg Med Chem 26(21):5718–5729. https://doi.org/10.1016/j.bmc.2018.10.022

Janczura KJ, Volmar C-H, Sartor GC, Rao SJ, Ricciardi NR, Lambert G, Brothers SP, Wahlestedt C (2018) Inhibition of HDAC3 reverses Alzheimer's disease-related pathologies in vitro and in the 3×Tg-AD mouse model. Proc Natl Acad Sci U S A 115(47):E11148–E11157. https://doi.org/10.1073/pnas.1805436115

Jiang Y, Li K, Li X, Xu L, Yang Z (2021) Sodium butyrate ameliorates the impairment of synaptic plasticity by inhibiting the neuroinflammation in 5XFAD mice. Chem Biol Interact 341:109452. https://doi.org/10.1016/j.cbi.2021.109452

Kazantsev AG, Thompson LM (2008) Therapeutic application of histone deacetylase inhibitors for central nervous system disorders. Nat Rev Drug Discov 7(10):854–868. https://doi.org/10.1038/nrd2681

Kilgore M, Miller CA, Fass DM, Hennig KM, Haggarty SJ, Sweatt JD, Rumbaugh G (2010) Inhibitors of class 1 histone deacetylases reverse contextual memory deficits in a mouse model of Alzheimer's disease. Neuropsychopharmacology 35(4):870–880. https://doi.org/10.1038/npp.2009.197

Krishna K, Behnisch T, Sajikumar S (2016) Inhibition of histone deacetylase 3 restores amyloid-β oligomer-induced plasticity deficit in hippocampal CA1 pyramidal neurons. J Alzheimers Dis 51(3):783–791. https://doi.org/10.3233/JAD-150838

Lane CA, Hardy J, Schott JM (2018) Alzheimer's disease. Eur J Neurol 25(1):59–70. https://doi.org/10.1111/ene.13439

Lattal KM, Barrett R, Wood MA (2007) Systemic or intrahippocampal delivery of histone deacetylase inhibitors facilitates fear extinction. Behav Neurosci 121(5):1125–1131. https://doi.org/10.1037/0735-7044.121.5.1125

Lee H-Y, Fan S-J, Huang F-I, Chao H-Y, Hsu K-C, Lin TE, Yeh T-K, Lai M-J, Li Y-H, Huang H-L, Yang C-R, Liou J-P (2018) 5-Aroylindoles act as selective histone deacetylase 6 inhibitors ameliorating Alzheimer's disease phenotypes. J Med Chem 61(16):7087–7102. https://doi.org/10.1021/acs.jmedchem.8b00151

Levenson JM, Sweatt JD (2005) Epigenetic mechanisms in memory formation. Nat Rev Neurosci 6(2):108–118. https://doi.org/10.1038/nrn1604

Levenson JM, O'Riordan KJ, Brown KD, Trinh MA, Molfese DL, Sweatt JD (2004) Regulation of histone acetylation during memory formation in the hippocampus*. J Biol Chem 279(39): 40545–40559. https://doi.org/10.1074/jbc.M402229200

Li L-H, Peng W-N, Deng Y, Li J-J, Tian X-R (2019) Action of Trichostatin A on Alzheimer's disease-like pathological changes in SH-SY5Y neuroblastoma cells. Neural Regen Res 15(2): 293–301. https://doi.org/10.4103/1673-5374.265564

Li Y, Lin S, Gu Z, Chen L, He B (2022) Zinc-dependent deacetylases (HDACs) as potential targets for treating Alzheimer's disease. Bioorg Med Chem Lett 76:129015. https://doi.org/10.1016/j.bmcl.2022.129015

Liang C-S, Li D-J, Yang F-C, Tseng P-T, Carvalho AF, Stubbs B, Thompson T, Mueller C, Shin JI, Radua J, Stewart R, Rajji TK, Tu Y-K, Chen T-Y, Yeh T-C, Tsai C-K, Yu C-L, Pan C-C, Chu C-S (2021) Mortality rates in Alzheimer's disease and non-Alzheimer's dementias: a systematic review and meta-analysis. Lancet Healthy Longev 2(8):e479–e488. https://doi.org/10.1016/S2666-7568(21)00140-9

Majid T, Griffin D, Criss Z, Jarpe M, Pautler RG (2015) Pharmacologic treatment with histone deacetylase 6 inhibitor (ACY-738) recovers Alzheimer's disease phenotype in amyloid precursor protein/presenilin 1 (APP/PS1) mice. Alzheimers Dement 1(3):170–181. https://doi.org/10.1016/j.trci.2015.08.001

McQuown SC, Wood MA (2011) HDAC3 and the molecular brake pad hypothesis. Neurobiol Learn Mem 96(1):27–34. https://doi.org/10.1016/j.nlm.2011.04.005

McQuown SC, Barrett RM, Matheos DP, Post RJ, Rogge GA, Alenghat T, Mullican SE, Jones S, Rusche JR, Lazar MA, Wood MA (2011) HDAC3 is a critical negative regulator of long-term memory formation. J Neurosci 31(2):764–774. https://doi.org/10.1523/JNEUROSCI.5052-10.2011

Meng J, Li Y, Zhang M, Li W, Zhou L, Wang Q, Lin L, Jiang L, Zhu W (2019) A combination of curcumin, vorinostat and silibinin reverses Aβ-induced nerve cell toxicity via activation of AKT-MDM2-P53 pathway. PeerJ 7:e6716. https://doi.org/10.7717/peerj.6716

Mielcarek M, Zielonka D, Carnemolla A, Marcinkowski JT, Guidez F (2015) HDAC4 as a potential therapeutic target in neurodegenerative diseases: a summary of recent achievements. Front Cell Neurosci 9:42

Misrani A, Tabassum S, Yang L (2021) Mitochondrial dysfunction and oxidative stress in Alzheimer's disease. Front Aging Neurosci 13:617588

Park S-Y, Kim J-S (2020) A short guide to histone deacetylases including recent progress on class II enzymes. Exp Mol Med 52(2):204–212. https://doi.org/10.1038/s12276-020-0382-4

Parra M (2015) Class IIa HDACs—new insights into their functions in physiology and pathology. FEBS J 282(9):1736–1744. https://doi.org/10.1111/febs.13061

Pascoal TA, Chamoun M, Lax E, Wey H-Y, Shin M, Ng KP, Kang MS, Mathotaarachchi S, Benedet AL, Therriault J, Lussier FZ, Schroeder FA, DuBois JM, Hightower BG, Gilbert TM, Zürcher NR, Wang C, Hopewell R, Chakravarty M, Savard M, Thomas E, Mohaddes S, Farzin S, Salaciak A, Tullo S, Cuello AC, Soucy J-P, Massarweh G, Hwang H, Kobayashi E, Hyman BT, Dickerson BC, Guiot M-C, Szyf M, Gauthier S, Hooker JM, Rosa-Neto P (2022) [11C]Martinostat PET analysis reveals reduced HDAC I availability in Alzheimer's disease. Nat Commun 13(1):4171. https://doi.org/10.1038/s41467-022-30653-5

Peleg S, Sananbenesi F, Zovoilis A, Burkhardt S, Bahari-Javan S, Agis-Balboa RC, Cota P, Wittnam JL, Gogol-Doering A, Opitz L, Salinas-Riester G, Dettenhofer M, Kang H, Farinelli L, Chen W, Fischer A (2010) Altered histone acetylation is associated with age-dependent memory impairment in mice. Science 328(5979):753–756. https://doi.org/10.1126/science.1186088

Penney J, Tsai L-H (2014) Histone deacetylases in memory and cognition. Sci Signal 7(355):re12. https://doi.org/10.1126/scisignal.aaa0069

Pulya S, Amin SA, Adhikari N, Biswas S, Jha T, Ghosh B (2020) HDAC6 as privileged target in drug discovery: a perspective. Pharmacol Res 163:105274. https://doi.org/10.1016/j.phrs.2020.105274

Pulya S, Mahale A, Bobde Y, Routholla G, Patel T, Swati, Biswas S, Sharma V, Kulkarni OP, Ghosh B (2021) PT3: a novel benzamide class histone deacetylase 3 inhibitor improves learning and memory in novel object recognition mouse model. ACS Chem Neurosci 12:883. https://doi.org/10.1021/acschemneuro.0c00721

Qing H, He G, Ly PTT, Fox CJ, Staufenbiel M, Cai F, Zhang Z, Wei S, Sun X, Chen C-H, Zhou W, Wang K, Song W (2008) Valproic acid inhibits Aβ production, neuritic plaque formation, and behavioral deficits in Alzheimer's disease mouse models. J Exp Med 205(12):2781. https://doi.org/10.1084/jem.20081588

Ricobaraza A, Cuadrado-Tejedor M, Pérez-Mediavilla A, Frechilla D, Del Río J, García-Osta A (2009) Phenylbutyrate ameliorates cognitive deficit and reduces tau pathology in an Alzheimer's disease mouse model. Neuropsychopharmacology 34(7):1721–1732. https://doi.org/10.1038/npp.2008.229

Ricobaraza A, Cuadrado-Tejedor M, Garcia-Osta A (2011) Long-term phenylbutyrate administration prevents memory deficits in Tg2576 mice by decreasing Abeta. Front Biosci (Elite Ed) 3(4): 1375–1384. https://doi.org/10.2741/e340

Ricobaraza A, Cuadrado-Tejedor M, Marco S, Pérez-Otaño I, García-Osta A (2012) Phenylbutyrate rescues dendritic spine loss associated with memory deficits in a mouse model of Alzheimer disease. Hippocampus 22(5):1040–1050. https://doi.org/10.1002/hipo.20883

Rivieccio MA, Brochier C, Willis DE, Walker BA, D'Annibale MA, McLaughlin K, Siddiq A, Kozikowski AP, Jaffrey SR, Twiss JL, Ratan RR, Langley B (2009) HDAC6 is a target for protection and regeneration following injury in the nervous system. Proc Natl Acad Sci U S A 106(46):19599–19604. https://doi.org/10.1073/pnas.0907935106

Rumbaugh G, Sillivan SE, Ozkan ED, Rojas CS, Hubbs CR, Aceti M, Kilgore M, Kudugunti S, Puthanveettil SV, Sweatt JD, Rusche J, Miller CA (2015) Pharmacological selectivity within class I histone deacetylases predicts effects on synaptic function and memory rescue. Neuropsychopharmacology 40(10):2307–2316. https://doi.org/10.1038/npp.2015.93

Sando R, Gounko N, Pieraut S, Liao L, Yates J, Maximov A (2012) HDAC4 governs a transcriptional program essential for synaptic plasticity and memory. Cell 151(4):821–834. https://doi.org/10.1016/j.cell.2012.09.037

Selenica M-L, Benner L, Housley SB, Manchec B, Lee DC, Nash KR, Kalin J, Bergman JA, Kozikowski A, Gordon MN, Morgan D (2014) Histone deacetylase 6 inhibition improves memory and reduces total tau levels in a mouse model of tau deposition. Alzheimers Res Ther 6(1):12. https://doi.org/10.1186/alzrt241

Simões-Pires C, Zwick V, Nurisso A, Schenker E, Carrupt P-A, Cuendet M (2013) HDAC6 as a target for neurodegenerative diseases: what makes it different from the other HDACs? Mol Neurodegener 8(1):7. https://doi.org/10.1186/1750-1326-8-7

Singh D, Gupta S, Verma I, Morsy MA, Nair AB, Ahmed A-SF (2021) Hidden pharmacological activities of valproic acid: a new insight. Biomed Pharmacother 142:112021. https://doi.org/10.1016/j.biopha.2021.112021

Soares Romeiro LA, da Costa Nunes JL, de Oliveira Miranda C, Simões Heyn Roth Cardoso G, de Oliveira AS, Gandini A, Kobrlova T, Soukup O, Rossi M, Senger J, Jung M, Gervasoni S, Vistoli G, Petralla S, Massenzio F, Monti B, Bolognesi ML (2019) Novel sustainable-by-design HDAC inhibitors for the treatment of Alzheimer's disease. ACS Med Chem Lett 10(4): 671–676. https://doi.org/10.1021/acsmedchemlett.9b00071

Stefanko DP, Barrett RM, Ly AR, Reolon GK, Wood MA (2009) Modulation of long-term memory for object recognition via HDAC inhibition. Proc Natl Acad Sci U S A 106(23):9447–9452. https://doi.org/10.1073/pnas.0903964106

Su Q, Li T, He P-F, Lu X-C, Yu Q, Gao Q-C, Wang Z-J, Wu M-N, Yang D, Qi J-S (2021) Trichostatin A ameliorates Alzheimer's disease-related pathology and cognitive deficits by increasing albumin expression and Aβ clearance in APP/PS1 mice. Alzheimers Res Ther 13: 7. https://doi.org/10.1186/s13195-020-00746-8

Sun J, Yuan B, Wu Y, Gong Y, Guo W, Fu S, Luan Y, Wang W (2020) Sodium butyrate protects N2a cells against $A\beta$ toxicity in vitro. Mediat Inflamm 2020:e7605160. https://doi.org/10.1155/2020/7605160

Sung YM, Lee T, Yoon H, DiBattista AM, Song JM, Sohn Y, Moffat EI, Turner RS, Jung M, Kim J, Hoe H-S (2013) Mercaptoacetamide-based class II HDAC inhibitor lowers Aβ levels and improves learning and memory in a mouse model of Alzheimer's disease. Exp Neurol 239:192–201. https://doi.org/10.1016/j.expneurol.2012.10.005

Tan J, McKenzie C, Potamitis M, Thorburn AN, Mackay CR, Macia L (2014) Chapter three—The role of short-chain fatty acids in health and disease. In: Alt FW (ed) Advances in immunology, vol 121. Academic, pp 91–119. https://doi.org/10.1016/B978-0-12-800100-4.00003-9

Tran AD-A, Marmo TP, Salam AA, Che S, Finkelstein E, Kabarriti R, Xenias HS, Mazitschek R, Hubbert C, Kawaguchi Y, Sheetz MP, Yao T-P, Bulinski JC (2007) HDAC6 deacetylation of tubulin modulates dynamics of cellular adhesions. J Cell Sci 120(8):1469–1479. https://doi.org/10.1242/jcs.03431

Tseng H-J, Lin M-H, Shiao Y-J, Yang Y-C, Chu J-C, Chen C-Y, Chen Y-Y, Lin TE, Su C-J, Pan S-L, Chen L-C, Wang C-Y, Hsu K-C, Huang W-J (2020) Synthesis and biological evaluation of acridine-based histone deacetylase inhibitors as multitarget agents against Alzheimer's disease. Eur J Med Chem 192:112193. https://doi.org/10.1016/j.ejmech.2020.112193

Turner BM (2000) Histone acetylation and an epigenetic code. BioEssays 22(9):836–845. https://doi.org/10.1002/1521-1878(200009)22:9<836::AID-BIES9>3.0.CO;2-X

Uliassi E, de Oliveira AS, de Camargo Nascente L, Romeiro LAS, Bolognesi ML (2021) Cashew nut shell liquid (CNSL) as a source of drugs for Alzheimer's disease. Molecules 26(18):5441. https://doi.org/10.3390/molecules26185441

Vecsey CG, Hawk JD, Lattal KM, Stein JM, Fabian SA, Attner MA, Cabrera SM, McDonough CB, Brindle PK, Abel T, Wood MA (2007) Histone deacetylase inhibitors enhance memory and synaptic plasticity via CREB: CBP-dependent transcriptional activation. J Neurosci 27(23): 6128–6140. https://doi.org/10.1523/JNEUROSCI.0296-07.2007

Vickers JC, Mitew S, Woodhouse A, Fernandez-Martos CM, Kirkcaldie MT, Canty AJ, McCormack GH, King AE (2016) Defining the earliest pathological changes of Alzheimer's disease. Curr Alzheimer Res 13(3):281–287. https://doi.org/10.2174/1567205013666151218150322

Wang J, Yu J-T, Tan M-S, Jiang T, Tan L (2013a) Epigenetic mechanisms in Alzheimer's disease: implications for pathogenesis and therapy. Ageing Res Rev 12(4):1024–1041. https://doi.org/10.1016/j.arr.2013.05.003

Wang Z, Zhang X, Li T, Li J, Tang Y, Le W (2013b) Valproic acid reduces neuritic plaque formation and improves learning deficits in APP Swe/PS1A246E transgenic mice via preventing the prenatal hypoxia-induced down-regulation of neprilysin. CNS Neurosci Ther 20(3): 209–217. https://doi.org/10.1111/cns.12186

Wang X-X, Xie F, Jia C-C, Yan N, Zeng Y-L, Wu J-D, Liu Z-P (2021) Synthesis and biological evaluation of selective histone deacetylase 6 inhibitors as multifunctional agents against Alzheimer's disease. Eur J Med Chem 225:113821. https://doi.org/10.1016/j.ejmech.2021.113821

Wang C, Zheng D, Weng F, Jin Y, He L (2022) Sodium butyrate ameliorates the cognitive impairment of Alzheimer's disease by regulating the metabolism of astrocytes. Psychopharmacology 239(1):215–227. https://doi.org/10.1007/s00213-021-06025-0

Wenk GL (2003) Neuropathologic changes in Alzheimer's disease. J Clin Psychiatry 64(Suppl 9): 7–10

Whittle N, Maurer V, Murphy C, Rainer J, Bindreither D, Hauschild M, Scharinger A, Oberhauser M, Keil T, Brehm C, Valovka T, Striessnig J, Singewald N (2016) Enhancing dopaminergic signaling and histone acetylation promotes long-term rescue of deficient fear extinction. Transl Psychiatry 6(12):e974. https://doi.org/10.1038/tp.2016.231

Xiao X, Liu X, Jiao B (2020) Epigenetics: recent advances and its role in the treatment of Alzheimer's disease. Front Neurol 11:538301

Xu K, Dai X-L, Huang H-C, Jiang Z-F (2011) Targeting HDACs: a promising therapy for Alzheimer's disease. Oxidative Med Cell Longev 2011:e143269. https://doi.org/10.1155/2011/143269

Xuan A-G, Pan X-B, Wei P, Ji W-D, Zhang W-J, Liu J-H, Hong L-P, Chen W-L, Long D-H (2015) Valproic acid alleviates memory deficits and attenuates amyloid-β deposition in transgenic mouse model of Alzheimer's disease. Mol Neurobiol 51(1):300–312. https://doi.org/10.1007/s12035-014-8751-4

Yang W, Chauhan A, Mehta S, Mehta P, Gu F, Chauhan V (2014a) Trichostatin A increases the levels of plasma gelsolin and amyloid beta-protein in a transgenic mouse model of Alzheimer's disease. Life Sci 99(1):31–36. https://doi.org/10.1016/j.lfs.2014.01.064

Yang W, Chauhan A, Wegiel J, Kuchna I, Gu F, Chauhan V (2014b) Effect of Trichostatin A on gelsolin levels, proteolysis of amyloid precursor protein, and amyloid beta-protein load in the brain of transgenic mouse model of Alzheimer's disease. Curr Alzheimer Res 11(10):1002–1011. https://doi.org/10.2174/1567205011666141107125531

Yoshida M, Kijima M, Akita M, Beppu T (1990) Potent and specific inhibition of mammalian histone deacetylase both in vivo and in vitro by Trichostatin A. J Biol Chem 265(28):17174–17179. https://doi.org/10.1016/S0021-9258(17)44885-X

Yu C-W, Chang P-T, Hsin L-W, Chern J-W (2013) Quinazolin-4-one derivatives as selective histone deacetylase-6 inhibitors for the treatment of Alzheimer's disease. J Med Chem 56(17):6775–6791. https://doi.org/10.1021/jm400564j

Zeng Q, Long Z, Feng M, Zhao Y, Luo S, Wang K, Wang Y, Yang G, He G (2019) Valproic acid stimulates hippocampal neurogenesis via activating the Wnt/β-catenin signaling pathway in the APP/PS1/nestin-GFP triple transgenic mouse model of Alzheimer's disease. Front Aging Neurosci 11:62

Zhang Z-Y, Schluesener HJ (2013) Oral administration of histone deacetylase inhibitor MS-275 ameliorates neuroinflammation and cerebral amyloidosis and improves behavior in a mouse model. J Neuropathol Exp Neurol 72(3):178–185. https://doi.org/10.1097/NEN.0b013e318283114a

Zhang X-Z, Li X-J, Zhang H-Y (2010) Valproic acid as a promising agent to combat Alzheimer's disease. Brain Res Bull 81(1):3–6. https://doi.org/10.1016/j.brainresbull.2009.09.003

Zhang L, Liu C, Wu J, Tao J, Sui X, Yao Z, Xu Y, Huang L, Zhu H, Sheng S, Qin C (2014) Tubastatin A/ACY-1215 improves cognition in Alzheimer's disease transgenic mice. J Alzheimers Dis 41(4):1193–1205. https://doi.org/10.3233/JAD-140066

Zhao L, Zhu L, Guo X (2018a) Valproic acid attenuates Aβ25-35-induced neurotoxicity in PC12 cells through suppression of mitochondria-mediated apoptotic pathway. Biomed Pharmacother 106:77–82. https://doi.org/10.1016/j.biopha.2018.06.080

Zhao W-N, Ghosh B, Tyler M, Lalonde J, Joseph NF, Kosaric N, Fass DM, Tsai L-H, Mazitschek R, Haggarty SJ (2018b) Class I histone deacetylase inhibition by tianeptinaline modulates neuroplasticity and enhances memory. ACS Chem Neurosci 9(9):2262–2273. https://doi.org/10.1021/acschemneuro.8b00116

Zhu X, Wang S, Yu L, Jin J, Ye X, Liu Y, Xu Y (2017) HDAC3 negatively regulates spatial memory in a mouse model of Alzheimer's disease. Aging Cell 16(5):1073–1082. https://doi.org/10.1111/acel.12642

SPRINGER NATURE

GPSR Compliance

The European Union's (EU) General Product Safety Regulation (GPSR) is a set of rules that requires consumer products to be safe and our obligations to ensure this.

If you have any concerns about our products, you can contact us on ProductSafety@springernature.com

In case Publisher is established outside the EU, the EU authorized representative is:

Springer Nature Customer Service Center GmbH
Europaplatz 3
69115 Heidelberg, Germany

The manufacturer's authorised representative in the EU is Springer Nature Customer Service Centre GmbH, Europaplatz 3, 69115 Heidelberg, Germany. If you have any concerns regarding our products, please contact ProductSafety@springernature.com

Printed and bound by CPI Group (UK) Ltd, Croydon, CR0 4YY
30/09/2026
02225322-0017